ENCYCLOPEDIA OF AUSTRALIAN ANIMALS

BIRDS

ENCYCLOPEDIA OF AUSTRALIAN ANIMALS

BIRDS

TERENCE R. LINDSEY

SERIES EDITOR RONALD STRAHAN

THE NATIONAL PHOTOGRAPHIC INDEX OF AUSTRALIAN WILDLIFE

THE AUSTRALIAN MUSEUM

Angus&Robertson

An imprint of HarperCollins*Publishers*

AN ANGUS & ROBERTSON BOOK
An imprint of HarperCollinsPublishers

First published in Australia in 1992 by
CollinsAngus&Robertson Publishers Pty Limited (ACN 009 913 517)
A division of HarperCollinsPublishers (Australia) Pty Limited
25-31 Ryde Road, Pymble NSW 2073, Australia

HarperCollinsPublishers (New Zealand) Limited
31 View Road, Glenfield, Auckland 10, New Zealand

HarperCollinsPublishers Limited
77-85 Fulham Palace Road, London W6 8JB, United Kingdom

National Library of Australia
Cataloguing-in-Publication data:

Encyclopedia of Australian animals.
    Includes index.
    ISBN 0 207 16976 4 (Birds).
    1. Zoology—Australia—Encyclopedias.
    2. Animals—Encyclopedias. I. Strahan,
    Ronald, 1922—

Layout by Sue Edmonds
Typeset by Midland Typesetters
Printed in Hong Kong

5 4 3 2 1
96 95 94 93 92

FOREWORD

The National Photographic Index of Australian Wildlife is a special project of the Australian Museum. Its initial purpose was to collect and preserve outstanding colour photographs of living animals, and thereby to supplement the Museum's collections of preserved specimens, but it soon became apparent that the photographs could be the basis of outstanding books. In turn, these books would preserve the photographic images, make them available to a wide public, and return revenue to the operation.

The Index served first as the core of the *Reader's Digest Complete Book of Australian Birds*, which has passed through numerous reprintings and appeared as a new edition in 1990. From its first appearance, it has been the most comprehensive and authoritative of popular references to the birds of Australia.

This was followed in 1983 by the *Australian Museum Complete Book of Australian Mammals*, published by Angus & Robertson. Filling a vacuum in both the professional and amateur literature of these animals, it is an even more significant work.

Meanwhile, steadily working away, the Index continues its association with Collins/Angus & Robertson in the production of ten volumes on the birds of Australia, seven of which have so far been published. Additionally, the collections of the Index, representing the work of the best of Australia's wildlife photographers, have been used by hundreds of authors and publishers to illustrate other works.

The four volumes of the *Encyclopedia of Australian Animals* document all Australian species of frogs, reptiles, birds and mammals and provide an invaluable reference for the increasing number of people in Australia, and around the world, who are concerned with our unique fauna.

The contributions of the Index represent but one aspect of the mission of the Australian Museum, which is to increase understanding of our natural environment and cultural heritage. We aim to be a catalyst in developing and changing people's attitudes through exhibitions, education programmes, lectures and publications. The *Encyclopedia of Australian Animals* is a valuable contribution to this mission and I am delighted by this latest product of the relationship between the Index and its publisher.

Des Griffin AM
Director
The Australian Museum

SPONSORS

The following have contributed to the cost of production of this work.

E.M. & M.J. Abrahams
J. & J. Allison
Arnott's Biscuits Ltd
A.W. Auldist
C.N. Banks
G.W.E. Barraclough
H. Barry
P.J. & J.Y. Bath
K.A. Blofeld
Dr. M. Bonnin
C. & T. Britton
G. Broinowski
J. & J. Broinowski
S. Broinowski
H.G. Brooks
P.J. Buckley
J., N., S., P., H. & H. Burton Taylor
A. Byrne
Sir Bede Callaghan
D.M. Carment
J. Champ
A. Charles
Clampett Properties Ltd
M.D. Cobcroft
B. & J. Coghlan
L.R. & P. Comben
K.H. Cousins
S. & M. Crouch
Mr and Mrs Cumbrae Stewart
Dalgety Farmers Ltd
P. & I. Davidson
S.G. Eagles
P. Edwards, E. & T. Frankland
J.O. Fairfax
Sir Vincent & Lady Fairfax
B.A. Farrell
C.A. Fay
Dr. M.A. Feilman OBE
R.A. Field

T. Florin
D. Ford
B. & P. France
K., R. & R. Frankland
K.I. Frecker
M.D. Frecker
E.W. Gibson
Sir Archibald Glenn
B. Goldrick
M.R. & B. Gordon
J.D. Gorter Pty Ltd
Gould League of NSW
Gould League of W.A.
I.G. Graham
R.W. Greaves
Sir David & Lady Griffin
R.M. Griffin
N. & E. Haines
C.M. Hall
J.D. Hamman
M. Hamilton
K.C. Hammer
R. & P. Harrison
H.R. Hawkeswood
Dr. J. & M. Hazel
S.M. Hicks
A. Holmes
R.M. Howarth
S. Hughes
Hunter Wetland Trust
T.D.E. Hyde
D. Inglis
M.M. Johnson
E. & T. Karplus
A.C. Keating
G.M. King
H.W. & M.A. Kinnersley
L.B. Kirk
L. Kramer

Y. Lewis
B. & N. Lithgow
A.F. Little
K. Lyall
K.A. Macpherson
H. & F. MacLachlan
A. McBride
F.L. Mahoney
M.J. Mashford
S. & R. Miles
P. Miles
L. & J. Minnett
B.N. Morrison
J. & M. Munro
Dr. M. Norst
Pancontinental Mining Ltd
L. Papi
Peddle, Thorpe & Walker P/L
Progressive Mortgage Co. Ltd.
Sir John & Lady Proud
Renison Goldfields Consolidated Ltd
N. Robinson
Royal Zoological Society of NSW
RSPCA (NSW)
C. & M. Ryman
P. & J. Sayers
R. & R. Schmidt
B.M. Scott
M. & R. Shepherd
R.A. Simpson
D. Solomon and family
South Australian Museum
Dr. S. Stevens
B. Stevenson
Sir Edward Stewart
Sunshine Foundation
Sir James Vernon
J.S. & M. Wilkey
R. & T. Yates

CONTENTS

For individual species listing,
see the Index of Common Names
or the Index of Scientific Names.

PREFACE

The National Photographic Index of Australian Wildlife was established by the Australian Museum in 1969, originally to create a set of reference photographs of all the Australian birds. These are being used in the production of a series of richly illustrated volumes, seven of which have so far been published by Collins/Angus & Robertson.

Expansion of the collection to include mammals provided a wealth of photographs to illustrate *The Australian Museum Complete Book of Australian Mammals*, the text of which was written by 110 experts. It is currently the standard work on Australian mammal species, accessible to the layman, but also referred to consistently by professional zoologists: no other source is more frequently cited in the current literature of Australian mammalogy.

In recent years, the Index has extended its scope to include the Australian frogs and reptiles. Thanks to the cooperation of hundreds of amateur and professional herpetologists, its collection of colour photographs of these animals is now the most comprehensive in existence. It therefore seemed appropriate to produce a series that would combine information and illustrations of all of these animals.

Zoologists tend to write about their areas of expertise but most members of the public have general interests. To meet this need, we have produced four volumes in which, for the first time, all of the basically four-legged Australian animals (tetrapods) have been treated in much the same way.

One need thumb through only a few pages to recognise that our knowledge of many species is remarkably slight. In some instances, we know almost nothing of the way of life of an animal, it being represented by a few museum specimens. It has been disappointing to the authors that we have been unable to provide an interesting account and colour photograph of every species, but the gaps may serve a useful purpose in drawing attention to the need for further research on the Australian fauna.

In a work of this nature, it is impossible to give credit for the sources of every item of information compressed into its pages: these represent the cumulative work of thousands of amateur and professional zoologists over the course of the past two centuries. However, since the work would not have been contemplated in the absence of the Index, it is pertinent to thank the hundreds of photographers who have contributed to the Index collections and Donald Trounson, who founded the Index and made its enterprises possible.

The greater part of this work was written by Terence Lindsey, whose elegance and authority is manifest. I take responsibility for the distribution maps and for the accounts of introduced species. Mr John Disney, assisted me in assessment of abundance and survival status of all the species. Mr John Waterhouse contributed greatly to the selection of illustrations and criticism of proofs.

The task of assembling the photographs for the four volumes was begun by the Archivist, Heather Lawrence, and carried through by the Collections Manager, Sally Bird. Joy Coghlan made sense of successively edited drafts of the text.

Finally, I acknowledge the many sponsors who contributed to the great cost of producing the work.

Ronald Strahan
Editor-in-Chief
National Photographic Index of Australian Wildlife

INTRODUCTION

This book is one of four volumes that comprise an encyclopedia of those Australian animals that have a backbone and four legs: the encyclopedia also includes all those animals such as snakes, birds and bats that have evolved from four-legged ancestors. These volumes, and their contents, are arranged according to zoological classification.

CLASSIFICATION

Zoological classification is based on a system of narrower and narrower categories, similar to those in the address of an overseas letter. Postal workers first sort letters in terms of continent and country, then cities or towns, street and number, and finally the surname and personal name of the addressee. This may be compared, for example, with the classification of the first species in this volume, the Emu. Having a backbone, it is a member of the Subphylum Vertebrata (not a worm, mollusc, arthropod, etc). Its wings are reduced to stubs but it nevertheless has four limbs and belongs within the Superclass Tetrapoda (not a fish). Its feathers place it in the Class Aves (not an amphibian, reptile or mammal). Reduced wings, lack of flight feathers and absence of a keel to the breastbone place it in the Order Struthioniformes (not one of the other two dozen or so orders of birds). A number of anatomical peculiarities mark it a member of the Family Dromaiidae (not an Ostrich or rhea). Its soft, hair-like plumage place it in the genus *Dromaius* (not a cassowary), which has only one living species, *novaehollandiae*.

Classification is a means of sorting and labelling animals but it has another important function in indicating evolutionary relationships. When a zoologist places two or more species in the same genus, this is the expression of an opinion that these are closely related to each other and share a common ancestor that is different from that of the species in other genera. Relationship is also implied when genera are placed in the same family, or families in the same order. If we had complete knowledge of all living and extinct species, we might be able to construct a classification that reflected the entire history of animal evolution, but lack of knowledge forces zoologists to make decisions on incomplete information. Consequently, schemes of classification are always open to revision in response to new information or opinions.

ZOOLOGICAL NAMES

Fundamental to the science of zoology is the binominal (two-name) system, according to which every species has a name composed of two parts, the genus name and the specific name. A genus name is a noun and can be used by itself, but a species name functions as an adjective and cannot stand alone.

A widespread species often includes populations that differ in size, proportions or coloration at the extremes of its range (north to south, east to west, wet to dry, or low to high altitude). Where the transition is continuous and gradual from one form to another, it is referred to as a cline. Where the variation is discontinuous and animals from one area differ recognisably from those in another, we refer to each form as a **subspecies**. Where one subspecies adjoins another, there are usually intermediate forms. Subspecies are named by adding a third (subspecific) name to that of the species.

PRONUNCIATION

Because most zoological names are constructed from Latin or Greek roots, they should be pronounced and stressed in such a way as to retain the identity of these elements. There are no absolute rules, but the pronunciation recommended in this encyclopedia is reasonably consistent and international.

Consonants are given their usual English values: a limited set of vowels is pronounced as below.

ay as in bay	*ie* as in die	*aw* as in law
a as in bat	*i* as in bit	*ue* as in sue
ah as in bah	*air* as in fair	*u* as in but
ee as in bee	*oh* as in doh	*oo* as in good
e as in bet	*o* as in dot	*ow* as in cow
er as in fern	*or* as in for	*oy* as in boy

A point to be noted is that proper names retain their identity, even when Latinised. Thus *burrelli* (after a Mr Burrell) is bu'-rel-ee, not bu-rel'-ee; *sladei* (after a Mr Slade) is slay'-dee; *godmani* (after a Mr Godman) is god'-mun-ee, not god-mah'-nee.

COMMON NAMES

Most Australian birds and mammals have agreed common names but only the more common frogs and reptiles are so blessed. The authors of this work have attempted to provide a consistent set of common names for every Australian tetrapod, choosing among alternatives when these are available and creating names where none previously existed.

PROSPECTS OF SURVIVAL

Human activities have led to the extinction of many vertebrate species and an even greater number are endangered or vulnerable. Nevertheless, many species are doing well at present. To deal with this range of situations, the encyclopedia introduces a three-fold system of categorisation of survival status comprising estimates of range and of abundance with a prognosis for each species.

Distribution

It is a reasonable assumption that a widely distributed species is more secure than one limited to a small area. Distribution is scored in six categories based on distribution maps and, since such maps always include areas where the species does not occur, these figures are always overestimates.

The principle underlying the scoring system for distribution is that the smaller the area occupied by a species, the greater the significance of that area. The scale therefore expands exponentially, beginning with small increases and ending with very large ones.

SCORING OF DISTRIBUTION

Less than 10,000 square kilometres ˙	10,000–30,000 square kilometres
30,000–100,000 square kilometres	100,000–300,000 square kilometres
300,000–1,000,000 square kilometres	More than 1,000,000 square kilometres

---

Abundance

In principle, abundance should be a measure of the number of individuals in a given area (and abundance multiplied by distribution would represent the population of a species). We seldom have the resources to conduct such counts but experienced observers can reach reasonable agreement on the six categories of abundance set out below.

Our best information on abundance comes from zoologists working in such institutions as natural history museums, national parks and wildlife services, the CSIRO, and universities. Such people are familiar with a range of species and are frequently in the bush. Amateur ornithologists and herpetologists are also valuable sources of information. The scoring system depends upon the opinions of experienced observers who are familiar with the species in question and judgment that a species is abundant does not necessarily mean that it will appear to be so to an untrained person. In the definition that follows, "locatable" means seen, heard, or judged to be present by nests, burrows, scrapings, dung, food scraps etc. "Appropriate times" may mean the time of day when the species is active or the time of year when its presence becomes obvious.

SCORING OF ABUNDANCE

VERY RARE	So infrequently located or trapped, despite considerable efforts, that the possibility of extinction cannot be excluded.
RARE	Known to be present within the distribution area but seldom located or trapped, despite considerable effort. May not be located over periods of several years.
VERY SPARSE	As for "sparse" but infrequently located, usually after considerable effort; apparently absent from many appropriate habitats.
SPARSE	Known to be present and frequently located at appropriate times in some appropriate habitats, but usually after some effort; if trappable, frequently absent from trap-lines set at such times and places.
COMMON	Usually locatable at appropriate times in most appropriate habitats; if trappable, usually represented to some extent in most trap-lines set at such times and places.
ABUNDANT	Locatable at all appropriate times in all appropriate habitats; if trappable, well represented in any trap-line set at such times and places.

Survival Status

Because conservation agencies have been mainly concerned with species at risk, evaluation of survival status has usually been restricted to these categories. The novelty in the approach taken here is recognition of two categories—"secure" and "possibly secure"—representing antitheses to "possibly extinct" and "presumed extinct". Before allotting a species to one of these categories, we have considered its past and present distribution and abundance, and those environmental and human factors that are currently acting to its detriment or benefit. Prognoses of survival status are scored as follows.

SCORING OF SURVIVAL STATUS

PRESUMED EXTINCT	No confirmation of the existence of the species in the wild for 50 years or more.
POSSIBLY EXTINCT	So infrequently located that the existence of the species in the wild has not been confirmed for up to 50 years.
ENDANGERED	Declining in distribution and/or abundance to such an extent that, without positive action to halt or reverse the trend, the species appears likely to become extinct in the near future.
POSSIBLY ENDANGERED	Declining in distribution and/or abundance but population still large and viable.
VULNERABLE	Currently of satisfactory distribution and/or abundance but foreseeable pressures could put the species at risk.
PROBABLY SECURE	Probably as below but insufficient data to be certain.
SECURE	No existing or foreseeable threat to continuance of the species.

Because this is the first attempt to survey the survival status of the Australian tetrapods, these evaluations will inevitably include some errors of fact or judgment. They should therefore be used as first approximations and as the basis for discussion.

In an overall sense, they must also be regarded as optimistic. A pessimistic approach, postulating unrestricted expansion of human populations, critical levels of environmental pollution, a severe "greenhouse" effect and possibly a "nuclear winter", would require the classification of virtually every tetrapod, including humans, as vulnerable or endangered.

CLASS
Aves

(ah'-vayz: "birds")

Birds are readily distinguishable from other tetrapods by their covering of feathers. The forelimbs are wings (degenerate in some non-flying birds); teeth are lacking; the upper and lower jaws are enclosed by a horny bill and the tail is reduced to a stub, familiar as the "parson's nose" of a roast fowl. The body of a typical (flying) bird is short and supported by a very firm skeleton that encloses the chest and almost all of the abdomen, resisting compression by the powerful flight muscles, most of which are attached to a deep keel on the underside of the sternum (breastbone). The rigidity of the body skeleton is compensated for by a very flexible neck, which may have up to 25 vertebrae.

The legs of a bird have the same basic structure as those of other tetrapods but the skeleton is simplified by the fusion of some bones. The ankle is elongated, forming a "third joint" of the leg, and there are never more than four toes, usually three directed forward and one backward. Some climbing birds have two forwardly directed and two backwardly directed digits. Emus have only three digits and the ostrich only two, all directed forward. The skeleton of the wing also is recognisably of the tetrapod pattern but only the second digit is well developed, the first and third being short stubs. The bigger bones are hollow and filled with air: in large birds, the skeleton may weigh less than the totality of the feathers.

Evidence of reptilian ancestry is seen in the feet and lower legs, which have a covering of scales. Except for the beak and, sometimes, areas on the head and neck, the rest of the body is covered by contour feathers, which define the shape of the bird; and flight feathers, arising from the wings and tail, which have an aerodynamic function. Flight and contour feathers have a central, hollow quill or shaft, supporting numerous stiff, parallel lateral filaments or barbs, each bearing minute, interlaced hook-like structures (barbules) that link the filaments together and give strength to this remarkably light structure. Down feathers, which lie under the contour feathers, have a short shaft and many flexible filaments which lack barbules and provide a wool-like insulating layer close to the skin. Among several other types of feathers are those which form sensory bristles (equivalent to the whiskers or vibrissae of mammals) around the mouth in some birds, particularly those which feed on flying insects.

Except at its base, a feather is composed of dead material, so damage to it cannot be repaired by healing. Compensation for wear and tear is provided by moulting, usually over an annual period, during which each feather is shed and replaced by a new one—without leaving the bird bare. Most species manage the replacement so well that they are not incommoded; others are unable to fly while moulting.

Although any feathered animal existing today is undoubtedly a bird, there are also some remarkably well preserved fossils, about 140 million years old, of crow-sized creatures known as *Archaeopteryx*. These had a covering of feathers, including flight feathers on the wings and tail, but a skeleton barely distinguishable from that of bipedal dinosaurs of the genus *Compsognathus*. The long, narrow jaws bore a series of sharp teeth and the skeleton of the tail was almost as long as that of the body and neck. In the absence of fossilised feathers, *Archaeopteryx* would have been classified as a dinosaur so the conclusion is inescapable that the first birds were "modified" dinosaurs. Some zoologists claim, quite reasonably, that it is therefore incorrect to say that dinosaurs are extinct: one highly specialised group of dinosaurs could be said to continue to exist, but is referred to as birds.

Birds with teeth persisted until about 90 million years ago. Thereafter, it seems, the jaws of all birds have been covered with horny bills. These have evolved into an immense variety of shapes, adapted to different types of food and to the manipulation of objects, as in nest-building. Although some beaks, such as those of parrots, provide a gnawing action, birds are incapable of chewing. However, food can be crushed and ground into small particles by the muscular gizzard, an organ derived from the stomach and best developed in seed-eating birds, less so in flesh-eaters and nectar-feeders.

A steady supply of readily digestible food is necessary for birds because, like mammals, they are warm-blooded. In fact, birds have a higher body temperature and rate of metabolism than mammals. The lungs of birds are not very large but they extend into a number of air-sacs in various parts of the body and even into the hollows of some of the larger bones. Very little gas exchange takes place in the air-sacs but, when breath is being exhaled, the sacs are compressed and sweep a large volume of (fresh) air through the lungs. Thus, whether inhaling or exhaling, the lungs of a bird are filled with air of near-atmospheric oxygen content, leading to far more efficient respiration than is possible in mammals or other tetrapods. When a bird is flying, the air-sacs are compressed with every wingbeat, so the flow of air through the lungs is related to the level of exertion.

As in mammals, the heart of a bird is effectively a double pump, one half supplying blood at moderate pressure to the lungs, the other receiving blood from the lungs and pumping it at high pressure to the rest of the body. Significant anatomical differences show that the similarity between the avian and mammalian hearts is the result of convergent evolution: a similar condition has also been achieved by the crocodiles.

The brain of modern birds is proportionately much larger than that of *Archaeopteryx*, probably reflecting a greater capacity for precise control of flight and much more complex behaviour. Of all the tetrapods, birds exhibit the most elaborate patterns of instinct, governing such intricacies as courtship, nest-building, care of the young and migration. The sense of taste and smell are less well developed in birds than in other tetrapods but hearing and vision are acute: the visual discrimination of birds is approximately four times greater than that of mammals and a considerable part of the brain is devoted to vision.

All birds lay eggs. Indeed, they are the only group of tetrapods without at least some species that give birth to live young. The eggs require warmth in order to hatch and this is usually provided by brooding on the part of one or both parents, the only exception being found in the megapodes, which lay eggs into a mound of soil and rotting vegetation or (geothermally) warm soil. Again, with the exception of megapodes, whose chicks must fend for themselves from the time of hatching, all young birds receive considerable protection from one or both parents—sometimes assisted by other relatives—until they are capable of independent life. The young of most species, which hatch in a virtually naked and almost helpless condition, are fed and brooded until they are able to fend for themselves. Others, such as chickens and ducklings, hatch with a covering of down and follow a parent on foot: they are not fed, but the parent may assist them in finding food and they are protected and brooded by the parent until they attain independence. In some species, the relationship between parents and young continues into adult life, giving rise to a social organisation based on an extended family.

The necessity to care for young underlies much of the mating behaviour of birds. Whereas, in most tetrapods, it is sufficient for two potential parents to reduce their normal antagonism or indifference to the point where copulation is tolerated, most birds engage in a quite elaborate courtship that creates a longer relationship between a male and female. Typically, but by no means universally, the two co-operate in establishing a territory from which they attempt to exclude other members of the species and within which they build a nest, into which the eggs are laid. They take turns in brooding the eggs and in brooding and feeding the young. The bond between the two parents usually dissolves at the end of a breeding season but in some species—usually long-lived birds—it may persist for years or even until the death of one of the partners. There are very many variations from this general pattern, including polygamy and polygyny, and care of the young by only one parent.

Many species are sedentary, spending their entire lives in a particular area. Some are nomadic, moving from one place to another in response to changes in the availability of food. Quite a few species engage in regular migrations, which may be as simple as a move towards the tropics in winter or as complex as that of the Short-tailed Shearwater which breeds on Bass Strait islands and migrates annually in a figure-eight path across the central and northern Pacific Ocean, so as to spend the height of the Australian winter in the Aleutian Islands. Just how birds navigate on such journeys, particularly over open oceans, remains one of the great puzzles of ornithology, hardly lessened by strong evidence that it can involve recognition of the position of stars in the sky and, possibly, a magnetic sense.

There are about 9000 species of living birds. This great diversity reflects the wide variety of habitats (polar to tropical, oceanic to desert) that are occupied by birds; their varied diet (nuts to nectar, insects to fishes); and, in general, a high degree of specialisation, permitting many species to live in more or less the same area, each exploiting a somewhat different food resource.

However, despite immense variation in the outward appearance of birds and in their ways of life, they are more uniform and exhibit less differences than are found in other tetrapods. For example, loss of limbs has not proceeded to the extent that we find in snakes or whales and, whereas mammals include both winged and four-footed forms, no bird has transformed its wings into feet. Nor, despite the facility with which penguins swim, have any birds become totally aquatic, like sea-snakes or the Dugong.

It seems that, having arisen as flying animals with highly specialised forelimbs, and having lost their teeth in order to reduce weight, birds closed off a number of evolutionary options. Essentially, they have been limited to variation in the shape and strength of the beak; length and flexibility of the neck; proportions of the wings, legs and feet; and—most significantly—variation in the size, structure, arrangement and colour of the feathers.

Birds have undergone so many "overlapping" evolutionary radiations that their external appearance and ways of life seldom provide a reliable guide to the relationships. Peculiarities of anatomy may serve to define a particular order or family of birds but their internal structure shows so little basic variation that it seldom offers clues to the relationships between such groups. Most higher-level classification tends therefore to be based upon (or reinforced by) biochemical evidence.

It is generally agreed that the "old" flightless birds, or ratites (emus, cassowaries, ostriches, rheas, kiwis and tinamous) are very different from the other modern birds but there is great disagreement on the relationship between each of these. In this book we have placed the emu and cassowaries together with the ostriches in the order Struthioniformes but some experts hold that they should be in a separate order, the Casuariformes. The remainder of the living birds are classified into about 25 orders, comprising some 170 families. One order, the Passeriformes (perching birds) include some 60 families and about half of the known species. Eighteen orders and about 87 families are represented in the native Australian avifauna, 16 being peculiar to the Australia–New Guinea region. Several of these have so evolved as to greatly resemble unrelated families in other parts of the world. Thus, members of the families Acanthizidae and Maluridae are similar to European wrens and warblers; members of the Epthianuridae to European chats; members of the Cinclorhamphidae to songlarks; members of the Neosittidae to nuthatches; members of the Cractidae to European magpies; and members of the Climacteridae are similar to European treecreepers.

Although birds are the most intensively studied of the Australian tetrapods, their classification is by no means settled, even at the family level, and experts continue to disagree on the genera to which some species should be assigned. The classification adopted here represents a general consensus at the present time.

Order STRUTHIONIFORMES

(Strue'-thee-on'-ee-for-mayz: "*Struthio*-order", after *Struthio*, the Ostrich)

The ratites are very large to huge birds which cannot fly, and accordingly lack the deep keel on the sternum or breast-bone that is otherwise such a characteristic feature of the avian skeleton (in other birds the keel serves as a point of attachment for the relatively enormous pectoral muscles mainly responsible for flight).

Surviving members of the group include the rheas of South America, the Ostrich of Africa, and the cassowaries and emus of Australasia, but the extinct moas of New Zealand and several other families known only from fossil remains are also included. Whether or not these groups are in fact closely related has been the subject of long-standing debate.

Family DROMAIIDAE

(droh-may'-i-dee: "*Dromaius*-family")

Emus constitute a group of large flightless birds confined to Australia and with only one surviving species. They are distantly related to the rheas of South America and the African Ostrich, but much more closely related to the cassowaries of New Guinea and Australia (many researchers combine emus and cassowaries in a single family).

Genus Dromaius

(droh-may'-us: "swift-foot")

The characteristics of the genus are essentially those of the single species.

Emu

Dromaius novaehollandiae (noh'-vee-hol-and'-ee-ee: "Australian swift-foot")

The Emu inhabits most of mainland Australia except rainforest and very arid deserts. Although sedentary in some places, it is generally nomadic or partly migratory, some populations moving seasonally over hundreds of kilometres. It is gregarious, and usually occurs in small groups, occasionally in mobs of thousands. It feeds mainly on green herbage but also eats insects and seeds.

An adult emu stands about 1.6 to 1.9 metres tall and may weigh up to 45 kilograms. Females outweigh males on average but the sexes are otherwise virtually identical.

Emus are sexually mature at about two years of age. Pairs form in late summer in a relationship that lasts about five months until five to 15 (usually seven to 11) eggs are laid, when the female wanders off to join wandering groups. Incubation (about 60 days) and care of young is by the male alone.

The Emu is common throughout Australia except in closely settled areas; it has been persecuted in some areas, reintroduced in others. Three insular populations (perhaps distinct species) once inhabited Tasmania, Kangaroo Island and King Island respectively, but were exterminated in the early years of European settlement.

HABITAT: grassland, savannah, open forest
HEIGHT: 1.6–1.9 m
DISTRIBUTION: more than 1 million km^2
ABUNDANCE: common
STATUS: secure

(G Little)

4

Family CASUARIIDAE

(kas'-ue-ah-ree'-id-ee: "*Casuarius*-family")

Three species of very large, flightless, heavy-bodied birds constitute the family Casuariidae, a group closely related to the emus (Dromaiidae). All inhabit New Guinea and one of them also occurs in north-eastern Australia. All live in rainforests.

In each species the plumage is glossy blue-black and the neck and head are naked, bright blue in colour and adorned with brightly coloured pendent wattles. The head is crowned with a large horny helmet or casque. The wings, as in the emus and other ratites, are reduced to mere vestiges—only the quills of the flight feathers are evident on superficial examination. The legs are stout and powerful, and the innermost toe is equipped with a large sharp claw—a kick from a cornered cassowary has on occasion proved fatal even to humans.

Cassowaries are wary, secretive and intensely solitary. They feed largely on fallen fruit.

Genus Casuarius

(kaz'-ue-ah'-ree-us: "cassowary", from a Malay name [*kesuari*] for the birds)

The characters of the single genus are those of the family.

Southern Cassowary

Casuarius casuarius (kaz'-ue-ah-ree-us: "cassowary cassowary")

The Southern Cassowary occurs in New Guinea and extreme north-eastern Australia, south to the vicinity of Cardwell, Queensland. It is solitary and territorial but individuals wander over an extensive home range. It is crepuscular in habits, hiding in dense thickets during the day. Normally shy and elusive, it occasionally becomes tame and comes to houses for food but it may behave aggressively and unpredictably in close contact with humans.

It is almost omnivorous, but feeds mainly on fallen fruit, swallowed whole (the seeds being voided undamaged). Females are larger and brighter than males and dominate them. Vocalisations include low hisses, booming and rumbling sounds.

Sexual maturity is reached at about three years. Peak breeding activity occurs from July to August. Females may pair with several males in a season, deserting each in turn after the eggs (usually four) are laid. The male incubates (about 60 days) and cares for the young, which are initially striped but develop dull brown juvenile plumage at about six months.

The Southern Cassowary remains locally common where unpersecuted, but it is declining in step with land clearing and destruction of rainforest.

HABITAT: tropical rainforest
HEIGHT: 1.5–2.0 m
DISTRIBUTION: 30,000–100,000 km^2
ABUNDANCE: very sparse
STATUS: possibly endangered to vulnerable

(HJ Pollock)

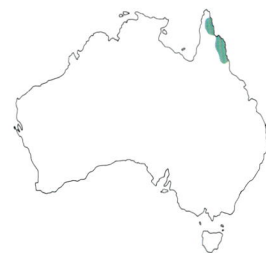

Order PODICIPEDIFORMES

(pod'-ee-sep'-id-ee-for'-mayz: "*Podiceps*-order", after a genus of grebes)

Members of this order have membranous lobes at the sides of the three forwardly directed toes. The tail is a downy stump without retrices (long tail feathers).

Family PODICIPEDIDAE

(pod'-ee-sep-id'-id-ee: "*Podiceps*-family", after a genus of grebes)

Some 20 to 22 species of foot-propelled diving birds found on all continents except Antarctica constitute this family which appears to lack close relatives, living or fossil. All are strictly aquatic and generally inhabit fresh standing waters, though many species congregate on sheltered coastal waters in winter. A conspicuous feature is the unusual structure of the feet, which are lobed, not webbed. Grebes are generally plain in plumage, but most have ornamental plumes on the head; these are shed in winter. Courtship displays are elaborate and varied. The sexes are similar but females are generally somewhat larger than males. Most species are no larger than a small duck.

Three species inhabit Australia.

Genus Podiceps

(pod'-ee-seps: "rump-foot")

This small group is distributed mainly in the Northern Hemisphere, and only one species occurs in Australia. They are medium-sized to large grebes characterised by the possession of showy crests, ruffs or tassels on the head, and notable for their spectacular and extremely intricate courtship behaviour.

Great Crested Grebe

Podiceps cristatus (kris-tah'-tus: "crested rump-foot")

The almost cosmopolitan Great Crested Grebe occurs throughout Australia but is most numerous in the Murray–Darling basin in the interior south-east. Its range has expanded since about 1950, especially in parts of northern Queensland, coastal New South Wales and Tasmania.

It is nomadic. Strongly aquatic and very seldom on land, it seldom flies and then does so mainly at night. Though not especially gregarious, it commonly forms winter flocks and sometimes breeds in dispersed colonies. The diet includes small fishes, tadpoles and aquatic invertebrates.

Breeding is mainly from November to March. Courtship includes spectacular and elaborate mutual aquatic displays. Sexes share nest-building, incubation and care of young, the usual clutch is four or five eggs.

HABITAT: cool temperate fresh waters, preferably extensive, deep and permanent
LENGTH: 46–51 cm
DISTRIBUTION: more than 1 million km^2
ABUNDANCE: very sparse
STATUS: probably secure

(P Slater)

Genus *Poliocephalus*

(poh'-lee-oh-sef'-ah-lus: "grey-headed")

A number of structural features, including a pronounced similarity in the arrangement of tendons in the hind limb, link these grebes with *Podiceps*, but, unlike *Podiceps*, they have very spartan courtship displays. There are only two species, one of which (the New Zealand Dabchick) seems to be a relatively recent derivative of the Hoary-headed Grebe of Australia. The latter differs strikingly from grebes of the genus *Tachybaptus* in its behaviour: it is a nomadic, gregarious bird which breeds in colonies and prefers extensive open waters. It seldom calls, flies readily, and its diet tends to focus upon aquatic insects rather than on fish.

Hoary-headed Grebe

Poliocephalus poliocephalus (poh'-lee-oh-sef'-ah-lus: "grey-headed grey-head")

The Hoary-headed Grebe occurs throughout Australia, but is much more common in the south than in the north, and rare in the tropics. It has recently established itself in New Zealand.

A nomadic and highly mobile species, its distribution, numbers and breeding cycle are heavily influenced by rainfall and associated flooding, and its ecology is closely linked with seasonal and ephemeral floodwaters. Strongly gregarious in all activities and at all seasons, it tends to congregate in winter in estuaries and sheltered coastal waters.

Pairing is brief, and courtship displays and vocal repertoire are relatively simple; the sexes share nest-building, incubation and care of young.

HABITAT: cool temperate to tropical areas of still water, usually with vegetated margins
LENGTH: 29–31 cm
DISTRIBUTION: more than 1 million km^2
ABUNDANCE: common
STATUS: secure

(CL Gill)

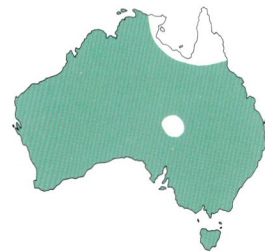

Genus *Tachybaptus*

(tak'-ee-bap'-tus; "fast-dipper")

Anatomically quite distinct from grebes of the genus *Podiceps* (especially in the structure of the hind limb), these small grebes also differ strikingly in their mating behaviour, in which a much reduced repertoire of display and courtship routines is associated with a greater dependence on a wider array of vocalizations. They are generally rather solitary, sedentary birds, living in dispersed pairs, foraging alone, and mainly inhabiting small sheltered bodies of water, although they frequently congregate in loose flocks in winter. They eat small fish and other aquatic animals. There are about four species, of which one occurs in Australia.

Australasian Grebe

Tachybaptus novaehollandiae (noh'-vee-hol-and'-ee-ee: "Australian fast-dipper")

The Australasian Grebe occurs in New Guinea, New Caledonia, Vanuatu and Australia; vagrants occur in New Zealand.

It inhabits wetlands of all kinds, but has a special fondness for small bodies of water—ornamental ponds, stock dams and the like—which are shunned by other grebes. It seldom occurs on salt water. It is mainly sedentary, or only locally dispersive, and lives in pairs. The diet includes small aquatic invertebrates and small fish.

It is strongly territorial when breeding; peak breeding activity takes place from January to April in the tropical north, from September to January in the south. Two or three successive broods may be raised. The usual clutch is four to six eggs.

HABITAT: cool temperate to tropical fresh water, often quite small areas
LENGTH: 23–26 cm
DISTRIBUTION: more than 1 million km²
ABUNDANCE: abundant
STATUS: secure

(AJ Olney)

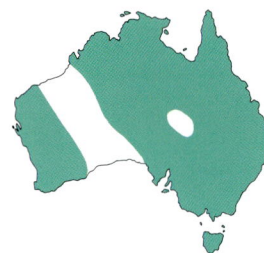

Order SPHENISCIFORMES

(sfen-is'-ee-for'-mayz: "*Spheniscus*-order", after a genus of penguins)

The characteristics of the order are essentially those of the single family.

Family SPHENISCIDAE

(sfen-is'-id-ee: "*Spheniscus*-family", after a genus of penguins)

About 17 species of penguins constitute this family of marine birds confined to the Southern Hemisphere. The wings are highly modified to serve as stiff flippers and the plumage is unusually tight and dense. Almost uniquely, the body is uniformly covered with feathers, there being no naked areas (apterya) between feather-tracts. Penguins cannot fly but propel themselves under water by means of a "flying" motion of the wings. They come ashore only to breed and moult.

Only one species breeds in Australia but several species visit at least occasionally, two with some frequency.

Genus Eudyptes

(ue-dip'-tayz: "good-diver")

The penguins belonging to this genus are all moderately large, with large, stout, deeply-grooved bills and relatively long tails; they are characterised especially by the possession of conspicuous yellow plumes or tassels on the sides of the head, the precise arrangement of these differing in detail from one species to another. The sexes are similar, except that males are somewhat larger and heavier than females.

Their distribution centres on the New Zealand region, extending to Macquarie Island and certain other subantarctic islands. There are four or five species, all of which have been recorded in Australia, all but two as very rare vagrants.

Rockhopper Penguin

Eudyptes chrysocome (krie'-soh-koh'-may: "golden-haired good-diver")

The Rockhopper Penguin breeds in small to enormous colonies on Heard, Macquarie and Marion Islands, on Tristan da Cunha and the Falkland Islands; and on various islands south of New Zealand. It does not breed in Australia but is a fairly frequent vagrant to Tasmania and elsewhere along the southern coast. Occurrences frequently involve individuals coming ashore to moult.

HABITAT: circumpolar coastal waters
LENGTH: *c* 61 cm
DISTRIBUTION: less than 10,000 km²
ABUNDANCE: very sparse
STATUS: secure

(G Robertson)

Fiordland Penguin

Eudyptes pachyrhynchus (pak'-ee-rink'-us: "thick-beaked good-diver")

(MF Soper)

The Fiordland Penguin breeds on Stewart Island and along the coast of the South Island of New Zealand. Nesting is solitary or in loose colonies; the site may be a cave or deep among the roots of trees in coastal forest. Adults seem to be largely sedentary but juvenile birds disperse widely and may not return to the breeding colonies until reaching sexual maturity at about five or six years of age.

The species occurs with some frequency as a vagrant to Victoria and Tasmania, often as individuals hauling out on some secluded beach to moult.

HABITAT: cool temperate to cold coastal waters
LENGTH: *c* 74 cm
DISTRIBUTION: not applicable
ABUNDANCE: rare to very sparse
STATUS: vulnerable

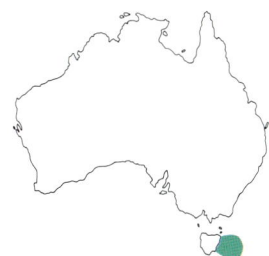

9

Genus Eudyptula

(yue-dip'-tue-lah: "little-good-diver")

The characteristics of the genus are those of the single species.

Little Penguin

Eudyptula minor (mie'-nor: "smaller little-good-diver")

Widely known as the Fairy Penguin, the Little Penguin inhabits the coast and offshore waters of southern Australia and New Zealand. Adults seem largely sedentary but immatures disperse widely.

Smallest of the penguins, adults weigh about 1 kilogram. The sexes are similar and immatures resemble adults; there is no seasonal variation.

Little Penguins are mainly solitary at sea but breed in colonies, usually on offshore islands, but also locally along the coast. Visited only at night, colonies vary in size from only a few pairs to several thousand.

The nest is always sheltered, usually in a burrow but sometimes in a mere crevice between rocks. Nesting usually begins in July, and two or three broods may be raised in succession. Two eggs form the clutch, incubated for 36 days. Chicks are brooded constantly for the first 20 days of life, then visited only to be fed; they fledge and go to sea at about 56 days of age. The average life-span is seven years, and many Little Penguins mate for life.

HABITAT: cool temperate coastal waters
LENGTH: 37–43 cm
DISTRIBUTION: 10,000–30,000 km²
ABUNDANCE: very sparse
STATUS: probably secure

(G Robertson)

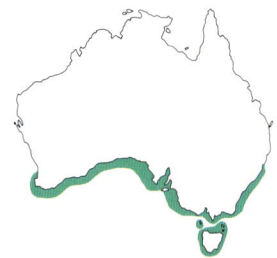

Order PROCELLARIIFORMES

(proh'-sel-ah-ree'-ee-for'-mayz: "*Procellaria*-order", after a genus of petrels)

Members of this order are immediately distinguishable at close quarters by having nostrils that are extended into two tubes lying on top of, or beside, the bill. The bill is hooked, and distinct grooves divide the horny covering into several distinct regions.

These so-called "tube-noses" comprise a group of strictly marine or pelagic species that normally come ashore only to nest, mostly on isolated oceanic islands. With few exceptions, the sexes are nearly identical in appearance (though males are generally slightly larger than females), there is no seasonal variation in plumage, and immatures closely resemble adults.

The pair bond is generally permanent, and both sexes cooperate in all phases of the nesting cycle. Colonial nesting is common, and many species—especially the smallest—nest in burrows, visited only at night. In all species the clutch consists of a single egg.

The order consists of four families (Diomedeidae, Procellariidae, Hydrobatidae, Pelecanoididae), all of which are represented in Australian waters.

Family DIOMEDEIDAE

(die'-oh-med-ay'-id-ee: "*Diomedea*-family", after a genus of albatrosses)

This family of very large seabirds has representatives in all oceans except the North Atlantic. The nasal tubes differ from those of other members of the order in being separate and relatively inconspicuous, situated at the sides of the upper mandible.

Squids figure prominently in the diet, but fish and crustaceans are also eaten, as well as galley refuse from ships. Most food is taken at the surface, and foraging is typically nocturnal.

There are 14 species in two genera (*Diomedea* and *Phoebetria*). Only one species breeds in Australia (several more on Macquarie Island), but nine species have been recorded in Australian seas, all persistently.

Genus Diomedea

(die'-oh-med-ay'-ah: after Diomede, a Trojan War hero)

These are small to very large albatrosses, characterised by a short, rounded tail, mainly white plumage (except the upperwing) and terrestrial courtship displays that involve elaborate rituals of bill clappering, bowing and dancing with wings outstretched. Several species show an intricate pattern of distinct successive plumages before maturing at five to nine years of age.

The smaller species are widely known as mollymawks, especially in New Zealand.

Buller's Albatross

Diomedea bulleri (bool'-er-ee: "Buller's Diomede", after W. Buller, New Zealand ornithologist)

This species breeds on Solander, Snares and Chatham Islands, New Zealand. Knowledge of its pelagic range is fragmentary, but it apparently disperses mainly eastwards into the southern Pacific.

It is an uncommon vagrant to seas off south-eastern Australia north to about Byron Bay, New South Wales, mostly from March to October.

Features important in field identification include the grey head, white forehead, and yellow ridge on the upper mandible.

The total population has been estimated to be about 30 000 breeding pairs. There are two subspecies, similar in appearance but with different life histories. *Diomedea b. bulleri*, which is rather sedentary, begins to breed in late summer at the Snares Islands, raising the young through autumn and winter. *D. b. platei* also begins breeding in late summer and then disperses across the Pacific as far as Chile.

HABITAT: southern oceans
LENGTH: *c* 76–81 cm
WINGSPAN: *c* 210 cm
DISTRIBUTION: not applicable
ABUNDANCE: rare
STATUS: secure

(*J Warham*)

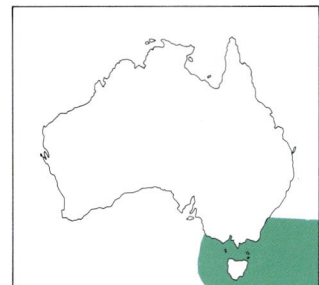

Shy Albatross

Diomedea cauta (kaw'-tah: "shy Diomede")

(*MF Soper*)

Also known as the White-capped Albatross, the Shy Albatross is generally solitary at sea. It follows ships but not persistently, and loiters about fishing vessels.

There are three distinct subspecies, regarded by some as species: *Diomedea c. cauta* breeds on islands off Tasmania and New Zealand; *D. c. salvini* breeds at the Bounty and Snares Islands and ranges eastward to South America; *D. c. eremita*, which breeds only at Pyramid Rock in the Chatham Islands, appears to be sedentary.

The Shy Albatross is common in south-eastern Australian waters from July to October. Numbers decline rapidly from about Sydney northward, and it is rare west of the Eyre Peninsula.

The single egg is laid in early October and hatches in late November; the chick fledges in mid-April when about four and a half months old.

HABITAT: southern oceans
LENGTH: *c* 99 cm
WINGSPAN: 198–256 cm
DISTRIBUTION: not applicable
ABUNDANCE: common
STATUS: secure

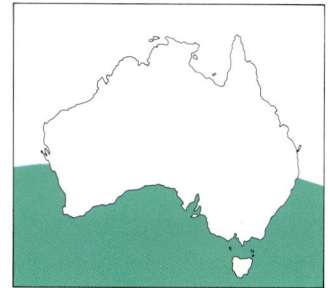

Yellow-nosed Albatross

Diomedea chlororhynchos (klor'-oh-rink'-os: "yellow-green-beaked Diomede")

The Yellow-nosed Albatross breeds at islands in the southern Atlantic and Indian Oceans. It favours distinctly warmer waters; it is also markedly less pelagic than other albatrosses, and is often common on shallower waters over the continental shelf.

It is common in Australian waters, from about Shark Bay in the west to about Brisbane in the east. It is much more numerous in the west and is most abundant from April to November. It is the smallest of the albatrosses, readily identified by the white head, mainly white underwing, and black bill with a yellow line along the upper ridge.

The nesting cycle lasts eight months, beginning in mid-August. The incubation period is about 78 days, the young fledging in about 130 days.

HABITAT: southern oceans
LENGTH: *c* 71–81 cm
WINGSPAN: 180–200 cm
DISTRIBUTION: not applicable
ABUNDANCE: common
STATUS: secure

(*T Howard*)

Grey-headed Albatross

Diomedea chrysostoma (kris'-oh-stoh'-mah: "golden-mouthed Diomede")

This albatross breeds on many subantarctic islands, including Macquarie and Campbell Islands in the Australasian region. Some 14 000 to 15 000 pairs breed at South Georgia, which is the largest breeding station; Campbell Island has a substantial breeding population, but few breed at Macquarie Island. Essentially a cold-water species, it favours oceans south of 35°S. It is fairly common in Australian seas, especially east and south of Tasmania and off the west and south coasts of the continent.

It is generally solitary at sea, and seldom follows ships. Crucial identification features include the grey head, a broad black leading edge to the underwing, and a black bill with a yellow line along both upper and lower ridges. Adult plumage is reached in five to seven years.

Breeding begins in October in dispersed colonies. The single egg is incubated for about 72 days; the chick is brooded for 18 to 28 days and fledges at about 140 days. At least at South Georgia, there is a two-year breeding cycle.

HABITAT: southern oceans
LENGTH: *c* 81 cm
WINGSPAN: *c* 220 cm
DISTRIBUTION: not applicable
ABUNDANCE: not applicable
STATUS: probably secure

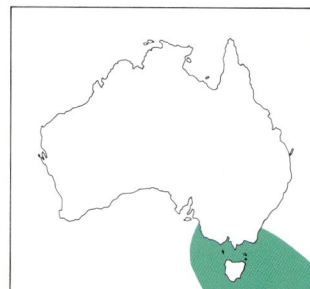

(MF Soper)

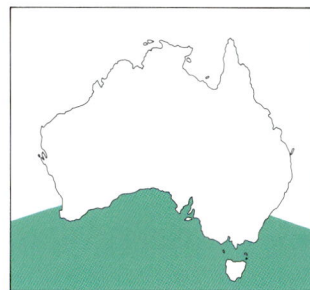

Royal Albatross

Diomedea epomophora (ep'-oh-moh-for'-ah: "shoulder-bearing Diomede")

The Royal Albatross breeds only on some islands off New Zealand and at Taiaroa Head near Dunedin on the South Island. Its pelagic range appears to be circumpolar, and it is seen with some frequency off south-eastern Australia.

The Royal Albatross closely resembles the Wandering Albatross in behaviour and appearance (the only field character useful at all ages is a narrow line of black along the cutting edges of both mandibles of the bill of the Royal Albatross).

The single egg is laid in late November or early December and hatches in about 79 days; the chick is brooded for about five weeks and fledges at about 240 days.

(J Warham)

HABITAT: southern oceans
LENGTH: *c* 120 cm
WINGSPAN: *c* 340 cm
DISTRIBUTION: not applicable
ABUNDANCE: rare to very sparse
STATUS: probably secure

Wandering Albatross

Diomedea exulans (ex'-ue-lanz: "wandering Diomede")

The Wandering Albatross breeds on many subantarctic islands, including Macquarie and (probably) Heard Islands. Its normal pelagic range encompasses the entire Southern Ocean, north to the Tropic of Capricorn and, rarely, beyond. It is markedly more numerous in winter and spring than at other seasons.

Sexual maturity is reached at about nine years after an intricate series of plumage changes; populations breeding on various islands display different patterns of progression to maturity.

The complete nesting cycle takes about 11 months, so successful pairs breed every second year. The nest, built mainly by the female from material fetched by the male, is a pedestal of mud, peat and vegetation up to half a metre high. The single egg is incubated by both sexes in alternate shifts lasting several weeks, until it hatches in 70 to 80 days. The chick fledges and goes to sea at an age of about 270 to 280 days.

HABITAT: southern oceans
LENGTH: *c* 120 cm
WINGSPAN: *c* 340 cm
DISTRIBUTION: not applicable
ABUNDANCE: common
STATUS: secure

(*J Warham*)

Black-browed Albatross

Diomedea melanophrys (mel'-an-of'-ris: "black-browed Diomede")

The Black-browed Albatross breeds on many subantarctic islands, including Heard and Macquarie Islands. Colonies vary greatly in size and are often associated with Grey-headed Albatrosses. The incubation period is about 68 to 71 days; the chick is brooded constantly for about 21 days, and fledges at about 120 days.

HABITAT: southern oceans
LENGTH: *c* 90 cm
WINGSPAN: *c* 230 cm
DISTRIBUTION: not applicable
ABUNDANCE: common
STATUS: secure

(*G Robertson*)

This species is probably the commonest albatross in Australian waters, north at least to about Fremantle in Western Australia and to Sydney in New South Wales, particularly from May to October. Its pelagic range is extensive, but it is markedly more abundant in the southern Atlantic than elsewhere.

It is not especially gregarious but hundreds of birds may form aggregations at occasional rich food sources such as the offal from fishing trawlers. It frequently follows ships. Adults are readily identified by the orange bill and the very small area of white in the otherwise mainly dark underwings.

14

Genus *Phoebetria*

(fee-bet'-ree-ah: "prophetess")

This genus consists of two similar and very closely related species, which differ chiefly in ecology: the Sooty Albatross is essentially a temperate species while the Light-mantled Albatross is associated with subantarctic waters and has the most 'southerly usual distribution of all albatrosses. The distinction is not clear-cut, and the two species breed together on several islands.

Usually silent and solitary at sea, *Phoebetria* albatrosses differ from *Diomedea* albatrosses most conspicuously in their dark, mainly sooty brown plumage, long wedge-shaped tails, and extraordinarily graceful, buoyant flight. The lower mandible has a long lateral groove (the sulcus) lined with a fleshy membrane.

Sooty Albatross

Phoebetria fusca (fus'-kah: "dusky prophetess")

The Sooty Albatross breeds on islands in the southern Atlantic and Indian Oceans, and its pelagic range is generally restricted to these oceans, between 30°S and 60°S latitude. It occurs fairly regularly from Western Australia across the Great Australian Bight to Tasmania.

Adult Sooty Albatrosses are relatively easily identified at sea, being completely dark slate-brown (including the mantle) and having a yellow sulcus.

The breeding cycle lasts about 10 months, beginning in August (mid-July on Tristan da Cunha). The incubation period is about 70 days; the chick is brooded almost constantly for its first 21 days and fledges at about 160 days.

(*JC Sinclair*)

HABITAT: southern oceans
LENGTH: 84–89 cm
WINGSPAN: *c* 203 cm
DISTRIBUTION: not applicable
ABUNDANCE: rare
STATUS: probably secure

Light-mantled Albatross

Phoebetria palpebrata (pal'-pe-brah'-tah: "eyelid-marked prophetess")

albatrosses, it tends to nest in dispersed single pairs, or as loose groups of pairs.

The breeding cycle lasts about 10 months, beginning in October (mid-September at the Crozets). The incubation period is about 66 to 69 days; the chick is brooded almost constantly for about 21 days, and fledges in 140 to 150 days.

HABITAT: southern oceans
LENGTH: 79–89 cm
WINGSPAN: 183–218 cm
DISTRIBUTION: not applicable
ABUNDANCE: rare
STATUS: secure

(*MF Soper*)

The Light-mantled Albatross breeds on many subantarctic islands (including Heard and Macquarie Islands) and has a circumpolar range, mainly south of the Antarctic Convergence. There are about 20 records in Australian territorial waters, mostly along the southern coast.

It closely resembles the Sooty Albatross in appearance and habits, but adults have a pale grey mantle and a blue (not yellow) sulcus on the lower bill. Unlike other

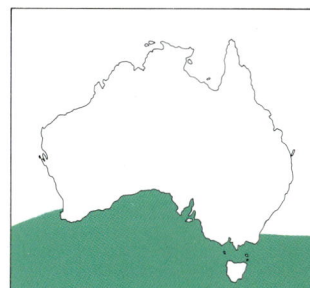

Family PROCELLARIIDAE

(proh'-sel-ah-ree'-id-ee: "*Procellaria*-family," after a genus of petrels)

This is the largest and most diverse family of tube-noses, consisting of about 62 species, of which about 41 have been recorded in Australian waters; nine breed in the region. Four groups may be recognised: fulmars (*Macronectes, Fulmarus, Daption*), gadfly-petrels (*Pterodroma*), prions (*Pachyptila*) and shearwaters (*Procellaria, Calonectris, Puffinus*); the monotypic genus *Halobaena* is commonly associated with the prions but has some characteristics linking it with gadfly-petrels.

Fulmars are gregarious and quarrelsome, and fly with a distinctive series of stiff "busy" flaps followed by extended glides. They follow ships and often attend fishing trawlers and whaling stations, squabbling in flocks for offal. The diet includes fish, squid, carrion and galley refuse from ships.

Gadfly-petrels are generally solitary at sea, and seldom form dense breeding colonies. Many are tropical in distribution. Polymorphism (the existence of more than one colour form) is common. Few species persistently follow ships, and most ignore them. The diet consists chiefly of squid and fish, gathered mainly at night.

Prions are small and gregarious. Plumage is mainly blue-grey above and white below, with a black-tipped tail. Their flight is distinctive: low, fast and erratic, twisting from side to side.

Shearwaters occur in all oceans. They are strongly gregarious. For many species (especially of *Puffinus*) the characteristic style of flight is low, fast and direct with rapid "stiff-armed" wing-beats. Most species feed largely on krill and fish, for which some regularly dive.

The pair bond is persistent, and both sexes cooperate at all stages of breeding. Nesting is colonial, sometimes mingled with other species; few species build nests, the single egg being usually laid at the end of a burrow, or in a crevice between boulders or on a cliff ledge. Pairs return to the nesting island for a period to establish ownership and refurbish their burrow, then return to sea while the female forms her egg (the so-called "honeymoon" period). The egg is incubated by both sexes in alternate shifts lasting several days. Once hatched, the chick is brooded almost constantly for several days, then abandoned, being visited only to be fed. Typically, visits to the nesting island take place only at night.

Genus Calonectris

(kal-oh-nek'-tris: "beautiful-swimmer")

This genus is in many respects intermediate between *Procellaria* and *Puffinus*. Both included species breed in the Northern Hemisphere, migrating southwards in winter. Only one, the Streaked Shearwater, occurs in Australian seas; the other is more or less restricted to the Atlantic Ocean.

Streaked Shearwater

Calonectris leucomelas (lue'-koh-mel'-as: "black-white beautiful-swimmer")

This large, unmistakable shearwater breeds on offshore islands in Japan, Korea, northern China, and the Soviet Union (near Vladivostok). After breeding it disperses southward into tropical waters, reaching Indonesia and New Guinea and, with apparently increasing frequency, the Coral Sea and coastal waters of northern and eastern Australia, south to Victoria (it was unrecorded in Australian waters before 1974). Most Australian records are between November and May.

It is gregarious at all seasons, and is most frequently encountered in parties or large flocks; even solitary birds readily join flocks of other shearwaters. The diet consists mainly of fish.

Very numerous in Japanese waters, it breeds in that country in large, dense colonies, generally in forested hills near the sea. Birds return to their colonies in March, most eggs being laid in May or June. The incubation period is about 54 days, and chicks fledge at about 66 days. At many islands, young are gathered for their meat and oil in much the same manner as Short-tailed Shearwaters are harvested in Australia.

(IJ Skira)

HABITAT: temperate and subtropical seas
LENGTH: *c* 48 cm
WINGSPAN: *c* 122 cm
DISTRIBUTION: not applicable
ABUNDANCE: rare
STATUS: secure

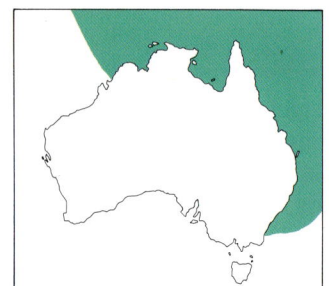

Genus *Daption*

(dap'-tee-on: anagram of *pintado*, Portuguese for "painted")

The characteristics of the genus are essentially those of the single species.

Cape Petrel

Daption capense (kah-pen'-say: "Cape [of Good Hope] *daption*")

Widely known as the Cape Pigeon or Pintado Petrel, the Cape Petrel is easily identified by its distinctive black-and-white "chequerboard" plumage.

It has a very extensive pelagic distribution, commonly penetrating into the Northern Hemisphere, but it is essentially a bird of the southern oceans, and its numbers decrease markedly north of about 25°S. In Australian waters it is very common at least as far north as Perth in the west and Sydney in the east; peak numbers are recorded between June and October.

The diet includes krill, squid and small fishes but it is an opportunistic feeder and many birds will trail in the wake of ships, scavenging for garbage and galley refuse. Noisy, gregarious and quarrelsome, they are usually encountered in small parties, but flocks of hundreds often congregate around fishing trawlers, groups of whales, or other local sources of abundant food.

Cape Petrels breed at a number of places along the shores of Antarctica and on many subantarctic islands including South Georgia and South Shetland Islands, the Crozets, and Kerguelen, Heard (possibly also Macquarie), Antipodes, Bounty, Snares and Campbell Islands. Colonies are variable in size but seldom very large. Nests are mere scrapes lined with chips of granite and a few wisps of grass, constructed on rock ledges, in caves or in crevices between jumbled rocks and boulders.

Birds begin breeding at about four years of age. Incubation lasts about 45 days; the chick is brooded continuously for the first 8 to 10 days, reaches adult weight at about 21 days and fledges at about 50 days.

(H & J Beste)

HABITAT: southern oceans
LENGTH: 38–40 cm
WINGSPAN: 81–91 cm
DISTRIBUTION: not applicable
ABUNDANCE: common to abundant
STATUS: secure

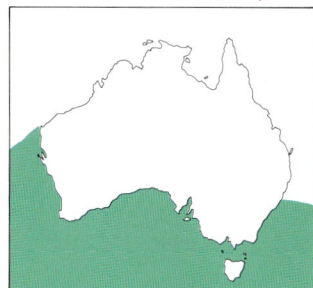

Genus *Fulmarus*

(fool-mar'-us: "foul-gull")

The genus *Fulmarus* has a bipolar distribution, with one species abundant in the northern Pacific and Atlantic Oceans and another in the Southern Hemisphere. Nesting is generally in dense colonies (visited by day) on steep rocky slopes or cliffs, and is highly synchronised. Courtship displays occur at the nest, which consists of a sheltered niche on rock face, lined with stone chips.

17

Southern Fulmar

Fulmarus glacialoides (glas'-ee-al-oy'-dayz: "like-*Fulmar glacialis* fulmar", a similar species in the Northern Hemisphere)

Also called the Antarctic Fulmar or Silver-grey Petrel, the Southern Fulmar breeds on South Sandwich, South Orkney, South Shetland, South Georgia, Bouvet and Peter Islands and at scattered sites of the Antarctic continent. Its pelagic range is circumpolar, lying mainly between the pack-ice zone and about 45°S. It is normally a rare vagrant in Australian waters but is prone to periodic wrecks, when beach-washed derelicts and sightings at sea may be recorded in large numbers off south-eastern South Australia, Victoria and Tasmania (where it is possibly regular). It is very rare in Western Australia and New South Wales.

Almost unmistakable, it is the only pale grey-and-white petrel in the Southern Hemisphere. It is generally silent at sea but noisy at its nesting colonies, where common calls include a rapid "kerk-kerk-kerk" and other shrill rattling cackling notes.

The season extends from October to March, and laying occurs in November and December. The egg is incubated for 43 to 44 days; the chick is brooded almost constantly for 15 to 18 days, then abandoned, visited only to be fed. It fledges in about 51 days.

HABITAT: southern oceans
LENGTH: 46–50 cm
WINGSPAN: 114–120 cm
DISTRIBUTION: not applicable
ABUNDANCE: very rare
STATUS: secure

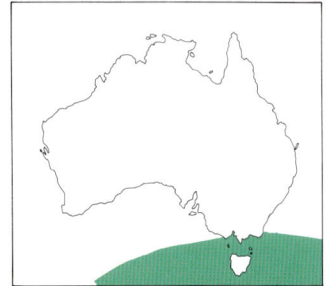

(R Puddicombe)

Genus *Halobaena*

(hal'-oh-bee'-nah; "sea-walker")

A monotypic genus of uncertain affinities, *Halobaena* seems most closely related to the gadfly-petrels (*Pterodroma*), but it closely resembles prions (*Pachyptila*) in general appearance and behaviour.

Blue Petrel

Halobaena caerulea (see'-rue-lay'-ah: "blue sea-walker")

The Blue Petrel breeds on South Georgia, Prince Edward and Marion Islands, Crozets, and Kerguelen and Macquarie Islands, and perhaps also on islands off Cape Horn. A bird of cold subantarctic waters, its pelagic range is circumpolar between the limits of pack-ice and about 40°S. It is an infrequent visitor to Australian waters, and then chiefly in the south-west, becoming gradually less frequent eastward along the southern coast; it is extremely rare in New South Wales and so far unrecorded in Queensland. It is subject to erratic wrecks, when large numbers may be cast ashore, dead or distressed, after severe winter storms. It is recorded most frequently in September and October. The white tip of the tail is diagnostic and a useful field character in identification.

The Blue Petrel is often seen in flocks of a hundred or so, but it also joins flocks of prions and other seabirds. It associates freely with whales and frequently follows ships. The diet is mainly krill.

Nesting colonies, in stretches of soft dry soil on coastal slopes, are occupied in early September. The egg is laid at the end of a burrow and hatches in about 46 days, and the young fledge in late February or early March; adults desert the colony some weeks later, but continue to visit sporadically throughout the winter.

HABITAT: subantarctic oceans
LENGTH: 28–30 cm
WINGSPAN: 58–66 cm
DISTRIBUTION: not applicable
ABUNDANCE: very rare
STATUS: secure

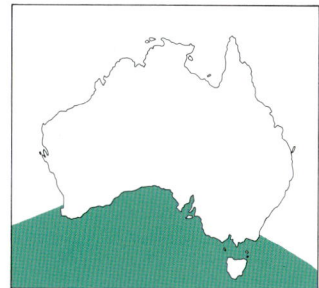

(MJ Carter)

Genus *Lugensa*

(lue-gen'-zah: "mourner", or sombrely clad)

Showing a number of features in common with the fulmars, the sole member of this genus differs from *Pterodroma* in its extremely laterally compressed bill, large head and very large eyes, and in its strikingly soft and fluffy plumage. It is strongly pelagic, and essentially a nocturnal predator on squids.

Kerguelen Petrel

Lugensa brevirostris (brev'-i-ros'-tris: "short-beaked mourner")

Ranging widely over the Southern Ocean from the limit of pack-ice to about 30°S, this petrel breeds at Tristan da Cunha, Gough, Marion and Prince Edward Islands, the Crozets and Kerguelen Island. Unknown in Australia before the first beach-washed derelict was found in 1926, it is still generally rare and seldom reported, but it is subject to periodic wrecks when hundreds may be recorded; two such events occurred in 1981 and 1984. It is most frequently reported in Western Australia, Victoria and Tasmania.

It nests in dispersed colonies, which are mainly occupied by early September, and most eggs are laid in mid-October. The incubation period is about 50 days and the nestling period is about 60 days, the chicks fledging in late January.

(JC Sinclair)

HABITAT: subantarctic oceans
LENGTH: *c* 36 cm
WINGSPAN: *c* 81 cm
DISTRIBUTION: not applicable
ABUNDANCE: very rare
STATUS: secure

19

Genus *Macronectes*

(mak'-roh-nek'-tayz: "large-swimmer")

This genus has two members, which are extremely similar and are by far the largest members of the family, approaching albatrosses in size. Both have a predominantly dark brown plumage. They are difficult to distinguish, the main difference between them lying in their ecological requirements: one is characteristic of temperate waters, the other of more southern, subantarctic waters.

Southern Giant-petrel

Macronectes giganteus (jee'-gan-tay'-us: "giant large-swimmer")

The Southern Giant-petrel breeds on many subantarctic islands, including Heard and Macquarie Islands, and at scattered sites on the Antarctic continent. Its has been recorded at sea over much of the Southern Hemisphere, north at least to 10°S. In Australian waters it is common off the southern coast, especially from June to September.

The unmistakable white morph makes up about 10 per cent of the population, but dark-plumaged birds are extremely difficult to distinguish from the Northern Giant-petrel: adult Southern Giant-petrels have mainly white heads and the bill has a pale green (not reddish) tip.

Southern Giant-petrels breed in loose colonies, sometimes involving several hundred pairs, on level ground on open coastal plateaus or headlands. Incubation lasts about 59 days, conducted in shifts (the male takes the first) of 2 to 12 days; the chick is brooded almost constantly for about 18 days, and fledges in about 115 days.

HABITAT: southern oceans
LENGTH: 86–99 cm
WINGSPAN: 185–205 cm
DISTRIBUTION: not applicable
ABUNDANCE: common
STATUS: secure

(G Robertson)

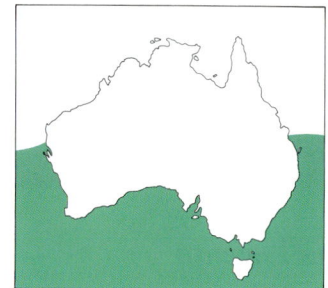

Northern Giant-petrel

Macronectes halli (haw'-lee: "Hall's large-swimmer", after Robert Hall, Australian ornithologist)

The Northern Giant-petrel breeds on many oceanic islands in the southern Indian Ocean and the New Zealand region, including Macquarie Island. Its pelagic range extends over most of the Southern Hemisphere but

(MF Soper)

mainly north of the Antarctic Convergence and south of the Tropic of Capricorn. It is common in Australian waters, though apparently somewhat less so than the Southern Giant-petrel, ranging north to Shark Bay in the west and Fraser Island, Queensland, in the east. It is most numerous in winter.

Identification at sea is extremely difficult: in mature Northern Giant-petrels the face is white, the crown dark, giving a distinct "dark-capped" appearance; the tip of the bill is dark red.

The breeding season is from August to February (beginning some five to eight weeks earlier than that of the Southern Giant-petrel). The incubation period is about 60 days;

the chick is brooded almost constantly for about 18 days; fledging takes about 108 days.

HABITAT: southern oceans
LENGTH: 81–94 cm
WINGSPAN: 180–200 cm
DISTRIBUTION: not applicable
ABUNDANCE: common
STATUS: secure

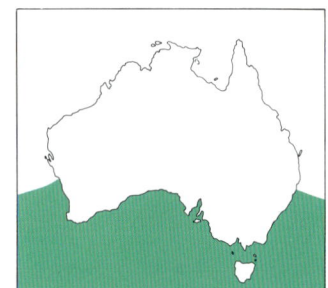

Genus *Pachyptila*

(pak'-ip-til'-ah: "thick-feather")

Pachyptila is a genus of six species of southern seabirds that are nearly identical in appearance. All have been recorded in Australian waters, several only as rare vagrants. The flight is brisk, with frequent bursts of stiff fluttering wing-beats. Strongly gregarious, prions are generally encountered in flocks. They ignore ships.

The diet consists mainly of small or very small pelagic crustaceans, and prions are peculiar in having a comb-like structure of lamellae along both sides of the palate (so well developed in some species that it is visible even when the bill is closed). This functions as a strainer during foraging, in a manner analogous to that of some whales: water is held in the mouth and expelled through the lamellae, trapping food items among the comb-like processes.

Slender-billed Prion

Pachyptila belcheri (bel'-cher-ee: "Belcher's thick-feather", after Sir Charles Belcher, Australian naturalist)

The Slender-billed Prion breeds at the Falklands, the Crozets and Kerguelen Island. It appears to be the most migratory of prions and (unlike other prions), it is very common in the southern Pacific. It is a frequent winter visitor to seas off southern Australia, especially in the south-west, but there are few sight records and beach-washed specimens are usually cast up only after winter storms.

It is scarce along the east coast although specimens have been found as far north as Fraser Island, Queensland.

HABITAT: southern oceans
LENGTH: *c* 26 cm
WINGSPAN: *c* 56 cm
DISTRIBUTION: not applicable
ABUNDANCE: very rare
STATUS: probably secure

(A Greensmith)

Antarctic Prion

Pachyptila desolata (de'-sol-ah'-tah: "Desolate [Island] thick-feather", referring to an earlier name of Kerguelen Island)

This prion has the most southerly breeding distribution of its group: it breeds at Cape Dennison, Antarctica; and on South Georgia, South Orkney, Kerguelen, Heard, Macquarie and Auckland Islands. Its pelagic range lies mainly in the southern Indian and Atlantic Oceans, and it is fairly common, at least as a beach-washed derelict, along the southern and eastern coastline of Australia between Perth and Sydney, erratically further north.

It breeds in dense colonies that are occupied in late October and early November. Eggs are laid throughout December; incubation shifts are rotated every 3 to 4 days throughout the incubation period of about 45 days. Most chicks hatch around the end of February, and fledge about 50 days later, leaving the colony in late March or early April.

HABITAT: southern oceans
LENGTH: *c* 25 cm
WINGSPAN: *c* 58 cm
DISTRIBUTION: not applicable
ABUNDANCE: very sparse
STATUS: probably secure

(MF Soper)

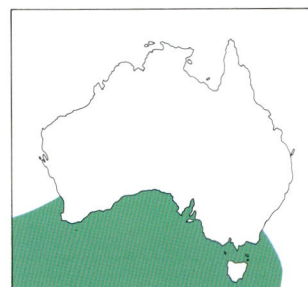

Salvin's Prion

Pachyptila salvini (sal-vin'-ee: "Salvin's thick-feather", after O. Salvin, ornithologist)

Salvin's (or the Lesser Broad-billed) Prion breeds on Marion and Prince Edward Islands and the Crozets in the southern Indian Ocean. Its pelagic range is uncertain, but includes the southern Atlantic and Indian Oceans north to about 10°S, and involves an eastwards dispersal into Australasian waters after breeding. It is a frequent beach-washed derelict in south-western Australia, becoming progressively less numerous along the southern coast.

The breeding season extends from October to April, with egg-laying occurring from November to January.

The species is often regarded as conspecific with the Broad-billed Prion, derelict specimens of which are occasionally found on southern Australian beaches.

HABITAT: southern oceans
LENGTH: *c* 25 cm
WINGSPAN: *c* 58 cm
DISTRIBUTION: not applicable
ABUNDANCE: sparse
STATUS: secure

(T Pescott)

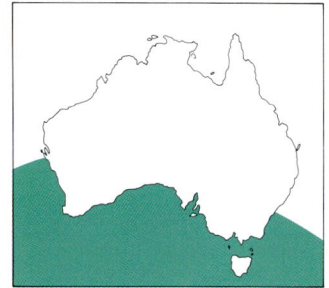

Fairy Prion

Pachyptila turtur (ter'-ter: "dove thick-feather")

The Fairy Prion breeds on Marion Island in the Indian Ocean; in Australia on a number of islands off Victoria, in Bass Strait and around Tasmania; and on various islands around New Zealand. It seems to be relatively sedentary, being common in all seasons in Australasian waters, north at least occasionally to the Tropic of Capricorn.

The breeding season extends from August to February, and eggs are laid mainly in November. Colonies are often large and dense. Incubation lasts about 56 days, conducted in shifts averaging 6 to 7 days; once hatched, the chick is brooded for 2 to 4 days, and fledges after about 44 to 45 days, leaving the colony in February.

(G Robertson)

HABITAT: subantarctic and southern temperate oceans
LENGTH: 25–28 cm
WINGSPAN: 56–60 cm
DISTRIBUTION: not applicable
ABUNDANCE: common
STATUS: secure

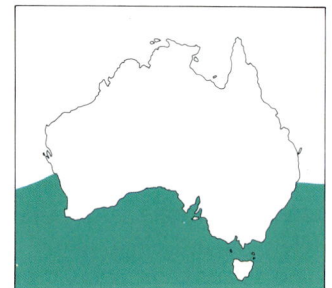

Genus *Procellaria*

(proh'-sel-ah'-ree-ah: "stormy[-bird]")

The four species in this genus are restricted to the Southern Hemisphere, and all have been recorded in Australian seas. The White-chinned Petrel occurs with some frequency but the status of the other three is uncertain: two (the Black and Westland Petrels) breed in New Zealand and are recorded with apparently increasing frequency at sea off the Australian east coast. They are very difficult to identify at sea.

They are big and burly and, unlike shearwaters, typically fly with slow, deliberate wing beats. The plumage is almost entirely dark brown or blackish, and the bill is short, stout and pale.

White-chinned Petrel

Procellaria aequinoctialis (ee'-kwin-ok'-tee-ah'-lis: "equinoctial storm-bird")

Otherwise known as the Shoemaker or Spectacled Petrel, this large, burly petrel breeds on several islands in the southern Atlantic and Indian Oceans, and near New Zealand (possibly including Macquarie Island). Its pelagic range extends around the Southern Ocean, mainly between 55°S and 30°S but regularly extending north along the Humboldt Current to Peru. Uncommon in Australian seas, it is most frequently recorded off the southern coast.

At sea, the white chin, together with the pale bill, is a good field character, but the first feature is very variable and may be lacking altogether; some individuals have irregular white patches elsewhere on the head, a characteristic that is most frequent in the population breeding on Tristan da Cunha (subspecies *P. a. conspicillata*).

The breeding season is variable and extended, its onset varying widely from colony to colony, but typically extending from about October to March. The egg is laid on a mud pedestal in a chamber at the end of a burrow 1 to 2 metres in length. The incubation period is about 60 days.

(MF Soper)

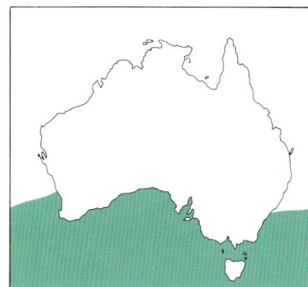

HABITAT: southern oceans
LENGTH: 51–58 cm
WINGSPAN: *c* 340 cm
DISTRIBUTION: not applicable
ABUNDANCE: rare
STATUS: secure

Westland Petrel

Procellaria westlandica (west-land'-ik-ah: "Westland storm-bird", referring to Westland district, South Island, New Zealand)

The Westland Petrel breeds in coastal mountain ranges on the South Island of New Zealand, but knowledge of its distribution at sea is vague. Since about 1980 it has been recorded with apparently increasing frequency off the coast of south-eastern Australia. It is very similar to the Flesh-footed Shearwater, and earlier occurrences may have been overlooked: it may prove to be a relatively frequent visitor to Australian waters.

HABITAT: cool temperate seas
LENGTH: *c* 51 cm
WINGSPAN: *c* 137 cm
DISTRIBUTION: not applicable
ABUNDANCE: very rare
STATUS: vulnerable

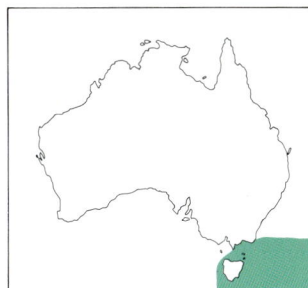
(IL McVinnie)

Genus *Pseudobulweria*

(sue'-doh-bool-wer'-ee-ah: "false-*Bulweria*", after a genus of petrels)

Members of this genus differ from *Pterodroma* (with which they have often been associated) in a range of anatomical, structural, behavioural and parasitological features, and they show a number of links with *Procellaria*. Their plumage is mainly dark, the nostrils are divided at the front by a bony septum, and the arrangement of their upper intestines is relatively simple. It is an essentially tropical group and only one species occurs in Australian waters with any frequency.

Tahiti Petrel

Pseudobulweria rostrata (ros-trah'-tah: "beaked false-*Bulweria*")

This is a bird of tropical seas which breeds in New Caledonia, the Society Islands and the Marquesas. Almost nothing is known of its distribution at sea or its life history, but it is recorded not infrequently just beyond the east coast continental shelf, south to about Wollongong in New South Wales. The breeding season is reportedly from January to May.

HABITAT: tropical and subtropical seas
LENGTH: 38–40 cm
WINGSPAN: c 84 cm
DISTRIBUTION: not applicable
ABUNDANCE: very rare
STATUS: vulnerable to probably secure

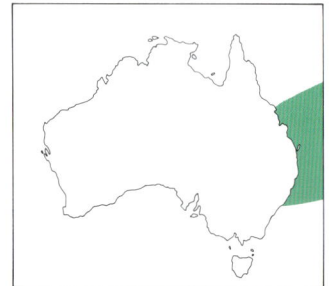

(F Hannecart)

Genus *Pterodroma*

(te'-roh-droh'-mah: "winged-runner")

Members of this genus are set apart from all other procellariiform birds by the coiled arrangement of the upper intestine, as well as a number of other anatomical features. Strongly pelagic, and largely solitary at sea, these birds generally ignore ships. Most are extremely difficult to identify, requiring specialised knowledge and experience. The diet consists mainly of squid and fish, gathered mainly at night. They are silent at sea, but often noisy at their breeding islands. They breed in colonies which, with some exceptions, are visited only at night; high-speed aerial chases low over the breeding grounds are often a conspicuous element of courtship. Adults share nest-building, incubation and care of young. Once hatched, the chick is brooded for several days then visited only to be fed.

Herald Petrel

Pterodroma arminjoniana (ar'-min-zhon'-ee-ah'-nah: "Arminjon's winged-runner", after Captain Arminjon, commander of the *Magenta*)

Also known as the Trinidade (or Trinidad) Petrel, this species breeds at a number of subtropical islands in the southern Atlantic, Indian and Pacific Oceans. It seems to be largely sedentary. The two subspecies (*P. a. arminjoniana* [Trinidade Petrel] of the South Atlantic and Indian Oceans, and *P. a. heraldica* [Herald Petrel] of the Pacific) are sometimes treated as distinct species.

There are few records in Australian waters (it was entirely unknown in Australia until 1959), but in 1980 it was discovered breeding on Raine Island off northern Queensland, where it is now known to breed regularly in small numbers.

Polymorphism is marked and intricate, with much individual variation; field identification is extremely difficult.

Little is known of the breeding cycle, but the season seems unusually extended, and colonies are visited throughout the year. Colonies are dispersed, and the egg is deposited in a crevice among rocks.

HABITAT: subtropical and tropical oceans
LENGTH: 35–39 cm
WINGSPAN: 88–102 cm
DISTRIBUTION: not applicable
ABUNDANCE: rare
STATUS: probably secure

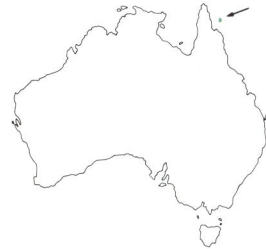

(*AD Forbes-Watson*)

White-headed Petrel

Pterodroma lessonii (les-son'-ee-ee: "Lesson's winged-runner", after R. P. Lesson, French zoologist)

Aptly named, the White-headed Petrel closely resembles the Great-winged Petrel in all but appearance. Breeding colonies are known on several islands in the southern Indian Ocean and around New Zealand, including Macquarie Island. Its pelagic range is circumpolar, generally south of 30°S latitude, but it is markedly more common in the southern Indian and Pacific Oceans than in the South Atlantic; it is generally uncommon in Australian seas, being most frequent off Tasmania. Recorded in all months, it is perhaps most numerous between June and September.

Breeding takes place in dispersed colonies. The egg, laid in late November or December, is deposited in a chamber at the end of a burrow a metre or more in length. Incubation takes some 58 days and the young spend about 102 days in the burrow before fledging sometime in May.

HABITAT: cool temperate southern oceans
LENGTH: 40–46 cm
WINGSPAN: c 109 cm
DISTRIBUTION: not applicable
ABUNDANCE: rare
STATUS: secure

(*T Palliser*)

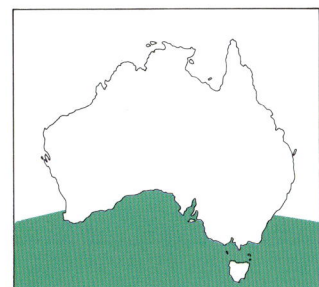

Gould's Petrel

Pterodroma leucoptera (lue-kop'-te-rah: "white-winged winged-runner")

Also called the White-winged Petrel, Gould's Petrel breeds in New Caledonia and also has a colony of an estimated 250 to 300 pairs on Cabbage Tree Island near Port Stephens, New South Wales. Its pelagic range is largely unknown, but it has been recorded at sea south to Tasmania, and New Zealand, the Galapagos Islands and near Tonga, and it is suspected of migrating to the eastern Pacific along the Subtropical Convergence.

At Cabbage Tree Island the colony is occupied from October to May. Egg laying takes place from late November to early December; the chicks hatch during December and fledge by April. Nests are occasionally in shallow burrows but mostly in crevices between boulders, or under tangles of fallen palm fronds.

HABITAT: tropical and subtropical oceans
LENGTH: *c* 30 cm
WINGSPAN: *c* 71cm
DISTRIBUTION: not applicable
ABUNDANCE: very sparse
STATUS: vulnerable

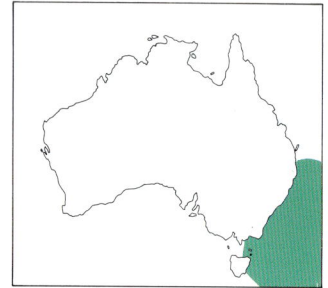

(PJ Fullegar)

Great-winged Petrel

Pterodroma macroptera (mak-rop'-te-rah: "long-winged winged-runner")

Also called the Grey-faced Petrel, this petrel breeds on many islands in the southern Atlantic and Indian Oceans and in the New Zealand region, as well as on offshore islands in south-western Australia. Its pelagic range is circumpolar in the subtropics, mainly between 25°S and 50°S latitudes. Birds of the New Zealand breeding population (subspecies *P. m. gouldi*) are common off the coasts of New South Wales, Victoria, Tasmania and (less so) south-eastern Queensland and South Australia.

Adults are generally sedentary, but immature birds disperse widely. Feeding is mainly at night, and cephalopods (squids, cuttlefishes, etc.) constitute the major item of diet.

Great-winged Petrels breed during the winter. The nests are either on the surface (in crevices between rocks, nestled between tree roots or beneath dense shrubs) or, more frequently, in a chamber at the end of a burrow. The breeding cycle occupies nine to 10 months. Some birds may return as early as late January, but maximum courtship activity occurs in April. Females leave the colony on a "honeymoon" exodus lasting two months, but males remain away only about seven weeks. The single egg hatches in about 55 days; the chick takes about 120 days or more to fledge.

HABITAT: cool temperate southern oceans
LENGTH: *c* 41 cm
WINGSPAN: *c* 109 cm
DISTRIBUTION: not applicable
ABUNDANCE: sparse to common
STATUS: probably secure

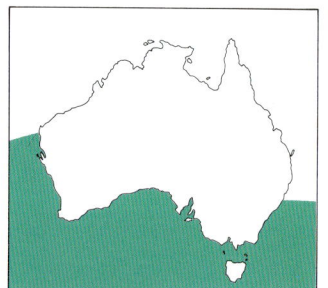

(D Garrick)

Soft-plumaged Petrel

Pterodroma mollis (mol'-is: "soft winged-runner")

The Soft-plumaged Petrel breeds at Gough Island, Tristan· da Cunha, Prince Edward and Marion Islands, the Crozets, and the Antipodes Islands south of New Zealand. Strongly pelagic, it ranges across the southern Atlantic and Indian Oceans, mainly between 25°S and 60°S. It is recorded with some frequency in south-western Australia, but occurrences decline markedly towards the south-east, and there are only a handful of records from the east coast.

It is mainly solitary at sea, but sometimes occurs in small parties. The diet is mainly krill, and foraging apparently largely diurnal; it occasionally approaches ships.

Breeding occurs in dispersed colonies, nests being in burrows, usually in dense vegetation. Colonies are occupied from September, and laying takes place in November.

HABITAT: southern oceans
LENGTH: 32–37 cm
WINGSPAN: 83–95 cm
DISTRIBUTION: not applicable
ABUNDANCE: very rare
STATUS: probably secure

(J Warham)

Kermadec Petrel

Pterodroma neglecta (neg-lek'-tah: "neglected winged-runner")

The Kermadec Petrel breeds at Balls Pyramid (Lord Howe Island), and at a number of islands in the New Zealand region and across the southern Pacific Ocean. Little is known of its pelagic distribution; adults may be sedentary but an annual transequatorial dispersal has been detected (perhaps involving mainly immatures) into the northern Pacific. It is almost unknown over Australian inshore waters (only three beach-washed derelicts have been documented, all from New South Wales), but it is recorded with some frequency just beyond the east coast continental shelf.

Polymorphism is marked, and individuals may be dark, light or intermediate in plumage pattern. Only one constant feature is diagnostic: the central shafts of the primaries are white, not dark as in related petrels.

HABITAT: temperate and subtropical oceans
LENGTH: *c* 38 cm
WINGSPAN: *c* 92 cm
DISTRIBUTION: not applicable
ABUNDANCE: sparse
STATUS: probably secure

(BD Bell)

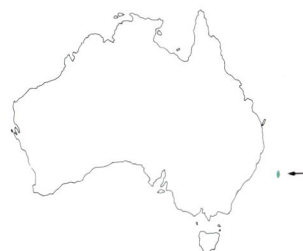

Black-winged Petrel

Pterodroma nigripennis (nig'-ri-pen'-is: "black-winged winged-runner")

This petrel breeds in New Caledonia, and at Lord Howe, and Norfolk Islands, the Kermadecs, the Chathams, and Three Kings and Austral Islands. Since about 1970 it has been expanding its range southward and westward, and is now frequently observed during the breeding season at Heron Island in Queensland and several other islands off the Australian mainland south to Muttonbird Island near Coffs Harbour in New South Wales, although breeding at these localities has yet to be confirmed.

Strongly migatory, it is absent from Australian waters between April and December, at which time it is recorded in numbers in the vicinity of Hawaii.

It nests in dispersed colonies, where it is active by day (except at the Chatham Islands); pairs indulge in spectacular high-speed aerial chases in courtship. Colonies are occupied from November to May; eggs are laid in December and January, and the young fledge in April.

HABITAT: temperate and subtropical oceans
LENGTH: *c* 30 cm
WINGSPAN: 63–71 cm
DISTRIBUTION: not applicable
ABUNDANCE: sparse
STATUS: vulnerable

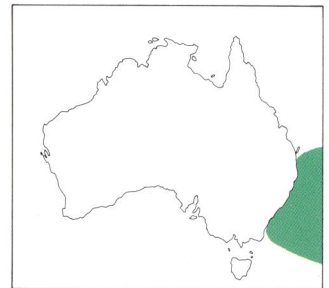

(*T & P Gardner*)

Providence Petrel

Pterodroma solandri (sol-and'-ree: "Solander's winged-runner", after D. C. Solander, Swedish botanist)

(*J Warham*)

Also known as the Brown-headed or Solander's Petrel, this species favours deep waters beyond the continental shelf. Relatively common off New South Wales and southern Queensland, it is widely dispersed over the western Tasman Sea. It is seldom reported beyond these limits, though there are scattered reports off Japan and (perhaps) near Hawaii and elsewhere in the Pacific; in recent years it has been reported with some frequency around Tasmania.

It formerly bred on Norfolk Island, where it was exterminated during the late 1700s, being slaughtered for food in the early days of the penal settlement on the island. About a century later it was recorded breeding on Lord Howe Island, where the population is currently estimated as about 20 000 breeding pairs. Recently a few pairs have been discovered breeding on Phillip Island near Norfolk Island.

The Providence Petrel breeds in winter, returning to its large, dense nesting colonies in March. The birds are noisy, tame and curious, and active about their colonies by day. The nest is a chamber lined with fragments of palm fronds at the end of a burrow about 1 metre long. Eggs are laid in mid-May and hatch in July; chicks fledge in November.

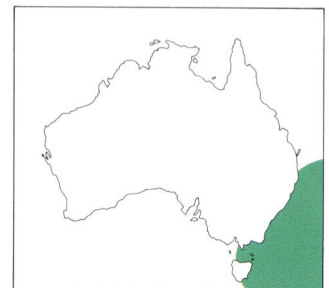

HABITAT: temperate oceans
LENGTH: *c* 40 cm
WINGSPAN: *c* 94 cm
DISTRIBUTION: not applicable
ABUNDANCE: rare
STATUS: possibly endangered

Genus *Puffinus*

(puf'-in-us: "puffin")

This is a diverse cosmopolitan genus of about 16 species, at least 11 of which have occurred in the Australasian region. The plumage is either entirely dark brown, or blackish above and white below. The bill is long and slender and, almost without exception, dark in colour. Most species are strongly gregarious.

Breeding is conducted in large (sometimes huge), dense colonies, visited only at night. Most species congregate in dense "rafts" on the water just offshore at dusk, waiting for darkness to visit their colonies. The nest is a chamber at the end of a burrow typically about a metre in length. Adults share nest building, incubation and care of young.

Little Shearwater

Puffinus assimilis (as-sim'-il-is: "similar puffin")

The Little Shearwater is widespread in the Atlantic, Indian and Pacific Oceans. In the Australasian region it breeds on islands near New Zealand, at the Kermadecs and Norfolk and Lord Howe Islands; and on a number of offshore islands of Western Australia from the Recherche Archipelago to the Abrolhos group. It is largely sedentary, but immature or non-breeding birds may range widely.

Its flight is low, rapid and direct, with brisk and shallow wing-beats interspersed with brief glides. Foraging singly or in small flocks, it generally feeds on the surface but also dives readily; occasionally it follows ships.

It breeds in relatively small, dispersed colonies. It has been estimated that nearly 2000 pairs breed on Eclipse Island in Western Australia, perhaps 1000 pairs on Norfolk Island, and about 4000 pairs on Lord Howe Island. Birds return to their colonies in early January, but eggs are not laid until June or July. The incubation period is about 55 days and the young fledge in about 70 to 75 days.

HABITAT: temperate southern seas
LENGTH: *c* 25–30 cm
WINGSPAN: 58–67 cm
DISTRIBUTION: not applicable
ABUNDANCE: very sparse
STATUS: probably secure

(PJ Fullagar)

Buller's Shearwater

Puffinus bulleri (bool'-er-ee: "Buller's puffin", after W. Buller, New Zealand ornithologist)

Also known as the Grey-backed or New Zealand Shearwater, Buller's Shearwater breeds at the Poor Knights Islands, New Zealand, where its numbers have increased dramatically in recent years. It is migratory, dispersing northwards into the North Pacific after breeding; it is numerous off the west coasts of North and South America. At all seasons, its movements and distribution accords closely with waters of the Subtropical Convergence. Unknown in Australian waters before 1954, it is now frequently recorded (mostly from October to June) off the east coast and has been found in burrows on offshore islands (though breeding remains unproven).

A conspicuous dark M-shaped band across the upperwings is a valuable field character. Buller's Shearwater is gregarious at sea; though seldom encountered in very large flocks, it frequently joins flocks of other shearwaters.

It returns to its large, dense breeding colonies in mid-September, and eggs are laid in November and December following a "honeymoon" absence at sea of about 30 days. The incubation period is about 51 days. Nesting completed, the colonies are deserted towards the end of May.

HABITAT: temperate and subtropical seas
LENGTH: *c* 46 cm
WINGSPAN: *c* 97 cm
DISTRIBUTION: not applicable
ABUNDANCE: rare
STATUS: secure

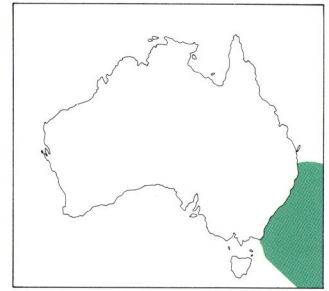

(*J Warham*)

Flesh-footed Shearwater

Puffinus carneipes (kar-nay'-ee-pez: "fleshy-footed puffin")

A bird of southern temperate waters, the Flesh-footed Shearwater breeds on St Paul Island in the Indian Ocean; on a number of offshore islands in Western Australia from about Cape Hamelin to the Recherche Archipelago; on Lord Howe Island; and on islands around northern New Zealand. The western population winters mainly in the Arabian Sea, dispersing southward to South Africa; the eastern population winters off Japan, dispersing eastward to the coast of North America. It is common in Australian waters from September to May, but numbers decline southwards and it is uncommon in Tasmania.

The large pale bill is a helpful fieldmark. The flight is distinctive, a series of long glides interspersed with unhurried, deep, stiff wingbeats. It often occurs in flocks of several hundred birds. It dives freely, but usually feeds on the surface, skimming the water with treading feet and shallow bellyflops.

Nesting colonies, which are reoccupied in late September, are large and dense; level sites close to the sea are preferred. Eggs are laid during November and early December and hatching is at its peak in late January; most chicks fledge during April and the colonies are deserted in May. Chicks fledge at about 92 days.

HABITAT: Temperate and subtropical seas
LENGTH: 41–45 cm
WINGSPAN: 100–105 cm
DISTRIBUTION: not applicable
ABUNDANCE: common
STATUS: secure

(*AEF Rogers*)

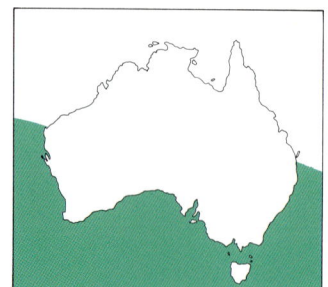

Fluttering Shearwater

Puffinus gavia (gah'-vee-ah: "gull-puffin")

This species breeds on many islands around the North Island and in Cook Strait, New Zealand. It does not breed in Australia but it is common offshore, especially between about June and October, from about Fraser

(*MF Soper*)

Island in Queensland, to the Eyre Peninsula in South Australia and, more rarely, further west to Western Australia.

It is usually encountered in flocks, sometimes quite large. The flight is low and direct, on brisk shallow wing-beats alternating with brief glides. It dives freely and expertly. It seldom approaches ships or fishing vessels.

Adults return to their colonies in August and most eggs are laid in September or early October. By February or March the young have fledged and the colonies are deserted.

HABITAT: temperate oceans
LENGTH: *c* 31–36 cm
WINGSPAN: *c* 76 cm
DISTRIBUTION: not applicable
ABUNDANCE: common
STATUS: probably secure

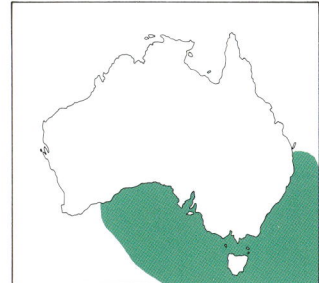

Sooty Shearwater

Puffinus griseus (griz-ay'-us: "grey puffin")

The Sooty Shearwater breeds on various islands in Chile and near Cape Horn, and in the vicinity of New Zealand; and in very small numbers on several offshore islands in south-eastern Australia. During summer it occurs southward to Antarctic waters and, after breeding, it migrates to the northern Pacific region, dispersing eastwards to the coast of North America and south to Peru and Ecuador. It also occurs in the Atlantic Ocean north to eastern Canada, Europe and the Mediterranean Sea.

Some New Zealand colonies are enormous, but relatively few birds breed in Australia (about 500 pairs in New South Wales, 1000 in Tasmania and about 3500 on Macquarie Island), always in scattered pairs in large colonies of Wedge-tailed Shearwaters or Short-tailed Shearwaters.

It is difficult to distinguish at sea from the much more abundant (in Australia) Short-tailed Shearwater, but it is frequently reported in the south-eastern States, though rare in

South Australia and Queensland. In Australian waters it usually associates with flocks of Short-tailed Shearwaters. It shows little interest in ships, but congregates enthusiastically around fishing boats for fish offal tossed overboard. The diet comprises krill, small fishes and squid, mostly obtained by shallow dives from the surface.

The breeding season extends from September to May. Most eggs are laid in mid-November and take about 53 days to hatch. The fledging period is about 97 days, the young leaving the colony during April; adults leave the colonies two or three weeks before the chicks fledge.

HABITAT: temperate and polar seas
LENGTH: 40–46 cm
WINGSPAN: 94–105 cm
DISTRIBUTION: not applicable
ABUNDANCE: abundant
STATUS: secure

(*J Warham*)

Hutton's Shearwater

Puffinus huttoni (hut'-on-ee: "Hutton's puffin", after F. W. Hutton, New Zealand ornithologist)

Hutton's Shearwater breeds only in the Seaward Kaikoura range of mountains on the South Island of New Zealand. Its pelagic range is uncertain, but young birds are suspected of undertaking a lengthy circumnavigation of Australia before returning to New Zealand and adopting a more or less sedentary habit as breeding adults.

It is extremely difficult to distinguish at sea from the closely related Fluttering Shearwater. Like most other shearwaters, it is strongly gregarious, and Hutton's and Fluttering Shearwaters frequently occur in mixed flocks. It frequents shallow inshore waters and is not uncommonly seen from shore.

The diet appears to consist mainly of small fishes, which are caught in shallow dives from the surface, generally with wings half-raised.

All known breeding colonies are at an altitude of 1200 metres or more, and some 9 to 24 kilometres inland. They are reoccupied in September when the ground is clear of snow, and eggs are laid in November. Neither incubation nor fledging periods have been reported.

HABITAT: southern oceans
LENGTH: *c* 38 cm
WINGSPAN: *c* 90 cm
DISTRIBUTION: not applicable
ABUNDANCE: sparse
STATUS: vulnerable

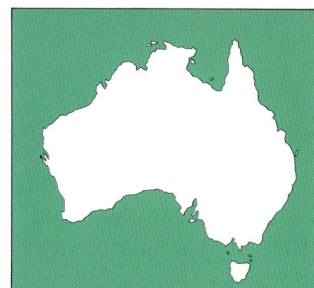

(*G Harrow*)

Wedge-tailed Shearwater

Puffinus pacificus (pah-sif'-ik-us: "Pacific puffin")

This abundant and widespread shearwater breeds on many islands in the tropical Indian and Pacific Oceans, from Madagascar to Mexico; in Australia it breeds on offshore islands around the western, northern and eastern coasts from Carnac Island near Fremantle in the west, to Montagu Island off Narooma, New South Wales, in the east. Population estimates include some 90 000 breeding pairs in New South Wales and 60 000 on Lord Howe Island.

Southern populations are migratory, but details are uncertain: birds banded at their breeding colonies in New South Wales have been recovered near the Philippines. The species is very rare in southern waters from June to August.

The Wedge-tailed Shearwater is polymorphic: one morph is entirely dark brown, the other extensively white below. The latter morph is relatively uncommon in the Australian region but occurs with some frequency in Western Australia.

Unlike that of many other shearwaters, the flight of the Wedge-tailed Shearwater is characteristically lazy, drifting and graceful, with frequent changes in height and direction. The species is gregarious, but birds usually gather in small parties rather than large flocks. However, large numbers congregate at transient rich food sources, often with other seabirds. They frequently attend fishing boats.

In New South Wales, colonies are reoccupied in early August and eggs are laid in late November. Eggs are laid over a period of about two weeks, and take about 52 to 54 days to hatch. Chicks are brooded for the first day, thereafter being visited only to be fed; they fledge in about 70 days. Adults leave the colony about a week before the young, and the colonies are deserted by the middle of May. Birds are normally at least three years old before they breed, but young non-breeders regularly visit the colonies.

HABITAT: temperate and tropical seas
LENGTH: 41–46 cm
WINGSPAN: 97–104 cm
DISTRIBUTION: not applicable
ABUNDANCE: common
STATUS: probably secure

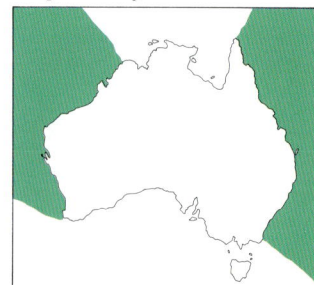

(*M Carter*)

Short-tailed Shearwater

Puffinus tenuirostris (ten'-ue-ee-ros'-tris: "slender-billed puffin")

The Short-tailed Shearwater breeds on islands of Bass Strait and Tasmania, and (in smaller numbers) in southern Western Australia, South Australia, Victoria and New South Wales. After breeding it migrates to the northern Pacific and it is absent from Australian seas from mid-May to late August. The total population is about 20 million pairs.

It is strongly gregarious, occurring sometimes in enormous flocks, especially when travelling. The diet is mainly krill but also includes cephalopods (squids, cuttlefishes, etc.) and small fishes. Its flight is brisk and direct; it seldom follows ships but eagerly attends fishing vessels.

The large, dense colonies are reoccupied in September. A "scratching-out" stage, lasting five to six weeks and spent mainly in refurbishing burrows, is followed by a "honeymoon" phase of about three weeks, spent at sea. Most eggs are laid during the last week in November. Incubation shifts, alternating between parents (the male taking the first) average about 14 days in duration, and the egg hatches in about 53 days. The fledgling period is about 94 days. Adults desert the colonies in mid-April, often within a period of a few days; the young leave several weeks later. Birds first breed at five or six years of age.

Chicks are harvested for flesh, oil and feathers in a small but well-established and tightly controlled industry that takes about 160 000 birds per year. The industry was founded last century by sealers who turned to "muttonbirds" after the seals had been almost exterminated. Aborigines did not exploit the birds.

HABITAT: temperate and polar seas
LENGTH: 41–43 cm
WINGSPAN: 97–100 cm
DISTRIBUTION: not applicable
ABUNDANCE: abundant
STATUS: secure

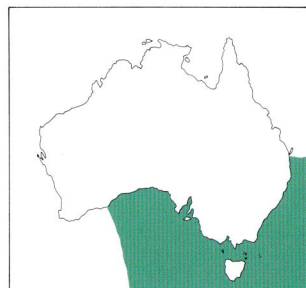

(*T & P Gardner*)

Family HYDROBATIDAE

(hie'-droh-bah'-tid-ee: "*Hydrobates*-family", after a genus of stormpetrels)

Stormpetrels are minute seabirds that in size, flight style and general behaviour may remind the casual observer of swallows. Not markedly gregarious except when breeding, they are strongly pelagic and far-ranging. Most are migratory.

There is no sexual or seasonal variation in plumage, and immatures closely resemble adults. The diet is mainly krill and surface plankton.

Stormpetrels breed in dense colonies, visited only at night. Both sexes prepare burrows, incubate and care for the young.

There are 21 species, seven of which have been recorded in Australia although only one breeds locally.

Genus Fregetta

(freg-et'-ah: "frigate [bird]")

The two members of this genus are mainly black above and white below, and extremely difficult to distinguish at sea. One breeds in subantarctic waters, the other in temperate or subtropical seas; both migrate to equatorial waters.

White-bellied Stormpetrel

Fregetta grallaria (gral-ar'-ee-ah: "stilt-like frigate [bird]")

A bird of temperate and subtropical seas, the White-bellied Stormpetrel breeds on Lord Howe, Kermadec, Austral, Juan Fernandez, Tristan da Cunha, Saint Paul and (possibly) Amsterdam Islands. It is strongly pelagic, avoiding inshore waters, and knowledge of its distribution at sea is fragmentary. It is not uncommon in the Coral and Tasman Seas and has been recorded off the coast of New South Wales.

The species is polymorphic and very difficult to identify at sea.

The breeding season appears to extend from December to May but very little is known of nesting behaviour.

HABITAT: temperate and subtropical oceans
LENGTH: *c* 20 cm
WINGSPAN: *c* 46 cm
DISTRIBUTION: not applicable
ABUNDANCE: very sparse to sparse
STATUS: probably secure

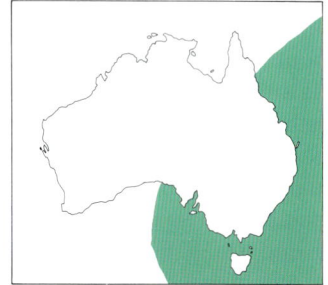

(*J Disney & P Fullegar*)

Black-bellied Stormpetrel

Fregetta tropica (trop'-ik-ah: "tropical frigate [bird]")

This stormpetrel breeds on South Georgia, South Orkney and South Shetland Islands, the Crozets, and Kerguelen, Auckland and Antipodes Isands, and perhaps also on Bouvet and Bounty Islands. Its main pelagic distribution lies between 30°S and 60°S latitudes, but it is strongly migratory and far-ranging and is fairly common in the Tasman Sea and the Southern Ocean south of Tasmania during winter.

The breeding season extends from November to April, egg-laying taking place in December and January. The incubation period is 38 to 44 days, and the young fledge at 65 to 71 days.

HABITAT: temperate and subantarctic oceans
LENGTH: *c* 20 cm
WINGSPAN: *c* 46 cm
DISTRIBUTION: not applicable
ABUNDANCE: very sparse to sparse
STATUS: secure

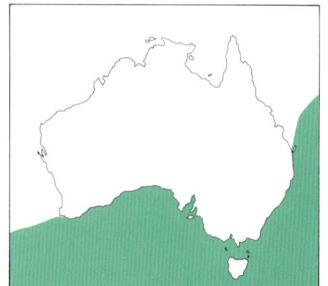

(*J Warham*)

Genus *Oceanites*

(oh'-see-ah-nee'-tayz: "oceanic [bird]")

This is a small group of mainly dark stormpetrels, members of which have a white rump and, sometimes, a white abdomen. They have relatively short, rounded wings and long slender legs. The legs and feet are black but the webs are often yellow, and the toes are more slender and the claws narrower and sharper than in many other stormpetrels. The significance of these characteristics is uncertain. Two species occur in Australian seas.

Grey-backed Stormpetrel

Oceanites nereis (ne-ray'-is: "Nereid oceanic [-bird]", the Nereids being Greek sea-nymphs)

The pelagic range of the Grey-backed Stormpetrel is circumpolar but unknown in detail. It is apparently largely sedentary, favouring cold subantarctic waters and avoiding shallow inshore waters, and is a regular, if uncommon, visitor to seas off South Australia and New South Wales but is so far unrecorded in Queensland or Western Australia.

With its distinctive black head, white underparts and grey rump, it is relatively easy to identify at sea. It occasionally occurs in small flocks and may follow ships.

The Grey-backed Stormpetrel breeds in dispersed colonies. Most eggs are laid in November or December, the young fledging in May; little more is known of the breeding cycle.

HABITAT: southern oceans
LENGTH: 16–19 cm
WINGSPAN: *c* 39 cm
DISTRIBUTION: not applicable
ABUNDANCE: sparse
STATUS: probably secure

(JC Sinclair)

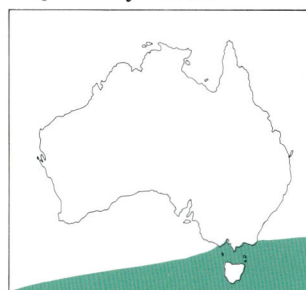

Wilson's Stormpetrel

Oceanites oceanicus (oh'-see-ah'-nik-us: "oceanic oceanic [-bird]")

Wilson's Stormpetrel breeds on rocky shorelines around the Antarctic continent and on many subantarctic islands in the southern Atlantic and Indian Oceans, including Heard Island. It does not breed on Macquarie Island or any of the islands south of New Zealand. Strongly migratory, its pelagic range is virtually cosmopolitan; major wintering grounds include the north-western Atlantic Ocean, the north-western Indian Ocean, and seas off north-western Australia. It is common in Australian seas, especially during its northward passage along the east and west coasts in spring and autumn.

It frequents shallow inshore waters as readily as the open ocean, and is usually encountered alone or in small groups. Its flight when feeding suggests that of a butterfly, fluttering and erratic with wings held high; it pats the water with its feet, dancing and skipping over the surface. It feeds mainly on surface plankton but readily follows ships for food scraps. The annual moult takes place at its northern wintering grounds.

Breeding takes place in loose colonies, generally from November to April (but varying substantially from colony to colony). Nesting burrows are often used by the same pair for several years. The single egg is incubated by both parents in daily shifts and hatches in 39 to 48 days. Chicks fledge at about 60 days.

HABITAT: all oceans
LENGTH: 15–19 cm
WINGSPAN: 38–42 cm
DISTRIBUTION: not applicable
ABUNDANCE: common
STATUS: secure

(M Price)

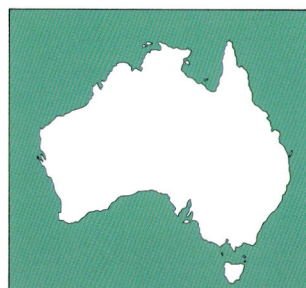

Genus *Pelagodroma*

(pel'-ah-goh-droh'-mah: "sea-runner")

This monotypic genus has grey and white plumage and has a virtually cosmopolitan distribution at low latitudes.

White-faced Stormpetrel

Pelagodroma marina (mah'-ree'-nah: "marine sea-runner")

This species ranges widely over the warmer regions of the Pacific, Indian, and Atlantic Oceans; breeding stations include the Cape Verde Islands, Tristan da Cunha, and several islands in the New Zealand region. It is the only stormpetrel to breed in Australia, on offshore islands along the southern coast from the Abrolhos in Western Australia to Broughton Island in New South Wales. It may be encountered at sea around Australia at any time of year.

Easily identified by the combination of grey upperparts and white face and forehead, it is pelagic, tending to avoid shallow inshore waters. It is usually solitary but may be encountered in small parties. It shows little interest in shipping.

The breeding season is protracted and non-synchronous. Colonies are reoccupied in September and most eggs are laid in November. Incubation takes about 56 days; fledging takes some 52 to 67 days, the young departing in March. Individuals begin breeding in their third year, and the pair bond is sustained over many seasons.

HABITAT: tropical and subtropical oceans
LENGTH: *c* 20 cm
WINGSPAN: *c* 42 cm
DISTRIBUTION: not applicable
ABUNDANCE: common
STATUS: secure

(CA Henley)

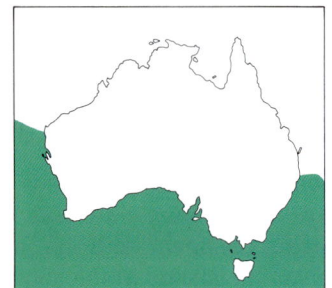

Family PELECANOIDIDAE

(pel'-ek-an-oy'-did-ee: "*Pelecanoides*-family', after the sole genus of diving-petrels)

Confined to the Southern Hemisphere, diving-petrels are small, tubby seabirds with very short wings. They fly seldom but with a distinctive action: very low, direct, and "buzzy", through air or water with equal facility. They are sedentary and gregarious, and breed in colonies.

There is no sexual or seasonal variation, and immatures differ only in their smaller, weaker bills. The diet is mainly krill.

The family includes only one genus with four, perhaps five, species. Only one of these breeds in Australia.

Genus *Pelecanoides*

(pel'-ek-an-oy'-dayz: "pelican-like")

The characteristics of the genus are those of the family.

Common Diving-petrel

Pelecanoides urinatrix (yue'-rin-ah'-trix: "diving pelican-like")

The Common Diving-petrel breeds on many islands in the southern Atlantic and Indian Oceans (including Heard Island) and off New Zealand, as well as on islands off Tasmania and in Bass Strait. It is mainly sedentary, but there is some evidence of dispersal, perhaps of immature birds.

Diving-petrels are difficult to detect at sea. They are usually encountered, alone or in flocks, on inshore waters close to their breeding islands. They ignore ships and fishing vessels.

In Australia the breeding season is generally from July to January, eggs being laid in August. The incubation period is about 53 or 54 days, and the young fledge and leave the colony at about 50 days of age.

HABITAT: southern oceans
LENGTH: 20–25 cm
WINGSPAN: 33–38 cm
DISTRIBUTION: not applicable
ABUNDANCE: sparse
STATUS: probably secure

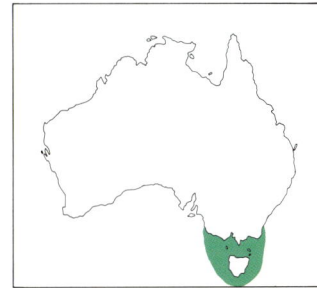

(P Harper)

Order PELECANIFORMES

(pel'-ek-ah'-nee-for'-mayz: "*Pelecanus*-order", after the single genus of pelicans)

A common characteristic of these large aquatic (and mainly marine) birds is that the four toes of the foot are joined by a single web (in other web-footed birds only the front three are so joined). The order includes six families: Pelecanidae (pelicans), Phaethontidae (tropicbirds), Sulidae (gannets and boobies), Phalacrocoracidae (cormorants and shags), Anhingidae (darters), and Fregatidae (frigatebirds). Representatives of all of these families occur in Australia or Australian waters.

Family PELECANIDAE

(pel'-ek-ah'-nid-ee: "*Pelecanus*-family", after the single genus of pelicans)

Pelicans are large aquatic birds mainly inhabiting warm temperate and tropical regions of all continents. They have a very long, robust bill, and leathery skin between the lower mandibles forms an enormously extensible pouch. There is only one genus with 7 or 8 species, one of which occurs in Australia.

The sexes are similar, there is little seasonal variation, and immatures closely resemble adults. Pelicans are strongly gregarious, nesting, foraging, roosting and travelling in flocks. Fish constitutes the main diet: feeding parties usually form a circle, facing inwards and dipping their bills in unison. They swim buoyantly and walk rather awkwardly on land; they are ponderous in take-off or landing on water, but they fly strongly and gracefully, often soaring to great heights.

Breeding is colonial, and parents cooperate in nest building, incubation and care of young.

Genus Pelecanus

(pel'-ek-ah'-nus: "pelican")

The characteristics of the genus are those of the family Pelecanidae.

Australian Pelican

Pelecanus conspicillatus (kon-spis'-il-ah'-tus: "spectacled pelican")

The Australian Pelican occurs throughout the continent, wherever there is water, but tends to breed mainly on the south and west coasts and the interior salt lakes. It also occurs in southern New Guinea, and has been recorded in Indonesia, New Zealand, and islands in the southwestern Pacific. It is strongly nomadic and subject to marked fluctuations in population, presumably related to drought.

Permanent colonies may involve about 1000 pairs, but often small transient breeding groups may form when and where conditions are appropriate; in some years no breeding occurs. Colonies are usually on secluded islets far from shore.

Nests, built mainly by the females, vary from mere scrapes to rather elaborate constructions of sticks, grass, and debris. The normal clutch is two eggs; incubation takes 32 to 35 days.

Naked and helpless, newly hatched chicks are brooded continuously by either parent. Once mobile and covered with down at about 25 days old, they congregate in creches of about 30 birds, remaining in such groups until they are able to fly at about three months.

HABITAT: cool temperate to tropical lakes (including salt lakes), rivers and estuaries
LENGTH: *c* 120 cm
WINGSPAN: *c* 340 cm
DISTRIBUTION: more than 1 million km^2
ABUNDANCE: common
STATUS: secure

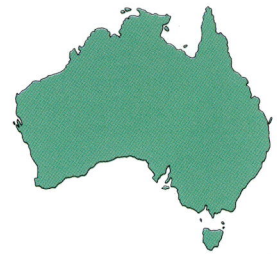

(*G Chapman*)

Family ANHINGIDAE

(an-hing'-id-ee: "*Anhinga*-family", after the sole genus)

Darters are closely related to cormorants, but differ in a number of anatomical and behavioural features. They are strongly aquatic, frequenting fresh waters almost exclusively, and largely tropical in distribution. There are two species in a single genus; one (the Anhinga) is American, the other is widespread from Africa to Australia.

Genus *Anhinga*

(an-hing'-ah: "darter", from Brazilian Amerindian name for darters)

The characteristics of the genus are essentially those of the single Australian species.

Darter

Anhinga melanogaster (mel'-ah-noh-gas'-ter: "black-bellied darter")

The Darter occurs across much of Africa and southern Asia to New Guinea and over most of Australia. It is very numerous in the far north, but it is also common in the extensive wetlands of the Murray-Darling basin in the interior south-east, and in the far south-west. It has occurred as a vagrant in Tasmania and also in New Zealand.

The adult male is mainly glossy black in plumage, while the female is greyish brown above and dull white below; immatures are similar to females but the head is pale. There is no seasonal variation.

The Darter is sometimes called "snakebird" from its habit of swimming with the body underwater and the long, slender, sinuous neck protruding above the surface. Unlike cormorants, which actively pursue their prey underwater, the Darter submerges silently and deliberately stalks unwary fishes. It is usually solitary, but small groups may congregate to loaf and preen on quiet river banks, submerged logs or similar sites.

Nesting is colonial, often in association with other waterbirds, and may occur at any time of year. Colonies are seldom large and solitary nesting is not uncommon. The male selects the nest-site, but the sexes cooperate in nest building, incubation and care of the young. The usual clutch is four eggs, incubated for 26 to 30 days; the young fledge at about 50 days.

HABITAT: extensive, still, deep fresh waters
LENGTH: 86–94 cm
DISTRIBUTION: more than 1 million km^2
ABUNDANCE: sparse to common
STATUS: secure

(*G Chapman*)

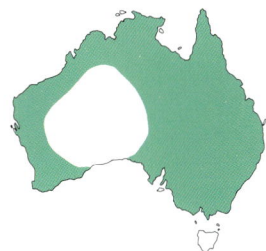

Family SULIDAE

(sue'-lid-ee: "*Sula*-family", after the genus of boobies)

Gannets and boobies are large, robust, exclusively marine seabirds with long pointed wings and tails and torpedo-shaped bodies. The bill is long and pointed. They forage by plunging on fish from a height in spectacular dives.

The sexes are similar and there is no seasonal variation, but juveniles are strikingly different from adults in appearance, and—especially in the genus *Morus*—go through an intricate series of intermediate plumages before reaching maturity.

Two eggs may be laid, but only one young is raised. There is no brood patch, the egg being incubated under the highly vascularised feet. Both sexes cooperate in nest building, incubation and care of young.

The family is traditionally regarded as comprising two genera, *Morus* (gannets) and *Sula* (boobies), but the differences between them are tenuous and the distinction is difficult to maintain, a taxonomic difficulty exacerbated by Abbott's Booby of Christmas Island, which in many respects is very distinct from both groups. In general, the gannets are larger than boobies and inhabit cool temperate rather than tropical seas, to which boobies are largely confined. All three species of gannets are very closely related and perhaps comprise a single species, inhabiting the North Atlantic, southern Africa, and Australasia. There are six species of boobies, three of which occur in Australian seas.

Genus Morus

(mor'-us: "fool")

The characteristics of the genus are discussed under the family heading.

Australasian Gannet

Morus serrator (se-rah'-tor: "sawyer fool")

The Australasian Gannet is common over inshore waters south of the Tropic of Capricorn. It breeds on a few islands off southern Victoria, Tasmania and New Zealand. Immature birds from New Zealand disperse to southern Australia, augmenting the local population between March and September.

Although exclusively marine, gannets seldom forage far out to sea, preferring coastal waters. With slow, stately wing-beats interspersed with lengthy glides, solitary birds or small parties patrol wide areas. When a shoal of fish is located, many birds may congregate, plunging into the sea on back-angled wings from 20 to 40 metres above the surface in spectacular dives. The targeted fish is captured from below, and swallowed before the bird surfaces.

Human persecution caused a severe decline in numbers around the turn of the century but, since about 1950, this trend appears to have been reversed. The total Australian breeding population is about 5000 to 6000 birds.

The breeding season is October to May. Gannetries are typically large and crowded. The egg is incubated for about 44 days by both parents in more or less daily shifts; the chick fledges at about 100 to 110 days. Fully adult plumage is attained during the third or fourth year, the birds first breeding at four to seven years.

HABITAT: cool to warm temperate coastal waters
LENGTH: 84–91 cm
WINGSPAN: 170–180 cm
DISTRIBUTION: not applicable
ABUNDANCE: common
STATUS: probably secure

(B Chudleigh)

Genus *Sula*

(sue'-lah: "plunderer")

The characteristics of the genus are discussed under the family Sulidae.

Masked Booby

Sula dactylatra (dak'-til-ah'-trah: "black-fingered plunderer")

The Masked Booby has a pantropical distribution, breeding on many oceanic islands in the Atlantic, Indian and Pacific Oceans. In the Australian region it breeds in the Cocos (Keeling) Islands, on Bedout and Adelie Islands, Western Australia; on Pandora, Raine and Swain Islands, Queensland; on many reefs and cays in the Coral Sea; and on Lord Howe and Norfolk Islands. It is largely sedentary, although generally absent from breeding colonies for about three months of the year. It is rare in coastal waters anywhere around the Australian mainland.

(J Hicks)

Adults superficially resemble the Australasian Gannet, but the head is pure white, and the tail entirely black. It forages far at sea, feeding mainly on flying fishes. It is usually solitary, and generally ignores ships.

The breeding season is extremely variable and very loosely synchronised: eggs and well-grown young may be found even within a single colony. Colonies, typically on grassy slopes above the sea, are generally rather small and dispersed. No nest is built, one to three eggs being laid in a rough scrape on the ground. Incubation takes 44 days; the chick is brooded until nearly a month old, and it fledges at about 120 days, reaching independence at about four months.

HABITAT: warm temperate and tropical oceans
LENGTH: 76–84 cm
WINGSPAN: 160–170 cm
DISTRIBUTION: not applicable
ABUNDANCE: sparse to common
STATUS: secure

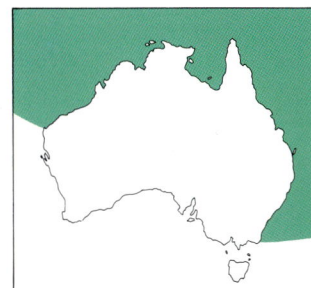

Brown Booby

Sula leucogaster (lue'-koh-gas'-ter: "white-bellied plunderer")

(*C Webster*)

The Brown Booby is common and widespread in tropical oceans. It breeds on many oceanic islands and, in the Australian region, on many offshore islands along the northern coast of Australia and along the Great Barrier Reef. It is common over inshore waters south to the Dampier Archipelago and about the Queensland–New South Wales border; it has been recorded as a vagrant in Victoria.

The deep brown upperparts and head and the pure white underparts of adults are diagnostic; juveniles are sufficiently similar to prevent confusion with other boobies.

Brown Boobies forage mainly over shallow inshore waters, usually in small parties. They spend much time loafing on channel markers, buoys, sandbanks and the rigging of ships.

Colonies vary greatly in size. Eggs may be laid in any month but with peaks of activity in spring and autumn. The nest may be a mere scrape in the ground or a substantial structure of sticks and vegetation. Two eggs are usually laid, but only one chick is successfully raised. The incubation period is about 43 days, chicks fledging at about 95 days.

HABITAT: tropical and subtropical oceans
LENGTH: 64–74 cm
WINGSPAN: 132–150 cm
DISTRIBUTION: not applicable
ABUNDANCE: common
STATUS: secure

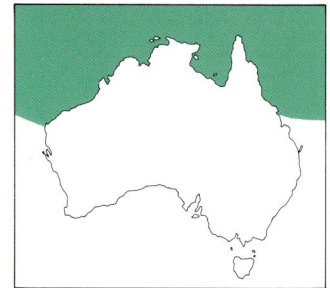

Red-footed Booby

Sula sula (sue'-lah: "plundering plunderer")

(*G Robertson*)

The Red-footed Booby, which is widespread in tropical oceans, breeds on very many small islands.

It is generally uncommon in the Australasian region, but it breeds at the Cocos (Keeling) Islands and Christmas Island, and on islands in the Coral Sea; vagrants are occasionally recorded in coastal Queensland, seldom elsewhere.

The Red-footed Booby occurs in a range of plumages too intricate to describe concisely: common plumages include a white form with black tail and flight feathers, and an ash-brown form, usually with a white tail. Diagnostic features include red feet and (when present) a white tail.

It feeds far out to sea, mostly at night and largely upon squid. Individuals often perch on the masts and rigging of ships.

It favours trees for nesting, resorting to the ground only when there is no available vegetation. The nesting season is extended and variable, and colonies vary greatly in size and density. The nest is a substantial structure of sticks, sometimes lined with a few leaves. Incubation takes about 46 days. Chicks fledge after 100 to 130 days but may be fed for several months thereafter.

HABITAT: tropical and subtropical oceans
LENGTH: 69–76 cm
WINGSPAN: 90–100 cm
DISTRIBUTION: not applicable
ABUNDANCE: very sparse to sparse
STATUS: secure

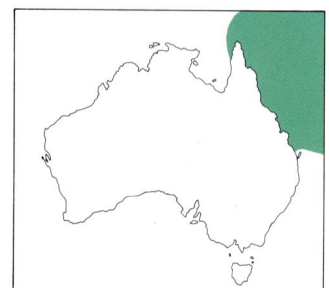

Family PHALACROCORACIDAE

(fal'-ak-roh-kor-as'-id-ee: "*Phalacrocorax*-family", after a genus of cormorants)

In the cormorants and shags, which comprise this family, the sexes are similar, the breeding plumage usually differs only slightly from the winter plumage, and immatures resemble adults.

The diet is mainly fish or other aquatic animals, pursued and captured underwater. The plumage quickly becomes saturated underwater, reducing the effort needed to remain submerged, but thereby requiring a bird to spend much time perched with its wings spread out to dry.

Nesting is colonial (often with other waterbirds). The nest is a rough bulky structure of sticks and twigs cemented with excreta. Adults share nest building, incubation and care of young.

Cormorants have a worldwide distribution. Some 33 species are usually grouped in a single genus, *Phalacrocorax*, but the Flightless Cormorant from the Galapagos Islands is often isolated in its own genus, *Nannopterum*.

Genus Phalacrocorax

(fal'-ak-roh-kor'-ax: "bald-crow")

The characteristics of the genus are essentially those of the family.

Great Cormorant

Phalacrocorax carbo (kar'-boh: "charcoal [-coloured] bald-crow")

Also widely known as the Black Cormorant, this species has an almost cosmopolitan, though disjunct, range across much of North America, Europe, Africa, Asia, Australia and New Zealand. It occurs throughout Australia, including Tasmania, but it is commonest south of about 20°S. It shows no detectable preference for fresh or salt water, but it prefers extensive wetlands to small ponds.

Birds often fish alone, but the species is generally gregarious in all activities, associating in casual groups of 10 or so; very large flocks are rare.

Nesting colonies, which may be very large, are typically situated in floodwaters or along river banks, and associated with other waterbirds. The breeding season is variable and protracted. The usual clutch is three or four eggs, which hatch in about 30 days. The young fledge at about 50 days but remain dependent for a month or more thereafter.

HABITAT: tropical to cool temperate fresh to salt waters
LENGTH: 71–92 cm
DISTRIBUTION: more than 1 million km²
ABUNDANCE: abundant
STATUS: secure

(D MacKenzie)

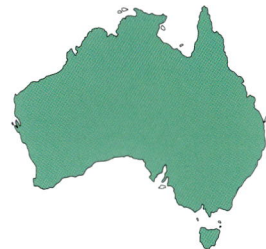

Black-faced Shag

Phalacrocorax fuscescens (fus-kes'-enz: "brownish bald-crow")

(D Watts)

HABITAT: cool temperate coasts
LENGTH: 61–69 cm
DISTRIBUTION: not applicable
ABUNDANCE: sparse to common
STATUS: probably secure

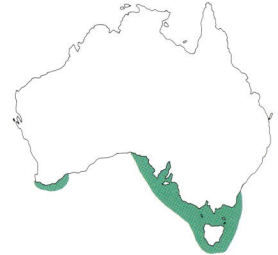

Also called the White-breasted Cormorant, this exclusively marine species occurs only off southern Australian coasts. Adults are largely sedentary.

It is readily confused with the Pied Cormorant, but its facial skin is black rather than yellow-orange; at a distance Pied Cormorants look mainly white-headed; Black-faced Cormorants look more black-headed.

Black-faced Shags are gregarious in all activities and at all seasons. They feed almost entirely on fish.

Nesting is colonial, mainly from August to December. Nests are substantial structures of seaweed, driftwood and other debris, built on rock ledges just above high tide level. A typical clutch consists of two eggs. Chicks congregate in creches once they are mobile.

Little Pied Cormorant

Phalacrocorax melanoleucos (mel'-ah-noh-lue'-kos: "black-white bald-crow")

Few ponds anywhere in mainland Australia are too small to harbour

(R Drummond)

one or two Little Pied Cormorants, at least as transient visitors; stock dams and ornamental waters in city parks are equally acceptable. The species is, however, uncommon in Tasmania. Elsewhere its distribution extends from Malaysia eastward to New Caledonia and New Zealand.

It tends to forage alone, feeding in shallow waters on yabbies and aquatic insects rather than fishes. It habitually loafs and roosts in groups, or in company with other species, especially the Little Black Cormorant.

It usually nests in colonies of about a dozen pairs, but isolated pairs are not unusual. The shallow nest usually overhangs water. The clutch varies from three to seven (usually three or four). Incubation and fledging periods are unknown.

HABITAT: cool temperate to tropical fresh waters, estuaries and bays
LENGTH: 58–63 cm
DISTRIBUTION: more than 1 million km²
ABUNDANCE: abundant
STATUS: secure

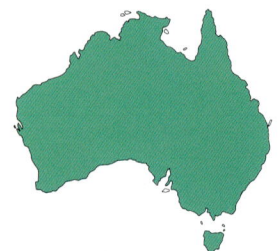

Little Black Cormorant

Phalacrocorax sulcirostris (sul'-kee-ros'-tris: "furrow-billed bald-crow")

This small cormorant has a distribution extending from central Indonesia eastward to New Caledonia and New Zealand. It occurs throughout Australia, wherever there is water, but it seldom breeds in Tasmania. It is strongly

(IR McCann)

nomadic, and few bodies of water throughout the continent—even very small ponds—are not visited at least occasionally.

A gregarious species, it prefers to forage, loaf and travel in small parties, but it occasionally forages alone, and very large flocks sometimes congregate. It feeds mainly upon fish and, although it sometimes forages in salt water, it has a marked overall tendency to avoid the coast.

Nesting colonies are typically rather small—around 100 birds. The nesting season is variable, and strongly dependent on local conditions. The usual clutch is four eggs.

HABITAT: tropical to cool temperate bodies of water
LENGTH: 58–63 cm
DISTRIBUTION: more than 1 million km^2
ABUNDANCE: abundant
STATUS: secure

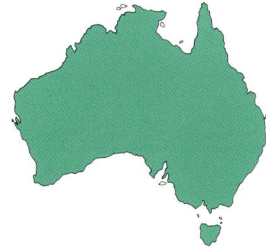

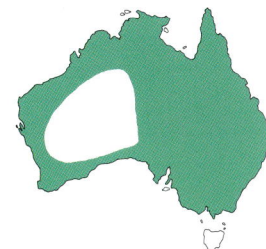

Pied Cormorant

Phalacrocorax varius (var'-ee-us: "multicoloured bald-crow")

(B Chudleigh)

It is gregarious but seldom forms large groups except in the south-west, where flocks of 10 000 to 20 000 have been reported. Like the Great Cormorant it prefers extensive, deep waters or marine environments, and is seldom encountered on small lakes or ponds. It often associates with other species of cormorants.

The breeding season is variable and extended, much influenced by local conditions. Colonies range from a few pairs to several thousand. The usual clutch is three eggs.

HABITAT: cool temperate to tropical bays, estuaries, mangrove swamps and inland waters
LENGTH: 66–81 cm
DISTRIBUTION: more than 1 million km^2
ABUNDANCE: abundant
STATUS: secure

This large cormorant is restricted to Australia and New Zealand. It occurs at least casually throughout Australia but it is most numerous on the wetlands of the Murray–Darling system as well as on the coast). It once occurred in Tasmania but seems to have become extinct there before white settlement; it is now only a vagrant. It is notably less nomadic than other Australian cormorants.

The Pied Cormorant is superficially similar to the much smaller Little Pied Cormorant, but has conspicuous black thighs and its bill is pale, slender and long (not yellowish and stubby).

Family FREGATIDAE

(freg-ah'-tid-ee: "*Fregata*-family", after the single genus of frigatebirds)

Frigatebirds obtain much of their food by harassing other seabirds returning to the breeding colony with food for their young. When the catch is regurgitated, the frigatebird snatches it in midair. Frequency and persistence of this habit varies from species to species, from colony to colony, and between sexes. Otherwise, food—mainly flying fishes—is snatched from the surface of the sea.

Courtship displays are spectacular. Males have an enormously distensible gular (throat) pouch, which in display is pumped full of air until it resembles a large scarlet balloon. The male displays this pouch at females flying overhead, simultaneously waving his long black wings, clapping his bill and whistling until a female descends to mate.

Breeding is colonial and parental duties are shared. One egg is laid and the chick is dependent for a long time.

There is no seasonal variation in plumage, but sexes differ markedly, and there is a complex sequence of plumages from juvenile to adult.

Five species, grouped in a single genus, constitute the family. Two are reasonably common in the Australian region, and a third (the Christmas Island Frigatebird) has been recorded as a vagrant.

Genus Fregata

(freg-ah'-ta: "frigate [-bird]")

The characteristics of the genus are those of the family.

Least Frigatebird

Fregata ariel   (ah'-ree-el: "fairy frigate [-bird]")

The Least Frigatebird is almost ubiquitous in tropical oceans. In the Australian region it breeds on many small islands in the Coral Sea, along the Great Barrier Reef, and along the coast of northern Australia west to Bedout Island, Western Australia, it also breeds on the Cocos (Keeling) Islands. Especially before or during cyclones, it is a common visitor along the northern and eastern coasts of Australia, south to Brisbane and casually even further south.

It spends most of its time soaring, alone or in groups, over seabird colonies or the sea nearby, often at very great heights. Flying fishes, young turtles and squids, deftly snatched from the surface, are frequent items of prey, but much food is also obtained by piracy.

Adults become sexually mature when five to seven years old. Breeding, generally in dispersed colonies, is extremely protracted, variable in timing and on a two-year cycle. Nests may be on the ground or in bushes. The incubation period is probably 44 to 66 days. The single chick fledges at about 24 weeks, but remains dependent on its parents for several months thereafter.

HABITAT: tropical oceans
LENGTH: 71–81 cm
WINGSPAN: 175–193 cm
DISTRIBUTION: not applicable
ABUNDANCE: common
STATUS: secure

(*I Morris*)

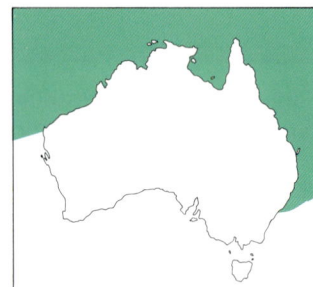

Greater Frigatebird

Fregata minor (mie'-nor: "smaller frigate [-bird]")

The Greater Frigatebird has a pantropical distribution and breeds mostly on oceanic islands. It is generally uncommon in the Australasian region, but breeds on several small islands in the Coral Sea; on Raine and Quoine islands on the outer Great Barrier Reef; and on the Cocos (Keeling) Islands and Christmas Island. It is a casual visitor to the tropical northern coast, west to about Point Cloates in Western Australia, and south to about Rockhampton in Queensland.

Nests, built by the female from material fetched by the male, are generally in the summits of bushes or mangroves. Both parents incubate the egg, which hatches at about 40 days. The chick remains dependent for food for several months after fledging at 20 to 24 weeks. A full nesting cycle may extend to 14 months.

HABITAT: tropical and subtropical oceans

LENGTH: 86–100 cm

WINGSPAN: 206–230 cm

DISTRIBUTION: not applicable

ABUNDANCE: rare to very sparse

STATUS: secure

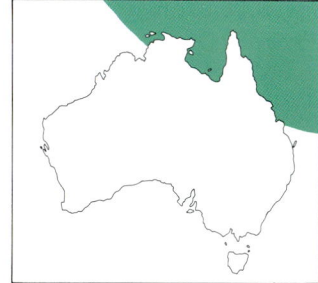

(V Serventy)

Family PHAETHONTIDAE

(fee-thon'-on'-tid-ee: "*Phaethon*-family", from the single genus of tropicbirds)

Tropicbirds (or bosunbirds) feed like gannets, plunging on prey from a height. Their diet consists mainly of fish (especially flying fishes) and squid. They fly well, occasionally swim, but can only shuffle awkwardly on land. The sexes are similar and there is no seasonal variation; immature birds differ from adults in being strongly barred.

Tropicbirds differ conspicuously from other pelecaniform birds in having marked and stone-coloured eggs (all other members of the order have plain chalk-white eggs). No nest is built, the single egg usually being laid in a sheltered niche or cleft in a rocky cliff, in the shelter of a bush, or sometimes in a tree. Breeding is semicolonial, and both parents cooperate in incubation and care of the young.

The family contains only one genus, *Phaethon*.

Genus Phaethon

(fee'-thon: "shining [bird]")

The characteristics of the genus are those of the family.

White-tailed Tropicbird

Phaethon lepturus (lep-tue'-rus: "delicate-tailed shining-bird")

The White-tailed Tropicbird has a pantropical distribution but does not breed in Australian seas, its closest breeding stations being Christmas Island and the Cocos (Keeling) Islands in the Indian Ocean and Walpole Island near New Caledonia. It is not uncommon in the Coral Sea, and it wanders with some frequency to the eastern coast of Australia, south to about Batemans Bay, New South Wales.

It differs from the Red-tailed Tropicbird in the bold black markings on its wings. It is also more delicately and slenderly built and more graceful and agile in flight. It further differs in being a nocturnal predator on squids, rather than fishes. It frequently follows ships. A subspecies restricted to Christmas Island is tinged a delicate apricot-gold shade.

HABITAT: tropical oceans

LENGTH: *c* 40 cm (not including tail streamers, which are up to 40 cm long)

WINGSPAN: *c* 95 cm

DISTRIBUTION: not applicable

ABUNDANCE: rare

STATUS: secure

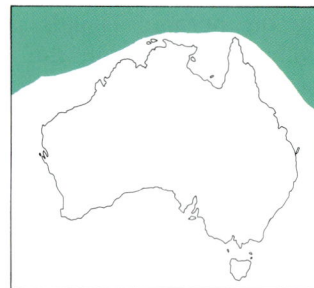

(G Robertson)

Red-tailed Tropicbird

Phaethon rubricaudus (rue'-bree-kaw'-dus: "red-tailed shining-bird")

The Red-tailed Tropicbird is widespread in the tropical Indian and western Pacific Oceans. It breeds on oceanic islands and a few offshore islands. In the Australian region breeding stations include Norfolk, Lord Howe, Raine and Herald Islands in the east and the Abrolhos group (now possibly abandoned) and Rottnest Island in the west; it also breeds on the Cocos (Keeling) Islands and Christmas Island. It is generally rare in Australian coastal waters, but it has occurred, at least as a vagrant, in all mainland States except South Australia.

It may be distinguished from the smaller White-tailed Tropicbird by its pure white wings (the other species has much black in the wing). It is strongly pelagic, avoiding inshore waters; it is usually solitary at sea, and shows little interest in ships. It feeds mainly by day, primarily on flying fishes.

The Red-tailed Tropicbird usually breeds in scattered pairs, which indulge in elaborate displays, flying in tandem and in turn beating their wings rapidly in a wide arc, causing the bird to rise in a "back-pedalling" effect.

The breeding cycle is variable and not always annual. The single egg is incubated by both parents and hatches at about 45 days. The chick fledges at about 40 days (when it weighs more than the adult) but remains at the nest, abandoned by its parents, until driven to flight by starvation some two months later.

HABITAT: tropical oceans

LENGTH: *c* 46 cm (not including tail streamers, which are up to 35 cm long)

WINGSPAN: *c* 104 cm

DISTRIBUTION: not applicable

ABUNDANCE: rare

STATUS: secure

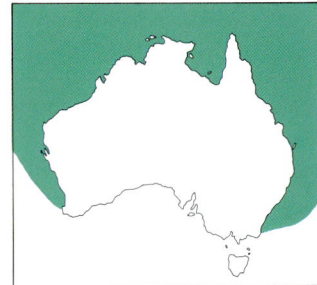

(B & B Wells)

Order CICONIIFORMES

(sis'-on-ee'-ee-for'-mayz: *Ciconia*-order", after a genus of storks)

Most members of this order are large to very large birds with long slender necks, long bills and long legs. Common to wetlands everywhere, they are adapted to wading in shallow water, feeding mainly on fish, amphibians and aquatic invertebrates. Many species breed in colonies but few are otherwise strongly gregarious; most mingle freely when foraging but seldom form flocks. Young are hatched helpless and nearly naked, and are fed and cared for by both parents for several weeks before fledging and leaving the nest.

Five families are usually recognised: Ardeidae (herons and bitterns), Scopidae (hammerkop), Ciconiidae (storks), Balaenicipitidae (shoebills), and Threskiornithidae (ibises and spoonbills). All but the Scopidae and Balaenicipitidae (both of which are exclusively African) are represented in Australia.

Family ARDEIDAE

(ar-day'-id-ee: "*Ardea*-family", after a genus of herons)

The herons constitute a family of some 64 species with a worldwide distribution. They have slim, slight bodies, long slender necks, long legs, and long, slender, dagger-like bills. They inhabit wetlands and obtain their food— mostly fish, molluscs, crustaceans and aquatic insects—by wading in shallow water along the margins.

The family can be divided into two groups: the bitterns, and the herons and egrets. Bitterns typically are stocky, are streaked and marbled in shades of brown, live in reeds and are intensely secretive in behaviour. Herons and egrets generally lack conspicuous colour pattern in their plumage, are gregarious at least in some activities, and nest in colonies. There is little substantial difference between herons and egrets, the latter term being usually reserved for those species that happen to be white in colour.

Genus Ardea

(ar-day'-ah: "heron")

Herons differ most conspicuously from egrets in being much more solitary in behaviour. Usually coloured in sober tones of grey, fawn or dull purple, the plumage lacks conspicuous elements of pattern. Females resemble males, there is little seasonal variation, and immatures resemble adults. Distribution is almost worldwide. Apart from several resident species, the Grey Heron (*A. cinerea*) of Eurasia has been recorded in Australia as a vagrant.

White-faced Heron

Ardea novaehollandiae (noh'-vee-hol-and'-ee-ee: "New Holland heron")

This heron occurs from eastern Indonesia eastward to New Guinea, New Caledonia, and New Zealand (where it has been resident since about 1940). The commonest and most widespread Australian heron, it is found throughout the continent, including Tasmania. In country areas it is often called the Blue Crane. It is easily identified by its plain grey plumage and white face and forehead.

It is generally rather solitary, but groups—sometimes quite large— may congregate where food is abundant, and it often loafs and roosts communally. The diet includes aquatic insects and their larvae, yabbies, molluscs, frogs, small reptiles and fish.

The breeding season is mainly from October to December, at least in the south, but this is strongly influenced by local conditions. Colonies are usually small—about five to 10 pairs—but may be quite large, and pairs sometimes nest alone or in mixed colonies with other waterbirds. The usual clutch is three or four eggs, which hatch in 24 to 25 days. Chicks fledge after 38 to 42 days.

HABITAT: tropical to cool temperate shallow fresh to salt water, from marine reefs to urban ponds and wet paddocks

LENGTH: 66–69 cm

DISTRIBUTION: more than 1 million km²

ABUNDANCE: abundant

STATUS: secure

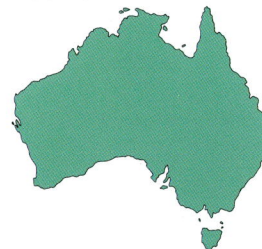

(G Chapman)

Pacific Heron

Ardea pacifica (pah-sif'-ik-ah: "Pacific [Ocean] heron")

The Pacific (or White-necked) Heron is almost exclusively Australian but it also occurs in southern New Guinea and has strayed to New Zealand. Although widespread, it is nowhere abundant, and is uncommon in Tasmania. It is strongly nomadic.

It frequents the margins of wetlands of all kinds. Generally rather solitary, it roosts and loafs in small loose companies, and numbers occasionally forage together without apparent conflict.

The nesting season is strongly influenced by local rainfall and other circumstances but is usually from September to January. Pairs nest alone or in small colonies. Four eggs constitute the usual clutch.

HABITAT: cool temperate to tropical, calm, shallow fresh water; rarely brackish or saline
LENGTH: 76–107 cm
DISTRIBUTION: more than 1 million km²
ABUNDANCE: common
STATUS: probably secure

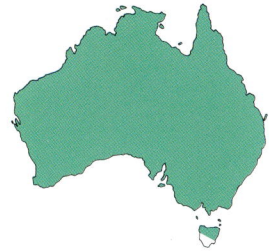

(H & J Beste)

Pied Heron

Ardea picata (pik-ah'-tah: "black-and-white heron")

This small tropical heron extends from Borneo eastward to New Guinea. In northern Australia, where it occurs from about Wyndham in Western Australia to about Ayr in Queensland, it is most common in coastal lowlands, especially deep, permanent lagoons with abundant waterlilies and other aquatic growth.

Seldom entirely alone, it generally loafs and forages in scattered groups, occasionally in flocks of a hundred or more birds. The diet includes aquatic insects and their larvae, frogs, crustaceans and small fish.

Breeding is colonial, generally in large mixed colonies, and the season is mainly from February to April. Two to four eggs form the usual clutch.

HABITAT: tropical mangrove swamps, lagoons, mudflats, sewage ponds, rubbish dumps; also freshly burned grassland
LENGTH: 43–49 cm
DISTRIBUTION: 300,000–1 million km²
ABUNDANCE: very sparse
STATUS: probably secure

(H & J Beste)

Great-billed Heron

Ardea sumatrana (sue'-mah-trah'-nah: "Sumatran heron")

The Great-billed Heron occurs from Burma through South-East Asia to New Guinea and northern Australia, where it occurs south to about Derby in the west and Rockhampton in the east. It is almost entirely confined to mangrove swamps so that its distribution is discontinuous.

Occasionally it also occurs in melaleuca thickets and it may wander upstream for some distance along major rivers.

It roosts in dense mangrove thickets, emerging to forage on mudflats at low tide, apparently feeding largely on crabs. It is solitary, wary and extremely secretive, and very little is known of its habits.

HABITAT: tropical to subtropical mangrove swamps bordering mudflats; less commonly, melaleuca swamps.
LENGTH: *c* 105 cm
DISTRIBUTION: 30,000–1 million km²
ABUNDANCE: very sparse
STATUS: probably secure

(CA Henley)

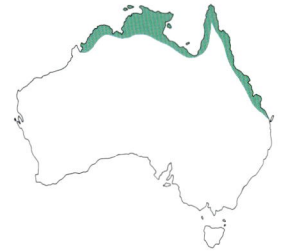

Genus *Botaurus*

(bot-or'-us: probably "cattle-bull", in reference to roaring call)

Bitterns stalk their prey with slow movements involving periods of immobility. The long, dagger-like beak is then employed, in an extremely rapid thrust, to catch or spear prey in or near the water. The rather protuberant eyes can be moved so as to scan in almost every direction while the head is held motionless.

When disturbed, bitterns take flight only as a last resort. Instead they extend their necks, pointing the bill straight up, and "freeze"; in this posture they are extremely difficult to discern in dense reeds. If a gentle breeze should sway the surrounding reeds the bird may even sway with them.

Brown Bittern

Botaurus poiciloptilus (poy'-sil-op'-til-us: "mottled-feathered cattle-bull")

Also called Australasian Bittern, this large brown species occurs in south-eastern and far south-western Australia as well as in New Zealand, New Caledonia and the Loyalty Islands. In Australia it is common in the wetlands of the Murray–Darling basin, but appears to be much less numerous in the south-east and Tasmania.

A solitary, secretive bird, it frequents dense stands of reeds at the margins of lakes, quiet rivers or swamps. It is partly nocturnal and very difficult to observe. The diet comprises frogs, small reptiles, yabbies and small fish. The characteristic call consists of several deep, deliberate booming notes.

Breeding is in solitary pairs. The nest is a platform of trampled reeds in dense cover. The usual clutch is four or five eggs. Incubation lasts about 25 days.

HABITAT: cool to warm temperate wetlands with dense reed beds
LENGTH: 66–76 cm
DISTRIBUTION: more than 1 million km^2
ABUNDANCE: sparse
STATUS: probably secure

(MF Soper)

Genus *Bubulcus*

(bue-bul'-kus: "ploughman")

The single species in this genus is a squat, relatively short-legged egret, which has evolved in conjunction with the large grassland mammal faunas of Africa and Asia.

The sexes are similar, and in the nesting season the head, breast and back are strongly washed with golden buff; immatures resemble winter adults.

Cattle Egret

Bubulcus ibis (ie'-bis: "ibis ploughman")

Breeding colonies are usually large and consist of several species of herons, egrets and other waterbirds. Breeding occurs mainly from November to February. The nest is a platform of sticks and twigs, placed in a tree. The clutch is three to six (usually four or five) eggs, which hatch in 22 to 26 days; the young fledge in 30 days.

HABITAT: cool temperate to tropical dairy pastures and grazing land, also margins of swamps
LENGTH: 48–53 cm
DISTRIBUTION: more than 1 million km²
ABUNDANCE: common
STATUS: secure

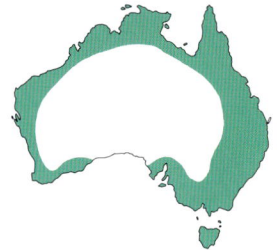

(G Weber)

The Cattle Egret, widespread elsewhere in the world, is now also widespread in Australia. Unknown before the 1900s (the details of its introduction or self-introduction are still obscure), it is most numerous in the north and east but has been recorded in all States including Tasmania and is apparently still spreading rapidly. It is common in New Zealand, and a regular trans-Tasman passage has been confirmed. The Cattle Egret does not breed in Victoria or Tasmania.

It is usually seen in small groups on grazing land, associating with cattle and horses, stalking through the grass for large insects. It frequently perches on fence posts or the backs of grazing animals.

Genus *Butorides*

(bue'-tor-ee'-dayz: "*Butor*-like", after a genus of herons)

This genus contains several populations of a small greenish heron living in tropical and warm temperate regions almost throughout the world. Formerly these populations were treated as three or more species but they are now generally regarded as constituting a single pantropical species.

Mangrove Heron

Butorides striatus (stree-ah'-tus: "striped *Butor*-like")

In Australia at least, this small squat heron is almost entirely restricted to coastal mangrove swamps, occurring wherever this environment is to be found around northern Australia, south to about Shark Bay in the west and Jervis Bay, New South Wales, in the east. It is largely sedentary in habit.

Also called the Striated Heron, it is generally solitary and is usually seen as it emerges onto tidal flats at low tide to forage. It stalks prey in a distinctive stealthy crouch, but may dash madly after it for the last few paces before capture. At high tide it roosts in mangroves.

The breeding season is mainly from September to January and two broods are sometimes raised in succession. The nest is a clumsy platform of sticks in mangroves, constructed by both parents; two to four eggs, which hatch in 21 to 24 days, form the usual clutch.

HABITAT: tropical to warm temperate mangrove swamps
LENGTH: 43–51 cm
DISTRIBUTION: 100,000–300,000 km²
ABUNDANCE: sparse
STATUS: probably secure

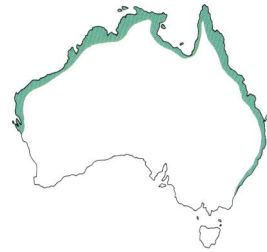

(N Chaffer)

Genus *Egretta*

(eg-ret'-ah: "egret")

Egrets are cosmopolitan white herons that breed in large colonies. In the breeding season the plumage is variously ornamented with plumes on the head, breast and lower back. Otherwise there is little seasonal variation, the sexes are similar, and immatures resemble adults.

Great Egret

Egretta alba (al'-bah: "white egret")

The range of the Great Egret is almost worldwide in temperate and tropical regions. It occurs in all Australian wetlands, but it is solitary by nature and nowhere really numerous. It is a casual visitor to Lord Howe, Norfolk and Macquarie Islands.

It usually wades quietly in search of fish, yabbies, frogs and other aquatic life, or stands silently in wait along the margins. It is generally silent, but frequently utters a deep rattling croak as it takes flight on being disturbed.

The nesting season is variable, but is mainly from October to December in the south, and March to May in the north. Nesting is colonial, usually in association with other waterbirds. The nest is a wide shallow platform of sticks and twigs. The three or four eggs constituting the usual clutch are laid at intervals of 48 hours and take 25 to 26 days to hatch. Hatching is asynchronous and the young take about 42 days to fledge.

HABITAT: cool temperate to tropical wetlands, sluggish rivers, estuaries, mudflats; also flooded pastures, rice paddocks
LENGTH: 41–49 cm
DISTRIBUTION: more than 1 million km²
ABUNDANCE: common
STATUS: secure

(J Yates)

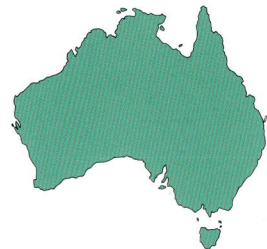

Little Egret

Egretta garzetta (gar-zet'-ah: "little-heron egret")

The Little Egret occurs across much of southern Europe, Africa and Asia. It is nowhere really common in Australia, but is very widespread in the east and in coastal regions of the north and west. It is rare but regular in Tasmania.

This species is generally rather solitary, but freely tolerates the presence of other herons or egrets when foraging. Unlike other egrets, it actively pursues fishes, amphibians, crustaceans and insects, often dashing madly to and fro in shallow water (other egrets stand and wait).

The main breeding season is from March to May in the tropical north and November to January in the south-east. It breeds for preference in large dense colonies with other waterbirds. The usual clutch is three to five eggs; the young fledge in about 40 to 45 days.

HABITAT: cool temperate to tropical, calm shallow water
LENGTH: 25–30 cm
DISTRIBUTION: more than 1 million km²
ABUNDANCE: sparse
STATUS: secure

(G Weber)

Plumed Egret

Egretta intermedia (in'-ter-med'-ee-ah: "intermediate egret")

(P Roberts)

The Plumed (or Intermediate) Egret occurs from Africa and southern Asia to Japan, New Guinea and northern and eastern Australia, west to the Kimberley and south (irregularly) to Tasmania and the vicinity of Adelaide.

It is often confused with the Great Egret but its neck is comparatively much shorter and lacks a distinct kink. It is more sociable than other egrets, and sometimes loafs and forages in flocks. Usual habitats include flooded grasslands, swamps, lagoons and the margins of lakes and rivers. Like the Cattle Egret, it sometimes associates with cattle.

The main breeding season is from March to May in the tropical north and November to January in the south-east. Breeding is usually in large dense colonies with other waterbirds. Three to five eggs form the usual clutch, incubated for 24 to 27 days; hatching is asynchronous, and the young fledge in about 35 days.

HABITAT: cool temperate to tropical, calm, shallow fresh water (infrequently brackish or saline); pastures and grassland
LENGTH: *c* 64 cm
DISTRIBUTION: more than 1 million km²
ABUNDANCE: common
STATUS: secure

Reef Heron

Egretta sacra (sah'-krah: "sacred egret")

The distribution of the Reef Heron extends from Burma eastward to Japan, Polynesia and New Zealand. Exclusively coastal in distribution, it occurs around Australia but it is very rare along the southern coast and merely a vagrant to Victoria and Tasmania. It is most numerous along the Great Barrier Reef.

It roosts and loafs communally but is otherwise rather solitary, actively defending feeding territories. It usually stalks prey in a stealthy hunched posture, hunting in rock pools, on coral reefs exposed at low tide, or in the tidewrack along beaches.

Nesting may occur in any month, but is chiefly from September to January. It usually nests singly but sometimes may form loose colonies. The nest, an untidy structure of sticks and debris, is placed in a tree, on the ground under a bush, or in a cave or rock ledge. The clutch is two or three eggs, which hatch in about 25 to 28 days.

HABITAT: tropical to cool temperate rocky coasts, coral reefs, occasionally mudflats and beaches
LENGTH: 61–66 cm
DISTRIBUTION: 10,000–30,000 km²
ABUNDANCE: sparse
STATUS: secure

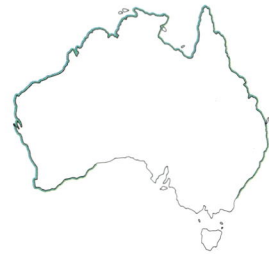

(D Watts)

Genus *Ixobrychus*

(ix'-oh-brik'-us: "reed-roarer")

This genus, containing the smallest members of the Ardeidae, has a worldwide distribution. Three species occur in Australia: two, the Little Bittern and the Black Bittern (which is often placed in the monotypic genus *Dupetor*), are common breeding residents, while the Yellow Bittern is generally regarded as an accidental visitor or vagrant to the tropical north.

Black Bittern

Ixobrychus flavicollis (flah'-vee-kol'-is: "yellow-necked reed-roarer")

This is a rather sedentary bird of coastal lowlands. It extends from India to the Bismarck Archipelago; in Australia, it is most common north of about Perth in the west and Sydney in the east.

Generally solitary, it especially favours quiet pools and backwaters of meandering, densely wooded coastal streams. It is very wary, secretive, and at least partly nocturnal, spending much of the day perched quietly low in leafy trees near water, or in dense reeds.

Nesting is in solitary pairs, usually from September to December. The nest is a loose untidy platform of sticks on a sheltered branch overhanging water. The usual clutch is four eggs.

HABITAT: tropical to warm temperate wetlands and the banks of sluggish rivers and estuaries, always with dense vegetation
LENGTH: 31–37 cm
DISTRIBUTION: more than 1 million km²
ABUNDANCE: sparse
STATUS: probably secure

(J Purnell)

Little Bittern

Ixobrychus minutus (mie-nue'-tus: "little reed-roarer")

(B & B Wells)

This species has an extensive distribution across much of the Old World. It is widespread in the better-watered parts of the mainland but does not occur in Tasmania. Apparently migratory, it is virtually unrecorded in the south during winter and is numerous in the tropical north of the continent at this season.

In the south it frequents dense stands of reeds—especially cumbungi—at the margins of lakes, quiet rivers and swamps. In the tropical north it often also occurs in sugarcane fields. It is secretive, elusive, and generally rather solitary. When foraging, it wades in shallow water or clambers nimbly through reed stems, feeding on aquatic insects and their larvae, frogs and small fish.

The nest, built mainly by the male, is a small shallow saucer of reeds in a dense clump of cumbungi or rushes. Four eggs, which take about 20 days to hatch, constitute the usual clutch. The young fledge at about 25 days.

HABITAT: cool temperate to tropical wetlands with dense reed beds
LENGTH: 25–36 cm
DISTRIBUTION: more than 1 million km²
ABUNDANCE: very sparse
STATUS: probably secure

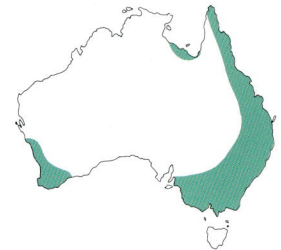

Genus *Nycticorax*

(nik'-tee-kor'-ax: "night-crow")

Representing a more or less cosmopolitan genus of three species, night-herons have a stocky build reminiscent of bitterns, but adults lack streaked plumage. (Juveniles are streaked, and are consequently often confused with bitterns.) They breed in colonies and, as the common name suggests, are largely nocturnal, roosting communally in large shady trees by day. Only one species occurs in Australia.

Nankeen Night-heron

Nycticorax caledonicus (kal'-ed-on'-ik-us: "[New] Caledonian night-crow")

Also called the Rufous Night-heron, this species occurs from Java eastward to Micronesia. It is common and widespread throughout mainland Australia and Tasmania wherever there is suitable habitat; it

(K Ireland)

has strayed to New Zealand. It appears to be largely sedentary, but visits temporary wetlands. It shows some preference for moderate cover at the margins of swamps and quiet rivers, where groups of up to 100 or so roost in favoured leafy trees (particularly willows) during the day. Birds feed singly at night on yabbies, frogs and small fish.

Nesting has been recorded in all months, but mostly from November to February. Two broods are occasionally raised in a single season. Pairs sometimes breed alone, but more usually in quite large colonies. Parents share nest building, incubation and care of young. Two or three eggs are laid and these take about 22 days to hatch. Youngsters fledge at 42 to 49 days.

HABITAT: Cool temperate to tropical wetlands, lagoons, dams, shallows of sluggish rivers, all with moderate cover
LENGTH: 56–66 cm
DISTRIBUTION: more than 1 million km²
ABUNDANCE: common
STATUS: secure

Family CICONIIDAE

(sis'-on-ee'-id-ee: "*Ciconia*-family", after a genus of storks)

Storks are very large wading birds with long slender necks and legs and long, heavy, dagger-like bills. The plumage is generally black, white or pied. The 17 included species occur in tropical and subtropical regions throughout the world; a single species occurs in Australia.

Genus Xenorhynchus

(ksee'-noh-rink'-us: "strange-beak")

The characteristics of the genus are essentially those of the species.

Black-billed Stork

Xenorhynchus asiaticus (ah'-zee-at'-ik-us: "Asian strange-beak")

The Black-billed Stork or Jabiru occurs from India eastward to southern New Guinea and Australia. It is common in the tropical north but numbers decline southwards. It has been recorded as a vagrant as far south as Victoria, but its occurrence south of the Clarence River in New South Wales is decidedly erratic.

It is usually encountered alone or in pairs, but small groups or flocks occasionally gather. It is wary and shuns cover, preferring extensive open shallows, where it stalks fishes, frogs and the larger crustaceans and insects. Normal flight appears ponderous and ungainly, but it often soars at great heights.

Nesting is solitary, usually from March to June. The nest is a huge wide platform of sticks and debris high in a living or dead tree in a secluded swamp. The usual clutch is two or three eggs. Adults maintain their pair bond for several successive years, perhaps for life.

HABITAT: tropical to warm temperate wetlands, lagoons, swamps, mudflats and irrigated cropland
LENGTH: 130–140 cm
DISTRIBUTION: more than 1 million km²
ABUNDANCE: very sparse
STATUS: probably secure

(H & J Beste)

Family THRESKIORNITHIDAE

(thres'-kee-orn-ith'-id-ee: "*Threskiornis*-family", after a genus of ibises)

This family incorporates the ibises and spoonbills: ibises have long downcurved bills while spoonbills have a much flattened spoon-shaped bill. The sexes are similar, there is little seasonal variation, and immatures resemble adults.

Spoonbills are much less gregarious than ibises, but both habitually breed in large, dense colonies. Nests are usually constructed in trees or bushes in swamps but sometimes on the ground. Parents share nest building, incubation and care of young.

The family is represented throughout the tropics and subtropics. Some 32 species are grouped in 18 or so genera, of which three are represented in Australia.

Genus *Platalea*

(plah'-tah-lay'-ah: "spoonbill")

The Roseate Spoonbill of tropical and warm temperate America is showy pink in colour and usually placed in its own genus, but all other spoonbills are included in *Platalea*. There are five species, widely distributed in tropical and warm temperate regions of the Old World; two species occur in Australia, where one is confined. All resemble ibises in general shape and proportions, but are characterised by the remarkably modified bill: long, flat, broad and spatulate. The plumage is white in all species, and most have display plumes in the form of nuchal crests or elongated feathers on the breast. They forage almost exclusively in shallow water, feeding on small aquatic invertebrates.

Yellow-billed Spoonbill

Platalea flavipes (flah'-vee-pez: "yellow-footed spoonbill")

The Yellow-billed Spoonbill is endemic to Australia but is nowhere numerous. It is strongly nomadic and is erratic in occurrence, but it is very rare on Cape York Peninsula and over much of the arid interior. It has strayed to New Zealand and Lord Howe Island.

It is much less gregarious than the Royal Spoonbill, and often forages alone on very small bodies of shallow water, such as stock dams, sweeping the bill rhythmically from side to side to collect the small aquatic insects that form the bulk of its diet. It often loafs conspicuously in high dead trees or on similar exposed perches.

Its nesting schedule and habits are little different from those of the Royal Spoonbill.

HABITAT: cool temperate to tropical shallow fresh to brackish water on margins of lagoons, swamps and dams, sometimes of small area
LENGTH: 76–92 cm
DISTRIBUTION: more than 1 million km²
ABUNDANCE: common
STATUS: secure

(IR McCann)

Royal Spoonbill

Platalea regia (re'-jee-ah: "regal spoonbill")

The Royal Spoonbill occurs across much of Australia, mainly in the north and east. It is rare in Western Australia south of the Kimberley and very erratic in occurrence over much of the arid interior. It also occurs in Indonesia and has been recorded in New Guinea and the Solomon Islands; it has bred in New Zealand since about 1950.

It is mildly gregarious, often foraging alone but more usually in small parties; large flocks are rare. It roosts communally in trees. Foraging is conducted almost exclusively in shallow water, the bill being swept rhythmically from side to side to sweep up small crustaceans, molluscs and other aquatic animals.

Local flooding strongly influences the nesting season, which is usually from July to November in the south-east, and from March to May in the tropical north. Nesting is colonial. Three eggs form the usual clutch.

HABITAT: tropical to cool temperate areas of shallow water, fresh to salt
LENGTH: 74–81 cm
DISTRIBUTION: more than 1 million km^2
ABUNDANCE: sparse
STATUS: secure

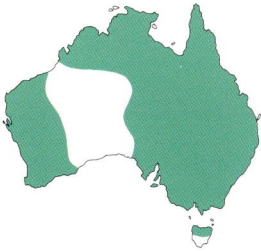

(TG Lowe)

Genus *Plegadis*

(pleg-ah'-dis: "sickle [-bill]")

This genus is usually regarded as consisting of a single almost cosmopolitan species, but the South American population is sometimes treated as a second distinct species. Plain deep brown in colour, both forms closely resemble *Threskiornis* in shape and proportions, but they are substantially smaller, somewhat more slender in build, and lack display plumes.

Glossy Ibis

Plegadis falcinellus (fal'-sin-ell'-us: "little-scythe sickle-bill")

The Glossy Ibis occurs in all continents. It is common in the northwest and the interior south-east of the Australian continent, but, being strongly nomadic, it is erratic and local in occurrence elsewhere. It does not normally occur in Tasmania.

It is gregarious, normally occurring in small groups or very large flocks. It roosts communally and associates freely with other waterbirds. It usually forages on mudflats or in shallow water for insects, molluscs, crustaceans, frogs and small fish.

Breeding is usually in response to local flooding. Breeding colonies are large and dense, often with other waterbirds in lignum or cumbungi swamps. The usual clutch of three or four eggs is incubated for about 21 days, and chicks take about 28 days to fledge.

HABITAT: cool temperate to tropical shallow fresh water in lagoons or flooded grasslands and pastures; less frequently mangrove swamps and estuarine mudflats
LENGTH: 48–61 cm
DISTRIBUTION: more than 1 million km²
ABUNDANCE: common
STATUS: secure

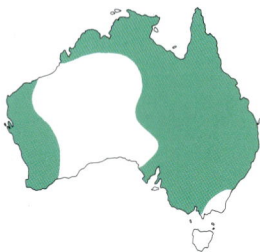

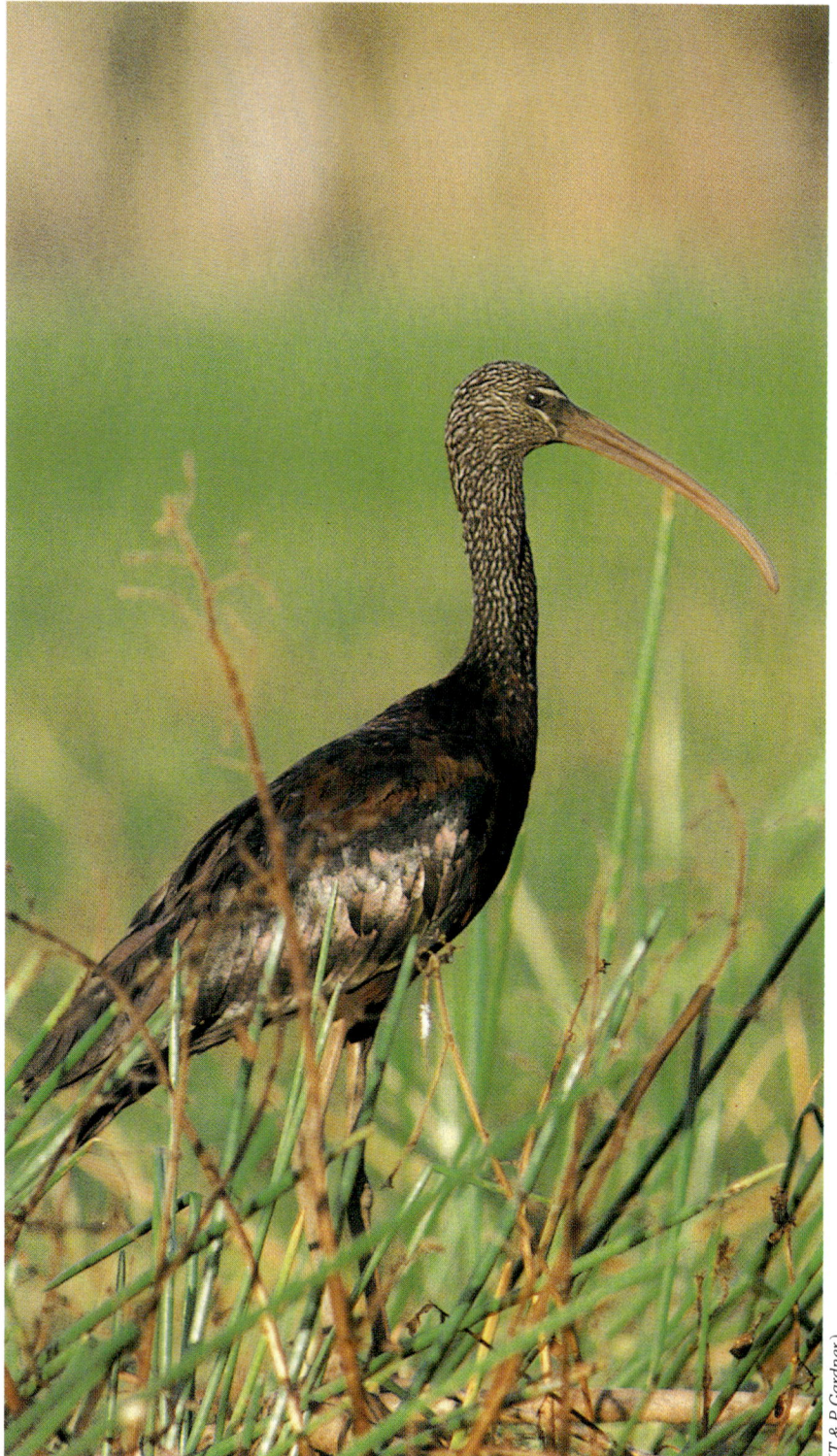

(*T & P Gardner*)

Genus *Threskiornis*

(thres'-kee-or'-nis: "sacred-bird")

The three species of ibis in this genus are widespread in tropical and temperate regions of the Old World; two species occur in Australia, where one is confined. They are rather large, stocky birds with long downcurved bills, naked black heads and stiff straw-like plumes on the upper breast. In Australia one species tends to favour grassland, feeding mainly on grasshoppers, whereas the other is most common in swamp and marshland and has a more generalised diet. Both are strongly gregarious in all activities and at all seasons. The sexes are similar.

Sacred Ibis

Threskiornis aethiopica (ee'-thee-oh'-pik-ah: "African sacred-bird")

Also called the White Ibis, this common species is found over much of northern and eastern Australia from the Kimberley to Eyre Peninsula, South Australia; it also occurs in the far south-west but not in Tasmania. Elsewhere it occurs from Africa across southern Asia to New Guinea and the Solomon Islands. It is strongly nomadic.

There has recently been marked expansion in range and numbers, especially in coastal regions of the south-east: it was virtually unknown in the vicinity of Sydney before 1950 but is now abundant in the city.

Strongly gregarious, it usually occurs in small groups to flocks of hundreds. It forages for insects, fishes, crustaceans and molluscs in swamps or on the margins of streams and lakes which adjoin grasslands, but it is also a frequent scavenger at many garbage tips, fowlyards, pigpens and city parks. It often associates with the Straw-necked Ibis.

Though strongly influenced by local conditions, breeding usually occurs between February and May in the tropical north and between August and November in the south-east.

It breeds in very large, dense colonies, often in company with other waterbirds: colonies may nest in mangrove or lignum swamps, or on the ground in reed beds. Sometimes two broods are raised in succession. The usual clutch is two to four eggs, which hatch in 20 to 25 days.

HABITAT: cool temperate to tropical grasslands
LENGTH: 69–76 cm
DISTRIBUTION: more than 1 million km²
ABUNDANCE: abundant
STATUS: secure

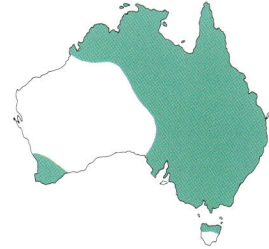

(G Chapman)

63

Straw-necked Ibis

Threskiornis spinicollis (spee'-nee-kol'-is: "spiny-necked sacred-bird")

The Straw-necked Ibis is endemic in continental Australia but it has occurred as a vagrant in Tasmania, southern New Guinea and Indonesia, and on Norfolk and Lord Howe Islands. It is especially abundant in the Murray–Darling basin, but it is strongly nomadic and few areas of the Australian continent are not visited at least occasionally by it.

It often occurs in mixed flocks with the Sacred Ibis but it prefers grassland and pastoral country rather than wetlands, which are the favoured habitat of the Sacred Ibis.

It usually occurs in very large flocks. The diet consists largely of grasshoppers and other large terrestrial insects.

Nesting occurs in large dense traditional colonies, often in company with Sacred Ibis and other waterbirds; breeding behaviour is strongly influenced by local flooding. The largest colonies may number tens or even hundreds of thousands of birds, and nests are often so close together they become trampled to a common platform as the season progresses. Sometimes two broods

are raised in succession. The usual clutch is two or three eggs, which hatch in 20 to 25 days; chicks fledge after about 35 days.

HABITAT: cool temperate to tropical grasslands
LENGTH: 59–76 cm
DISTRIBUTION: more than 1 million km²
ABUNDANCE: common
STATUS: secure

(T & P Gardner)

Order ANSERIFORMES

(an'-se-ree-for'-mayz: "*Anser*-order", after a genus of geese)

Members of this order are moderately small to very large aquatic birds which lay plain unmarked eggs. The newly hatched young are clad in dense fluffy down and are more or less independent on hatching, requiring only guardianship from the parents.

The two families which comprise the order are similar in their basic anatomy but very different in general appearance and modes of life. In general, the waterfowl (Anatidae) have webbed feet, a reduced hind toe, and a highly specialised bill that is extremely distinctive in shape; males have a copulatory organ approximating in function the mammalian penis; almost all species

moult their flight feathers simultaneously.

About 130 species form an almost cosmopolitan group. The three species of screamers (Anhimidae) vaguely resemble large geese. They are exclusively South American in distribution, have unspecialised chicken-like bills, only partly webbed feet with a well developed hind toe, and lack a copulatory organ and do not moult the wing feathers simultaneously. They are unique among birds in lacking uncinate processes (tab-like structures extending backward from each rib to overlap the one behind, an almost diagnostic feature of the avian skeleton).

Family ANATIDAE
(ah-nah'-tid-ee: "*Anas*-family", after a genus of ducks)

Ducks, geese and swans are among the most widespread and best-known of the birds that frequent wetlands. There is little scientific significance in these names but "duck" is generally used for the smaller species, "goose" for the larger (usually grazing) species, and "swan" for those that are large and have exceptionally long, graceful necks.

They are mainly aquatic and spend little time ashore. Most fly readily and well but a few species are flightless or nearly so. Many are strongly migratory. Some species are accomplished divers, others forage mainly by dabbling and upending in the shallows. The diet is extremely varied.

All but the Magpie Goose moult all their flight feathers at once, rendering them flightless for several weeks while the new feathers grow back.

There are about 150 species, of which 23 occur in Australia; two of these are casual vagrants and one is a largely unsuccessful introduction.

Genus Anas
(ah'-nahs: "duck")

With a worldwide distribution and comprising about 40 species, this is by far the largest of the anatid genera. The genus encompasses the typical dabbling ducks, which feed by patrolling shallow water, taking food from the surface or upending for it. Males have striking plumages worn during courtship and usually moulted shortly thereafter; at other seasons they resemble females. Immatures resemble adult females.

Chestnut Teal
Anas castanea (kah'-stan-ay'-ah: "chestnut [-coloured] duck")

(*P. Klapste*)

lagoons. It usually forages in the shallows, dabbling and upending for seeds, shoots, tubers and roots of a variety of aquatic plants as well as small invertebrates.

August to November are peak breeding months, and nesting is in dispersed single pairs. The nest, of down only, is placed in cover on the ground or in a tree cavity. Ten eggs form a typical clutch, hatching in 26 days. The male takes little part in nesting activities but remains close by.

HABITAT: cool to warm temperate coastal mangrove swamps, marshes, mudflats
LENGTH: 38–46 cm
DISTRIBUTION: more than 1 million km²
ABUNDANCE: common
STATUS: secure

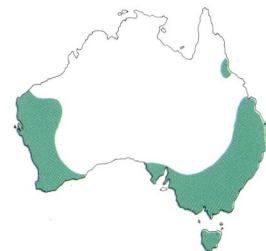

The Chestnut Teal occurs through much of southern Australia and along the eastern and western coasts to about the Tropic of Capricorn: in Tasmania it is probably the most numerous duck.

Adults are fairly sedentary and live permanently in small groups which may coalesce into much larger flocks, especially in winter. The Chestnut Teal is mainly nocturnal or crepuscular, spending the day loafing on mudbanks, in mangrove swamps, or on secluded coastal

65

Grey Teal
Anas gibberifrons (gib'-er-ee-fronz: "humped-forehead duck")

(CA Henley)

It breeds in dispersed single pairs but, during major floods, these may be so closely packed as to constitute, in effect, a vast single colony. Several broods may be raised in succession. Tree cavities are favoured as nest-sites, but almost any convenient site may be chosen. The male takes little part in nesting activities. The usual clutch is seven or eight eggs, which hatch in 24 to 26 days.

HABITAT: cool temperate to tropical wetlands; bodies of water of almost any size but infrequently brackish or saline
LENGTH: 41–48 cm
DISTRIBUTION: more than 1 million km²
ABUNDANCE: abundant
STATUS: secure

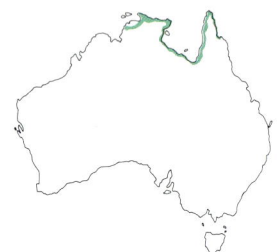

This species extends from the Andaman Islands eastward to the Solomons and is perhaps the most numerous and widespread of the ducks that occur in Australia. Its stronghold is the wetlands of the Murray–Darling basin, but it is also abundant in Tasmania and the south-east, the far south-west, and the tropical north. Highly mobile, it may occur in numbers anywhere else across the continent whenever local rainfall permits.

It is strongly gregarious and often occurs in very large flocks. The diet is varied, and feeding birds dabble in the shallows for seeds and shoots of aquatic plants, insects and their larvae, and minute crustaceans and molluscs.

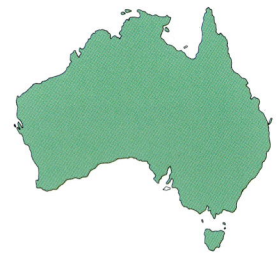

Garganey
Anas querquedula (kwer-kwed'-ue-lah: "duck-duck")

The Garganey has an extensive distribution across Eurasia, wintering in Africa, India, South-East Asia and New Guinea. Before the 1960s it was known in Australia only on the basis of three old specimens from Victoria, but it is now reported with apparently increasing frequency in the Northern Territory and northern Queensland. It may prove to be an uncommon annual visitor.

HABITAT: cool temperate shallow fresh waters; sewage farms, gravel pits, floodlands
LENGTH: 38–41 cm
DISTRIBUTION: 10,000–30,000 km²
ABUNDANCE: rare
STATUS: secure

(MD England)

Blue-winged Shoveler
Anas rhynchotis (rink-oh'-tis: "beaked duck")

The Blue-winged (or Australasian) Shoveler occurs in Australia and New Zealand. Nowhere numerous, it is most abundant in the far south-west of Western Australia, in the lower Murray–Darling Basin and in Tasmania; elsewhere in the south-east it is distinctly uncommon. Its occurrence in northern Australia is very rare and erratic.

Not markedly gregarious, it seldom forms large flocks. It is extremely wary, avoiding the margins of waterbodies and tending to remain far out from shore, mingling inconspicuously with other ducks. It feeds at night, mostly on aquatic insects and their larvae.

The main nesting season is from August to November, but this is influenced by local flooding. Two broods may be raised in succession, but breeding may not be attempted in drought years. Nests are on the ground near water, well hidden in dense vegetation.

HABITAT: cool temperate to subtropical wetlands, particularly shallow lagoons with abundant vegetation; cumbungi and melaleuca swamps
LENGTH: 45–53 cm
DISTRIBUTION: more than 1 million km²
ABUNDANCE: sparse
STATUS: probably secure

(*G Chapman*)

Pacific Black Duck
Anas superciliosa (sue'-per-sil-ee-oh'-sah: "eyebrowed duck")

The Pacific Black Duck occurs from Indonesia eastward to Polynesia and New Zealand. It tolerates a great range of environments and is probably the most widely distributed duck in Australia. It appears to prefer a sedentary way of life but populations engage in unpredictable movements: some marked groups have remained on one small wetland for several years, but at other times birds scatter and range widely, some moving across the continent or even to New Zealand.

Normally alert and very wary, it quickly becomes tame where unmolested. It lives in small groups, seldom in very large flocks, dabbling in the shallows for seeds, roots and shoots of water plants; occasionally it grazes ashore.

It will breed at any time in response to major local flooding but consistent peaks in activity occur from July to October in the south and March to May in the north. Sometimes two broods are raised in succession. The nest is very variable in site, materials and construction. The usual clutch is eight to 10 eggs, hatching in 26 to 28 days. Chicks are led from the nest immediately on hatching, and fledge in 47 to 59 days. Males take little active part in nesting.

HABITAT: cool temperate to tropical bodies of water of any area, fresh to saline, but with preference for deep, permanent fresh water with abundant aquatic and waterside vegetation
LENGTH: 51–61 cm
DISTRIBUTION: more than 1 million km²
ABUNDANCE: abundant
STATUS: secure

(*R Whitford*)

Genus *Anseranas*

(an'-ser-ah'-nahs: "goose-duck")

The characteristics of the genus are essentially those of the single species.

Magpie Goose

Anseranas semipalmata (sem'-ee-pal-mah'-tah: "half-webbed goose-duck")

The Magpie Goose is abundant in northern Australia and southern New Guinea but its distribution has been much reduced since white settlement: once locally common in New South Wales, it is now rare.

It is strongly migratory and nomadic, moving regularly from flooded black-soil plains during the wet season to deep, permanent tropical lagoons during the dry. It feeds mainly on bulbs, roots and seeds of various aquatic plants, especially sedges, rushes and rice.

The basic social unit is a family party, but these freely congregate in flocks of many thousands of birds. Magpie Geese wade along muddy margins, upend in deep water, perch and loaf freely in trees, and roost communally.

Nesting is colonial, mostly from March to May, coinciding with monsoonal rains; if these are delayed, breeding may not be attempted. Nests are bulky mounds of trampled reeds in swamps.

Often the breeding unit is a trio rather than a pair. Clutch size varies from one to 16, two females often laying in the same nest. The incubation period is 24 to 25 days. Young are led from the nest on hatching, and fledge in about 77 days. The pair bond is persistent, lasting probably for life.

HABITAT: tropical to subtropical wetlands, black-soil flood plains, wet grasslands
LENGTH: 75–92 cm
DISTRIBUTION: 300,000–1 million km^2
ABUNDANCE: common
STATUS: probably secure

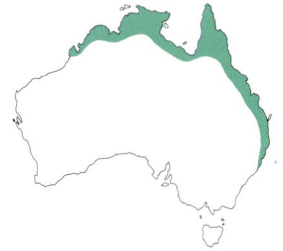

(T & P Gardner)

Genus *Aythya*

(ay'-thee-ah: "diving-bird")

Aythya is a genus of some 14 species of diving ducks best represented in the Northern Hemisphere; only one species occurs in Australia. Males have a plumage distinct from that of females.

Hardhead

Aythya australis (os-trah'-lis: "southern diving-bird")

The Hardhead (or White-eyed Duck) is largely confined to Australia although vagrants frequently reach Indonesia, New Guinea, New Zealand and various islands in the south-west Pacific. In Australia it is most common in the east, the lower Murray–Darling basin and in the extreme south-west, but it wanders widely in response to local flooding. It does not breed in Tasmania.

It is strongly gregarious and often occurs in very large flocks. It dives freely to feed mostly on aquatic plants and invertebrates, especially molluscs.

Breeding is mainly from October to December, in dispersed single pairs. The nest is a compact bowl of reeds, thickly lined with down, in a lignum bush or reed clump near water. The usual clutch is nine to 12 eggs which take about 25 days to hatch. The male takes little part in nesting activities.

HABITAT: cool temperate to tropical wetlands, preferably deep permanent swamps and lakes with abundant fringing and emergent vegetation
LENGTH: 42–59 cm
DISTRIBUTION: more than 1 million km²
ABUNDANCE: common
STATUS: secure

(P Klapste)

Genus *Biziura*

(biz'-ee-ue'-rah: meaning unknown)

This a monotypic genus, restricted to southern Australia is related to *Oxyura*. The characteristics of the genus are those of the single species. There is little seasonal or sexual variation (although females are much smaller than males and lack the gular pouch) and immatures resemble adult females. No pair bond is formed and, after courtship and copulation, males take no further part in nesting activities.

Musk Duck

Biziura lobata (loh-bah'-tah: "lobed *biziura*")

The Musk Duck is common in south-western and south-eastern Australia but rare in Queensland. It frequents deep fresh waters of almost any size and the fact that it quickly reaches transient deep waters following local flooding indicates some degree of nomadism. Occasionally it occurs on the sea.

Completely aquatic, it swims very low with the body almost awash, and is very clumsy on land. It flies rarely and then mainly at night. Although territorial and rather solitary, it often gathers in large flocks in winter. It dives for its food, mainly bottom-living aquatic invertebrates.

The male's courtship display is remarkable: the bird floats or swims slowly with tail arched well over the back and bill pouch fully inflated, uttering a persistent piercing whistle in rhythm with vigorous kicks resulting in sideways splashes of water.

Breeding is mainly from September to December. The nest is a loose bowl of woven and trampled reeds, often with a loose woven canopy overhead and well hidden in dense cover. Two or three eggs form the usual clutch. Eggs are sometimes laid in the nests of other ducks.

(G Chapman)

HABITAT: cool to warm temperate deep permanent fresh water; casually in almost any waterbody, fresh to saline
LENGTH: 61–73 cm
DISTRIBUTION: more than 1 million km²
ABUNDANCE: sparse
STATUS: probably secure

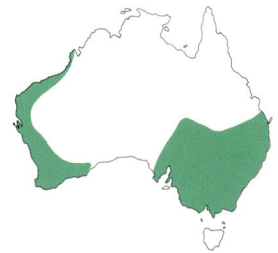

Genus Cereopsis

(se'-ray-op'-sis: "waxy-face")

The characteristics of the genus are essentially those of the single species.

Cape Barren Goose

Cereopsis novaehollandiae (noh'-vee-hol-an'-dee-ee: "New Holland waxy-face")

(CA Henley)

The Cape Barren Goose breeds on small offshore islands in Bass Strait, the vicinity of Spencer Gulf, and the Recherche Archipelago along the southern coast of Australia. After breeding there is some dispersal to the adjoining mainland.

Strongly territorial when breeding, Cape Barren Geese live at other times in family parties, which graze over exposed grasslands and the margins of ponds, lagoons and swamps. The diet consists mainly of grasses and sedges.

Birds first breed when four or five years old, and probably mate for life. Parents share all nesting activities. The nest is a rough pad of grass beside a tussock or rock or under a bush. Four or five eggs, which hatch in about 35 days, form the usual clutch.

HABITAT: cool temperate grasslands on small offshore islands
LENGTH: 75–90 cm
DISTRIBUTION: 100,000–300,000 km²
ABUNDANCE: sparse
STATUS: vulnerable

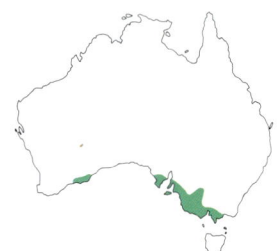

Genus *Chenonetta*

(ken'-on-et'-ah: "goose-duck")

The characteristics of the genus are essentially those of the single species.

Maned Duck

Chenonetta jubata (jue-bah'-tah: "maned goose-duck")

The Maned (or Wood) Duck occurs more or less throughout Australia but it is most common in the east and the far south-west. It is vigorously expanding in both range and numbers, having benefited from the provision of stock dams and other recent land management practices, and is now common over much of the arid interior, the east coast and Tasmania.

Mildly gregarious, it lives mainly in small groups but large flocks sometimes congregate during winter. It forages, largely by night, for seeds and fresh green shoots, spending much of the day camped quietly by water. It tends to avoid salt or brackish water.

The breeding season is variable, but mostly from September to November in the south, January to March in the north. Parents share most nesting activities but only the female incubates. The nest is a pile of down in a tree cavity, not necessarily near water. The usual clutch is nine to 12 eggs, incubated for 28 days.

HABITAT: cool temperate to tropical bodies of fresh water with vegetated edges, usually in open woodland or savanna
LENGTH: 46–51 cm
DISTRIBUTION: more than 1 million km^2
ABUNDANCE: abundant
STATUS: secure

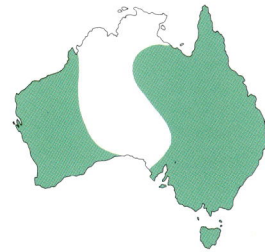

(B Chudleigh)

Genus *Cygnus*

(sig'-nus: "swan")

Swans are best represented in the Northern Hemisphere. Largest and most majestic of waterfowl, all five members of this genus have slender, graceful necks surpassing the body in length, and short, stout legs and large feet. The bill is broad, high and arched, and often brightly coloured or boldly patterned. The plumage of the sole Australian species is black, but others are pure white, with the exception of a South American species which has a black head and neck. Immatures are dingy grey or brown. Swans are mainly vegetarian and strongly aquatic. Some breed in colonies, others in dispersed strongly territorial pairs, but most congregate in flocks in winter. Adults mate for life. They have deep, resonant, bugling calls.

Black Swan

Cygnus atratus (ah-trah'-tus: "blackened swan")

The Black Swan is endemic in Australia, but it has been introduced to New Zealand and elsewhere in the world. It occurs throughout the continent, but is rare in the tropics and breeds only south of about North West Cape in the west and the Atherton Tableland in the east. It seems to be gradually extending its range northward.

It is nomadic and dispersive, often travelling by night. It is strongly gregarious when not breeding, and loafing flocks congregate on extensive sheltered waters. It occasionally forages on land, but is mostly aquatic in habit. The diet is mainly water plants, obtained from the surface or by upending.

Nesting may be solitary or colonial, conducted by pairs or trios, and is very flexible in timing. The nest is a large untidy heap of reeds and grasses, often moored in shallow water. Incubation of the five or six eggs that constitute the usual clutch takes 35 to 45 days; chicks take about 113 to 160 days to fledge and remain with their parents for some time thereafter.

HABITAT: cool temperate to tropical wetlands with extensive areas of shallow, vegetated water, fresh to salt; flooded pastures; ponds in parks
LENGTH: 112–140 cm
DISTRIBUTION: more than 1 million km²
ABUNDANCE: common
STATUS: secure

(P Klapste)

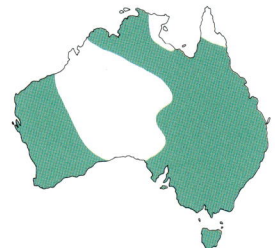

Genus Dendrocygna

(den'-droh-sig'-nah: "tree-swan")

This very distinct genus of eight species is widespread in the tropics and extends into temperate regions in the Southern Hemisphere. Two species of *Dendrocygna* (whistling-ducks) occur in Australia. All are of moderate size, with long necks normally held erect, broad, rounded wings, long legs and large feet. The hind toe is well developed, the webs indented, and the claws fairly long and sharp. Most species have elongated, lanceolate plumes of some contrasting colour on the flanks. The plumage pattern is elaborate, and the sexes are similar in appearance. The colour pattern of the downy young is very distinctive.

Whistling-ducks are noisy and intensely gregarious. Mainly vegetarian and partly nocturnal or crepuscular habits, many species congregate by day in densely packed "camps" in secluded swamps. In flight their wings produce a characteristic whistling noise, blending and merging with the generally whistling timbre of their calls. Adults mate for life.

Wandering Whistling-duck

Dendrocygna arcuata (ark'-ue-ah'-tah: "bowed tree-swan")

The Wandering (or Water) Whistling-duck occurs from Borneo eastward to Fiji. In Australia it is most abundant in the coastal lowlands of

(CA Henley)

the Kimberley and the Top End, but it has been reported in all States except Tasmania.

It is strongly gregarious and usually inhabits deep, permanent tropical lagoons during the dry season, dispersing during the wet to forage over flooded grasslands. It feeds, mainly at night, on bulbs, roots and seeds of various aquatic plants and grasses. During the day it camps in deep water or a secluded corner of a swamp. In deep water it dives freely for food.

Breeding is usually from January to March in dispersed pairs, but nesting may not be attempted if monsoon rains are delayed. Pairs probably mate for life, and share all nesting activities. The nest is a scant concealed pad of grasses in or near a swamp, often some distance from water. Eight to 10 eggs make up the usual clutch; incubation takes 28 to 30 days; and the young fledge in about 90 days.

HABITAT: mainly deep, permanent freshwater swamps and lagoons
LENGTH: 55–61 cm
DISTRIBUTION: more than 1 million km²
ABUNDANCE: common
STATUS: secure

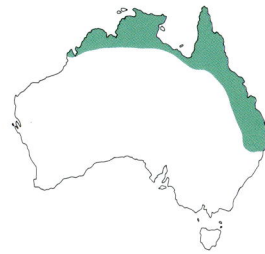

Plumed Whistling-duck

Dendrocygna eytoni (ay'-ton-ee: "Eyton's tree-swan", after T. C. Eyton, English ornithologist)

Also called Grass Whistle-duck, the Plumed Whistling-duck occurs, at least casually, almost throughout Australia, but mostly in the lush subinterior grasslands and associated wetlands.

It usually lives in large, compact flocks of up to several thousand birds. It feeds largely on grass seeds gathered at night; during the day it camps in dense flocks in a secluded swamp, often far from the feeding grounds. It perches freely in trees. Flocks maintain an almost constant twittering whistle.

It beeds in dispersed single pairs, mainly from January to March, but the schedule is strongly influenced by local conditions. The nest is a scant concealed pad of grasses in or near a swamp, often some distance from water. Clutch size ranges from eight to 14; eggs hatch in 28 days, and the young are led from the nest immediately on hatching. Parents, which share nest-building, incubation and care of young, perhaps mate for life.

HABITAT: mainly deep, permanent tropical lagoons and marshes
LENGTH: 44–62 cm
DISTRIBUTION: more than 1 million km²
ABUNDANCE: common
STATUS: secure

(LC Llewellyn)

Genus *Malacorhynchus*

(mal'-ah-koh-rink'-us: "soft-beak")

The characteristics of the genus are essentially those of the single species.

Pink-eared Duck

Malacorhynchus membranaceus (mem'-brah-nah'-say-us: "membraned soft-beak")

Highly mobile, the endemic Pink-eared Duck is characteristic of ephemeral floodwaters across Australia. Its stronghold is around the lower Murray–Darling basin and the far south-west of Western Australia. It is rare on the east coast and is seldom reported from Tasmania.

It is strongly gregarious, living in small to very large flocks, often associating with Grey Teal and other wildfowl. It seldom comes ashore but groups often perch above the water on snags and similar objects. Its foraging is highly specialised: groups swim through the shallows filtering minute aquatic animals from the water.

Breeding is in response to floods. Nests are usually dispersed but, at the peak of breeding activity, may be very close together. The six to eight eggs which form the usual clutch are buried in a heap of down that serves as a nest, jammed into any convenient cavity in a bush, fence post or tree hollow.

HABITAT: cool temperate to tropical shallow open residual floodwaters; lignum swamps; infrequently in permanent fresh water, coastal lagoons, mangrove swamps
LENGTH: 38–45 cm
DISTRIBUTION: more than 1 million km^2
ABUNDANCE: sparse
STATUS: probably secure

(G Chapman)

Genus *Nettapus*

(net'-ah-poos: "duck-foot")

Smallest of waterfowl, the pygmy-geese are characterised by their short, blunt, goose-like bills, compact silhouettes, and elaborate plumage patterns in which white and glossy bottle-green are prominent. The sexes are generally similar but females are much duller than males. Pygmy-geese inhabit quiet, deep, permanent lagoons that have an abundance of aquatic vegetation, especially water-lilies. Rather quiet and inactive waterfowl, they spend most of their time in deep water and seldom come ashore. They feed mainly on the seeds of aquatic plants. There are three species in the genus, which is restricted to the Old World tropics. Two species occur in Australia.

White Pygmy-goose

Nettapus coromandelianus (ko'-roh-man'-del-ee-ah'-nus: "Coromandel [southern Indian] duck-foot")

Widespread from India to southern China and New Guinea, the White (or Cotton) Pygmy-goose is restricted in Australia to the eastern coastal lowlands from far north-eastern Queensland to the Clarence River in New South Wales. However, it was extirpated by drainage and agriculture in New South Wales during the 1930s. It is now a rare vagrant in that State and its current stronghold is in the Ayr–Townsville–Charters Towers region of Queensland.

It is sedentary and gregarious, living mostly in small flocks. Like the Green Pygmy-goose it is strictly aquatic but, unlike that species, it seldom dives for food. It feeds mainly on water plants and insects.

HABITAT: tropical to subtropical coastal lagoons with abundant vegetation; also large bodies of calm fresh water
LENGTH: 35–38 cm
DISTRIBUTION: 100,000–300,000 km²
ABUNDANCE: sparse
STATUS: vulnerable

(D & M Trounson)

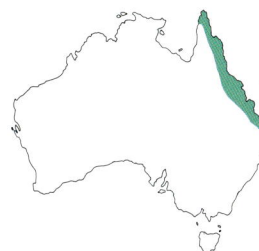

Green Pygmy-goose

Nettapus pulchellus (pool-kel'-us: "pretty little duck-foot")

The Green Pygmy-goose occurs in Indonesia, New Guinea and tropical northern Australia. Its distribution coincides closely with the occurrence of deep, permanent tropical lagoons with abundant aquatic vegetation, upon which it is dependent for refuge during the dry season, although it may disperse widely at other times.

Strictly aquatic and almost never ashore, it feeds mainly on waterlilies and the seeds of aquatic plants, and dives freely for food. It lives in pairs or small groups, but flocks of several hundred birds sometimes congregate. Its breeding behaviour has not been closely studied.

HABITAT: tropical lagoons and swamps with abundant vegetation
LENGTH: 30–36 cm
DISTRIBUTION: 300,000–1 million km²
ABUNDANCE: sparse
STATUS: vulnerable

(G Anderson)

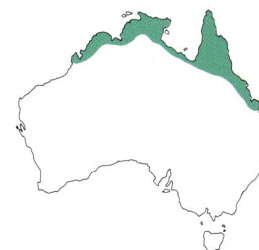

Genus Oxyura

(ox'-ee-ue'-rah: "sharp-tail")

Oxyura is a genus of about six species widespread in North and South America, Africa and Eurasia; one species is endemic in Australia. They are small rotund ducks that dive for food. Males have a distinct breeding plumage but at other seasons resemble females; immatures resemble adult females. Males take no part in nesting activities.

Blue-billed Duck

Oxyura australis (os-trah'-lis: "southern sharp-tail")

The Blue-billed Duck occurs mainly in the lower Murray–Darling basin and in far south-western Australia. It occurs casually in southern Queensland, New South Wales, Victoria and Tasmania.

It is strongly territorial when breeding but otherwise lives in small groups, although quite large flocks congregate on occasion. It is completely aquatic, dives freely and almost never comes ashore. It is mainly nocturnal or crepuscular, feeding on aquatic plants and insect larvae.

The breeding season is from September to December. The nest is a deep bowl of grass and reeds lined with down, well hidden in dense reeds over water. Five or six eggs form the usual clutch, incubated for 26 to 28 days.

HABITAT: cool to warm temperate deep, permanent freshwater lakes, lagoons and swamps, preferably with extensive reed beds
LENGTH: 35–44 cm
DISTRIBUTION: more than 1 million km^2
ABUNDANCE: sparse
STATUS: vulnerable

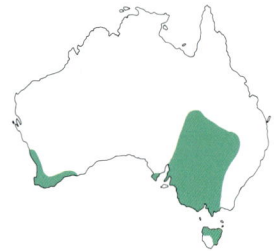

(R Whitford)

Genus Stictonetta

(stik'-toh-net'-ah: "freckled-duck")

The characteristics of the genus are essentially those of the single species.

Freckled Duck

Stictonetta naevosa (nee-voh'-sah: "spotted freckled-duck")

The Freckled Duck breeds most commonly in the lower Murray–Darling basin, in the channel country surrounding Lake Eyre and in the far south-west of Western Australia. It occurs from time to time over most of the continent.

It is secretive in behaviour, and spends most time in small groups, although quite large flocks have been reported. It is seriously endangered: a census in 1983 indicated a total population of not more than 19 000 birds, possibly far fewer. Despite protection, many are shot by duck hunters.

(E Beaton)

HABITAT: cool temperate to sub-tropical swamps (cumbungi, lignum or melaleuca, usually of large extent); also lakes and floodwaters
LENGTH: 52–60 cm
DISTRIBUTION: more than 1 million km^2
ABUNDANCE: rare
STATUS: vulnerable

Genus *Tadorna*

(tad-or'-nah: "duck")

Moderately large but unremarkable in proportions, shelducks are characterised by their bold plumage patterns, slight sexual dimorphism, and aggressive disposition. Strongly territorial when breeding, they often form small groups in winter but seldom congregate in large flocks. Most forage for preference along muddy or grassy margins of wetlands or, sometimes, the seacoast. They feed on aquatic plants, but such invertebrates as worms and molluscs are prominent in the diet. There are eight species in the group, which is widespread in temperate and tropical regions across the Old World, but not in the Americas. Two species occur in Australia.

Burdekin Shelduck

Tadorna radjah (rah'-jah: "shelduck duck", supposedly from an Asian native name for the species)

The Burdekin Shelduck is found in Indonesia, New Guinea and coastal northern Australia, west to the vicinity of Broome in Western Australia, and south to about Rockhampton in Queensland; in the early years of European settlement it probably occurred at least as far south as Brisbane.

It retreats to deep permanent coastal lagoons during the dry season and disperses widely during the wet. It lives mainly in small groups and feeds principally on molluscs; foraging is largely nocturnal.

Most breeding occurs from April to June but the season is strongly influenced by local rainfall. No nest is built, the six to 12 eggs being laid on the floor of a tree cavity, scantily lined with down. The selected tree is usually in or near water. Only the female incubates but the sexes share other nesting activities.

HABITAT: tropical shallow brackish water, particularly mudflats and mangrove swamps
LENGTH: 50–56 cm
DISTRIBUTION: 300,000–1 million km^2
ABUNDANCE: sparse
STATUS: probably secure

(T & P Gardner)

77

Mountain Shelduck

Tadorna tadornoides (tad'-or-noy'-dayz: "duck-like duck")

The Mountain (Australian or Chestnut-breasted) Shelduck occurs in south-western Australia from about Shark Bay to Esperance and, in the east, from about Spencer Gulf to Sydney, including Tasmania. Vagrant individuals or groups frequently occur further north.

When breeding, it is strongly territorial. At other times it occurs in small groups, feeding mainly upon algae and other aquatic vegetation, supplemented by insects, crustaceans and molluscs. It feeds mainly at night. After breeding, it resorts in flocks to large bodies of fresh or brackish water in order to moult.

The main breeding season is from June to September. Pairs probably mate for life, sharing most nesting activities except incubation, which is conducted by the female. The nest is typically in a cavity 2 metres or more up in a tree, often far from water, but ground-nesting is not uncommon. The usual clutch is 10 to 14 eggs, which hatch in 30 to 35 days.

HABITAT: cool temperate to subtropical estuarine mudflats and muddy shores of other bodies of water (fresh to saline); flooded pastures, irrigation crops
LENGTH: 59–72 cm
DISTRIBUTION: more than 1 million km²
ABUNDANCE: common
STATUS: secure

(CL Gill)

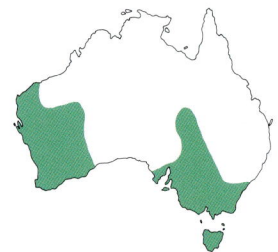

Order ACCIPITRIFORMES

(ak'-sip-it'-ree-for'-mayz; "*Accipiter*-order", after a genus of hawks)

The birds included in this order vary enormously in size, but most are notable for their well developed powers of flight, extraordinarily acute vision, powerful hooked bills and taloned feet, and their essentially carnivorous diet. With few exceptions they are active and hunt by day. Females are usually substantially larger than males. Five groups may be distinguished within the order as traditionally listed: Cathartidae (New World vultures and condors), Pandionidae (osprey), Accipitridae (kites, hawks and eagles), Sagittariidae (secretary-bird), and Falconidae (falcons), but their relationships with each other are much disputed and it seems doubtful that the group is a natural one. In particular, the New World vultures are very distinct in many ways and are now often isolated in their own order. Similarly, the falcons seem less closely related to the Accipitridae than their superficial similarities might suggest and they, too, are sometimes isolated in their own order, the Falconiformes. The families Pandionidae and Sagittariidae consist of only one species each, the former almost cosmopolitan, the latter confined to Africa.

Family PANDIONIDAE

(pan'-dee-on'-id-ee: "*Pandion*-family", after the single Osprey genus)

The characteristics of the family are those of the single genus and species.

Genus Pandion

(pan-dee'-on: "osprey")

The characteristics of the genus are those of the species.

Osprey

Pandion haliaetus (hal'-ee-ee'-tus: "sea-eagle osprey")

The Osprey has an almost worldwide distribution. In Australia it occurs around almost the entire coastline, though it is far more numerous in the north than in the south. It is very rare in Tasmania and does not breed there.

It feeds almost entirely on fish, for which it hunts in steady systematic flights over water at a height of some 20 metres. Once a fish is sighted, the bird hovers briefly, then plunges feet-first into the water to capture it. Prey is usually taken to an exposed high perch, such as a dead tree, for consumption. Despite its predominantly coastal distribution, the Osprey will hunt over fresh, salt or brackish water.

The nest is a large pile of sticks high in a tree, usually dead, and often conspicuous. On offshore islands the birds sometimes nests on rock ledges or cliff tops. Two or three eggs form the usual clutch. Both sexes share all nesting activities.

HABITAT: tropical to cool temperate coasts and margins of estuaries, rivers and lakes
LENGTH: 50–63 cm
DISTRIBUTION: 300,000–1 million km²
ABUNDANCE: common
STATUS: probably secure

(*V. Fazio*)

Family ACCIPITRIDAE

(ak'-sip-it'-rid-ee: "*Accipiter*-family", after a genus of hawks)

About 218 species are included in this cosmopolitan family, which has representatives in almost all environments from arid desert to rainforest. The range in size is great, some members being small and others being among the largest of all flying birds, several exceeding six kilograms in weight. Variously known as hawks, eagles, buzzards, kites, and by several other names, almost all are carnivorous and diurnal. They show a very wide range of diets and hunting strategies, ranging from snails, insects and other invertebrates to fairly large mammals, captured by browsing in foliage, low hovering flight, swift dashes through dense cover, or spectacular dives from great heights. Many species feed on carrion.

The Accipitridae superficially resemble falcons (Falconidae) but differ in a number of anatomical peculiarities, in the structure of their eggshells, and in their patterns of moult, as well as in a number of behavioural peculiarities: for example, hawks usually build nests, whereas falcons do not; falcons also habitually kill their prey by severing the neck vertebrae with a quick bite.

The family may be broadly divided into eight or nine groups but many genera resist generalised treatment. Kites (*Milvus, Elanus, Haliastur* and several other genera) are cosmopolitan but best represented in the tropics; they patrol open country at no great height and feed mainly in invertebrates and carrion. Sea eagles (*Haliaeetus*) show some similarities to kites but are much larger, feed largely on fish, and are generally found near water. Vultures, which are essentially African and tropical Asian in distribution, scan for food while soaring at very great heights, and feed on carrion. Sparrowhawks and goshawks (*Accipiter*) are cosmopolitan, broad-winged and long-tailed, swift and dashing, and are primarily bird-catchers in dense cover. Buteos, best represented in the Northern Hemisphere, scan for small prey from moderate heights over open country. Eagles (*Aquila*) are large, powerful, cosmopolitan, and feed on fairly large mammals attacked from a great height in open country. Broad-winged and long-tailed, harriers (*Circus*) fly slowly low over open country and feed largely on small mammals. Several tropical genera hunt in rainforest and jungle, often specialising on prey ranging from stick insects to bats and monkeys.

Genus Accipiter

(ak-sip'-it-er: "hawk")

This cosmopolitan genus of about 50 species is by far the largest in the family and notable especially for the extreme difference in size between the sexes: in some species the female weighs nearly twice as much as the male. Supremely adapted for swift and agile flight in dense cover, these hawks have rather short, broad and blunt wings and long, rounded or slightly notched tails. The legs are generally long and slender, and the talons well developed. The male usually differs from the female in plumage as well as in size: the pattern varies widely, but in many species the upper parts of the male are more or less plain grey and the underparts narrowly barred in pale rufous, whereas females are often heavily streaked below. Many species are small and dashing, mostly catching birds by means of a hunting technique that combines sudden surprise attack and outstanding dexterity in flight; some are larger and prey more on mammals.

Collared Sparrowhawk

Accipiter cirrhocephalus (si'-roh-sef'-ah-lus: "tawny-headed hawk")

The Collared Sparrowhawk occurs in Indonesia and New Guinea, and virtually throughout Australia, including Tasmania.

It lives in pairs that apparently maintain permanent territories. A remarkably quick and agile hunter, it feeds mainly on small birds captured on the ground or in the canopy. It also eats nestling birds and large insects, rarely small mammals and reptiles.

Most breeding takes place from September to October. The nest is a small rough platform of sticks and twigs, lined with leaves, placed up to 30 metres high in a tree. Often the abandoned nest of some other bird is used, and the nest may be refurbished and used again in subsequent nestings. Three eggs form the usual clutch.

(R Garstone)

HABITAT: tropical to cool temperate dense eucalypt forest
LENGTH: 28–39 cm
DISTRIBUTION: more than 1 million km^2
ABUNDANCE: common
STATUS: probably secure

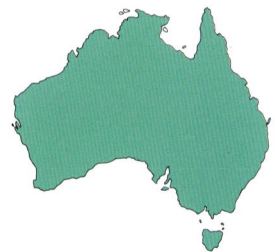

Brown Goshawk

Accipiter fasciatus (fas'-ee-ah'-tus: "banded hawk")

Also called the Australian Goshawk or Chicken-hawk, this common species occurs in Indonesia and New Guinea and throughout Australia, including Tasmania.

(P Klapste)

Adults live in pairs which appear to maintain permanent territories, but immatures wander widely. The diet is mainly small mammals, reptiles and birds, captured on or near the ground; rabbits are prominent in the diet. Young and inexperienced birds frequently take to raiding poultry farms, and many are shot for taking chickens and racing pigeons.

The nesting season is mainly from August to January in the south, September to November in the north. The nest is a rough platform of sticks and twigs, lined with green leaves, high in a tree; it may be refurbished and used again in subsequent seasons. The usual clutch is two or three eggs. The incubation period is 35 to 36 days and the young fledge after about 35 to 36 days in the nest. The female undertakes incubation and most of the care of the young, assisted by the male who regularly brings food to the vicinity of the nest.

HABITAT: tropical to cool temperate wooded country, preferably riverine woodland and dense forests
LENGTH: 43–56 cm
DISTRIBUTION: more than 1 million km^2
ABUNDANCE: common
STATUS: secure

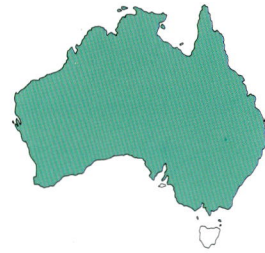

Grey Goshawk

Accipiter novaehollandiae (noh'-vee-hol-an'-dee-ee: "New Holland hawk")

The Grey Goshawk occurs from Indonesia eastward to the south-west Pacific region. In Australia it inhabits the northern and eastern coastal regions and nearby highlands. Adults are sedentary, maintaining permanent territories, but immatures disperse widely.

The species occurs in two distinct forms—white and grey—which freely interbreed. The ratio of one form to another varies geographically for unknown reasons: individuals of the white form dominate the local population in Tasmania and the Kimberley but are virtually unknown in northern Queensland. The sexes are similar in appearance and immatures resemble adults.

Small birds make up the bulk of the diet, which occasionally also includes small mammals, reptiles and large insects. Most prey is captured by sudden flight from a secluded perch.

The nest is a rough platform of sticks and twigs, usually at a considerable height in an angophora or eucalypt; two or three eggs form the normal clutch. Incubation, lasting about 35 days, is shared by both parents, but the female undertakes most care of the young, with food being fetched by the male. Young fledge in about 37 days but are fed by their parents for some time thereafter.

(G May)

HABITAT: tropical to cool temperate dense forests with high rainfall; seldom far from the coast
LENGTH: 35–54 cm
DISTRIBUTION: 300,000–1 million km^2
ABUNDANCE: very sparse
STATUS: probably secure

Genus *Aquila*

(ak-wil'-ah: "eagle")

Rivalling the sea eagles in size, the nine or 10 species in this genus have rather long, broad wings and medium to long tails which vary greatly in shape between species. The genus is best represented in Africa and Eurasia, but one species occurs in western North America and another is widespread in Australia. Most live largely on rabbit-sized mammals captured on more or less open terrain after a dive from a great height, but diet and hunting techniques vary and some of the smaller species live mainly on carrion. Eagles are much given to soaring effortlessly at great heights, and are commonest in mountainous country. Most are brown or black in colour, and few have striking plumage patterns.

Wedge-tailed Eagle

Aquila audax (aw'-dax: "bold eagle")

The Wedge-tailed Eagle occurs in southern New Guinea and throughout Australia, including Tasmania. It is most numerous in the interior and avoids coastal areas. Adults are sedentary, living in pairs that occupy permanent territories.

The large, wedge-shaped tail is unmistakable in silhouette, but colour varies from pale sandy brown in juveniles to near-black in fully adult birds. Much time is spent soaring, often at great heights. The species is generally solitary but groups often congregate at a carcass. The diet consists mainly of mammals (especially rabbits), reptiles and carrion.

Breeding begins in July or August. The nest is a huge structure of sticks and twigs, usually in a tree with a commanding view over surrounding countryside. Two eggs form the usual clutch, incubated almost entirely by the female. The incubation period is about 45 days. Young fledge in about 70 days but are dependent on their parents for several months thereafter.

HABITAT: tropical to cool temperate wooded country with adjacent open areas
LENGTH: 90–100 cm
DISTRIBUTION: more than 1 million km²
ABUNDANCE: common
STATUS: secure

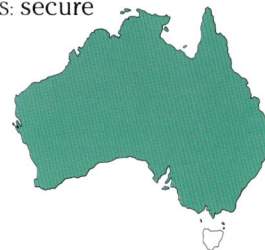

(G Steer)

Genus *Aviceda*

(ah'-vee-say'-dah: meaning unknown but derived from Latin *avis*, bird)

This genus of medium-sized hawks is widespread in tropical forests in Africa, southern Asia and Australasia. Known as cuckoo-falcons or bazas, they are arboreal hunters which take most of their prey, consisting of small birds, reptiles and large insects such as mantids, in the foliage of trees, hunting within the canopy rather than from above it. They have rather long, moderately pointed wings and tails of medium length. The bill is rather small and there are two tooth-like denticles on the upper mandible; the legs are short and stout and the talons well-developed. Several feathers at the nape are prolonged to form a crest. The plumage is strongly patterned, and the underparts are often barred; the sexes are similar. Bazas are mildly gregarious, sometimes occurring in loose parties. There are five species, one of which occurs in north-eastern Australia.

Crested Hawk

Aviceda subcristata (sub'-kris-tah'-tah: "slightly crested *Aviceda*")

The Crested Hawk (or Pacific Baza) occurs from Indonesia to the Solomon Islands. In Australia it occurs mainly in coastal lowlands and associated highlands. It is mainly sedentary except for the post-fledging dispersal of young. Uniquely among the Australian hawks, it has an obvious crest.

It lives in pairs or family groups which are thought to maintain permanent territories. It feeds mainly on large arboreal insects and small reptiles, captured by snatching them from dense foliage in the upper canopy. It often soars in wide circles high above forest; courtship displays involve spectacular tumbling flights.

Nesting takes place from October to February. The nest is a small, flimsy, shallow platform of sticks, lined with green leaves and placed on the horizontal limb of a tree, at least 15 metres from the ground. The clutch consists of two or three eggs, which hatch in about 33 days. The young fledge after 32 to 35 days in the nest.

HABITAT: tropical to warm temperate rainforest; wet sclerophyll forest and its margins
LENGTH: 35–43 cm
DISTRIBUTION: 300,000–1 million km^2
ABUNDANCE: very sparse
STATUS: vulnerable

(JM Cupper)

Genus Circus

(ser'-kus: "hawk")

The harriers form an almost cosmopolitan and very distinct genus of about 10 or 12 species with no obvious links with other hawks; two species occur in Australia. All are medium sized, with long, blunt wings and long, narrow, rounded tails. The bills are rather small and the legs long and slender, but the most striking characteristic is the owl-like ruff of stiff feathers surrounding the ears. Females are substantially larger than males, and usually differ also in being dull brown and streaked: adult males of many species are white and pale grey. Most species have a white rump. All have in common a distinctive hunting technique, in which the bird methodically patrols grassland, reedbeds or similar open country at a height of about 10 metres, often sailing briefly on slightly upswept wings, while scanning the ground below. The diet consists mainly of mice and other mammals of similar size. All but one species nest on the ground.

Swamp Harrier

Circus approximans (ap-rox'-im-anz: "approaching [i.e., similar] hawk")

The Swamp Harrier occurs in New Guinea, New Zealand, and many islands of the south-western Pacific region. In Australia it is most common in the east and the far south-west, and it seldom breeds north of 33°S. Most southern individuals migrate northward to spend the winter, returning south in spring.

It occurs singly or in pairs. The hunting style is distinctive: it flies slowly and methodically low over the tops of reeds and crops, frequently sailing on slightly upswept wings and rocking slightly from side to side, occasionally hovering clumsily. The diet includes birds, snakes and other reptiles, and rabbits; occasionally it eats carrion.

Breeding is mainly from October to January. The nest is a rough platform of reeds and grass on the ground, usually well hidden in a reed bed or crops. Parents share nest building, incubation and care of young. The clutch varies from one to five eggs, usually three or four, but seldom are more than three young raised. The incubation period is 33 days and the young fledge in about 41 to 50 days.

Some authorities regard the Swamp Harrier as conspecific with the Marsh Harrier of Africa and Eurasia.

(*P Klapste*)

HABITAT: tropical to cool temperate reed beds, lush pastures and croplands, seldom far from water
LENGTH: 50–58 cm
DISTRIBUTION: more than 1 million km²
ABUNDANCE: sparse
STATUS: vulnerable

Spotted Harrier

Circus assimilis (as-sim'-il-is: "similar hawk")

The Spotted Harrier occurs in western Indonesia and throughout mainland Australia, but it is a very rare vagrant in Tasmania.

Strongly nomadic, it occurs singly or occasionally in small groups. It systematically covers its hunting area, flying low and slowly, periodically hovering in a characteristically clumsy fashion, and pouncing on its prey on the ground. Rabbits, pipits and quails figure largely in the diet, which also includes other small terrestrial birds, mammals, reptiles, and small insects.

Alone among harriers, it usually nests in a tree (all others nest on the ground). The nest is a rough, bulky structure of sticks and twigs, lined with green leaves. Two or three eggs constitute the usual clutch. The breeding season is strongly influenced by local conditions, but is usually from August to October.

(*J McCann*)

HABITAT: tropical to cool temperate open plains and grasslands; also cereal crops
LENGTH: 53–60 cm
DISTRIBUTION: more than 1 million km²
ABUNDANCE: common
STATUS: secure

Genus *Elanus*

(el-ah'-nus: "kite")

This is an almost cosmopolitan genus of small kites in which the bill and feet are rather small, the wings long and pointed, and the tail moderately long, square-ended and slightly notched. The plumage is white below and grey above, with a large black panel on the upperwing.

Black-shouldered Kite

Elanus notatus (noh-tah'-tus: "marked kite")

The Black-shouldered Kite is common across much of the continent but is a casual visitor to Tasmania.

It lives in pairs and feeds mostly on small mammals and large insects.

(E Hosking)

When hunting, it patrols at a height of 10 metres or so, frequently pausing to hover. It chooses an exposed perch: the top of a dead tree, a telephone pole, a wire fence or some similar situation.

Most breeding is from May to September but the season is heavily influenced by local conditions. Several broods may be raised in succession. The nest is a deep cup of sticks and twigs, lined with leaves, usually 10 metres or higher in a eucalypt; often the deserted nest of a crow or magpie is used. Three or four eggs constitute the usual clutch, which is incubated for about 28 days. Fledging is variable but usually takes about 42 to 50 days, and the young remain with their parents for several weeks after leaving the nest.

HABITAT: well-watered, lightly timbered grassland, crops, pastures
LENGTH: 35–38 cm
DISTRIBUTION: more than 1 million km²
ABUNDANCE: sparse
STATUS: probably secure

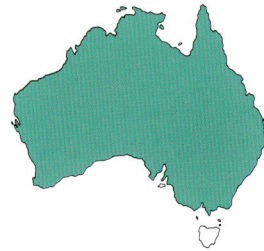

Letter-winged Kite

Elanus scriptus (skrip'-tus: "written kite", referring to black W marking under wings)

The endemic Letter-winged Kite breeds mainly from Lake Eyre northward into south-western Queensland but individuals or flocks may appear almost anywhere in eastern Australia. It resembles the closely related Black-shouldered Kite but is easily distinguished in flight by the presence of a conspicuous black bar extending the

(H & J Beste)

length of the underwing.

Its ecology is intimately linked with that of the Long-haired Rat, numbers of which increase enormously in response to successive years of rainfall in the inland deserts. When rats are abundant the kites breed rapidly, but when the return of the usual drought conditions leads to a collapse of the rat populations, large numbers of kites must starve or leave the region. There is a tendency for such displaced birds to move south and east.

The Letter-winged Kite is strongly gregarious, hunting, roosting and breeding in loose flocks. It feeds at night.

The nest is a small, flimsy bowl of sticks and twigs lined with fur and usually situated in dense foliage in a tree. Clutch size ranges from three to six eggs, usually five. Young leave the nest at about 35 days.

HABITAT: warm temperate semiarid to arid woodland and grassland, varying considerably with cycles of rainfall and drought
LENGTH: 33–38 cm
DISTRIBUTION: more than 1 million km²
ABUNDANCE: common
STATUS: secure

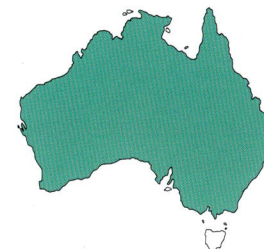

Genus *Erythrotriorchis*

(e-rith'-roh-trie-ork'-is: "red-raptor")

The characteristics of the genus are essentially those of the single species.

Red Goshawk

Erythrotriorchis radiatus (rah'-dee-ah'-tus: "barred red-raptor")

The Red Goshawk inhabits northern and eastern Australia, from the Kimberley to north-eastern New South Wales.

It is extremely rare and thinly distributed; very little is known of its biology. It inhabits forest and woodland, and feeds mainly on large birds, including cockatoos, pigeons and kookaburras.

HABITAT: tropical to warm temperate well-watered forest and woodland
LENGTH: 45–61 cm
DISTRIBUTION: 300,000–1 million km²
ABUNDANCE: rare
STATUS: vulnerable

(JM Cupper)

Genus *Haliaeetus*

(hal'-ee-ee-et'-us: "sea-eagle")

Eight species of eagles constitute this genus, some members of which are among the largest in the family. Usually known as sea-eagles, they feed largely on fish (either caught by plunge-diving into water from a height, or scavenged on beaches or lake shores). Everywhere they are strongly associated with the seacoast or major inland rivers and lakes. They occur in North America, Eurasia, Africa and Australasia. The wings are very long; the tail is rather short, rounded or wedge-shaped; and the moderately short legs have very powerful feet and talons. The bill is large, deep, and strongly curved. Many species have striking plumage patterns.

White-bellied Sea-eagle

Haliaeetus leucogaster (luke'-oh-gas'-ter: "white-bellied sea-eagle")

The White-bellied Sea-eagle occurs from India through south-eastern Asia to Australia, where it occurs around the coast, along major inland rivers, and on large bodies of water—natural or artificial—throughout the country. Grey above and white below, with very broad wings and a small diamond-shaped tail, adults are unmistakable.

It lives in pairs. In many areas these maintain permanent territories; elsewhere, both adults and young disperse after breeding. The diet includes birds, small mammals and fishes (taken by plunging into water from flight) but it also scavenges along beaches for carrion. It is often seen wheeling in high majestic circles in the sky, but it also spends long periods perched in high, prominent trees or other vantage points.

Breeding generally begins in August. The nest is usually in the highest available tree, a huge mass of sticks and twigs accumulated over many years; on offshire islands it may be placed on the ground or on a cliff ledge.

Two eggs form the usual clutch. Young leave the nest after 70 to 78 days and depart from the area a few weeks afterward.

HABITAT: tropical to cool temperate coasts and extensive bodies of inland water
LENGTH: 75–85 cm
DISTRIBUTION: more than 1 million km^2
ABUNDANCE: common
STATUS: secure

(TV Modra)

Genus *Haliastur*

(hal'-ee-as'-ter: "sea-goshawk")

Two rather large kites make up this genus, which has a distribution from India to the Solomon Islands and New Caledonia: both species occur in Australia. They are most abundant in the tropics but one species extends into southern Australia. The wings are rather long and moderately broad, the tail is medium to quite long and slightly rounded, and the legs are short and stout. One species is largely dull brown in plumage, variably mottled and streaked, but the other (in adult plumage) is strikingly coloured in chestnut and white. They feed on insects, small vertebrates (including fish) and carrion. The genus is very closely related to *Milvus*, and the two genera might comfortably be merged.

Brahminy Kite

Haliastur indus (in'-dus: "Indian sea-goshawk")

The Brahminy Kite is widespread from India to the Solomon Islands. It inhabits the warmer coasts of Australia, seldom wandering any distance inland, and it appears to be sedentary.

Adults (essentially bright red-brown with white head and breast) are unmistakable, but immatures can be easily mistaken for several other raptors, especially the Whistling Kite.

The Brahminy Kite lives in pairs and small groups. The diet includes fish, crustaceans and small reptiles, but it is mainly a scavenger, usually seen patrolling along tidelines for carrion.

It builds a rough, bulky nest of sticks and twigs, lined with leaves and grass and often decorated with seaweed, high in a tree. Two eggs form the usual clutch. The nesting season is from August to October in the east, May to September in the west.

HABITAT: tropical to warm temperate beaches, estuaries, coastal lagoons, mangrove swamps, coastal woodland, offshore islands
LENGTH: 48–51 cm
DISTRIBUTION: 300,000–1 million km²
ABUNDANCE: common
STATUS: secure

(H & J Beste)

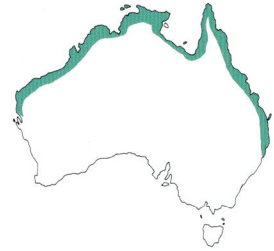

Whistling Kite

Haliastur sphenurus (sfen-ue'-rus: "wedge-tailed sea-goshawk")

One of the most abundant and widespread of Australian hawks, the Whistling Kite occurs across Australia except Tasmania where it is a casual vagrant, though it occasionally breeds there. Elsewhere it occurs in New Guinea, the Solomon Islands, Vanuatu and New Caledonia.

It usually lives in pairs, occasionally flocks, and spends much time wheeling and soaring high in the sky. It has a distinctive shrill whistling cry, audible at a considerable distance. Like the Brahminy Kite, it often scavenges along beaches in coastal districts, but its diet also includes rabbits and other small mammals, birds, lizards, carrion, and insects.

Breeding may take place at any time of year. The nest is a large untidy platform of sticks and twigs high in a tree. The clutch is two or three eggs.

HABITAT: tropical to cool temperate wooded country, particularly eucalypt forest near water; avoids rainforest and treeless regions
LENGTH: 53–59 cm
DISTRIBUTION: more than 1 million km²
ABUNDANCE: abundant
STATUS: secure

(T & P Gardner)

Genus *Hamirostra*

(ham'-ee-ros'-trah: "hook-bill")

The characteristics of the genus are essentially those of the single species.

Black-breasted Buzzard

Hamirostra melanosternon (mel'-ah-noh-stern'-on: "black-breasted hook-bill")

The Black-breasted Buzzard is essentially a bird of the arid interior and the far north. It is very rare in the south-east and the south-west and seems to have declined markedly in the lower Murray–Darling region during the 20th century.

Generally uncommon and apparently sedentary, it is usually seen alone, soaring high. Little is known of its biology but the diet includes reptiles (chiefly lizards), mammals (mostly rabbits), and nestling birds and eggs.

The breeding season is from August to November. The nest is a large, loose structure of sticks that is usually built in a dead tree. The usual clutch is two eggs, and the incubation period is about 36 days.

(D & V Blagden)

HABITAT: tropical to warm temperate arid to semiarid, lightly timbered country
LENGTH: 53–60 cm
DISTRIBUTION: more than 1 million km²
ABUNDANCE: sparse
STATUS: probably secure

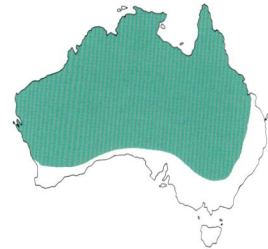

Genus *Hieraaetus*

(hie'-ree-et'-us: "eagle-hawk")

Five species, including one inhabiting Australia and New Guinea, are usually placed in this genus, but it is difficult to characterise and may not be a natural group. Some species are clearly very closely related to eagles of the genus *Aquila* but the placement of others is problematic: in general they are smaller and less powerful than *Aquila* eagles, with smaller bills and somewhat longer legs. Most species occur in Asia and Africa.

Little Eagle

Hieraaetus morphnoides (morf-noy'-dayz: "eagle-like eagle-hawk")

The Little Eagle inhabits New Guinea and mainland Australia but it has not been recorded in Tasmania. It is largely sedentary and inhabits most kinds of wooded country. It is most easily recognised by its distinctive underwing pattern. The species is polymorphic, with pale, dark and many intermediate forms. It feeds largely on rabbits, pounced on from flight.

Breeding is variable in timing. The nest is a rough bulky mass of sticks and twigs, usually in one of the largest trees available but seldom an isolated one; one or two eggs form the usual clutch.

HABITAT: tropical to cool temperate forests and woodlands
LENGTH: 48–55 cm
DISTRIBUTION: more than 1 million km²
ABUNDANCE: common
STATUS: secure

(T & P Gardner)

Genus *Lophoictinia*

(loh'-foh-ik-tin'-ee-ah: "crested-kite")

The characteristics of the genus are essentially those of the single species.

Square-tailed Kite

Lophoictinia isura (ee-sue'-rah: "square-tailed crested-kite")

The Square-tailed Kite is very rare in South Australia and much of the arid interior, but is otherwise widespread across Australia except Tasmania. It seems to be migratory, moving south in winter.

It is seldom seen. It may not be a very numerous species, but its habits also contribute to the paucity of observations: it spends much time sailing through woodland, either in the canopy or just above it, where it is difficult to observe.

Little is known of its biology. When available, nestling birds figure prominently in the diet, but it also feeds on young rabbits, small reptiles, large insects and carrion.

The breeding season is from August to December. Usually placed on a horizontal limb some 15 to 20 metres from the ground, the nest is a substantial structure of sticks, lined with green leaves. The nest may be reused for several successive seasons. The usual clutch is two or three eggs.

HABITAT: tropical to cool temperate woodland and heath
LENGTH: 50–56 cm
DISTRIBUTION: more than 1 million km^2
ABUNDANCE: sparse
STATUS: probably secure

(DGW Hollands)

Genus *Milvus*

(mil'-vus: "kite")

A genus of two or possibly three species, *Milvus* has an extensive distribution across the Old World, especially in tropical and warm temperate regions. These kites are rather large, with relatively short legs, long and moderately narrow wings, and a long tail which is either notched or distinctly forked. The talons are relatively small and straight. Opportunistic foragers, they feed on carrion as well as insects and small vertebrates. One species occurs in Australia.

Black Kite

Milvus migrans (mee'-granz: "wandering kite")

The Black (or Fork-tailed) Kite has an extensive distribution through the Old World. In Australia it occurs throughout the continent, although it is far more numerous in the north than in the south. There is some evidence that it is expanding its range southward.

The forked tail is distinctive but in flight may appear to be briefly square-cut. The Black Kite is chiefly a scavenger of the arid interior, feeding on carrion, garbage, and any insects, small mammals or reptiles that it can capture. It is strongly gregarious, and very large flocks often congregate. The Black Kite spends most time in flight, wheeling high and scanning the ground for food.

Bushfires attract birds from afar.

Nesting is variable but is mainly from August to November. The nest is an untidy bowl of sticks and twigs, often lined with cattle dung, and usually prepared by renovation of an old nest of a crow or hawk. Two or three eggs, incubated for about 35 days, constitute the usual clutch.

HABITAT: tropical to warm temperate semiarid to arid country; congregating around homesteads, airfields, rubbish dumps, slaughter yards
LENGTH: 52–55 cm
DISTRIBUTION: more than 1 million km²
ABUNDANCE: common
STATUS: secure

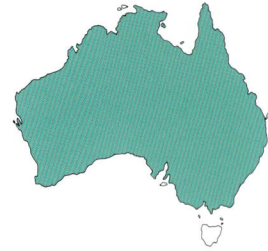

(J Daley)

Family FALCONIDAE

(fal-kon'-id-ee: "*Falco*-family", after a genus of falcons)

Falcons comprise a worldwide family of some 61 species, generally regarded as related to the Accipitridae. They are chiefly diurnal, have outstanding powers of flight and vision, formidable talons and hooked beaks, and feed mainly on vertebrates.

Although superficially similar to hawks, falcons differ in a number of trenchant characteristics of anatomy and physiology. Although some species have catholic diets, most specialise in catching birds and are spectacular, pouncing on their prey in full flight after an electrifying dive from a great height.

Falcons seldom if ever build their own nests, depositing their eggs instead on a cliff ledge or similar site, or in the abandoned nest of some other bird. Courtship displays are spectacular and are conducted in flight. Eggs are laid at intervals of 2 to 3 days; unlike the situation in many hawks, falcon chicks typically show little aggression to each other, and often the entire clutch is successfully raised.

Genus *Falco*

(fal'-koh: "falcon")

This genus is widespread around the world, with many species, six of which occur in Australia. The characteristics of the genus are as outlined for the family.

Brown Falcon

Falco berigora (be'-ree-gor'-ah: "berigora falcon", after supposed Aboriginal name for the species)

The Brown Falcon, which also inhabits New Guinea, is Australia's most numerous and widespread falcon, although populations have suffered some decline over recent decades in some coastal regions of the south-east. Adult Brown Falcons apparently live in sedentary pairs but

(*P Klapste*)

immature birds disperse widely.

The Brown Falcon exists in light, dark and intermediate forms; some individuals may be as dark as the Black Falcons and are easily mistaken for this species. Unusually long-legged for a falcon, the Brown Falcon moves easily on the ground and spends much time perched on fence and telegraph posts and similar low, exposed perches. It hunts low to the ground and eats a wide variety of prey: grasshoppers, small reptiles, mammals and birds.

It seldom if ever builds its own nest, adopting instead the abandoned nest of some other bird, but it is said to sometimes line the nest with green leaves and twigs. Two or three eggs form the usual clutch.

HABITAT: virtually every Australian environment except rainforest and some dense wet sclerophyll forest
LENGTH: 41–50 cm
DISTRIBUTION: more than 1 million km²
ABUNDANCE: abundant
STATUS: secure

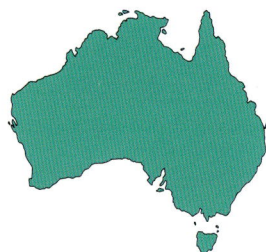

Nankeen Kestrel

Falco cenchroides (sen-kroy'-dayz: "speckled-hawk-like falcon")

Two characters make the Nankeen (or Australian) Kestrel unmistakable: the upperparts are an unusual shade of gingery rufous and the bird persistently hovers when hunting for prey. It is a small, lightly built falcon that is abundant and widespread in open country throughout mainland Australia, though rare in Tasmania.

It also occurs in Indonesia and New Guinea, and in recent years has become established on Norfolk and Lord Howe Islands.

It lives in pairs that may be resident, but many individuals wander northwards in winter, and the species is nomadic over much of the interior. The diet consists mainly of

grasshoppers and other large terrestrial insects and mice.

No nest is built, and the site is very variable: buildings, cliff ledges, tree cavities or the old nests of other birds are often utilised. The usual clutch consists of four eggs, but may vary from three to five. The incubation period is 28 days and young fledge after 28 to 33 days in the nest.

HABITAT: almost any mainland Australian environment with open space, including urban areas
LENGTH: 30–36 cm
DISTRIBUTION: more than 1 million km²
ABUNDANCE: abundant
STATUS: secure

(*H & J Beste*)

Grey Falcon

Falco hypoleucus (hie'-poh-lue'-kus: "under-white falcon")

The endemic Grey Falcon is a bird of the arid interior. It lives in pairs and feeds almost exclusively on birds. Rare and elusive, it is Australia's least numerous falcon, and very little is known of its movements or biology.

It usually breeds in the abandoned nest of some other bird, often a crow or raven. The clutch is two to four eggs. Incubation is reportedly 35 days, fledging 42 to 49 days.

(*A Eames*)

HABITAT: tropical to cool temperate arid desert, tussock grassland, spinifex and scrubland
LENGTH: 32–45 cm
DISTRIBUTION: more than 1 million km^2
ABUNDANCE: sparse
STATUS: probably secure

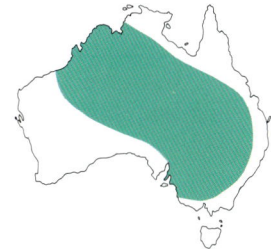

Little Falcon

Falco longipennis (lon'-jee-pen'-is: "long-winged falcon")

The Little Falcon (or Australian Hobby) occurs in Indonesia and New Guinea and throughout mainland Australia, but it is not numerous in Tasmania. In some areas it is migratory or nomadic, in others resident. It lives alone or in pairs.

Small, dark, swift and dashing, it feeds mainly on bats and small birds as well as insects, especially swarming termites or flying grasshoppers. It is often seen at dusk, patrolling quiet reaches of rivers and streams for early bats; or streets in towns and cities in search of sparrows, starlings and pigeons going to roost.

The breeding season is from August to October in the south, somewhat earlier in the tropical north. The Little Falcon seldom, if ever, builds its own nest, preferring to use the abandoned nest of some other bird of prey or of a raven. This is neither repaired nor lined, but it may be reused over several successive seasons. The usual clutch is three eggs; the young fledge at about 35 days.

(*T Howard*)

HABITAT: tropical to cool temperate forest and woodland
LENGTH: 30–35 cm
DISTRIBUTION: more than 1 million km^2
ABUNDANCE: common
STATUS: secure

Peregrine Falcon

Falco peregrinus (pe'-re-gree'-nus: "wandering falcon")

This is one of the most widespread of all birds, with populations on all continents except Antarctica. It is widely but somewhat thinly distributed in Australia, reaching its greatest abundance in the south-east. It is generally sedentary.

Relying upon speed and flying ability to overcome its prey, its hunting techniques place it among the most dramatic of the raptors, a character long recognised in the sport of falconry. Typically it soars high, watching the space below and pouncing on birds in mid-flight after a high-speed dive (the momentum of the talon-first collision often knocking the victim to the ground). It preys almost exclusively on birds, especially starlings, mynas, pigeons and others of similar size.

The Peregrine Falcon is especially sensitive to environmental pollution resulting from the worldwide use of chlorinated hydrocarbons since the 1940s, one effect of which is the production of thin-shelled, fragile eggs that break during incubation. Entire populations have been decimated (in North America the species was reduced to the brink of extinction in the 1960s and 1970s), but the effect is less in Australia, which is now the stronghold of the species: the entire world population has been estimated at 12 000 to 18 000 pairs, of which about 3000 to 5000 are in Australia.

In the south-east the breeding season is regular, beginning in early August. No nest is built, the two or three eggs forming the usual clutch being laid on a cliff ledge (or occasionally on a tall building); sometimes a tree cavity is used, or (rarely) the abandoned nest of some other hawk. Cliff sites, in particular, are often traditional. The incubation period is about 36 days, and the young fledge after about 39 or 40 days in the nest.

HABITAT: usually heavily timbered and ruggedly mountainous country
LENGTH: 38–48 cm
DISTRIBUTION: more than 1 million km²
ABUNDANCE: sparse
STATUS: vulnerable

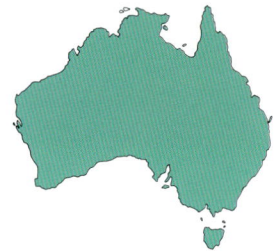

(*D & M Trounson*)

Black Falcon

Falco subniger (sub-nie'-jer: "somewhat black falcon")

The endemic Black Falcon is widespread over mainland Australia, but it is rare in Western Australia and the tropical north; breeding is almost confined to the interior south-east from about Lake Eyre to the lower Murray–Darling basin. It is nomadic and strongly dispersive.

The Black Falcon is aggressive and active, spending much of its time in flight and often challenging other hawks and chasing other birds. It lives alone or in pairs, feeding on rabbits caught on the ground and birds (especially parrots and pigeons) captured in flight, often after a spectacular stoop. Unlike other falcons, it sometimes eats carrion.

The usual breeding season is from June to September, but it is strongly influenced by local conditions. Nests are usually in trees along watercourses. The Black Falcon seldom or never builds its own nest, usually adopting the abandoned nest of some other hawk—or sometimes that of a crow or raven—without any attempt at repair. Two to four eggs constitute the usual clutch.

HABITAT: cool to warm temperate semiarid to arid woodland
LENGTH: 45–55 cm
DISTRIBUTION: more than 1 million km²
ABUNDANCE: sparse
STATUS: probably secure

(D Holland)

Order GALLIFORMES

(gal'-ee-for'-mayz: "*Gallus*-order", after a genus of fowl)

The ordinary domestic chicken is typical of a cosmopolitan group of small to large, mainly terrestrial, granivorous or insectivorous birds that together constitute the order Galliformes. They have short, broad, rounded wings but the tail varies in length from very short to very long. The bill is short, deep, and curved, and the legs and feet stout and strong; all species have a well-developed hind toe. Males often have elaborate plumage patterns, bright colours, and flamboyant ruffs, capes or other display features, but females are generally brownish and nondescript. The newly hatched young are clad in dense fluffy down and are quite indepen-dent, requiring only guardianship from their parents.

Four families are included: Cracticidae (curassows and guans); Opisthocomidae (hoatzin); Megapodiidae (megapodes); and Phasianidae (pheasants, partridges and quails). The first two families are exclusively American (mainly South American), and differ strikingly from the other groups in being essentially arboreal. Megapodes are essentially Australasian. The Phasianidae dominates the order in number of species (more than 200) and in distribution (virtually worldwide), and the group includes chickens, turkeys, pheasants, and most birds popularly known as gamebirds.

Family MEGAPODIIDAE

(meg'-ah-poh-dee'-id-ee: "*Megapodius*-family", after a genus of scrubfowl)

The megapodes or mound-builders are notable for using a breeding strategy resembling that of some reptiles but unique among birds: the eggs are not incubated by either parent but are buried in mounds of earth, leaf litter or sand until they hatch. The heat necessary to hatch the eggs comes from geothermal activity, from the heat of fermentation produced by decaying vegetable matter, or from sunlight. Megapodes carefully engineer their mounds to take advantage of one or other of these sources, the choice varying with the species.

Some 15 species comprise the family, which is almost exclusively Australasian. All are large, chicken-like terrestrial birds which (with one exception) inhabit rainforest. They are mostly confined to the tropics, but the Malleefowl inhabits semiarid woodland in southern Australia

Genus *Alectura*

(al'-ek-tue'-rah: "cock-tail")

The characteristics of the genus are essentially those of the single species.

Brush-turkey

Alectura lathami (lay'-tham-ee: after J. Latham, English ornithologist)

The Brush-turkey inhabits eastern Australia from Cape York Peninsula south to the Illawarra district of New South Wales but it is now seldom recorded south of the Hawkesbury River. It was introduced to Kangaroo Island in the 1930s.

It is most common in rainforest near the coast but it also inhabits more open environments at some distance inland. It spends almost all of its waking hours on the ground but roosts at night in trees. The diet comprises fallen fruit, insects, snails and other invertebrates.

Generally shy and solitary, it soon becomes tame around forest picnic areas and can sometimes be fed by hand.

The nest mound, built and maintained by the male, may reach 7 metres in diameter and 2 metres in height, though new mounds are much smaller.

The male allows the female onto the mound only for mating and egg laying. Captive females have laid as many as 80 eggs in a season but the natural clutch size is unknown—perhaps 18 to 24, laid at intervals of 2 to 3 days.

HABITAT: tropical to warm temperate coastal rainforest and lantana thickets
LENGTH: 59–75 cm
DISTRIBUTION: 300,000–1 million km^2
ABUNDANCE: common
STATUS: possibly endangered

(*N Chaffer*)

Genus Leipoa

(lie-poh'-ah: "egg-leaver")

The characteristics of the genus are those of the single species.

Malleefowl

Leipoa ocellata (oh'-sel-ah'-tah: "eye-marked egg-leaver" [i.e., marked with eye-like spots])

The Malleefowl inhabits semiarid districts of southern Australia. It is strongly sedentary and territorial: males, in particular, are effectively restricted to the vicinity of their mounds throughout their lives.

It is a large bird but, except when occupied at the mound, extraordinarily wary, elusive and difficult to observe. Although strictly terrestrial when active, it roosts at night in low bushes.

The diet of the Malleefowl consists largely of insects, flowers and green shoots, and fruits and seeds of acacias and other shrubs. Numbers have been seriously reduced by land clearing, competition from rabbits and sheep, and predation by foxes, feral cats and shooters.

The pair bond is permanent, but the sexes are seldom together. When ready to lay, the female approaches the mound and the male digs down through the mound to expose the egg chamber, refilling it after an egg is laid. Clutch size and frequency of laying depends on the season, varying from five to 33 eggs (usually 15 to 24) and two to 17 days respectively. The incubation period is about 49 days. The breeding season is generally from September to April.

HABITAT: cool to warm temperate mallee and other semiarid eucalypt woodland

LENGTH: 55–61 cm

DISTRIBUTION: 300,000–1 million km²

ABUNDANCE: very sparse

STATUS: endangered

(T & P Gardner)

Genus Megapodius

(meg'-ah-poh'-dee-us: "large-foot")

These are small megapodes, rather chicken-like in general appearance, with feathered heads and short crests. They have sombre, dull brown or grey plumage, inhabit rainforest, and are very vocal, often calling by night as well as by day. All members are similar and closely related, and species limits are controversial, some researchers listing three species, others as many as nine. The genus is the most widely distributed of the megapode genera, ranging from eastern Indonesia through New Guinea to Micronesia and northern Australia.

Orange-footed Scrubfowl

Megapodius reinwardt (rine'-vart: "Reinwardt's large-foot", after a Dutch collector)

The Orange-footed Scrubfowl (or Jungle Fowl) occurs in Indonesia, New Guinea and tropical Australia. It is common and conspicuous within its distribution, though sometimes wary and difficult to observe. Noisy by day and night, it utters a wide range of loud raucous chortling notes, crows and screams.

It is sedentary, maintaining permanent territories occupied by an unknown number of individuals. The diet consists of fallen fruits and berries, seeds and green shoots.

Both sexes maintain the mound, which may be shared by several pairs. Newly made mounds are often small but, constantly added to, mounds may ultimately reach 10 metres or more in diameter and several metres in height. Little is known of breeding biology.

HABITAT: rainforest, especially coastal lowland rainforest; also mangrove swamps
LENGTH: 45–47 cm
DISTRIBUTION: 300,000–1 million km²
ABUNDANCE: sparse
STATUS: vulnerable

(H & J Beste)

Family PHASIANIDAE

(faz'-ee-ah'-nid-ee: "*Phasianus*-family", after a genus of pheasants)

About 213 species around the world constitute the family Phasianidae. They vary markedly in size, colour and ornamentation: some species—such as quail—are small and obscurely marked in plain brown, while others—such as peacocks—are large, brightly coloured and equipped with flamboyant tail feathers, capes or other adornments. In the latter case, males generally differ conspicuously from females in appearance. The best-known member of the group is the domestic fowl.

Only three species occur naturally in Australia, several exotic species (including the Common Peafowl, California Quail and Common Pheasant) have, however, been introduced at various times, seldom with any great success.

The Australian representatives are all quails of the genus *Coturnix*, very small birds, cryptically coloured in dull browns, which inhabit grasslands of various kinds. They are secretive and reluctant to fly, preferring to run when disturbed. The nest is a shallow scrape in the soil, sparsely lined with grass and generally well hidden in a tussock or similar site. Most species lay large and variable clutches, typically of about seven to 14 eggs.

Genus Coturnix

(koh-ter'-nix: "quail")

Quails are very small, almost tail-less chicken-like birds with plump bodies, shortlegs and very short necks. Several are plain brown in colour, intricately streaked and barred, but some are boldly marked in slate blue, black, and white. They live in grassland and seldom emerge from dense cover. Their flight is low and swift but seldom sustained. The genus is widespread in Africa, Eurasia and Australasia: there are about nine species, of which three occur in Australia.

Brown Quail

Coturnix australis (os-trah'-lis: "southern quail")

This species occurs naturally in Indonesia, New Guinea and Australia and has been introduced to Fiji and New Zealand. In Australia its distribution is mainly coastal, especially in the north and east; it is common in Tasmania. Some researchers have recognised two species: the Brown Quail *C. australis* and the Swamp Quail *C. ypsilophorus*, but the status of these populations is still unclear.

Brown Quail usually live in small groups. They tend to be relatively sedentary, but will disperse far if forced by adverse conditions. Seeds and insects constitute the usual diet.

HABITAT: tropical to cool temperate grassland, especially dense grasses at the margins of swamps and lagoons; also lush pastures, heaths and bogs
LENGTH: 17–19 cm
DISTRIBUTION: more than 1 million km^2
ABUNDANCE: common
STATUS: probably secure

(G Rogerson)

King Quail

Coturnix chinensis (chin-en'-sis: "Chinese quail")

The King Quail extends from India eastward through southern Asia and to the Bismarck Archipelago. In Australia the King Quail occurs in coastal lowlands from the Kimberley to Cape York Peninsula and along the east coast; it is seldom recorded south of about Sydney. Males are unmistakable, but females are generally brown like other quail and are best identified by their very small size.

The King Quail is extremely shy, secretive and difficult to flush. It is often active by night, feeding on small seeds and insects. It seems largely sedentary, but local populations fluctuate markedly: in any given locality it may be very common one year, scarce or absent the next.

HABITAT: tropical to warm temperate boggy heaths, swamps, margins of lagoons; also cereal crops, rice stubble and rank pastures
LENGTH: 13–14 cm
DISTRIBUTION: 300,000–1 million km^2
ABUNDANCE: sparse
STATUS: probably secure

(J McCann)

Stubble Quail

Coturnix novaezealandiae (noh'-vee-zee-land'-ee-ee: "New Zealand quail")

The Stubble Quail once occurred in both Australia and New Zealand but became extinct in New Zealand before 1900, and in Tasmania between 1910 and 1960. It became established in Western Australia early in the 20th century. It is generally widespread and common in the south-east and south-west, rare in the tropical north, and patchily distributed in the arid interior.

Females differ from males chiefly in that the throat is off-white, not light chestnut; otherwise the sexes resemble each other and immatures resemble adult females.

It is gregarious and usually encountered in coveys in open country. Nomadic and erratic in occurrence, it may be abundant one year but rare or absent the next. The diet consists of seeds, green shoots, and insects.

Breeding usually begins in August, most clutches hatching in October or November, although nesting may continue for many months in good seasons. The usual clutch is seven or eight eggs.

HABITAT: cool to warm temperate grasslands and croplands; generally avoids dense grass near water
LENGTH: 18–19 cm
DISTRIBUTION: more than 1 million km^2
ABUNDANCE: abundant
STATUS: secure

(AY Pepper)

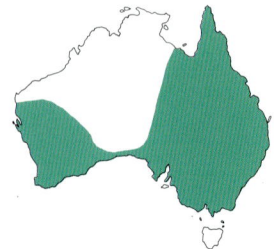

Order GRUIFORMES

(grue'-ee-for'-mayz; "*Grus*-order", after a genus of cranes)

This order contains 11 families: Turnicidae (buttonquails); Pedionomidae (plains-wanderer); Rallidae (rails, crakes, moorhens and coots); Aramidae (limpkin); Psophiidae (trumpeters); Gruidae (cranes); Heliornithidae (finfoots); Rhynochetidae (kagu); Eurypygidae (sun-bitterns); Cariamidae (seriemas); and Otididae (bustards). The families Rallidae and Gruidae are almost cosmopolitan; the Turnicidae and Otididae are widespread in warmer regions of the Old World (all four families are represented in Australia). The remainder, however, comprise only one or several species each, and have very restricted distributions, mainly in warm temperate or tropical regions.

The order is superficially an extremely heterogeneous group, difficult to characterise succinctly, but the various families are linked by a number of anatomical peculiarities (it has recently been shown, however, that Pedionomidae almost certainly does not belong here). The included species vary in size from very small (buttonquails) to extremely large (cranes), but most have in common a predominantly terrestrial habit: most feed and nest on the ground, and several species cannot fly. Many live in swamps and marshes. Insects are prominent in the diet. The young are hatched covered in dense down (often black) and are usually led from the nest almost immediately after hatching. They are fed by one or both parents at first but soon reach functional independence.

Family TURNICIDAE

(ter-nis'-id-ee: "*Turnix*-family", after a genus of buttonquails)

The Turnicidae or buttonquails constitute a group of about 16 species widespread in the Old World. Seven species occur in Australia.

Buttonquails superficially resemble true quails (family Phasianidae) but are not related, differing in many features of anatomy, physiology and behaviour; one external distinction is their lack of a hind toe. Like quails, they are small, chicken-like birds that inhabit dense grass and similar places. Extremely secretive, they are reluctant to fly when disturbed, and very difficult to observe.

The female is larger and more brightly coloured than the male and takes the initiative in courtship. The female defends the breeding territory against other females, and may take several mates in a season, leaving incubation and care of the young to the male in each case (although in some species she does not entirely abandon them). The nest is a mere scrape in the ground, lined with grass and well hidden in deep cover.

All but one of the species is included in the genus *Turnix*, which has characteristics typical of the family. The sole exception, the Lark-quail *Ortyxelos meiffrenii* of central Africa, is very distinct. It inhabits tropical grassland, has boldly marked black and white wings, and on the ground somewhat resembles a small pratincole.

Genus Turnix

(ter'-nix: "quail")

The characteristics of the genus are essentially those of the family.

Chestnut-backed Buttonquail

Turnix castanota (kas'-tah-noh'-tah: "chestnut-backed quail")

The endemic Chestnut-backed Buttonquail inhabits northern Australia from the Kimberley to Arnhem Land.

(J. Barnett)

It is similar to the Buff-breasted and Painted Buttonquails in appearance, behaviour and ecology (some researchers regard all forms as conspecific), but little is known of its biology. It lives in coveys of six to 20 birds.

Breeding occurs mainly from December to May. Built in a shallow scrape in the ground in the shelter of a bush, the nest is woven of grass and leaves and often domed or partly domed. The clutch consists usually of four eggs, which hatch in about 14–15 days. Incubation is by the male.

HABITAT: tropical stony hillsides and ridges, arid scrublands and the grassy understorey of open eucalypt woodland
LENGTH: 14–20 cm
DISTRIBUTION: 100,000–300,000 km²
ABUNDANCE: sparse
STATUS: probably secure

Red-backed Buttonquail

Turnix maculosa (mak'-ue-loh'-sah: "spotted quail")

The Red-backed Buttonquail extends from Indonesia eastward to the Philippines and the Solomon Islands. In Australia it occurs from the Kimberley to Arnhem Land and from Cape York Peninsula along the east coast (including several islands of the Great Barrier Reef) to about the Manning River, New South Wales.

It is subject to marked fluctuations in numbers and periodic irruptions that result in its occurrence far from its usual range: it has, for example, occurred (and bred) in the Riverina. It lives in pairs or small coveys and feeds on insects and seeds.

Breeding occurs mainly between October and July. The nest is of woven grass, placed in a shallow depression in the ground, and usually screened by interwoven surrounding vegetation. The clutch consists usually of four eggs, which hatch in about 14 days.

HABITAT: tropical to warm temperate rank vegetation at the margins of creeks, lagoons and marshes; crops and coastal scrubs, especially areas prone to seasonal flooding
LENGTH: 11–15 cm
DISTRIBUTION: 100,000–300,000 km^2
ABUNDANCE: sparse
STATUS: probably secure

(ER Josika)

Black-breasted Buttonquail

Turnix melanogaster (mel'-ah-noh-gas'-ter: "black-bellied quail")

The endemic Black-breasted Button-quail inhabits extreme north-eastern New South Wales and eastern Queensland, north to about Gladstone. It was formerly common north at least to the Atherton Tableland, but it is now very rare, its range having declined severely since European settlement.

Extremely secretive and elusive, it is sedentary and lives in pairs or coveys of four to 10 birds.

Breeding occurs mainly between September to March, but may occur at any time. The nest is a shallow depression in the ground in the shelter of a bush or shrub, and lined with leaves. The clutch consists of three or four eggs, which hatch in about 16 days.

(AT Foster)

HABITAT: warm temperate to subtropical dense rainforest with abundant dry leaf litter; also lantana thickets and hoop pine plantations
LENGTH: 15–18 cm
DISTRIBUTION: 30,000–100,000 km^2
ABUNDANCE: very sparse
STATUS: vulnerable

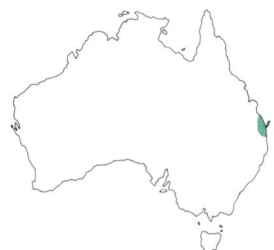

Buff-breasted Buttonquail

Turnix olivei (ol'-iv-ee: "Olive's quail", after E. Olive, a Queensland naturalist)

This species is restricted to Cape York Peninsula, where it inhabits heaths and grassy clearings in rainforest and open eucalypt woodland. It is seldom reported and little is known of its habits.

Breeding occurs mainly between December and May. Sometimes partly domed, the nest is a loose woven structure of grass in a depression in the ground, sheltered by a bush or shrub. Three or four eggs constitute the clutch.

HABITAT: tropical heaths; grassy clearings in rainforests; open eucalypt woodland
LENGTH: 18–20 cm
DISTRIBUTION: 100,000–300,000 km²
ABUNDANCE: rare
STATUS: possibly endangered

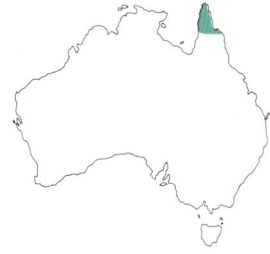

Red-chested Buttonquail

Turnix pyrrhothorax (pi'-roh-thor'-ax: "fire-chested quail")

The Red-chested Buttonquail occurs in eastern Australia generally, from Arnhem Land and Cape York Peninsula south to Victoria and west to Yorke Peninsula, South Australia, mainly west of the Great Dividing Range. It has also been recorded at scattered localities across northern Australia to the Kimberley.

It is generally strongly nomadic and irruptive but it seems to be resident in some areas. Very secretive and difficult to observe, it lives in pairs or small coveys, eating seeds and insects.

Breeding occurs mainly between September and March. Usually domed, the nest is built loosely of woven grass in a depression in the ground in the shelter of a bush or shrub. Three or four eggs form the clutch, and incubation takes 13–14 days.

HABITAT: tropical to cool temperate woodland, mulga and callitris scrubs; also pastures, stubble and crops
LENGTH: 12–16 cm
DISTRIBUTION: more than 1 million km²
ABUNDANCE: sparse
STATUS: probably secure

(FE Lewitzka)

103

Painted Buttonquail

Turnix varia (var'-ee-ah: "different quail")

The Painted Buttonquail inhabits New Caledonia and southern and

(*M Seyfort*)

eastern Australia, where it is fairly common from about Cooktown in Queensland, to Tasmania and the Eyre Peninsula in South Australia, and in south-western Western Australia, north to the Houtman Abrolhos Islands.

It seems relatively sedentary but may disperse far if forced by adverse conditions. It is partly nocturnal, and eats insects and seeds. The call is a low hollow booming note.

Breeding occurs mainly between August and March. The nest is a depression in the ground, lined with grass and leaves and sometimes partly domed. The clutch of four eggs is incubated by the male and hatches in about 13 or 14 days.

HABITAT: tropical to cool temperate heath, open woodland and lightly timbered ridges
LENGTH: 16–21 cm
DISTRIBUTION: 300,000–1 million km^2
ABUNDANCE: sparse
STATUS: probably secure

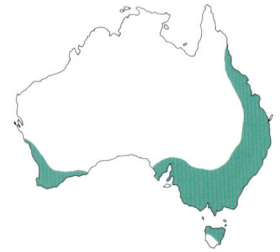

Little Buttonquail

Turnix velox (vel'-ox: "swift quail")

The Little Buttonquail occurs almost throughout mainland Australia except for the tropical north and most of the eastern coastal strip. It is migratory, rare in the south in winter but returning every spring. In any locality its numbers may fluctuate dramatically from year to year in response to severe drought or unusually high rainfall.

It lives in large or small coveys which generally scatter if forced into flight.

Breeding may take place in any month. Both sexes build the nest, which is a shallow depression lined with grass in the shelter of a tussock. Four or five eggs form the usual clutch; they hatch in 13 to 14 days.

HABITAT: tropical to cool temperate tussock grassland, spinifex and woodland with a grassy understorey; crops
LENGTH: 13–15 cm
DISTRIBUTION: more than 1 million km^2
ABUNDANCE: common
STATUS: secure

(*J McCann*)

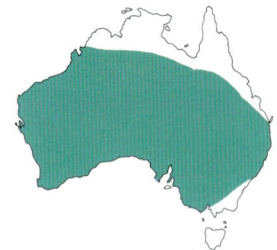

Family PEDIONOMIDAE

(ped'-ee-oh-noh'-mid-ee: "*Pedionomus*-family", after the single genus of the family)

The enigmatic Plains-wanderer, sole member of this family, lacks close relatives. It has many features in common with buttonquails (family Turnicidae), especially in behaviour, ecology and general appearance. As in the buttonquails, the female is larger and brighter than the male and takes the initiative in courtship, leaving the male to incubate and care for the young. It differs from the buttonquails in many respects (e.g., in possession of a hind toe) and recent studies indicate that the two families may not be even distantly related.

Genus Pedionomus

(ped'-ee-oh-noh'-mus: "plains-wanderer")

The characteristics of the genus are those of the single species.

Plains-wanderer

Pedionomus torquatus (tor-kwah'-tus; "collared plains-wanderer")

The Plains-wanderer is restricted to the interior south-east of Australia. It has declined in range and number since European settlement and is now very rare. Most recent records are in the Riverina and northern central Victoria. It seems to be largely sedentary.

Living alone, in pairs or in family parties, it feeds on insects and seeds. It is mainly nocturnal, roosting by day in a shallow scrape in open ground. It flies only under compulsion, preferring when disturbed to scurry away. It has the distinctive habit of frequently raising itself on tiptoe, as if to peer about over the top of grass cover. The most typical call is a soft, penetrating lowing note.

Most breeding occurs between September and December, but occasionally at any time of year. The nest is a scrape in the ground, lined with grass. The usual clutch is of four eggs, but may vary between two and five. Incubated mainly by the male, they hatch in about 23 days. Led from the nest soon after hatching, the young are fed at first on insects, then weaned onto seeds.

HABITAT: cool temperate semiarid plains with a mixture of grasses and herbs; also stubble and other dry croplands
LENGTH: 16–18 cm
DISTRIBUTION: 300,000–1 million km²
ABUNDANCE: very sparse
STATUS: possibly endangered

(TG Lowe)

Family RALLIDAE

(ral'-li-dee: "*Rallus*-family", after a genus of rails)

The rails, crakes, moorhens and coots constitute a family of some 140 species found on all continents except Antarctica. Many species live in forests (especially in the tropics) but, in general, rails inhabit dense vegetation at the margins of wetlands. They are generally common and widespread, though notoriously shy and difficult to observe.

The sexes are similar, there is little seasonal variation, and immatures generally resemble adults except in being duller; downy young are sooty black. In some (not all) species only the female incubates but both sexes share in caring for the young. The chicks are led from the nest on hatching and are fed by both parents for the first few days. Thereafter they feed independently but remain with their parents for several months.

Genus Eulabeornis

(yue'-lab-ay-or'-nis: "cautious-bird")

The characteristics of the genus are essentially those of the single species.

Chestnut Rail

Eulabeornis castaneiventris (kas'-tah-nay'-ee-vent'-ris: "chestnut-bellied cautious-bird")

The Chestnut Rail lives only in coastal mangroves in tropical northern Australia from about Derby, Western Australia, to the head of the Gulf of Carpentaria, and also in the Aru Islands, Indonesia. It is noisy but wary and extremely difficult to observe except at low tide, when it emerges to feed over exposed mudflats, capturing crabs and other small invertebrates.

It is sedentary and strongly territorial. The breeding season is mainly from November to January. The nest is a large flattish structure of sticks, usually situated on a horizontal mangrove branch out of reach of tides. The usual clutch consists of four eggs.

HABITAT: tropical mangroves
LENGTH: 42–44 cm
DISTRIBUTION: 10,000–30,000 km²
ABUNDANCE: rare
STATUS: possibly endangered

(AL Hertog)

Genus Fulica

(fuh-li-ca: "coot")

Coots are strongly aquatic rails that differ strikingly from other members of the family in their gregariousness and their habit of diving for their food. They are largely vegetarian but their varied diet also includes molluscs and other aquatic invertebrates. Like gallinules they are mostly slate grey in colour, and have rather short, pointed bills and conspicuous frontal shields, which are usually white. Their toes are lobed like those of grebes. There are eight species, most of which occur in South America, but one species, the Common or Eurasian Coot, is widespread in the Old World, including Australia.

Eurasian Coot
Fulica atra (ah'-trah: "black coot")

A common and familiar bird across much of Europe, Africa and Asia, the Eurasian Coot also occurs on suitable lakes, lagoons and marshes across Australia, but it is not common in Tasmania. Unlike other rails, it dives for most of its food (mainly aquatic plants), and favoured wetlands are, in general, deep, extensive and permanent, though isolated pairs are not uncommon on small farm dams and similar situations. It rarely occurs on salt water but it often congregates on brackish coastal lagoons. It is nomadic.

Strongly gregarious, Eurasian Coots often form large rafts far out on open water, but are strongly territorial and aggressive when breeding. The vocabulary includes a variety of loud, strident and abrupt notes.

The main breeding season is from September to December in the south, January to April further north. The nest is a bulky platform of reed stems, often with an entrance ramp, and often with a sheltering woven hood. The usual clutch is five or six eggs, which hatch in 23 to 26 days.

HABITAT: mostly permanent fresh waters with ample cover of rushes, reeds, cumbungi or scrub; often in artificial ponds of city gardens
LENGTH: 32–39 cm
DISTRIBUTION: more than 1 million km²
ABUNDANCE: common
STATUS: secure

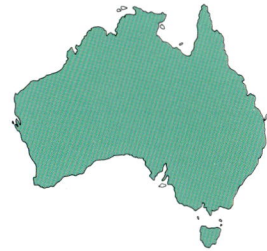

(P Klapste)

Genus Gallinula

(gah'-lin-yue'-lah: "small-fowl")

The gallinules comprise a group of about 14 species widespread on all continents except Antarctica. Four species occur in Australia. Generic limits are vague and debatable, and the group is sometimes placed entirely in the genus *Gallinula* or distributed in various combinations amongst the genera *Amaurornis, Tribonyx, Gallinula* and several others. In general, gallinules are medium-sized to rather large rails with long legs, short, stout bills, and a well-developed and conspicuous horny shield (usually red) on the forehead. Many species are plain slate grey in colour, with white undertail coverts, but some are brown. The sexes are similar. Not necessarily restricted to swamps and marshes, gallinules frequent the margins of freshwater wetlands of most kinds, including ornamental waters in city parks, but some inhabit tropical rainforests. They have loud, strident calls, persistently flick their tails and, while swimming, bob their heads back and forth in a distinctive manner.

Tasmanian Native-hen

Gallinula mortierii (mor'-tee-e'-ree-ee; "Dumortier's small-fowl", after Dumortier, Belgian zoologist)

The Tasmanian Native-hen cannot fly. It is now endemic in Tasmania, but fossil remains show that it once occurred on the mainland as well. Locally very common, it is a shy bird which inhabits the margins of marshes, lagoons, lakes and streams.

(J Fennell)

It is largely vegetarian, grazing over waterside lawns and pasture and disappearing into dense cover at the least disturbance. Apart from seeds and green shoots, insects and other small invertebrates are also eaten.

The Tasmanian Native-hen lives in sedentary groups, typically composed of a female with one or two males together with the young of the previous season, which jointly defend permanent territories some 5000 to 15 000 square metres in extent. Territorial defence is vigorous, involving loud calls (including a distinctive duet between male and female) and occasional fighting. Breeding involves pairs or trios in about equal numbers.

Most breeding takes place between August and November. The nest, often canopied, is well hidden in dense cover; an additional structure close by is usually built in which the young are brooded once hatched. Six eggs form the usual clutch.

HABITAT: mainly boggy pastures and adjacent reedbeds and the margins of lagoons and swamps
LENGTH: *c* 45 cm
DISTRIBUTION: 30,000–100,000 km^2
ABUNDANCE: abundant
STATUS: secure

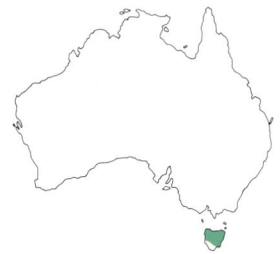

Bush-hen

Gallinula olivacea (ol'-iv-ah'-say-ah: "olive small-fowl")

The Bush-hen occurs in the Philippines, eastern Indonesia, New Guinea, the Bismarck Archipelago and, in Australia, in the Top End and eastern Queensland and north-eastern New South Wales south to about Ballina. Inhabiting dense rank vegetation near water, it is extremely elusive and difficult to observe, though noisy. It is mainly nocturnal.

Its vocabulary is varied; common calls include a loud, harsh "knee-you", often in series, and a persistent "tok".

Breeding usually occurs between October and April. The nest is a shallow bowl of woven grass in dense cover. The usual clutch is six or seven eggs, and is incubated by both parents.

HABITAT: tall rank grass and dense vegetation beside forest streams; lantana thickets
LENGTH: *c* 26 cm
DISTRIBUTION: 100,000–300,000 km^2
ABUNDANCE: sparse
STATUS: probably secure

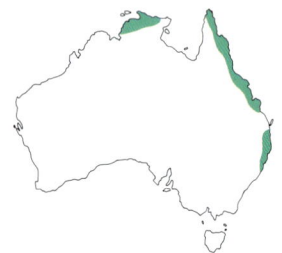

(L Robinson)

Dusky Moorhen

Gallinula tenebrosa (ten'-eb-roh'-sah: "dark small-fowl")

(*G Chapman*)

The breeding season is variable, but mainly from January to May in the north, September to December in the south; two broods are sometimes raised in succession. The nest is a bulky saucer of broken reeds near water, often with an entrance ramp of trampled reeds. The normal clutch consists of six to nine eggs.

HABITAT: mainly shallow lagoons and marshes fringed with reeds and sedges
LENGTH: *c* 35 cm
DISTRIBUTION: more than 1 million km²
ABUNDANCE: common to abundant
STATUS: secure

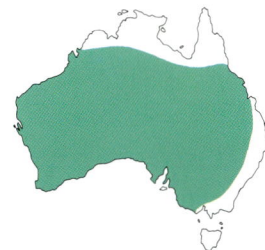

The Dusky Moorhen inhabits Indonesia, New Guinea and Australia, where it is largely confined to the east and south-east and the far south-west: however, it has only recently extended its range to include Tasmania, where the first record was in 1976. It is common in most kinds of wetlands margined with reeds, and displays a marked preference for fresh water over salt.

On ornamental ponds in city parks it often becomes extremely tame, but it is otherwise shy and wary.

It lives in groups that together defend permanent territories. Unlike other members of its family (with the exception of the coots) it spends a great deal of its time swimming, and its diet consists mainly of aquatic plants obtained by dipping, or sometimes by up-ending like a duck.

Black-tailed Native-hen

Gallinula ventralis (ven-trah'-lis: "[notable-] bellied small-fowl")

(*B & B Wells*)

only when hard pressed, but runs nimbly. The diet consists of seeds, aquatic plants, and insects and other invertebrates.

The breeding season is extremely erratic, and strongly influenced by local flooding. Several broods may be raised in rapid succession. Four to nine eggs (usually five or six) constitute the usual clutch.

HABITAT: mainly interior wetlands fringed with lignum and canegrass; temporary floodwaters
LENGTH: *c* 35 cm
DISTRIBUTION: more than 1 million km²
ABUNDANCE: sparse
STATUS: secure

Also called Waterhen and Swamp-hen, the Black-tailed Native-hen occurs throughout Australia, but it is only a rare vagrant to the Top End and the entire east coast, and there is only a single record from Tasmania. Strongly nomadic and gregarious, the Native-hen inhabits the reedy margins of swamps, lakes, billabongs and quiet backwaters of rivers, moving out, often in huge numbers, to ephemeral floodwaters in time of flood; lignum swamps are especially favoured. It normally flies

Genus *Poliolimnas*

(poh'-lee-oh-lim'-nas: "grey-marsh [-bird]")

The characteristics of the genus are essentially those of the single included species.

White-browed Crake

Poliolimnas cinereus (sin'-er-ay'-us: "ashy grey-marsh-bird")

The White-browed Crake is widely distributed in South-East Asia, New Guinea and islands of the south-western Pacific, but in Australia it occurs only in the Top End and north-eastern Queensland. It inhabits marshes and lagoons, especially those with abundant floating vegetation, upon which it forages, feeding on aquatic insects and other invertebrates, seeds and green shoots. It is sedentary and mainly crepuscular in habits.

The breeding season extends from September to May. The nest, constructed of reed stems and leaves, is well hidden in dense vegetation; three to six eggs are laid.

HABITAT: tropical lagoons with dense aquatic vegetation
LENGTH: *c* 19 cm
DISTRIBUTION: 30,000–100,000 km²
ABUNDANCE: sparse
STATUS: probably secure

(RD Mackay)

Genus *Porphyrio*

(por-fi'-ree-oh: "purple [-bird]")

About four species of large, brightly coloured rails widespread in warmer parts of the Old World comprise the genus *Porphyrio*. Except for their deep blue or purplish colouration they resemble gallinules and, like these, have large conspicuous frontal shields and white undertail coverts. However, they are much larger and have deeper, blunter bills. Largely vegetarian, they inhabit the margins of wetlands of all kinds, living in dense cover but frequently emerging to graze on short grass or partly submerged aquatic vegetation. The sexes are similar.

Purple Swamphen

Porphyrio porphyrio (por-fi'-ree-oh: "purple purple-bird")

The Purple Swamphen has an extensive distribution from southern Europe and Africa across southern Asia to New Zealand and islands of the south-western Pacific. In Australia it occurs in wetlands more or less across the continent; like the Dusky Moorhen it is often common on ornamental ponds in urban parks, where it is easily observed grazing on grassy margins and waterside lawns.

In such situations it is often so tame that it can be fed by hand. It eats mainly aquatic plants, but it sometimes does damage to crops; its varied diet also includes frogs, stranded fish and, on occasion, eggs and chicks of other birds.

It is mainly sedentary, but undertakes wide movements under some circumstances. A gregarious species, it lives in pairs or flocks and often breeds communally. Calls are extremely loud and resonant. It swims and flies readily and is also able to perch in trees; when walking it flicks its tail continually.

The breeding season is very variable but July to December represent peak breeding months in the south-east. The nest is a bulky platform of reeds near water; surrounding vegetation is often interwoven to form a loose overhead hood. Five to seven eggs, which hatch in 23 to 24 days, form the usual clutch.

HABITAT: temperature to tropical reed beds by lakes and rivers
LENGTH: 44–48 cm
DISTRIBUTION: more than 1 million km^2
ABUNDANCE: common
STATUS: secure

(*TR Lindsey*)

Genus *Porzana*

(por-zah'-nah: "crake")

The members of this genus, generally known as crakes, comprise an almost cosmopolitan genus of about 13 species, three of which occur in Australia. They are very small, chicken-like birds with rather long legs and short, stout bills. The plumage patterns are intricate and many species are spotted with white, especially on the back. The sexes are similar in appearance. They live mainly in dense reedbeds in swamps and marshes, where they are usually common but, being intensely secretive and largely nocturnal or crepuscular, they are very difficult to observe. When disturbed they fly only with reluctance, but several species are migratory.

Australian Crake

Porzana fluminea (flue-min'-ay-ah: "riverine crake")

Also called Spotted Crake, this species is endemic in Australia, where it appears to be commonest and most widespread in the interior south-east.

It inhabits densely-vegetated wetlands of most kinds, including mangroves, but it is more tolerant of saline conditions and ephemeral floodwaters than other crakes and rails, and often forages in the open.

Its diet includes aquatic insects and other small invertebrates. The Australian Crake is nomadic.

The breeding season is from August to February. The nest is a cup or saucer of reeds hidden in dense vegetation, often with a woven canopy. Five eggs constitute the usual clutch.

HABITAT: mainly inland swamps and lagoons; the muddy margins of temporary floodwaters
LENGTH: *c* 20 cm
DISTRIBUTION: more than 1 million km²
ABUNDANCE: sparse
STATUS: probably secure

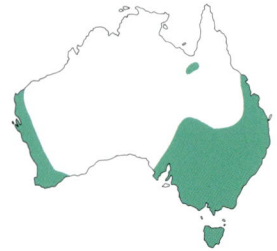

(*JN Yates*)

Marsh Crake

Porzana pusilla (pue-sil'-ah: "very-small crake")

Also called Baillon's Crake and Little Crake, the Marsh Crake has an extensive distribution across Africa, Europe and Asia; it is widespread in Australia, especially in the south-

(*B & B Wells*)

east, but it is very rare in Tasmania. Little is known of its movements, but available information suggests a regular north–south migration. It is common in wetlands generally, but it especially favours secluded swamps and marshes with abundant floating aquatic vegetation, upon which it often forages far from cover. It eats aquatic insects and other small invertebrates.

The breeding season extends from September to February. The nest is a saucer of woven reed stems and leaves, often sheltered above by a canopy. The usual clutch is five or six eggs, though from four to eight have been reported.

HABITAT: tropical and temperate freshwater swamps and lagoons with abundant aquatic vegetation; reedbeds
LENGTH: *c* 16 cm
DISTRIBUTION: more than 1 million km²
ABUNDANCE: sparse
STATUS: probably secure

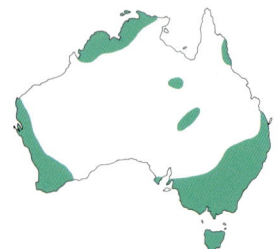

Spotless Crake

Porzana tabuensis (tah'-bue-en'-sis: "Tongan crake")

Perhaps the most secretive and elusive of Australian crakes, the Spotless Crake occurs in eastern and

(D Hadden)

south-eastern Australia, and elsewhere is widespread in South-East Asia, New Guinea, New Zealand, and islands of the south-western Pacific. There is some evidence to suggest migratory habits, but little is known of its movements. It inhabits wetlands generally (and in some areas dry land, as on coral atolls) but in Australia it shows a distinct preference for deep, permanent cumbungi swamps. It always forages in dense cover, feeding on aquatic insects and other small invertebrates; it has also been seen to eat carrion and fruit.

The breeding season is generally from September to January. The nest, of woven reed stems and leaves, is extremely well hidden, and the usual clutch consists of four to six eggs.

HABITAT: tropical and temperate freshwater swamps and lagoons with abundant aquatic vegetation; reedbeds
LENGTH: *c* 18 cm
DISTRIBUTION: more than 1 million km²
ABUNDANCE: sparse
STATUS: probably secure

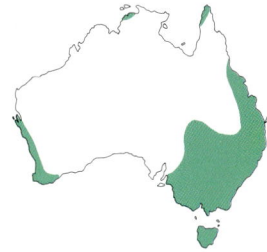

Genus Rallina

(ral-lee'-nah: "little-rail")

The four species in this genus somewhat resemble rails of the genus *Rallus* except that their colour patterns are simpler (in most species the head and neck is deep rufous, the back uniformly dark and the underparts barred), they have shorter, somewhat stouter bills, and live mainly in rainforest. The genus is restricted to south-eastern Asia and the Australasian region, and only one species occurs in Australia.

Red-necked Rail

Rallina tricolor (trie'-kol-or: "three-coloured little-rail")

The Red-necked Rail inhabits dense tropical rainforests in coastal and near-coastal districts of north-eastern Queensland, favouring especially the vicinity of creeks and streams. Elsewhere it occurs in Indonesia, New Guinea and the Bismarck Archipelago.

It is mainly crepuscular and

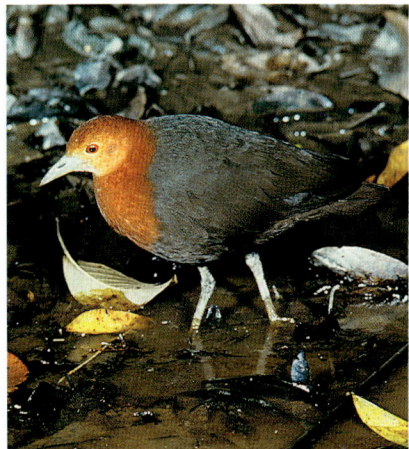

(H & J Beste)

nocturnal, and therefore difficult to observe, but it is very noisy. Little is known of its movements but some individuals apparently migrate to and from New Guinea. Foraging, it rakes through leaf litter and wades along shallow streams; its varied diet includes seeds as well as insects and other invertebrates. The most characteristic call is a monotonous "tok-tok-tok", uttered in lengthy series.

Adults maintain permanent territories. Most eggs are laid from December to March. The nest is a shallow cup of vine stems, twigs and leaves, secluded in the buttresses of a rainforest tree or hidden in a vine tangle or a fern up to two metres from the ground. The usual clutch is five eggs, which hatch in 18 to 22 days. Chicks are fed by their parents for the first three to five days and fledge at four to six weeks.

HABITAT: the margins of rivers and streams in tropical rainforest
LENGTH: 28–30 cm
DISTRIBUTION: 30,000–100,000 km²
ABUNDANCE: sparse
STATUS: probably secure

Genus *Rallus*

(ral'-us: "rail")

Most of the approximately 18 species in this almost cosmopolitan genus are rather large to medium-sized rails with long bills and relatively long legs. They have intricate plumage patterns but are generally coloured in sober shades of brown, grey, white and dull rufous; in particular, the underparts are often barred. Most live in swamps and marshes, and have loud, strident calls. Some are migratory but a few are flightless. The sexes are similar, there is no seasonal variation, and immatures resemble adults.

Lewin's Rail

Rallus pectoralis (pek'-tor-ah'-lis: "[notable-] chested rail")

(J Purnell)

Lewin's Rail inhabits New Guinea and eastern and south-eastern Australia, including Tasmania, mainly in coastal and subcoastal regions. A population once existed in far south-western Australia, but it is known from only four specimens and has not been seen since 1932.

Among the most elusive of rails, it inhabits dense vegetation along streams, river meadows and the margins of coastal lagoons. Extremely reluctant to fly, it avoids disturbance by following a network of tunnels it maintains under the vegetation. Often its presence is detectable only by calls, and knowledge of its movements is fragmentary; there is nevertheless some reason to suspect it is migratory, at least at the southern extremities of range. In an extensive vocal repertoire, one especially distinctive call resembles the sound of two large coins being rapped together. It forages along secluded muddy margins, and tends to probe rather than peck for food, eating worms and molluscs as well as insects.

The breeding season is from August to December. A clutch of four to six eggs is laid in a well-hidden, woven cup of grass and reeds, often with an entrance ramp; the eggs hatch in about 20 days.

HABITAT: dense vegetation at the margins of forest streams and lagoons
LENGTH: 20–23 cm
DISTRIBUTION: 300,000–1 million km^2
ABUNDANCE: sparse
STATUS: probably secure

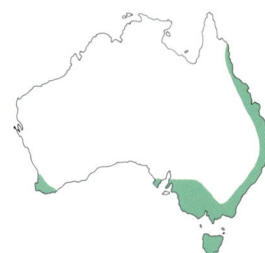

Banded Landrail

Rallus philippensis (fil'-ip-en'-sis: "Philippines rail")

The Banded Landrail, or Buff-banded Rail, occurs in Australia (mainly in the east and south-east and far south-west but also at scattered localities in the interior). It is also widespread in the Indo-Pacific region. It seems to be migratory in the south, resident in the north.

It lives singly or in pairs in a variety of swampy or vegetation-choked environments, mainly at the margins of streams, quiet rivers, marshes and lagoons, also in mangrove and melaleuca swamps and, along the Great Barrier Reef and elsewhere, on coral sand cays. Although a secretive and elusive bird, it is often common near habitation.

It often feeds in the open, in the early morning and evening. It has a variety of calls, most of which are loud, sharp and penetrating; they include an extraordinary note that has been described as a "deep thudding grunt". The diet includes insects and other aquatic invertebrates.

The nest is a flimsy cup woven of grass and reeds under a tussock or in a similar well-hidden situation. The usual clutch is five or six eggs, sometimes as many as 11. The chicks are led from the nest on hatching, and fed by both parents for the first week. If the brood is large it is sometimes divided by the parents, each independently assuming sole care of several, but perhaps more often the male loses interest and wanders off after the first week or two. Several broods may be raised in succession.

HABITAT: densely vegetated swamps and lagoons; mangroves; coral cays
LENGTH: 29–32 cm
DISTRIBUTION: more than 1 million km^2
ABUNDANCE: sparse
STATUS: probably secure

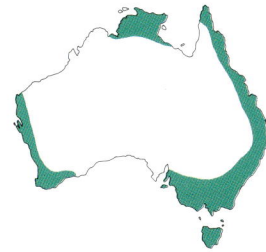

(B & B Wells)

Genus *Tricholimnas*

(trik'-oh-lim'-nas: "marsh-thrush")

The characteristics of the genus are those of the single species.

Lord Howe Woodhen

Tricholimnas sylvestris (sil-ves'-tris: "woods marsh-thrush")

The reduced wings of this rail render it incapable of flight. It lives on the ground in dense vegetation, feeding upon insects, worms and other invertebrates taken from leaf litter and decaying wood with a long, powerful bill. It originally had no fear of humans so European settlers found it a convenient food source. Introduction of rats, cats, pigs and goats led to direct predation on the Woodhen and destruction of its habitat, so that in 1972 the population was estimated to be fewer than 20 birds. Since then, a program of captive breeding and release, together with reduction of the introduced mammals, has led to an increase in the population (estimated in 1989 to be about 200).

The sexes are similar. Breeding occurs from October to January. Pairs appear to be permanent and territorial. The nest is a shallow depression in the ground. One to four eggs comprise the clutch, which is incubated by both parents.

HABITAT: rainforest with dry leaf litter
LENGTH: *c* 40 cm
DISTRIBUTION: less than 10,000 km^2
ABUNDANCE: rare
STATUS: endangered

(GA Hoye)

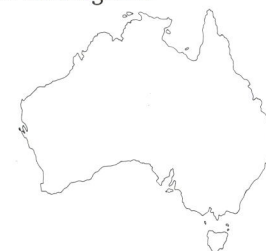

Family GRUIDAE

(gru-i-dee: "*Grus*-family", after a genus of cranes)

Cranes constitute a family of 14 or 15 species found on all continents except South America and Antarctica. They are an unusually homogeneous group, all species being large, elegant wading birds (some standing up to 1.5 metres tall) that inhabit grasslands or extensive shallow wetlands. Most species belong to the genus *Grus*, which is widespread in North America, Eurasia, Africa and Australasia; three other genera are confined to Africa.

Cranes are typically either grey or white in colour, often with small areas of black, and with wattles or patches of naked pink or yellow skin on the head and neck. The sexes are similar (though males are somewhat larger than females); there is no seasonal variation; and immatures closely resemble adults. Wary birds, they are strongly gregarious in habit and omnivorous in diet.

Cranes are unusually long-lived (captive birds have reached 60 years) and mate for life.

Genus Grus

(groos: "crane")

The characteristics of this genus are essentially those of the family.

Sarus Crane

Grus antigone (an-tig'-oh-nay: "Antigone-crane", after Greek mythological daughter of Oedipus)

The origin of the Australian population of the Sarus Crane is mysterious. It is so similar to the native Brolga (most readily distinguished by pink, not slate-grey, legs and a red collar on the upper neck) that its presence was not detected until the late 1960s. It is now known to be fairly common in north-eastern Queensland and the vicinity of the Gulf of Carpentaria (with scattered reports in the Top End and Kimberley), but it remains uncertain whether it is a native species or a recent immigrant: its numbers seem to be expanding rapidly. Its distribution outside Australia extends from India to Indonesia.

In its habits and behaviour it closely resembles the Brolga: it is omnivorous in diet; when not breeding it roosts, forages and travels in flocks; and it maintains extensive nesting territories and long-lasting pair bonds by means of elaborate dancing displays and loud bugling calls. The two species occur widely together in north-eastern Queensland. They interbreed freely in captivity but any interaction in the wild is uncertain.

The breeding season extends from January to March. Typical breeding territories are 50 to 60 hectares in extent. Both sexes build the nest (a rough platform of sedges about two metres in diameter), incubate the two eggs (which hatch in about 34 days) and care for the young, which remain with their parents until the following breeding season.

HABITAT: swamps, shallow lagoons, floodwaters, wet grassland, salt marsh, stubble and croplands
LENGTH: 110–120 cm
DISTRIBUTION: 100,000–300,000 km²
ABUNDANCE: sparse
STATUS: probably secure

(*K. Ireland*)

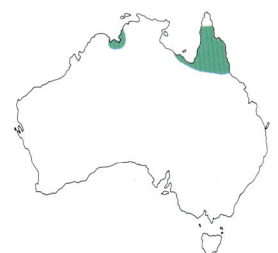

Brolga

Grus rubicundus (rue'-bee-kun'-dus: "reddish crane")

The Brolga, or Native Companion, occurs almost throughout mainland Australia but it is rare in the west, being most numerous in the east and the tropical north; it has been seriously reduced in the interior south-east since European settlement. It also occurs in New Guinea, where it probably breeds.

Strongly gregarious when not breeding, the Brolga is not migratory, but wanders in flocks (sometimes large) in winter, visiting crops and flooded grassland and sometimes causing damage. Such flocks represent aggregations of family groups,

each led by a dominant male. In the tropical north, flocks numbering 1000 or more are not infrequent. Its varied diet includes tubers, seeds, grain, insects and other invertebrates, and small vertebrates. It generally forages by day, roosting communally on shallow floodwaters by night.

Permanent pair bonds are thought to be maintained by dramatic, stately dancing displays: birds face each other, wings half-spread and shaking, each bowing and shaking its head during alternate advances and retreats, with many exaggerated

leaps and parachuting descents, the birds periodically pausing to utter loud, trumpeting calls.

The breeding season is generally from September to December in the south, February to June in the north. Nesting territories are usually in extensive swampy grasslands, and the nest is a platform of grasses and sedges about 1.5 metres in diameter. Both sexes incubate the clutch of two eggs (which hatch in 28 to 30 days) and care for the young, which are mobile within hours of hatching but remain with their parents for a year or more.

HABITAT: Swamps; shallow lagoons; floodwaters; grassland; saltmarsh; croplands
LENGTH: 95–125 cm
DISTRIBUTION: more than 1 million km²
ABUNDANCE: common
STATUS: secure

(*J McCann*)

Family OTIDIDAE

(oh-tid'-id-ee: "*Otis*-family", after a genus of bustards)

The bustards, family Otididae, comprise about 22 to 24 species which live mainly in Africa and southern Asia. They are medium-sized to large birds, generally inhabiting arid plains, semidesert and grassland—although two or three live in forest. Males are larger and more brightly coloured than females, and courtship displays are complex and sometimes spectacular. Only one species occurs in Australia.

Genus Ardeotis

(ar'-day-oh'-tis: "heron-bustard")

Members of this genus are widespread in Africa, southern Asia, and Australasia. They are rather large bustards, inhabiting open arid grassland and feeding mainly on large, ground-living insects and small vertebrates. Males are substantially larger than females. Slightly crested, and with dense long plumage on the upper breast, they also possess an inflatable gular sac: these features are used in combination in a spectacular display in which the male struts about with his tail cocked, wings drooped, breast feathers spread to form a large, drooping fan, and gular sac inflated while uttering deep bellowing and booming sounds. Members of two other genera of bustards use elements of this display, but are relatively silent: the use of vocalizations seems restricted to *Ardeotis*.

Australian Bustard

Ardeotis australis (os-trah'-lis: "southern heron-bustard")

The Australian Bustard occurs throughout the Australian mainland.

(R Drummond)

It does not tolerate intensive land management for agriculture and other human activities, and has become extremely rare over most of south-eastern Australia since European settlement. It is still common in northern Australia, but suffers persecution from shooters and predation from foxes.

Strongly nomadic, it usually occurs alone or in small parties (generally consisting of birds of a single sex), but occasionally in large flocks. It eats seeds, insects and small reptiles.

In northern Australia the breeding season is usually from January to March. The male has an impressive courtship display in which he struts about, booming loudly, with wings drooped, tail spread and cocked, and long feathers of the upper breast spread into a swaying fan. The clutch is one, occasionally two, eggs, incubated by the female.

HABITAT: tropical to warm temperate semiarid plains, grassland and lightly timbered country
LENGTH: 80–120 cm
DISTRIBUTION: more than 1 million km²
ABUNDANCE: common
STATUS: vulnerable

Order CHARADRIIFORMES

(kah-rad'-ree-oh-for'-mayz: "*Charadrius*-order", after a genus of plovers)

This large and cosmopolitan order encompasses about 340 species of birds, distributed among 18 families. There are a number of small peripheral groups (including, of those represented in Australia: Jacanidae (jacanas); Rostratulidae (painted-snipe); Haematopodidae (oystercatchers); Recurvirostridae (stilts and avocets); Burhinidae (stone-curlews); Glareolidae (pratincoles) and Stercorariidae (skuas and jaegers). In terms of number of species, extent of distribution, and relative abundance, however, the order is dominated by three families: Charadriidae (plovers), Scolopacidae (sandpipers) and Laridae (gulls and terns). All three are represented in Australia by many species.

Almost all are associated with water and most forage by wading in or near shallow water, feeding on a variety of invertebrates and small aquatic vertebrates, especially fishes. A number of species inhabit freshwater marshes and even interior deserts, but the Charadriiformes dominate coastal avifaunas to the extent that few birds seen commonly along beaches, coastal mudflats and tidal estuaries are not members of this order.

In structure and mode of life they divide into three very distinct groups, often treated as suborders. Both of the families Charadriidae and Scolopacidae are included in the suborder Charadrii. Generally rather small and soberly plumaged, with long narrow wings and rather short tails, these birds are notable in general for their spectacular migrations, often involving entire populations in an annual cycle of movement between the Northern and Southern Hemispheres.

The Laridae are characteristic of the suborder Lari, to which group the skuas and jaegers also belong. These birds are often white below and pale grey above and have long narrow wings and well-developed powers of sustained and effortless flight. The tail is usually either moderately short and rounded, or long and deeply forked. Gulls and terns are common along seacoasts throughout the world. Most scavenge along beaches or dive for fish from moderate heights.

The suborder Alcae contains the auks, puffins and guillemots, a small group of exclusively marine birds found only in colder parts of the Northern Hemisphere.

Usually both parents are involved at all stages of the nesting cycle. The nest is commonly a scrape in the ground, variably lined with grass or other vegetation. The young, which are hatched alert, mobile, and covered in dense fluffy down, are led from the nest soon after hatching.

Family JACANIDAE

(jah-kah'-nid-ee: "*Jacana*-family", after a genus of lily-trotters)

About eight species of small wading birds constitute the family Jacanidae, found in South and Central America, Africa, Asia and Australasia. All live exclusively on tropical and warm temperate freshwater wetlands, and have conspicuous adaptations (notably the extremely elongated toes) for walking on floating vegetation. For all species, the typical habitat is extensive, deep permanent lagoons and swamps with abundant aquatic vegetation. One species occurs in Australia.

The sexes are similar in plumage, and immatures generally resemble adults; in some species there is marked seasonal variation in plumage.

Females are larger than males, and several species are polyandrous. Generally, the female defends the territory; but the bulk of nest building, incubation and care of the young is undertaken by the male. Very little is known of the breeding behaviour of the Australian species.

Genus *Irediparra*

(ie'-red-ip'-a-rah: "Iredale's-bird", a clumsy name based on that of T. Iredale, Australian ornithologist)

The characteristics of the genus are essentially those of the single species.

Lotusbird

Irediparra gallinacea (gal'-in-ah'-say-ah: "fowl-like Iredale's-bird")

(G Threlfo)

over lily pads and other floating vegetation, flicking its tail and bobbing its head, frequently leaning forward to peer under the margins of lily pads, and snatching at surface invertebrates. The flight is swift, low, and brisk, with shallow wing-beats and trailing feet.

Breeding is strongly influenced by local conditions, especially water levels, but generally occurs from January to May in the north, September to January in the east. The nest is a fragile raft of grass and plant stems, moored amongst submerged vegetation; four eggs form the usual clutch. Sometimes two broods are raised in succession, but little is known of the nesting cycle.

The Lotusbird or Comb-crested Jacana is uncommon to locally abundant in fresh waters of the coastal lowlands of northern and eastern Australia, south on the east coast commonly to about Grafton in New South Wales, sporadically to the Hawkesbury River; its status in the Kimberley is uncertain. Elsewhere it occurs in Indonesia, New Guinea and the Philippines.

Its extraordinarily long toes and persistent habit of walking on floating vegetation render it unmistakable, but the pinkish comb on the head is a valuable field mark. Females are somewhat larger than males, but the sexes are otherwise similar.

It is mainly sedentary, but it undertakes local movements in response to changing water levels. It is gregarious outside the breeding season, sometimes congregating in large numbers at favoured ponds and lagoons, but it is strongly territorial when breeding. Its calls include high-pitched chittering notes, often uttered in flight. It feeds on insects and the seeds and shoots of aquatic plants.

Almost never encountered ashore, it is nevertheless an active and conspicuous species, walking easily

HABITAT: deep, permanent, vegetation-choked tropical and warm temperate wetlands (large or small); sewage farms
LENGTH: 19–23 cm
DISTRIBUTION: 300,000–1 million km²
ABUNDANCE: sparse to common
STATUS: probably secure

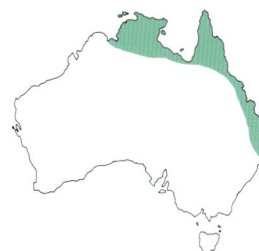

Family BURHINIDAE

(bue-rie'-nid-ee: "*Burhinus*-family", after a genus of stone-curlews)

Variously known as stone-curlews, thick-knees or dikkops, some nine species constitute this family, found on all continents except North America. There are two genera (*Burhinus* and *Esacus*); Australia has a single representative of each.

Most species are about the size of a Masked Lapwing (but members of *Esacus* are substantially larger), and the plumage is an unobtrusive blend of shades of grey, fawn and dull rufous. At least partly nocturnal, they have abnormally large yellow eyes and a distinctively hunched posture and furtive mannerisms. They fly well but reluctantly, relying generally on camouflage to evade disturbance. Except for *Esacus*, the various species inhabit mainly scrub and dry, open, broken country. They are strongly vocal, and are notable for their complex, eerie calls, often uttered in chorus.

Genus *Burhinus*

(bue-rie'-nus: "bull-nose")

Members of this genus are found on all continents except North America. Mainly nocturnal in habits and skulking in behaviour (though noisy at night), they inhabit scrub, open woodland and rough farmland. Though territorial when breeding, most are otherwise gregarious, congregating in flocks or "clans" during the winter and, under some circumstances, even when nesting. They feed mainly on terrestrial insects and other small invertebrates, but they are omnivorous and forage opportunistically. Most are sedentary.

The sexes are similar in appearance, immatures resemble adults, and there is no seasonal variation.

Bush Stone-curlew

Burhinus grallarius (gral-ar'-ee-us: "stilt-like bull-nose")

(AJ Olney)

eats insects and other arthropods.

It is strongly vocal, with a complex repertoire which includes an eerie, long-drawn, ululating "whistling scream" that is perhaps the most familiar call, well known in quiet country districts at night. It often calls in chorus.

Breeding occurs mainly between July and January. One to three (usually two) eggs form the clutch, which is laid in a shallow scrape in the ground. Parents share incubation and care of young. The eggs hatch in 28 to 30 days, and the young fledge at about 47 to 50 days.

HABITAT: open woodland, savannah, dune scrub, fringes of mangroves; sometimes golf courses, orchards
LENGTH: 52–58 cm
DISTRIBUTION: more than 1 million km²
ABUNDANCE: sparse to common
STATUS: vulnerable

The Bush Stone-curlew (or Southern Stone-Curlew or Bush Thick-knee) is widespread in Australia and southern New Guinea, but it has not been recorded in Tasmania since 1895, and its Australian mainland distribution has been drastically reduced since European settlement, especially in the south-east; it is now most common in the far north. The introduction of foxes, changes in fire regimes, land clearing and human persecution have all been implicated in its decline.

Males are slightly larger than females but the sexes are otherwise similar, and immatures resemble adults. There is some subtle regional variation in the shade of grey or rufous in the plumage but there are no subspecies.

The Bush Stone-curlew is terrestrial, sedentary, mainly nocturnal in habits, and extremely furtive in behaviour, relying heavily on camouflage. It is strongly territorial when breeding (the pair bond is apparently permanent), but at other times it often forms loose clans, which roost together and scatter to feed.

It is almost omnivorous but mainly

Genus Esacus

(es-ah'-kas: meaning unknown)

This genus contains one or two species, which closely resemble stone-curlews of the genus *Burhinus* but differ most conspicuously in habitat, being characteristic of tropical and warm temperate coral reefs and cays and undisturbed beaches in the Oriental and Australasian regions respectively. Both forms may prove to be conspecific. One of them occurs in Australia.

Beach Stone-curlew

Esacus magnirostris (mag'-nee-ros'-tris: "great-billed *esacus*")

The Beach Stone-curlew (or Beach or Reef Thick-knee) occurs in northern Australia from about Red Rock in New South Wales northward and westward to Point Cloates in Western Australia, and from New Caledonia north to the Philippines and west through Indonesia to the Andaman Islands. It is sedentary, generally uncommon, and extremely sensitive to human disturbance. It is usually encountered alone or in pairs.

Almost unmistakable, the Beach Stone-curlew is a large, dignified, deliberate bird with a long stout bill and a conspicuously patterned head. The sexes are similar, and immatures resemble adults.

Little is known of its behaviour, but it eats mainly crustaceans and other small invertebrates. Breeding occurs mostly between October and December, and the single egg is usually laid on bare sand close to the high tide line in the shelter of flotsam, driftwood, or a small bush. Both parents share incubation and care of the young.

HABITAT: remote and secluded beaches, coral reefs and cays, mangrove fringes, estuarine mudflats
LENGTH: 53–57 cm
DISTRIBUTION: 10,000–30,000 km^2
ABUNDANCE: sparse
STATUS: vulnerable

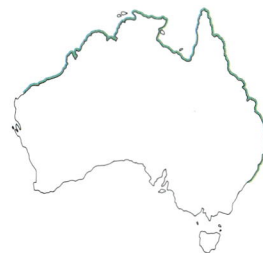

(G Chapman)

Family ROSTRATULIDAE

(ros-trah-tue'-lid-ee: "*Rostratula*-family", after a genus of snipes)

Two species, each in a distinct genus, constitute this family, which lacks close relatives but shows some similarities with the jacanas. Very little is known of the South American species, but the other has an extensive distribution in tropical and warm temperate wetlands across the Old World.

Painted Snipes resemble true snipes (family Scolopacidae) in size, general appearance, and skulking habits but have broader, more rounded wings and a more colourful plumage pattern. Females are more brightly coloured than males, exceed males in size by about 10 per cent, and take the initiative in courtship, subsequently leaving the male to raise the brood alone.

121

Genus Rostratula

(ros-trah'-tue-lah: "small-beak")

The characteristics of the genus are essentially those of the single species.

Painted Snipe

Rostratula benghalensis (ben'-gah-len'-sis: "Bengal small-beak")

(TG Lowe)

The Painted Snipe has an extensive distribution in Africa and across southern Asia to southern Japan, the Philippines and Indonesia. In Australia it is most numerous in the Murray–Darling basin and elsewhere in the interior south-east: it has not been reported in Western Australia since last century, its status in the Northern Territory is uncertain, it is rarely reported from northern Queensland, and there is only a single record in Tasmania. Possibly migratory, its movements are unknown and its occurrence is erratic and unpredictable: it seldom remains long in any locality.

Its smoky brown hood with conspicuous white head markings and "horse collar" around the neck are distinctive. Males resemble females but are very much duller and less distinctly marked, and somewhat smaller. There is no seasonal variation.

Intensely secretive, crepuscular, and mainly solitary, the Painted Snipe skulks in dense vegetation by day, emerging at dusk (but occasionally during the day) to feed over adjacent mudflats and other open areas. The varied diet includes seeds as well as insects and other small aquatic invertebrates. Calls include a harsh, abrupt "kek!" when flushed, and a lengthy series of soft, mellow notes. Flight is normally low and feeble, with legs trailing.

Breeding is strongly influenced by local conditions, but occurs mainly from September to December. The species is polyandrous, and the female has an intricate courtship display involving spread wings and fanned and raised tail; the pair bond with each of several males in turn lasts only until the eggs are laid. The male builds the nest, incubates, and rears the young alone. The nest is a shallow waterside scrape scantily lined with grass; the four eggs hatch in about 21 days.

HABITAT: swampy vegetation bordering tropical and warm temperate freshwater wetlands
LENGTH: 23–26 cm
DISTRIBUTION: more than 1 million km^2
ABUNDANCE: sparse
STATUS: probably secure

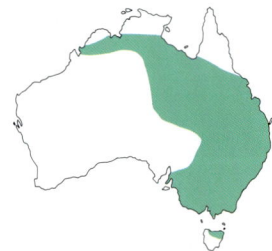

Family HAEMATOPODIDAE

(heem'-ah-toh-poh'-did-ee: "*Haematopus*-family", after a genus of oystercatchers)

Oystercatchers are large waders found along coastlines on every continent except Antarctica. About 12 species, grouped in a single genus, divide into two types: those with plumage entirely black and those that are black and white; all have bright red bills, legs and feet. There is a marked tendency for black species to inhabit rocky coastlines, pied species to favour beaches. In some areas (especially Europe) certain species occupy moorland and other non-coastal habitats. Most are sedentary.

The sexes are similar in appearance, immatures resemble adults, and seasonal variation is negligible.

Australia has two endemic species: one black and one pied.

Genus Haematopus

(hee'-mah-toh-poos: "[blood-] red-foot")

The characteristics of the genus are those of the family.

Sooty Oystercatcher

Haematopus fuliginosus (fool'-i-jin-oh'-sus: "sooty red-foot")

(*G Rogerson*)

January. Both sexes vigorously defend their territory against trespassers, and cooperate in nest building, incubation and care of the young. Two eggs (rarely three) are laid in a shallow scrape in loose sand between rocks, mostly on offshore islands.

HABITAT: littoral, favouring (but not restricted to) rock ledges and reefs, tide pools
LENGTH: *c* 50 cm
DISTRIBUTION: 30,000–100,000 km²
ABUNDANCE: sparse
STATUS: probably secure

Sedentary in habits and exclusively coastal in distribution, this species occurs around Australia, though it is somewhat more numerous in the south than in the north.

It strongly resembles the Pied Oystercatcher in habits and behaviour, and the two species occasionally occur together in mixed flocks, but there is a strong (though not invariant) tendency for them to occupy different habitats: the Pied Oystercatcher being seen on sand beaches, the Sooty Oystercatcher along rocky shores.

Breeding occurs from October to

Pied Oystercatcher

Haematopus longirostris (lon'-jee-ros'-tris: "long-billed red-foot")

(*JR Napier*)

either on the ground or in the air.

October to January is the usual breeding season (somewhat earlier in the tropics). Two or three eggs form the clutch, laid in a shallow scrape in the sand, usually between high tide mark and the first line of dunes. Incubation, by both sexes, takes about 28 days; the young associate with their parents for several months after fledging.

HABITAT: ocean beaches and estuarine flats
LENGTH: 48–51 cm
DISTRIBUTION: 30,000–100,000 km²
ABUNDANCE: sparse
STATUS: secure

A conspicuous, noisy and wary shorebird, the Pied Oystercatcher occurs along the entire Australian coastline, but it is most common in the Tasmanian region and its numbers generally decline northwards. It is mainly sedentary.

It is strongly territorial when breeding but tends to associate in flocks of about 10 to 50 during winter; it may flock casually with other shorebirds. At low tide it forages on exposed sand or mudflats, roosting on nearby reefs, rock shelves or sandbars at high tide. The diet is mainly bivalve molluscs but includes worms, crustaceans, and other small littoral invertebrates. Courtship involves several birds in "piping ceremonies" and noisy chases,

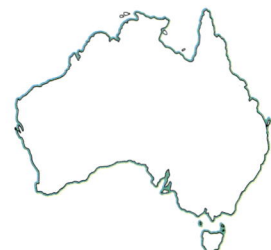

Family GLAREOLIDAE

(glar'-ay-oh'-lid-ee: "*Glareola*-family", after a genus of pratincoles)

Two subfamilies make up the family Glareolidae, a group of 17 species of arid zone waders found across the Old World. Nine species of Cursoriinae, or coursers, are mainly Indian and African in distribution, but two species of Glareolinae, or pratincoles, extend to Australia. The pratincoles comprise two genera: *Stiltia* has only one species, while the remaining eight are grouped in *Glareola*.

As a breeding bird, *Stiltia* is confined to Australia. Long-legged and with long slender wings, it shows some affinities with the coursers. Pratincoles of the genus *Glareola* are short-legged and have deeply forked tails, and are essentially African and Asian in breeding distribution; one species migrates to Australia.

Pratincoles are unusual among land birds in possessing salt glands in the head. They are gregarious in all activities, nesting in colonies and migrating in flocks. They feed on insects captured on the ground or in the air. Females resemble males and immatures resemble adults. *Glareola* has a pectinated middle toe, *Stiltia* does not. In general, *Stiltia* shows more adaptations to a desert existence than does *Glareola*.

Genus Glareola

(glah'-ray-oh'-lah: "gravel [-bird]")

The characteristics of the genus are discussed under the family heading.

Oriental Pratincole

Glareola maldivarum (mahl'-dee-var'-um: "Maldives gravel-bird")

The Oriental Pratincole breeds on the vast, arid plains and steppes of Central Asia, from Mongolia, Manchuria and central Siberia south to India, Thailand and southern China. In winter it migrates southward to Indonesia, New Guinea and Australia, where it is most numerous and most frequently recorded (though extremely variable in numbers from year to year) in the Kimberley and the Top End in November and December. Soon after arriving in Australia, it tends to move inland in flocks of thousands, which spend the next few months roaming the southern fringe of the region of summer monsoon rainfall, seldom remaining more than a week or so in any one spot. Most leave again in April. Occasional stragglers reach New South Wales and South Australia.

Very active and graceful both on the ground and on the wing, the Oriental Pratincole feeds mainly on insects captured in flight: it attends plagues of grasshoppers and swarming termites. It feeds mainly at dawn and dusk, and often spends the day loafing in dense flocks on claypans and similar open sites.

Little is known of its breeding behaviour.

HABITAT: mainly arid interior; breeds on steppes and dry grasslands; at other seasons open plains, claypans, mudflats, airfields
LENGTH: *c* 23 cm
DISTRIBUTION: more than 1 million km²
ABUNDANCE: sparse
STATUS: secure

(*N Yamagata*)

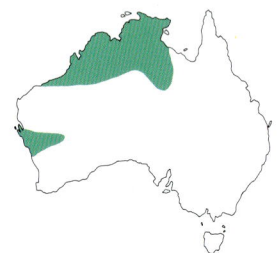

Genus Stiltia

(stil'-tee-ah: "stilt")

The characteristics of the genus are essentially those of the single species.

Australian Pratincole

Stiltia isabella (iz'-ah-bel'-ah: "greyish-yellow stilt")

This elegant and graceful wader occurs across mainland Australia, but it avoids south-eastern coastal districts and is rare or absent over much of Western Australia. It is most common in the north and interior south-east, where it mainly breeds (few pratincoles breed north of about 20°S) and where it is largely migratory, being present in numbers only between about October and March (though recorded in all months). Elsewhere across the continent it is strongly nomadic. In winter substantial numbers visit Indonesia and New Guinea.

Strongly gregarious in all activities, the Australian Pratincole feeds mainly on swarming insects, captured in buoyant hawking flight or in nimble sprints on the ground. It drinks frequently but withstands high temperatures with little apparent discomfort. Most feeding is at dawn or dusk. It is fond of loafing on outback gravel roads.

It breeds in loose colonies, raising a single brood per season, generally between October and December. Colonies are most often on arid gibber flats, though seldom far from water. Two eggs are laid on bare earth and incubated by both parents in shifts averaging about 90 minutes; the eggs hatch in about 21 days and the young reach independence in about 28 days.

HABITAT: mainly arid interior: gibber flats, open plains, claypans, airfields
LENGTH: 19–22 cm
DISTRIBUTION: more than 1 million km²
ABUNDANCE: sparse
STATUS: probably secure

(R & D Keller)

Family CHARADRIIDAE

(kar'-ad-ree'-id-ee: "*Charadrius*-family", after a genus of plovers)

Plovers are small to medium-sized shorebirds with long wings, short tails and legs of moderate length. Plovers have relatively large eyes, and the bill is fairly uniform in structure: it is straight, rather short and fairly stout, showing little of the wide range of modifications characteristic of the related family Scolopacidae.

Plovers are generally gregarious, wary and active, and have well-developed powers of flight. Some are strongly migratory, others sedentary. Some are active by night as well as by day. Plumage colours are usually subdued, but there is often a distinct breeding plumage, brighter and more colourful than the winter plumage. The sexes are usually similar or nearly so, and immatures usually resemble adults, especially the winter plumage.

There are about 50 species grouped in nine or 10 genera, and the family can be conveniently, if crudely, divided into two major groups—the lapwings and the plovers, with one genus and five genera respectively—and several aberrant forms, of which the Australian *Peltohyas* is a conspicuous example. Plovers are found on all continents except Antarctica.

Genus *Charadrius*

(kar-ah'-dree-us: "valley [-bird]")

This is a cosmopolitan, relatively homogeneous genus of about 30 species. Most are small, and have plain brown upperparts and white underparts crossed by one or several breast bands of black or chestnut. Many frequent seashores, and most have in common a characteristic foraging idiosyncrasy: they run on twinkling legs for some distance, then stop abruptly, bob the head, stab at the ground when prey is located, then run again.

Double-banded Plover

Charadrius bicinctus (bie-sink'-tus: "double-banded valley-bird")

The Double-banded Plover breeds in New Zealand. It is a trans-Tasman migrant, most individuals wintering in south-eastern Australia between January and July. It is most numerous on the coast, but it is not uncommon in the interior. Some birds wander north to Cairns and west to Perth, and occasional vagrants reach Fiji, Vanuatu, Norfolk Island and Lord Howe Island.

(J Fennell)

It is gregarious on its wintering grounds, usually occurring in flocks of a few dozen, aloof from other waders. It eats mainly insects on its breeding grounds, marine worms in its winter quarters.

In breeding plumage it is unmistakable by virtue of its double breast band, but at other seasons it strongly resembles other small plovers in both appearance and behaviour. A pronounced ochre wash on the head and nape and (usually) a faint suggestion of twin bars remaining on the sides of the breast usually aid identification.

Breeding occurs from August to December, generally in loose colonies. The male constructs the nest but both parents share the burden of incubation and care of the young. The nest is a shallow scrape lined with animal dung, shell fragments and scraps of local vegetation. The usual clutch consists of three eggs, which hatch in about 27 days. Fledging takes about 40 days.

HABITAT: breeds in New Zealand on consolidated shingle banks along riverbeds, or in ploughed fields, dunes or upland tussocks; at other seasons mainly coastal: beaches, mudflats, sewage farms
LENGTH: 17–19 cm
DISTRIBUTION: 30,000–100,000 km^2
ABUNDANCE: sparse
STATUS: probably secure

Little Ringed Plover

Charadrius dubius (due'-bee-us: "doubtful valley-bird")

The Little Ringed Plover is a casual winter visitor to the tropical north of Australia. It is widespread across Europe and Asia to Japan, wintering south to Africa, India, and Indonesia. Unknown in Australia before 1972, it is now recorded with increasing frequency, and has even occurred in South Australia and Tasmania.

It favours disturbed areas and waste ground, and seems to have profited from human construction and development, expanding its range substantially in Europe and perhaps elsewhere during the 20th century. In habits it resembles other small plovers, and feeds mainly on insects.

HABITAT: mainly fresh waters and their margins; gravel pits, shingle banks, sewage farms
LENGTH: 14–15 cm
DISTRIBUTION: less than 10,000 km^2
ABUNDANCE: rare
STATUS: secure

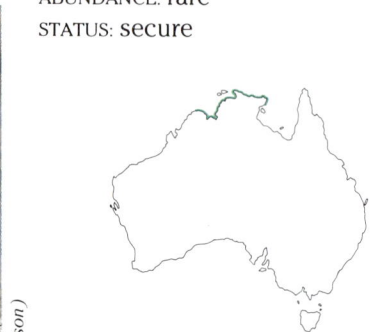

(L Jonsson)

Large Sand Plover

Charadrius leschenaultii (le'-shen-awl'-tee-ee: "Leschenault's valley-bird", after J. B. Leschenault de la Tour, French botanist)

The Large Sand Plover breeds in Central Asia from Armenia to the Aral Sea, and locally and perhaps erratically in Afghanistan, Turkey, Jordan and Egypt. Like the Mongolian Plover, it winters around the coasts of the Indian Ocean from South Africa eastward to the Philippines, New Guinea, New Zealand and Australia. It is generally uncommon in Australia except in the far north-west, and it seldom forms large flocks. Often it is solitary. It is recorded in all months but is most numerous between September and April.

It closely resembles the Mongolian Plover except in being somewhat longer-legged and possessing a larger, heavier bill; in breeding plumage both species have a rich suffusion of rufous on the head and upper breast. It feeds mainly on insects while nesting; at other seasons it eats small crustaceans (especially crabs), molluscs, and polychaete worms.

The Large Sand Plover breeds in scattered pairs during a season lasting from April to August. Both parents incubate and feed the young; three eggs form the clutch.

HABITAT: breeds on steppe, desert and other barren environments; otherwise almost exclusively coastal, favouring extensive sand flats
LENGTH: 22–25 cm
DISTRIBUTION: 10,000–30,000 km²
ABUNDANCE: sparse
STATUS: secure

(*N Chaffer*)

Black-fronted Plover

Charadrius melanops (mel'-ah-nops: "black-faced valley-bird")

This small plover, often common on farms and in other familiar situations, has a red eye ring, a black-tipped red bill, and a bold black Y-shaped mark across an otherwise white breast. It is widespread throughout Australia, both in coastal regions (although it is rare beside salt water) and in the interior. It is expanding in range and numbers in Tasmania, and around 1950 reached New Zealand, where it is now common. Tolerant of extremely small wetlands, it often occurs in roadside ditches or beside small stock dams, so long as there is at least a small stretch of open mud or rough gravel on which to forage. It occurs mainly in pairs, and is usually sedentary.

The diet consists of aquatic and terrestrial insects and their larvae, small molluscs, and seeds. The most common call is a distinctive, abrupt "tip", but it also often utters a series of dry rattling or churring notes. It is often active at night.

Breeding occurs from April to September in the far north, September to December in the south. It is territorial, and breeds in dispersed pairs. Sometimes two broods are raised in succession. The nest is a shallow scrape in loose shingle or gravel, and two or three eggs form the clutch. Incubation, shared by both parents, takes about 26 days.

HABITAT: shallow margins of wetlands, usually with pebbles, shingle or mud: lakes, rivers, marshes, stock dams, sewage farms; seldom on the seashore
LENGTH: 16–18 cm
DISTRIBUTION: more than 1 million km²
ABUNDANCE: abundant
STATUS: secure

(IP Rowles)

Mongolian Plover

Charadrius mongolus (mon'-gol-us: "Mongolian valley-bird")

The Mongolian Plover is very difficult to distinguish from the Large Sand Plover but has a shorter, slighter bill, slightly shorter legs and (in breeding plumage) a narrower breast band. On its breeding grounds it eats mainly insects, but in its winter quarters it feeds on small crustaceans, molluscs and marine worms.

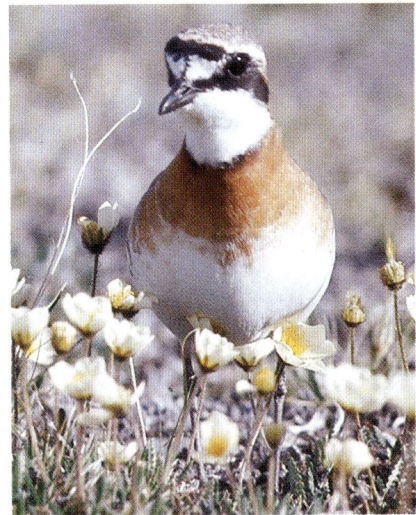

(B Chudleigh)

In Australia it is sometimes solitary but usually occurs in flocks, sometimes numbering several hundred. It associates freely with other shorebirds. It forages over bare sand or mud exposed at low tide, in the typical plover fashion of quick spurting runs interrupted by abrupt pauses to lunge at the surface.

The Mongolian Plover breeds in Asia in two widely separated clusters of populations: in Tien Shang, Tibet and Mongolia; and in north-eastern Siberia. Strongly migratory, it is widespread in winter around the shores of the Indian Ocean from South Africa to Australia, extending rarely eastwards to New Zealand. In Australia it is one of the most numerous and widespread of the migratory plovers, but it seems most abundant in Queensland and least so in the far south-west. It is most common from September to April, but substantial numbers remain through the winter.

Breeding occurs from April to August, usually in small scattered colonies. The male is said to carry out most of the incubation (which takes about 22 to 24 days) and care of the young (which fledge in about 30 to 35 days). The nest is a shallow scrape in the ground. Three eggs usually form the clutch.

HABITAT: breeds on steppes and deserts, mainly at high altitudes; at other seasons mainly coastal: beaches, mudflats, mangroves
LENGTH: 19–21 cm
DISTRIBUTION: 10,000–30,000 km²
ABUNDANCE: sparse
STATUS: secure

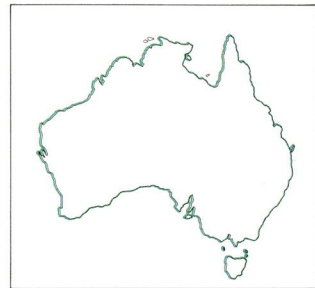

Hooded Plover

Charadrius rubricollis (rue'-bree-kol'-is: "red-necked valley-bird")

The Hooded Plover is easily identified as a small bird the colour of light sand, with a black head and a white nape, strongly associated with open surf beaches of southern Australia from about Jervis Bay in the east to Bunbury in the west. It seems unusually sensitive to human disturbance and is declining. The Hooded Plover is mainly sedentary, and usually occurs in pairs, but flocks sometimes congregate in winter.

It forages in the intertidal zone, feeding on small invertebrates such as insects, polychaete worms, isopods, amphipods, and small bivalve molluscs. It has the typical plover trick of running some distance, then stopping abruptly, bobbing the head.

Breeding occurs between September and January. The nest is a shallow scrape in the sand, often in the lee of a rock or piece of driftwood, and sometimes sparsely lined with grass or weed. Two or three eggs form the clutch, and both sexes incubate and care for the young.

HABITAT: ocean beaches and dunes or (in Western Australia) inland salt lakes
LENGTH: 19–22 cm
DISTRIBUTION: 100,000–300,000 km²
ABUNDANCE: sparse
STATUS: possibly endangered

(M Bonnin)

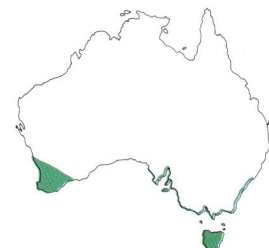

Red-capped Plover

Charadrius ruficapillus (rue'-fee-kah-pil'-us: "red-haired valley-bird")

A pale reddish crown renders this small plover unmistakable, but it otherwise closely resembles other plovers in size, appearance and behaviour. Where it occurs on the coast, it generally favours drier substrates (e.g., sand above high tide mark) than those habitually frequented by other waders. It very

(C. Seller)

seldom wades. It rarely congregates in large flocks but is nevertheless gregarious, and usually encountered in small parties. It feeds mainly on small insects.

The Red-capped Plover is common and widespread almost throughout Australia, both along the coast and in the interior; it is a vagrant to New Zealand and has been reported in southern New Guinea. It is mainly sedentary in coastal areas, nomadic in the interior.

The breeding season is very variable, but nesting occurs mainly between September and November, either in dispersed pairs or loose colonies. Several broods may be raised in succession. The nest varies from a shallow scrape in the ground, sometimes lined with shell fragments or small pebbles, to a substantial pad of grass and local vegetation. The usual clutch consists

of two eggs but varies from one to three. Hatching takes about 31 to 33 days. Both parents incubate and care for the young.

HABITAT: beaches, sand flats and mudflats, claypans; coast and interior
LENGTH: 14–16.5 cm
DISTRIBUTION: more than 1 million km²
ABUNDANCE: common
STATUS: secure

Oriental Plover

Charadrius veredus (ve'-red-us: "fleet valley-bird")

The Oriental Plover breeds in eastern Mongolia and Manchuria and migrates southward to winter in Indonesia, New Guinea and Australia, where it is widespread but most numerous in the far north-west: numbers decline sharply southward, and it is only a rare vagrant in New

South Wales, Victoria and Tasmania. It is present in Australia mainly between September and February.

Unlike other migratory plovers it seldom occurs on the coast, frequenting instead arid grasslands, where it may be solitary or in small flocks, but often occurs in flocks of

thousands. It often associates with pratincoles but seldom with other waders.

Almost nothing is known of its nesting habits.

HABITAT: breeds in deserts, steppes, saltflats and other sterile environments; at other seasons mainly arid or overgrazed grasslands; mainly interior, seldom coastal
LENGTH: 22–25.5 cm
DISTRIBUTION: more than 1 million km²
ABUNDANCE: sparse
STATUS: secure

(T Ishii)

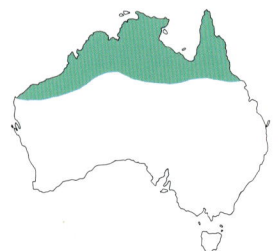

Genus *Erythrogonys*

(e-rith'-roh-gon'-is: "red-knee")

The characteristics of the genus are essentially those of the single species.

Red-kneed Dotterel

Erythrogonys cinctus (sink'-tus: "banded red-knee")

The Red-kneed Dotterel occurs in southern New Guinea and over much of Australia, but it seems most numerous in the interior south-east. Since about 1950 it has spread to south-eastern coastal regions and to Tasmania, where it was formerly unknown.

It is a bird of the margins of wet-lands of all kinds, wading in shallow water for its food, which consists of aquatic insects and their larvae, and seeds (especially those of legumes). It is nimble in the ground and in the air, and it occasionally swims; like other plovers, it frequently bobs its head. It resembles the Black-fronted Plover in much of its behaviour, but is nomadic rather than sedentary, gregarious rather than solitary. Its ecology is intimately linked with the erratic cycles of drought and transient floodwaters that are characteristic of the interior.

The onset of breeding is strongly influenced by local flooding. Pairs breed in loose colonies and both sexes share in incubation and care of the young, raising a single brood in a season. The nest is a mere shallow scrape in damp ground. Four eggs form the usual clutch.

HABITAT: temporary floodwaters, quiet open shallows of inland freshwater lagoons and grassy swamps, especially those with a substrate of mud or fine sand
LENGTH: *c* 18 cm
DISTRIBUTION: 300,000–1 million km²
ABUNDANCE: common
STATUS: secure

(D & M Trounson)

Genus *Peltohyas*

(pel'-toh-hie'-as: "shielded-plover")

The characteristics of the genus are those of the single species.

Inland Dotterel

Peltohyas australis (os-trah'-lis: "southern shielded-plover")

The Inland Dotterel is widespread across mainland Australia, but reaches its greatest abundance in the southern interior; it is seldom recorded on the east coast or in the tropical north. It is nomadic or partly migratory, and strongly gregarious except when breeding.

Largely independent of surface water, it appears to obtain most of its moisture needs from the leaves and shoots of succulent herbs; otherwise it mostly eats insects. It is mainly nocturnal.

Breeding occurs in solitary pairs or loose colonies between August and October. The nest is a shallow scrape in loose soil, sometimes rimmed with tiny pebbles or scraps of local vegetation. Three eggs form the clutch.

HABITAT: arid, stony and sterile plains, claypans, gibbers
LENGTH: 19–23 cm
DISTRIBUTION: more than 1 million km²
ABUNDANCE: sparse
STATUS: secure

(B & B Wells)

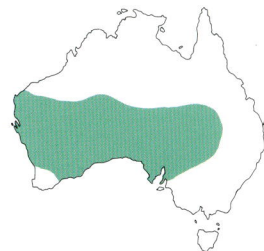

Genus *Pluvialis*

(plue'-vee-ah'-lis: "rain [-bird]")

Three or four species constitute this genus of medium-sized plovers, characterised mainly by their spangled upperparts and black underparts (in breeding plumage) and their enormous migrations: all breed in the high Arctic and winter almost throughout the world. In their annual migrations certain populations undertake some of the longest uninterrupted journeys attempted by birds—for example, from Alaska to Hawaii, or from Nova Scotia in Canada to Argentina via a great circle route over the open Atlantic.

Lesser Golden Plover

Pluvialis dominica (dom-in'-ik-ah: "San-Domingo rain-bird")

The Lesser Golden Plover breeds on the Arctic tundra of the Soviet Union, Alaska and Canada (eastward to Baffin Island); in autumn it migrates southward to winter in South America and across the islands of the Pacific Ocean to India, Indonesia and Australia. Vagrant birds sometimes reach Europe and Africa.

It differs from the Grey Plover most conspicuously in its spangled brownish ("golden") upperparts (the Grey Plover is spangled grey and white), but in flight it reveals more or less plain brown upperwings and rump. As in the Grey Plover, breeding birds are boldly black below, but in Australia this plumage is likely to be seen only in the final few weeks before the northward migration in February or March.

In Australia it is common and widespread, both inland and on the coast, though seldom in large numbers. It is most numerous from September to April, but substantial numbers remain through the winter. The most characteristic call is a clear, melodious "quee". It feeds mainly on terrestrial insects (especially grasshoppers), worms and small molluscs.

It breeds in dispersed pairs between June and September, laying a clutch of four eggs in a shallow scrape on open ground. Both sexes cooperate in incubation and raising the young; incubation takes 26 to 27 days, fledging about 22 days.

HABITAT: breeds on well-vegetated slopes on high Arctic tundra; on migration and wintering grounds, mainly grasslands bordering marshes, ploughed or stubble fields, sports grounds, airfields; also salt marshes, mangroves, estuarine mudflats
LENGTH: 23–26 cm
DISTRIBUTION: 300,000–1 million km²
ABUNDANCE: sparse to common
STATUS: secure

(*W Lankinen*)

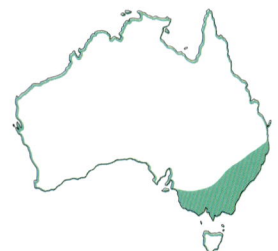

Grey Plover

Pluvialis squatarola (skwah'-tah-roh'-lah: "plover rain-bird")

The Grey Plover breeds on the Arctic tundra of northern Canada, Alaska and the Soviet Union, and migrates southward for the northern winter, in which season it is virtually cosmopolitan. In Australia it is widespread but almost exclusively coastal in distribution, reaching maximum numbers in the far north-west; it is rather uncommon in the south. Most individuals arrive in September and leave in February.

The Grey Plover is shy and wary, and on its wintering grounds usually rather silent and solitary. In breeding plumage (seldom seen in Australia), birds are boldly black below, but this feature is lost in winter. In flight, the Grey Plover is easily distinguished from the Lesser Golden Plover by its white rump, white wing bar, and black axillaries, but at rest the two species are very similar.

While breeding the Grey Plover eats mainly insects, but in its winter quarters it feeds largely on molluscs, crustaceans and polychaete worms.

It breeds in dispersed pairs between June and September, laying a clutch of four eggs in a shallow depression in dry open ground, often on a slight mound. Both parents share in incubation (which takes about 26 to 27 days) and the care of the young (which fledge at about 35 to 45 days).

HABITAT: breeds on lowland tundra in the high Arctic; otherwise almost restricted to coastal environments, especially extensive intertidal mudlfats, sandy beaches
LENGTH: 27–30 cm
DISTRIBUTION: 30,000–100,000 km²
ABUNDANCE: sparse
STATUS: secure

(RE Webster)

Genus *Vanellus*

(vah-nel'-us: "winnowing fan", referring to the action of the wings)

The lapwings are generally larger than other plovers, have relatively broad, blunt wings and bold conspicuous patterns of plumage, and inhabit mainly grasslands of various kinds. They are usually active, excitable and noisy. Most species have a sharp horny spur at the bend of the wing. Lapwings are found on all continents except North America; Australia has two species.

Masked Lapwing

Vanellus miles (mee'-layz: "soldier winnowing-fan")

The Masked Lapwing (or Spur-winged Plover) is widespread in Indonesia, southern New Guinea, and (self-introduced around 1940) New Zealand; in Australia it is a common and familiar species throughout most of the north and east, from the Kimberley to Tasmania, where it has expanded vigorously over the past few decades. It also appears to be expanding in central Australia. Vagrants have reached Lord Howe and Christmas Islands.

Though they hybridise freely, northern and eastern populations are so different that they were formerly considered to be separate species: northern birds are longer-legged, are much more extensively white about the head and neck, and have larger and longer yellow facial wattles than eastern birds.

The Masked Lapwing is mainly sedentary, but individuals and small parties may wander widely, especially in winter. While breeding it lives in pairs, which defend territories with conspicuous aggression, but at other times it is strongly gregarious. It often calls and travels at night. The diet consists of terrestrial invertebrates, especially those (such as earthworms and beetle larvae) found just below the surface. The common call is a distinctive grating staccato rattle.

The breeding season varies according to region and local climatic conditions, but is mainly from July to October in the south-east. Often several broods are raised in succession. The nest is merely a shallow depression, sparsely lined, on open ground well away from cover. Usually four eggs constitute the clutch; incubation takes about 27 days. Sexes share nest building, incubation and care of young.

HABITAT: short, open grassland, usually near water: river flats, margins of swamps and marshes, golf courses, playing fields, often urban nature strips
LENGTH: 35–39 cm
DISTRIBUTION: more than 1 million km²
ABUNDANCE: abundant
STATUS: secure

(*AF Evans*)

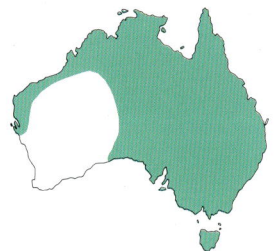

Banded Lapwing

Vanellus tricolor (trie'-kol-or: "three-coloured winnowing-fan")

The Banded Lapwing is widespread in Australia but it is much more common in the south than in the north, and all but absent from the Kimberley, the Top End, and northern Queensland. It is nomadic, and sometimes congregates in large flocks.

It closely resembles the Masked Lapwing in appearance and behaviour. However, it can be identified by the bold band of black across the breast, and it tends to replace the Masked Lapwing on more sterile grasslands of the arid interior. It is wary, noisy and excitable and, like the Masked Lapwing, frequently active at night. It often occurs far from water, but will travel many kilometres in search of shade. It feeds on insects and other terrestrial invertebrates, differing from the Masked Lapwing in its preference for small animals living on the surface rather than just beneath it. It also eats the seeds and green shoots of various grasses and herbs.

Most breeding occurs in spring, but it is very variable, and the Banded Lapwing differs little from the Masked Lapwing in details of its nesting cycle.

HABITAT: arid grasslands, stony plains, bare paddocks, airfields and other areas of short grass; usually far from cover
LENGTH: 25–29 cm
DISTRIBUTION: more than 1 million km²
ABUNDANCE: abundant
STATUS: secure

(H & J Beste)

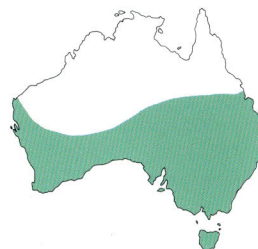

Family RECURVIROSTRIDAE

(ree-kur'-vee-ros'-trid-ee: "*Recurvirostra*-family", after the genus of avocets)

Avocets and stilts are medium-sized wading birds with conspicuous black and white plumage, very long legs (among birds only flamingos have longer legs in proportion to body length) and long slender bills that are upturned in the avocets, straight in stilts. Some species have patches of chestnut in the plumage. The sexes are similar, immatures resemble adults except in being duller, and there is no seasonal variation. All are strongly gregarious and usually breed in colonies.

There are three genera: *Recurvirostra* (avocets) comprises four species, one each in North America, South America, Eurasia, and Australasia. Notable for their upswept bills, these species inhabit brackish wetlands and feed largely on brine shrimps. The stilts (*Himantopus*) are common and conspicuous inhabitants of tropical and temperate wetlands around the world and are variously regarded as constituting a single cosmopolitan species or four or five geographically distinct species. *Cladorhynchus* is uniquely Australian and contains only one species; it shares so many anatomical, behavioural and ecological features with flamingos that some researchers have suggested a relationship between the two groups.

Genus Cladorhynchus

(clah'-doh-rink'-us: "curved [branch]-beak")

The characteristics of the genus are essentially those of the single included species.

Banded Stilt

Cladorhynchus leucocephalus (lue'-koh-sef'-ah-lus: "white-headed curved-beak")

(*G Rogerson*)

The Banded Stilt occurs only in Australia, where it is largely restricted as a breeding species to remote salt lakes in the arid south-western interior. After breeding it becomes strongly nomadic, dispersing erratically but with some frequency northwards to the Pilbara and eastwards to southern Victoria, occasionally Tasmania and New South Wales. Its breeding behaviour and subsequent movements are strongly influenced by local rainfall: nesting occurs in vast colonies only when and where unusual rainfall has flooded normally dry salt flats to produce huge standing crops of brine shrimps, on which the bird mainly feeds. So remote and inhospitable is its preferred breeding environment that only about 20 nesting colonies have ever been discovered (the first in 1930), and very little is known of the nesting cycle. Most breeding records have been from July to September.

Adults are almost entirely white in plumage, except for black wings (with a white trailing edge) and a broad chestnut band across the breast, extending and shading down to the blackish abdomen. Some individuals lack the chestnut band and these are generally considered immature, but they may represent an undescribed non-breeding plumage; the point has yet to be established.

Banded Stilts are strongly gregarious in all activities, and even solitary vagrants associate freely with other wading birds. They usually forage in densely packed mobs in quite deep water, frequently wading belly-deep and often swimming.

HABITAT: mainly arid interior salt lakes; disperses after breeding to coastal salt marshes, sewage farms, estuarine mudflats
LENGTH: 36–45 cm
DISTRIBUTION: more than 1 million km^2
ABUNDANCE: rare to sparse
STATUS: probably secure

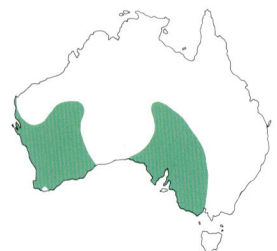

Genus *Himantopus*

(him-an'-toh-poos: "thong-foot")

The characteristics of the genus are discussed in the family account.

Black-winged Stilt

Himantopus leucocephalus (lue'-koh-sef'-ah-lus: "white-headed thong-foot")

The Black-winged (or Pied or White-headed Stilt) occurs from the Philippines south through eastern Indonesia, New Guinea and the Solomon Islands to Australia and New Zealand; it is merely a casual visitor to Tasmania. Many researchers combine it with populations in tropical America, Africa, Asia and elsewhere in a single cosmopolitan species, *H.*

himantopus. Abundant and conspicuous, it is essentially sedentary except when compelled to move in response to fluctuating water levels.

Black and white, with extraordinarily long pink legs, it can be confused only with the Banded Stilt, which differs in having an entirely white head, a white back, and a white trailing edge to the wing (the wings

of the Black-winged Stilt are entirely black). The sexes are similar, and immatures differ mainly in their smudgy grey heads. Conspicuous in behaviour as well as appearance, the Black-winged Stilt constantly flies about a human intruder onto its territory, uttering persistent yapping calls.

Though often alone or in pairs on small ponds, it is markedly gregarious, congregating in large loose flocks on favoured wetlands. Feeding birds wade in shallow water, gathering small aquatic insects, molluscs, crustaceans and diatoms. It seldom swims.

Breeding (in solitary pairs or loose colonies) is extremely erratic in timing, but usually occurs from August to November: several broods may be raised in a season. The nest may be a simple shallow scrape or a substantial structure of grass and other local vegetation, at the muddy margins or on small islands. Three to four eggs form the clutch, incubated for 22 to 25 days. The young leave the nest within hours of hatching, and fledge at about 30 days. Both parents incubate and care for the young.

HABITAT: shallow tropical and temperate fresh waters; brackish swamps, lakes, estuarine mudflats, sewage farms
LENGTH: 35.5–38 cm
DISTRIBUTION: more than 1 million km²
ABUNDANCE: abundant
STATUS: secure

(G Weber)

Genus *Recurvirostra*

(ree-ker'-vee-ros'-trah: "recurved-beak")

The characteristics of the genus are discussed under the family heading.

Red-necked Avocet

Recurvirostra novaehollandiae (noh'-vee-hol-an'-dee-ee: "New-Holland recurved-beak")

The Red-necked Avocet is restricted to Australia. Though transient individuals or small groups may occur anywhere, the species is almost unknown in Tasmania and normally very rare east of the Great Dividing Range, and breeding seldom occurs north of about 23°S Vagrants have reached New Zealand. It is strongly nomadic, its movements heavily influenced by local rainfall.

The reddish head and slender upturned bill render it unmistakable. The sexes are similar and immatures resemble adults except that the head is paler; there is no seasonal variation.

The Red-necked Avocet is markedly gregarious but flocks are seldom very large (typically 40 or 50) except just after the breeding season. It feeds mainly on aquatic insects and crustaceans, for which it wades in shallow water, rhythmically sweeping the bill from side to side just below the surface. It occasionally swims, or upends like a duck.

Breeding is in dispersed colonies and occurs mainly from August to November, but the timing is strongly influenced by local rainfall. The nest is a shallow scrape in damp ground, sparsely or substantially lined with samphire or similar vegetation. The normal clutch consists of four eggs.

(*G Chapman*)

HABITAT: temperate shallow wetlands (fresh, brackish or salt); floodwaters; estuarine mudflats
LENGTH: 43–45 cm
DISTRIBUTION: more than 1 million km²
ABUNDANCE: sparse
STATUS: probably secure

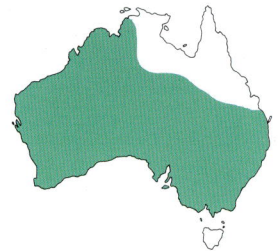

Family SCOLOPACIDAE

(skol'-oh-pah'-sid-ee: "*Scolopax*-family", after a genus of woodcocks)

About 85 species grouped in 27 genera constitute the family Scolopacidae, a group of birds mainly inhabiting the intertidal zone between land and sea (except when breeding) and notable for their spectacular migrations. Although there are a number of exceptions, a typical member of the family breeds on tundra at high Arctic latitudes and migrates to the Southern Hemisphere to spend the winter. Only the Charadriidae and some seabirds rival the distance and scope involved in these annual journeys, and some species are virtually worldwide in distribution.

Generally (if somewhat loosely) known as waders or shorebirds, members of the Scolopacidae are small to medium in size, with long, pointed wings and relatively short tails. The bill and legs may be short or long, and the hallux (hind toe) is short or even absent. The sexes are usually similar in appearance but many have breeding plumages conspicuously brighter and more intricate than the plumage generally worn on the wintering grounds, at which time their appearance differs only in detail (apart from size and structure), most species being pale grey or dull brown above and white below. Many are extremely difficult to identify in the field.

Most are strongly gregarious, and many species commonly occur in vast flocks involving tens or even hundreds of thousands of individuals. Typically, shorebirds congregate in massed roosts at high tide and disperse over nearby exposed mud and sand flats to feed at low tide.

Most breed in dispersed territorial pairs, but some breed in colonies. Four eggs usually constitute the clutch, deposited in a simple grass-lined depression in the ground. The young are hatched down-covered and relatively independent, and are led from the nest by the parents almost immediately on hatching.

Genus *Arenaria*

(ah'-ray-nah'-ree-ah: "sandy [-place-dweller]")

Apart from the virtually cosmopolitan Ruddy Turnstone this genus contains only one species, the Black Turnstone (*Arenaria melanocephala*) of north-western North America. The two species are very similar, and the characteristics of the genus are essentially those of the Ruddy Turnstone.

Ruddy Turnstone

Arenaria interpres (in'-ter-prays: "go-between sand-dweller")

(*AT Foster*)

The Ruddy Turnstone breeds around the high Arctic in Alaska, northern Canada, Greenland, Scandinavia and the Soviet Union. At other seasons it occurs on beaches and seacoasts virtually throughout the world as either a winter resident or passage migrant. In Australia it is common around the entire coastline, and there are a few scattered records from the interior, especially in the south-east; it is also common on Lord Howe, Norfolk and Christmas Islands, and has been reported at Macquarie Island. It is most numerous from September to April, but substantial numbers remain through the austral winter.

Its moult cycle is complex, and it appears in various plumages which differ much in detail. Nevertheless it is one of the easiest of waders to identify: consistent fieldmarks include a tubby build, short stout bill, short orange legs, and a black or smudgy grey breast cut off abruptly from white underparts. In flight it reveals a white rump and bold white wing bars. Birds in breeding plumage are marbled black, white and bright chestnut on the upperparts.

It is consistently gregarious but seldom occurs in large flocks, though it associates freely with other waders. It is generally alert, active, bustling and pugnacious, and it has a distinctive habit of using its bill to bulldoze through piles of drifted seaweed or to flip over pebbles and seashells. While breeding, it eats mainly insects, but its winter diet includes small arthropods, molluscs and crustaceans, and occasionally birds' eggs and carrion.

It breeds from May to August in dispersed pairs. The female builds the nest but both parents participate in incubation and rearing the young (the female often abandoning the young to the exclusive care of the male before they fledge). The nest is a shallow sheltered depression lined with a few leaves. The clutch usually consists of four eggs, which hatch in about 22 to 24 days; the young fledge at about 19 to 21 days.

HABITAT: breeds on Arctic tundra; otherwise seacoast generally, but mainly coral cays, reefs, sand or shingle beaches in preference to mudflats; occasionally saltfields and sewage farms
LENGTH: 22–24 cm
DISTRIBUTION: 300,000–1 million km²
ABUNDANCE: common
STATUS: secure

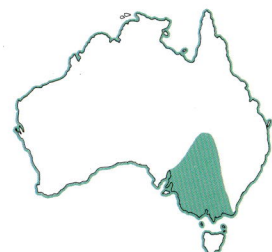

139

Genus *Calidris*

(kah-lid'-ris: Greek name for a shorebird, the identity of which is uncertain)

Sometimes known as stints, members of this genus are very small shorebirds with bills and legs of short to moderate length. Many are intricately patterned in shades of brown and rufous when breeding but in winter plumage are generally pale grey or brown above and white below. Most are extremely difficult to identify. All are strongly migratory and most are intensely gregarious, often occurring in huge, densely packed flocks notable for their impressive massed aerial manoeuvres. There are 19 species, most of which have been seen in Australia at least once; several are among the most numerous and widespread of migratory shorebirds in Australia.

Sharp-tailed Sandpiper

Calidris acuminata (ah-kue'-min-ah'-tah: "pointed [-tailed] shorebird")

Commuting annually from its breeding grounds in north-eastern Siberia, this small unobtrusive bird is the most abundant and widespread of the migratory waders in Australia, arriving in August and departing in March (very few remain through the austral winter). Its numbers generally increase to the south, and it seems that the greater concentration occurs in the interior south-east of the continent. Elsewhere it is widespread from Indonesia and New Guinea to the Solomon Islands, New Caledonia, and New Zealand.

It is strongly gregarious and associates freely with other waders, but tends to occur in small parties rather than large flocks. It often roosts or shelters in flooded tussock grassland or on samphire flats, scattering to forage in shallow water and feeding on polychaete worms, small molluscs and crustaceans. It is generally quiet, but birds taking flight frequently utter a soft, dry "trit-trit".

Breeding occurs from June to August, but little is known of the nesting cycle except that the female incubates and raises the young unaided by the male.

HABITAT: breeds on high Arctic tundra; habitat flexible at other seasons but especially the grassy margins of interior shallow fresh waters; also flooded fields, sewage farms, estuarine mudflats, mangroves, reefs, beaches
LENGTH: 18–26 cm
DISTRIBUTION: more than 1 million km²
ABUNDANCE: abundant
STATUS: secure

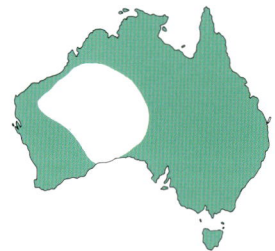

(*G Chapman*)

Sanderling

Calidris alba (al'-bah: "white shorebird")

The Sanderling breeds in the high Arctic in northern Canada, Greenland and the Soviet Union: it occurs virtually worldwide as a winter resident or passage migrant. In Australia it is almost exclusively coastal and is generally local and uncommon in occurrence. Peak numbers are reached between September and May, but many birds remain through the austral winter.

In breeding plumage its underparts are strongly tinged with rufous, but in winter it is mainly very pale grey, with white underparts, a conspicuous black mark at the shoulder, and bold white wing bars. It otherwise strongly resembles other small *Calidris* sandpipers in appearance.

Although it often associates with other waders, especially at high-tide roosts, it seldom occurs in large flocks and is usually encountered in parties of about 10 to 20. It prefers exposed surf-struck sand beaches, and its normal foraging technique is strongly characteristic, scuttling after receding waves, snatching hurriedly at prey on or near the surface, then running nimbly ("like little clock-work toys") back up the beach in advance of the next wave. It feeds mainly on small crustaceans.

It breeds in dispersed pairs from June to August. The nest is merely an exposed depression in the ground, and four eggs constitute the usual clutch. Incubation takes 24 to 27 days, and the young fledge at about 17 days. Both parents participate in incubation and rearing the young, but "double clutching", a procedure in which the female lays two clutches in quick succession, abandoning one to the exclusive care of the male and raising the other herself, has also been reported.

HABITAT: breeds on coastal tundra in the high Arctic; at other seasons strongly associated with open sand beaches with heavy surf; occasionally mudflats, reefs, salt marshes
LENGTH: 20–21 cm
DISTRIBUTION: 10,000–30,000 km^2
ABUNDANCE: common
STATUS: probably secure

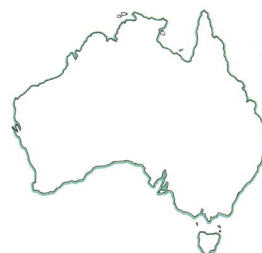

(*JP Myers*)

Baird's Sandpiper

Calidris bairdii (bair'-dee-ee: "Baird's shorebird", after S. F. Baird, Australian ornithologist)

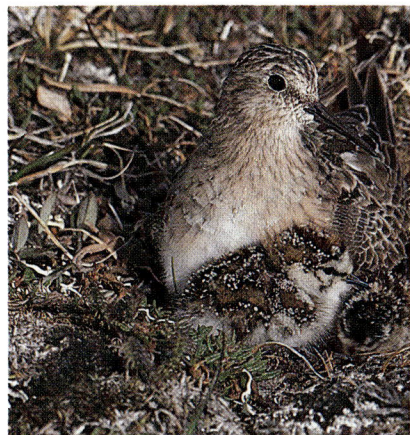

(*F Bruemmer*)

Baird's Sandpiper breeds in the high Arctic in far north-eastern Siberia, Alaska, northern Canada, and Greenland; it winters mainly in South America. It is generally considered to be a very rare vagrant in Australia, but it has been recorded in all States except Queensland and it may prove to be more regular in occurrence than is indicated by present data. Extremely difficult to distinguish from other small *Calidris* sandpipers, it is quiet, inconspicuous and rather secretive, and tends to favour the shelter of flooded vegetation in preference to open, exposed environments such as beaches and mudflats.

HABITAT: breeds on dry, arid upland tundra; at other seasons mainly grassy margins of lagoons and marshes, flooded fields
LENGTH: 14–18 cm
DISTRIBUTION: less than 10,000 km^2
ABUNDANCE: very rare
STATUS: probably secure

Red Knot

Calidris canutus (kah-nue'-tus: "[King-] Canute shorebird")

The Red Knot breeds around the shores of the Arctic Ocean in Alaska, northern Canada, Greenland and the Soviet Union: at other seasons it occurs almost throughout the world as either a winter resident or a passage migrant. In Australia it is almost exclusively coastal in distribution, but is common and widespread. Numbers decline southward and it is most numerous in the far north-west. Its abundance reaches a peak between September

(B Chudleigh)

and April, but substantial numbers remain through the austral winter.

Among the most strongly gregarious of small waders, it often occurs in very large, dense flocks, associating freely with other waders and frequently indulging in impressive massed aerial manoeuvres. Like other waders, it congregates to roost at high tide on reefs, shingle banks and similar sites, at low tide dispersing over adjacent mudflats to feed. In its foraging it is intense, concentrated and active, pecking rapidly and methodically at the surface with the head held low and the bill nearly vertical. On its breeding grounds it feeds on insects, berries and seeds, but in its winter quarters it eats mainly small molluscs and other marine invertebrates.

Breeding occurs in dispersed territorial pairs from June to August. The nest is a shallow depression in open ground, lined with grass. The usual clutch consists of four eggs,

which hatch in 21 to 22 days. Both parents participate in incubation and in rearing the young, but the female frequently abandons the chicks to the exclusive care of the male before they fledge at about 18 to 20 days.

HABITAT: breeds on tundra or barren stony slopes in the high Arctic; at other seasons mainly coastal, especially on extensive estuarine mudflats and beaches
LENGTH: 23–25 cm
DISTRIBUTION: 300,000–1 million km²
ABUNDANCE: common to abundant
STATUS: secure

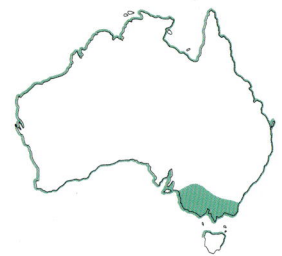

Curlew Sandpiper

Calidris ferruginea (fe'-rue-jin'-ay-ah: "rusty shorebird")

The Curlew Sandpiper breeds in the high Arctic in the Soviet Union and has an enormous wintering distribution extending across Africa and southern Asia to Australia and New Zealand. It is one of the commonest and most widespread migratory waders in Australia, where it seems most numerous in the south-east and north-west; it is common both on the coast and at wetlands in the interior. Peak numbers are reached between September and

(G Moon)

April but substantial numbers remain through the austral winter. Nationwide counts since the 1970s suggest numbers may be increasing.

A small pallid wader with a slender downcurved bill and a white rump, it differs from most *Calidris* sandpipers in being relatively easy to identify. In breeding plumage the white underparts are replaced with deep chestnut.

It usually occurs in large compact flocks, often associated with other waders, but it is also often encountered in small parties. Light, active and agile both on the ground and in flight, it congregates like other small waders at massed roosts at high tide, dispersing at low tide over adjacent exposed sand and mud flats to feed. It generally forages in shallow water, pecking and probing at the surface and feeding on small marine invertebrates, especially polychaete worms.

Breeding occurs from June to August. The pair bond is very brief,

and the female incubates and raises the young alone, hatching them from a clutch of four eggs laid in an exposed, grass-lined depression in the ground.

HABITAT: breeds on lowland tundra in the high Arctic; flexible at other seasons, favouring extensive estuarine mudflats but also frequenting margins of shallow wetlands, fresh or salt, coastal or interior; also beaches, reefs
LENGTH: 18–19 cm
DISTRIBUTION: more than 1 million km²
ABUNDANCE: abundant
STATUS: secure

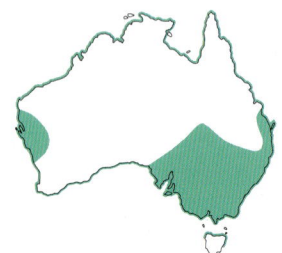

White-rumped Sandpiper

Calidris fuscicollis (fus'-kee-kol'-is: "brown-necked shorebird")

The White-rumped Sandpiper breeds in the Canadian high Arctic and winters mainly in South America. Like Baird's Sandpiper it is presently known only as a very rare vagrant to Australia, but it is recorded with apparently increasing frequency and may in fact be regular in occurrence. Unlike other *Calidris* sandpipers, it has a narrow patch of white across the rump, but it is otherwise difficult to identify.

It associates freely with other waders but is often solitary.

HABITAT: breeds on Arctic tundra; at other seasons favours muddy margins of shallow swamps, salt marshes, floodwaters, sand or mud flats
LENGTH: 15–17 cm
DISTRIBUTION: not applicable
ABUNDANCE: rare
STATUS: probably secure

(DM Cottridge)

Pectoral Sandpiper

Calidris melanotos (mel'-ah-noh'-tos: "black-backed shorebird")

The Pectoral Sandpiper breeds in the high Arctic in Alaska, northern Canada, and the Soviet Union: it winters mainly in South America. Formerly considered to be a very rare vagrant to Australia, its status remains uncertain because of the difficulty in distinguishing it from the more abundant Sharp-tailed Sandpiper, but it is recorded annually in small numbers, mostly in the southeast and generally between September and April.

It closely resembles the Sharp-tailed Sandpiper in behaviour but is somewhat less gregarious: it associates freely with Sharp-tailed Sandpipers but seldom with other waders, and is often solitary. It is quiet, inconspicuous, wary and rather shy. The characteristic call is a hoarse dry reedy "trrt", usually uttered by birds taking flight.

Breeding occurs from June to August. Males congregate at communal display grounds, visited by females who, after mating, retire to lay, incubate and rear the young unaided. The nest is a well-hidden and substantial structure of grass and local vegetation in a depression in the ground. Four eggs form the clutch, and incubation takes about 21 to 23 days.

HABITAT: breeds on Arctic tundra; at other seasons favours grassy margins of shallow fresh waters; also sewage farms, salt marshes, floodwaters; occasionally estuarine mudflats, beaches
LENGTH: 19–23 cm
DISTRIBUTION: 300,000–1 million km^2
ABUNDANCE: sparse
STATUS: secure

(B Chudleigh)

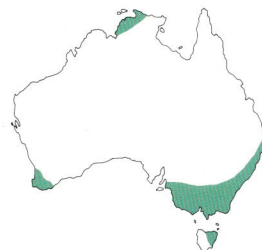

143

Little Stint

Calidris minuta (mee-nue'-tah: "little shorebird")

It is extraordinarily difficult to distinguish the Little Stint from the Red-necked Stint in the field. Unknown in Australia before 1977, the Little Stint has now been recorded a number of times, and it seems increasingly likely that its occurrence is regular but mostly undetected. It breeds in the far north of the Soviet Union and winters mainly from India westward to Africa and the Mediterranean region.

Intensely gregarious, it closely resembles the Red-necked Stint in behaviour, congregating in roosts at high tide and scattering to feed over adjacent mud or sand flats at low tide. It feeds actively, pecking rather than probing at the surface, and taking worms, tiny molluscs and crustaceans.

It breeds in dispersed pairs from June to August. Four eggs are deposited in a shallow grass-lined depression in the ground, and incubated for about 20 to 21 days. The female often lays two separate clutches in quick succession, abandoning one to the care of the male and rearing the other herself.

HABITAT: breeds on coastal lowland tundra in the high Arctic; at other seasons favours extensive estuarine mudflats or sand flats, reefs, salt marshes; occasionally inland fresh waters
LENGTH: 12–14 cm
DISTRIBUTION: not applicable
ABUNDANCE: very sparse
STATUS: probably secure

(B Chudleigh)

Cox's Sandpiper

Calidris paramelanotos (pah'-rah-mel'-ah-noh'-tos: "*melanotos*-like shorebird")

Almost nothing is known of this recently discovered and enigmatic species. It presumably breeds somewhere in Siberia, but its existence is so far known only from a few specimens and a number of sight records, all in Australia with the exception of one report from Hong Kong and another from Massachusetts, USA. In appearance it closely resembles the Pectoral Sandpiper but has a longer, blacker and slightly decurved bill. It has been tentatively described as a species but may prove on further investigation to be a hybrid.

HABITAT: breeding habitat unknown; most Australian records on estuarine mudflats and shores of shallow lakes
LENGTH: uncertain
DISTRIBUTION: insufficient data
ABUNDANCE: rare
STATUS: insufficient data

(DW Eades)

Red-necked Stint

Calidris ruficollis (rue'-fee-kol'-is "rufous-necked shorebird")

The Red-necked Stint breeds in northern Siberia and Alaska; it winters in southern China, South-East Asia, Indonesia, New Guinea, Australia, and New Zealand. One of the most abundant and widespread of the migratory waders in Australia, it is mainly coastal in occurrence but is also a common visitor to wetlands across the interior. Maximum concentrations occur in the far north-east, where flocks exceeding 100 000

(B Chudleigh)

individuals have been recorded. Peak numbers are reached between September and April but substantial numbers remain through the austral winter.

It is a small and tubby wader, with a comparatively short black bill. In breeding plumage it is strongly tinged with rufous around the head and neck, but in winter plumage it is plain and nondescript, mousy grey-brown above and white below, and extremely difficult to distinguish from other *Calidris* sandpipers.

Usually in large dense flocks and often in the company of other waders, it congregates at tightly packed high-tide roosts and often indulges in impressive massed aerial manoeuvres. Its flight is swift and agile, and it tends to flutter on landing. At low tide, flocks scatter to feed over adjacent exposed mud or sand flats, foraging in a characteristically avid, concentrated and bustling manner, pecking or

probing rapidly at the surface and frequently uttering soft twittering notes. It feeds on small worms, molluscs and crustaceans.

Little is known of its breeding behaviour.

HABITAT: breeds on high Arctic tundra; at other seasons mainly coastal, favouring extensive estuarine mudflats, beaches, salt marshes
LENGTH: 13–16 cm
DISTRIBUTION: more than 1 million km^2
ABUNDANCE: abundant
STATUS: secure

Long-toed Stint

Calidris subminuta (sub'-mee-nue'-tah: "rather-small shorebird")

This little-known wader breeds locally across central and eastern Siberia: it winters from southern China west to India and south to Australia, where it occurs mostly from August to March and mainly in the interior. It seems especially numerous in Western Australia, and has been recorded only once in Tasmania.

It is only mildly gregarious, and

often solitary, though it sometimes associates with other small waders. It favours the grassy margins of swamps, where it tends to skulk in the shelter of grass tussocks and other vegetation, relatively seldom venturing onto exposed muddy flats. It is very agile in flight and, when disturbed, often towers high, uttering a distinctive trilled "trring-trring". It feeds mainly on insects, small

molluscs and crustaceans.

Very little is known of the nesting cycle, except that it breeds in dispersed pairs in June and July. The nest is a well-hidden, grass-lined depression in the ground, and four eggs constitute the usual clutch.

HABITAT: breeds mainly in boggy clearings in taiga forest; at other seasons mainly muddy margins of shallow lagoons, marshes and floodwaters; also rice paddies
LENGTH: 13–14 cm
DISTRIBUTION: more than 1 million km^2
ABUNDANCE: sparse
STATUS: probably secure

(T Loseby)

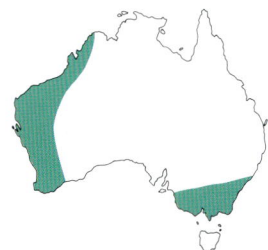

Great Knot

Calidris tenuirostris (ten'-ue-ee-ros'-tris: "narrow-billed shorebird")

The Great Knot breeds in far north-eastern Siberia and winters from southern China west to Pakistan and south to Australia, where it occurs mainly from October to March. It is numerous in the far north-west, where flocks exceeding 20 000 individuals have been recorded, but numbers decline sharply southwards, and it is rare and irregular in Tasmania. Substantial numbers remain through the austral winter.

It is unmistakable in breeding plumage, but in winter it is not easy to distinguish from the Red Knot except in being substantially larger: it also has a relatively longer, more slender bill, a distinct white rump, and inconspicuous pale wing bars. It is strongly gregarious and freely associates with other waders, especially favouring the company of Bar-tailed Godwits. It resembles other small waders in behaviour, congregating at massed roosts at high tide and scattering to forage over adjacent mudflats at low tide, feeding on small marine invertebrates.

Very little is known of its breeding behaviour.

HABITAT: breeds on barren alpine slopes; at other seasons mainly coastal, especially on estuarine mudflats and open beaches
LENGTH: 26–28 cm
DISTRIBUTION: 10,000–30,000 km^2
ABUNDANCE: very sparse to common
STATUS: probably secure

(*T & P Gardner*)

Genus *Gallinago*

(gah'-lin-ah'-goh: "gamebird")

Snipes of one species or another are found on all continents except Antarctica. They are stout-bodied shorebirds with very long, straight, slender bills (which, like those of other shorebirds, can be opened only at the tip). Their characteristic stance is a crouch. All are similar in plumage and many are extremely difficult to identify. Some are sedentary and others migratory. Unlike most shorebirds, they are rather solitary and intensely secretive, spending most of their lives in swamps, bogs or dense tussock grassland, emerging to feed on soft ground, probing the mud for earthworms and similar invertebrates. Breeding males conduct display flights over their territories, diving from considerable heights while laterally extending the modified outermost tail feathers, allowing the rush of air through them to produce a distinctive winnowing note. Three species are known to occur in Australia.

Japanese Snipe

Gallinago hardwickii (hard'-wik-ee-ee: "Hardwicke's gamebird", after C. Hardwicke, Australian naturalist)

The Japanese (or Latham's) Snipe breeds mainly in Japan (though some breed in southern Sakhalin and the Kurile Islands in the Soviet Union): it winters almost entirely in Australia. It is most numerous in the south-east, including Tasmania, and very rare in the west. Very few birds remain through the austral winter, and it is generally present only from about August to April or May.

It is crepuscular and mainly solitary but loose flocks sometimes congregate on migration or in especially favoured localities. It is very wary and secretive, and usually encountered only when inadvertently flushed from dense grass or swamp vegetation, whereupon it zigzags rapidly away, uttering an abrupt, grating "krek!" and often flying wildly and erratically about before settling back in dense cover a few hundred metres away. It forages in soft mud, probing deeply for worms and other small invertebrates.

It breeds in dispersed territorial pairs from May to July. Sheltered by a bush or grass tussock, the nest is a shallow depression in the ground, lined with grass and leaves. Four eggs form the normal clutch, incubated apparently only by the female.

HABITAT: breeds in **grassy clearings in forest**; at other seasons, **alpine bogs, rough damp tussock pasture, margins of swamps and marshes**
LENGTH: 24–26 cm
DISTRIBUTION: more than 1 million km²
ABUNDANCE: sparse to common
STATUS: probably secure

(J Purnell)

Swinhoe's Snipe

Gallinago megala (meg'-ah-lah: "large gamebird")

Swinhoe's Snipe breeds in central Siberia and Mongolia and winters from India across South-East Asia to western Micronesia, New Guinea and northern Australia, where it appears to be common in the Top End and the Kimberley from November to February. Its Australian status is, however, very uncertain because of the extreme difficulty in distinguishing it in the field from the Pintail Snipe and the Japanese Snipe.

It closely resembles other snipes in general behaviour, but little is known of its biology or breeding behaviour.

HABITAT: breeds in **forest glades, extensive meadows**; at other seasons, **muddy margins of wetlands, flooded rice paddies, sewage farms**
LENGTH: *c* 29 cm
DISTRIBUTION: 30,000–100,000 km²
ABUNDANCE: very sparse to common
STATUS: probably secure

(S Ishii)

Pintail Snipe

Gallinago stenura (sten-ue'-rah: "narrow-tailed gamebird")

(*F van Gessel*)

The Pintail Snipe breeds in north-eastern Siberia from the Urals to Manchuria and winters from India across south-eastern Asia to Indonesia and the Philippines. It is almost impossible to distinguish in the field from the Japanese Snipe or Swinhoe's Snipe, and in consequence its Australian status is very uncertain, but it is reported with some frequency from the Pilbara region of northern Western Australia, and it is probably more numerous and regular in occurrence in the far north-west than presently available data might indicate.

Like other snipes it is mainly crepuscular, solitary, secretive, and extremely difficult to observe, usually being encountered only when accidentally flushed from dense cover. It feeds mainly on earthworms.

It breeds in dispersed pairs from May to August. Four eggs are laid in a shallow grass-lined depression in the ground, and incubated for about 20 days.

HABITAT: breeds in bogs, rank meadows, marshes; flexible at other seasons; margins of wetlands, stubble and fallow pasture, paddyfields
LENGTH: 25–27 cm
DISTRIBUTION: 10,000–30,000 km²
ABUNDANCE: rare
STATUS: probably secure

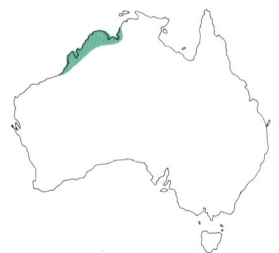

Genus *Limicola*

(lim'-ee-koh'-lah: "mud-dweller")

The characteristics of the genus are essentially those of the single species.

Broad-billed Sandpiper

Limicola falcinellus (fahl'-sin-el'-us: "little-scythe mud-dweller")

The Broad-billed Sandpiper breeds in northern Scandinavia and across the Soviet Union; it winters mainly in India and South-East Asia, extending in lesser numbers to Indonesia and Australia, where it is scarce and irregular, almost exclusively coastal in occurrence, and most numerous in the far north-west. It occurs mainly from October to March, records during the austral winter being exceptional.

It is gregarious, but usually occurs in small parties rather than large flocks, and is often encountered alone. Markedly less active than other small waders, it forages deliberately over soft ooze, probing deeply with the bill held vertically downwards. It feeds on worms, small molluscs and crustaceans.

Breeding occurs from June to August, in loose colonies. The male builds several nests, one of which is then selected by the female; both parents share in incubation and rearing the young. The nest is a shallow saucer of grass, hidden in a grass tussock or similar site. The normal clutch consists of four eggs, which hatch in about 21 days.

HABITAT: breeds in bogs in montane subarctic zone; at other seasons favours estuarine mudflats, salt marshes, reefs; occasionally sewage farms, shallow freshwater lagoons
LENGTH: 16–17 cm
DISTRIBUTION: 10,000–30,000 km²
ABUNDANCE: sparse to common
STATUS: probably secure

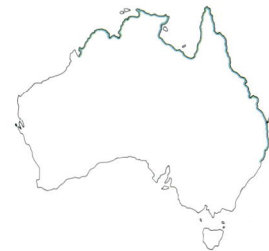

(T Loseby)

Genus *Limnodromus*

(lim'-noh-droh'-mus: "marsh-runner")

The characteristics of this genus are essentially those of the single species.

Asian Dowitcher

Limnodromus semipalmatus (sem'-ee-pahl-mah'-tus: "half-webbed marsh-runner")

This wader has been recorded with increasing frequency since the 1970s, particularly in the far north-west, where it may well prove to be regular and common.

It closely resembles the Bar-tailed Godwit, with which it often associates, but it is somewhat smaller and has a long, straight, entirely black bill, which is slightly swollen at the tip. It is strongly gregarious, and typically forages by thrusting its bill deep into soft mud, with a distinctive stiff-necked, wooden or puppet-like action. Its diet includes worms and insect larvae. Away from its breeding grounds it is generally rather silent and undemonstrative.

Very little is known of its breeding behaviour.

(R Drummond)

HABITAT: breeds at margins of floodwaters, wetlands; at other seasons mainly coastal, especially on estuarine mudflats and beaches
LENGTH: 28–35 cm
DISTRIBUTION: less than 10,000 km²
ABUNDANCE: rare
STATUS: probably secure

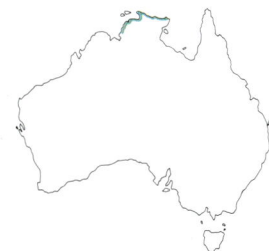

Genus *Limosa*
(lim-oh'-sah: "muddy [-bird]")

Godwits are large shorebirds which superficially resemble curlews except in the structure of their bills, which are long, slender, and very slightly upcurved. Intensely gregarious, they usually occur in large dense flocks and associate freely with other waders. On their wintering grounds they tend to favour the seacoast but are not unknown on inland swamps and other fresh waters. There are four species, three of which have occurred in Australia.

Bar-tailed Godwit
Limosa lapponica (lah-pon'-ik-ah: "Lapland muddy-bird")

(AF Evans)

The Bar-tailed Godwit breeds from Scandinavia across northern Asia to Alaska and winters across Africa and southern Asia to Australia and New Zealand. It is one of the most widespread of migratory waders in Australia, common around the entire coastline but most abundant in the far north-west; its occurrence is rare and erratic in the interior. Substantial numbers remain through the austral winter, even in southern Australia.

A large plain wader with a long, slightly upturned bill, it closely resembles the Black-tailed Godwit in appearance and behaviour, and the two often occur in mixed flocks. It is immediately identifiable in flight by its plain upperwings (the Black-tailed Godwit has a bold white wing bar) but at rest the two species are not easy to distinguish.

It is strongly gregarious, usually feeding, roosting and travelling in large dense flocks and associating freely with other waders, especially other godwits. Deliberate in most of its movements, it usually walks rather than runs, and forages by advancing steadily over exposed mudflats or wading in shallow water, using its bill in rapid vertical probes. In Australia its diet consists largely of worms, molluscs and small crustaceans.

It breeds in dispersed pairs from May to August. The nest is a shallow depression in the ground, lined with grass and leaves. The clutch consists usually of four eggs, which hatch in about 20 to 21 days. Both sexes incubate and care for the young.

HABITAT: breeds on moorland, peat bogs, heath and tundra; at other seasons mainly coastal, especially on estuarine mudflats; also beaches, mangroves, lagoons, sewage farms
LENGTH: 37–39 cm
DISTRIBUTION: 10,000–30,000 km^2
ABUNDANCE: sparse to abundant
STATUS: secure

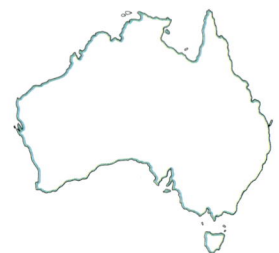

Black-tailed Godwit

Limosa limosa (lim-oh'-sah: "muddy muddy-bird")

The Black-tailed Godwit breeds across Eurasia from Iceland to Kamchatka and winters in the Mediterranean region and across Africa and southern Asia to New Guinea, Australia and (erratically) New Zealand. In Australia it is common, but numbers decline sharply southwards and its occurrence inland is erratic; it is most numerous in the Gulf of Carpentaria. Peak numbers are reached between September and March but substantial numbers overwinter, especially in the north.

Strongly gregarious, it usually occurs in large flocks and associates freely with other waders, but it is sometimes solitary. It is wary and excitable but seldom calls on its wintering grounds. It usually forages by wading slowly in shallow water, using its bill in rapid vertical exploratory probes, with a sudden deep thrust of the entire bill (often completely immersing the head) when prey is encountered. Its diet includes aquatic insects and their larvae (when breeding) and worms, molluscs and crustaceans (at other seasons).

Breeding occurs from April to August, in dispersed pairs or sometimes loose colonies. The nest is a shallow depression in the ground, lined with grass and leaves. Usually four eggs constitute the clutch. Incubation takes 22 to 24 days, and the young fledge at about 25 to 30 days. Both parents participate in nest construction, incubation and care of the young.

HABITAT: breeds on marshes, moorland, bogs and wet marginal farmland; at other seasons wetlands generally, coastal or inland; especially estuarine mudflats, beaches, reefs, sewage farms

LENGTH: 40–44 cm

DISTRIBUTION: more than 1 million km^2

ABUNDANCE: sparse to common

STATUS: secure

(E Soothill)

Genus Numenius

(nue-may'-nee-us: "curlew")

Typical shorebirds in most other respects, members of this genus differ in their relatively large size, their mostly brownish colouration, and their long, slender, and strongly decurved bills. They are also notable for their loud, ringing, haunting cries, uttered especially in display over the breeding grounds. There are about eight species, three of which occur commonly in Australia.

Eastern Curlew

Numenius madagascariensis (mad'-ah-gas'-kar-ee-en'-sis: "Madagascar curlew")

Almost nothing is known of the nesting behaviour of the Eastern Curlew: even its breeding distribution is uncertain, although it is known to lie within the Soviet Union and includes Kamchatka Peninsula and eastern Manchuria. During the austral summer it occurs in New Zealand and is recorded eastward casually as far as Fiji and Samoa, but the bulk of its total population of about 6000 individuals winters in Australia, where it is common on many feeding grounds around the entire coastline. Reaching its greatest abundance between September and May, it is most numerous in the far north-west, least so in the south and south-west, and it is very seldom recorded in the interior. Substantial numbers remain through the austral winter. Recent counts indicate that it is declining.

It usually occurs in flocks or small parties, but it is often solitary. Wary, aloof and deliberate in general comportment, it forages mainly by probing in soft mud, taking polychaete worms, small crustaceans and molluscs.

HABITAT: breeds in peat bogs and tussock grassland; almost exclusively coastal at other seasons, favouring extensive tidal flats
LENGTH: 53–61 cm
DISTRIBUTION: 10,000–30,000 km^2
ABUNDANCE: sparse to common
STATUS: secure

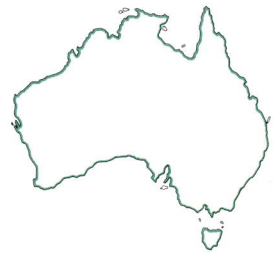

(*G Chapman*)

Little Curlew

Numenius minutus (mee-nue'-tus: "little curlew")

The Little Curlew breeds in central and north-eastern Siberia and migrates southward through the Philippines and Indonesia to winter mainly in Australia, and possibly southern New Guinea. It is a casual visitor to New Zealand. In Australia it occurs mainly from October to April, winter records being very few. It is only a passage migrant on the north coast, apparently wintering mainly on the grasslands of the northern interior. Its numbers decline very abruptly southward.

It is strongly gregarious, often occurring in very large flocks and freely associating with other waders. It feeds mainly on insects but its diet also includes berries.

Breeding occurs from May to August, generally in loose colonies. Both parents incubate the clutch of four eggs, laid in an exposed shallow depression lined with grass, until they hatch in about 22 to 23 days.

HABITAT: breeds in fire-damaged clearings in dry subarctic taiga forests; at other seasons mainly dry open grasslands including sports grounds, airfields
LENGTH: 28–30 cm
DISTRIBUTION: more than 1 million km^2
ABUNDANCE: sparse to common
STATUS: secure

(*M Carter*)

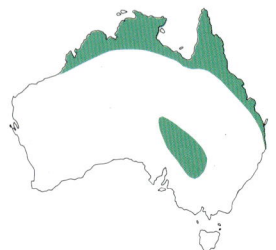

Whimbrel

Numenius phaeopus (fee'-oh-poos: "grey-footed curlew")

During its breeding season the Whimbrel is a common and familiar inhabitant of moors and peat bogs almost throughout Alaska, northern Canada, Scandinavia and the northern Soviet Union. At other seasons it occurs virtually worldwide as either a winter resident or a passage migrant. In Australia it occurs along the entire coastline but numbers decline southward and it is relatively rare in the far south. It is most numerous from September to May, but substantial numbers remain through the austral winter, especially in the north.

In appearance the Whimbrel closely resembles the Eastern Curlew but it is very much smaller, with a relatively shorter, straighter bill; it differs also in its conspicuous head pattern and dingy white rump.

Active and wary, it usually occurs in small parties or occasionally large flocks, but it is sometimes solitary. It often associates with other waders. Its varied diet includes (while breeding) berries, snails, and insects, and (on its wintering grounds) molluscs and small crustaceans.

Its vocabulary includes a high, ringing, rippling call.

Breeding occurs from May to August, in dispersed pairs, and involves both parents at all stages. The nest is a shallow scrape in the ground, lined with grass. The usual clutch consists of three or four eggs, which hatch in about 27 to 28 days. The young fledge at about 35 to 40 days.

HABITAT: breeds on tundra, moorland, peat bogs; at other seasons mainly coastal: estuarine mudflats, coral reefs, mangrove swamps, beaches
LENGTH: 40–42 cm
DISTRIBUTION: 10,000–30,000 km²
ABUNDANCE: sparse to common
STATUS: secure

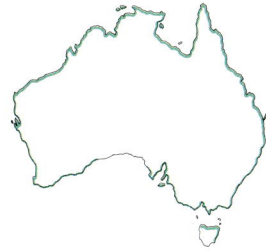

(B Chudleigh)

Genus *Phalaropus*

(fah'-lar-oh-poos': "coot-foot")

This is a genus of three small waders notable for their partial reversal of the role of the sexes in breeding. The female is larger and more brightly coloured than the male and takes the initiative in mating. After laying her eggs, the female then abandons them to the exclusive care of the male. Phalaropes are also remarkable in their pelagic habit when not breeding, spending the winter in flocks far out in the oceans of the Southern Hemisphere. Even on their breeding grounds they swim rather than wade when foraging, spinning around like tops on the surface of the water, stirring small invertebrates to the surface with the action of their unusually lobed toes. Brightly coloured in breeding plumage, all three species are similar in winter, being generally pale grey above and white below. All have been recorded in Australia, but only the Red-necked Phalarope with any frequency.

Red-necked Phalarope

Phalaropus lobatus (loh-bah'-tus: "lobed coot-foot")

The Red-necked Phalarope breeds in Alaska, northern Canada, Greenland, Europe and across the Soviet Union. It winters at sea in the Southern Hemisphere, especially important wintering areas being located off western South America, off Arabia, and in the seas north of New Guinea. It is merely a casual visitor to Australia but is recorded regularly and with apparently increasing frequency, especially in the south-east.

It is strongly gregarious except when breeding, and is sometimes encountered in huge rafts at sea. Like other phalaropes it forages mainly by spinning, top-like, on the water, rapidly pecking at the surface for small aquatic invertebrates stirred up by the action of the feet. Generally quiet and unobtrusive, it is often absurdly tame and can sometimes almost be caught by hand.

Breeding occurs from June to August, in dispersed pairs or small loose colonies. The nest site is selected by the female, but the male is left to incubate and rear the young alone. The nest is a shallow grass-lined depression in the ground, often in the shelter of a grass tussock.

Four eggs constitute the clutch. Incubation takes about 17 to 21 days, and the young fledge at about 20 days.

(*MW Grosnick*)

HABITAT: breeds on most freshwater wetlands, including ditches, small pools in peat bogs, shallow bays in lakes and rivers; pelagic at other seasons
LENGTH: 18–19 cm
DISTRIBUTION: not applicable
ABUNDANCE: rare
STATUS: probably secure

Genus *Philomachus*

(fil'-oh-mah'-kus: "battle-lover")

The characteristics of the genus are essentially those of the single species.

Ruff

Philomachus pugnax (pug'-nax: "pugnacious battle-lover")

The Ruff breeds across northern Eurasia and winters mainly in Africa and south-western Asia. Unknown in Australia before 1962, it is now recorded almost annually in small numbers, and has occurred in all States. Most reports span the period from September to March.

Adorned with flamboyant ruffs and crests, breeding males are unmistakable, and are among the most remarkable of birds in that no two individuals are precisely alike: the ruff may be almost any colour from white through yellow, red and brown to black, and in a range of patterns that may be barred, mottled, spotted or plain. Females are plain dull brown, differing little in summer or winter plumage, and are also substantially smaller.

The Ruff usually occurs alone or in small parties, seldom in large flocks, but it associates freely with other waders. Generally quiet and undemonstrative, it seldom calls; it forages slowly and methodically on mudflats or flooded grassy margins, probing and pecking and feeding mainly on insects and their larvae.

Breeding occurs from May to August. Males congregate at communal arenas, displaying with spectacular antics to itinerant females who retire after mating to incubate and rear the young unaided. The nest is a well-hidden grass-lined depression in the ground, and usually four eggs form the clutch. Incubation takes 20 to 23 days, and the young fledge at about 25 to 28 days.

HABITAT: breeds on lowland forest tundra wetlands, damp meadows, wet heaths and moors; at other seasons, favours muddy margins of shallow fresh wetlands, floodwaters
LENGTH: 26–32 cm
DISTRIBUTION: 10,000–30,000 km²
ABUNDANCE: rare
STATUS: secure

(A Lindau)

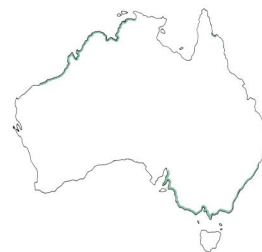

Genus *Tringa*

(tring'-ah: "sandpiper")

Shorebirds of the genus *Tringa* are generally slender-bodied and have long legs and long slender bills that are straight or very slightly upturned. Though gregarious, they are usually markedly less so than some other shorebirds, especially those of the genus *Calidris*, and are often encountered alone. Alert, wary and excitable, many have loud ringing calls. On the whole, they tend to favour fresh water over salt, and some are more common in the interior than on the coast. There are 15 species, of which at least nine are known to migrate to Australia.

Grey-tailed Tattler

Tringa brevipes (brev'-ee-pez: "short-footed sandpiper")

The Grey-tailed Tattler breeds in north-eastern Siberia from about Lake Baikal to Kamchatka: it has an extensive wintering distribution in South-East Asia and through Indonesia to New Guinea, Australia and New Zealand. It is almost exclusively coastal when not breeding, and is generally common and widespread along Australian shores, though rare and erratic in the far south. It is almost numerous between September and April, but substantial numbers remain through the austral winter.

Active, wary and nervous, it occurs alone or in small parties, and associates freely with other waders. It forages on mudflats, or fossicks busily in tide pools or piles of wave-washed seaweed, feeding on small crabs and other marine invertebrates. It often roosts in mangrove swamps but tends to avoid sandy beaches.

Almost nothing is known of its breeding behaviour.

HABITAT: breeds near mountain streams at high altitudes; at other seasons almost exclusively coastal, favouring mangroves, estuarine mudflats, coral cays, beaches, surf-washed rocks and reefs
LENGTH: *c* 25 cm
DISTRIBUTION: 10,000–30,000 km²
ABUNDANCE: common
STATUS: secure

(BJ Coates)

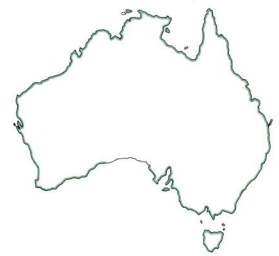

Wood Sandpiper

Tringa glareola (glah'-ray-oh'-lah: "gravelly sandpiper")

The Wood Sandpiper breeds across northern Eurasia from Scotland to Kamchatka, and on the Aleutian Islands, Alaska. It winters in Africa, India, South-East Asia and Australia, where it is common in the north but rather rare in the south. It favours interior wetlands rather than the coast, and is most numerous from August to April, seldom occurring during the austral winter.

It sometimes congregates in flocks and associates freely with other waders, but it is often solitary. Nervous, wary and excitable, it has a very distinctive alarm call, a shrill "chiff-if-if", uttered chiefly in flight. Its flight is strong and erratic, and it zigzags away with impressive acceleration on being disturbed. It forages on open mudflats or by wading in shallow water, probing or pecking at the surface and feeding mainly on aquatic insects and their larvae.

Breeding occurs in dispersed pairs from May to August. The nest is a shallow scrape in the ground, lined with grass and leaves. Both parents incubate the clutch of four eggs, which hatch in 22 to 23 days. The young fledge at about 30 days.

HABITAT: breeds on tundra, marshes, peat bogs, and clearings in taiga woodland; at other seasons mainly in freshwater wetlands, especially flooded margins with drowned trees, debris and other cover nearby; occasionally estuarine mudflats
LENGTH: 19–21 cm
DISTRIBUTION: more than 1 million km²
ABUNDANCE: very sparse
STATUS: secure

(DM Cottridge)

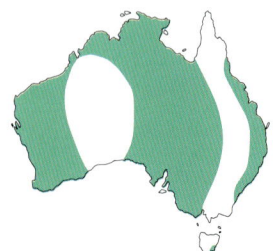

Common Sandpiper

Tringa hypoleucos (hie'-poh-lue'-kos: "white-below sandpiper")

The Common Sandpiper breeds across Europe and Asia from Spain and Britain to Japan, and winters across Africa and southern Asia to New Guinea and Australia, where it is rather uncommon but widespread. It is most numerous in the north, where it tends to be seen from September to April, though some individuals remain all year. It is a casual vagrant in New Zealand.

It is usually solitary, and tends to avoid the company of other waders. Quiet and unobtrusive when foraging, it constantly teeters and bobs its head as it fossicks busily about boulders and flood debris, and along narrow muddy margins. Its diet includes aquatic insects and their larvae, as well as small molluscs and crustaceans. Its flight is very distinctive: direct and very low over water, on shallow, clipped wing beats, interspersed with glides on stiffly downcurved wings.

It breeds in dispersed territorial pairs from May to August. Usually well hidden, the nest is a variable structure of grass and leaves placed in a shallow depression in the ground. The four eggs which constitute the clutch hatch in about 21 to 22 days, and the young fledge at about 26 to 28 days. The female builds the nest but both parents incubate and care for the young.

HABITAT: breeds near the margins of clear lakes, streams and rivers; at other seasons tends to avoid open, exposed environments and favours muddy or pebbly margins of waterways, coast or interior, salt marshes, mangroves; drains, sewage farms; rarely estuarine mudflats, coral reefs, beaches
LENGTH: 19–21 cm
DISTRIBUTION: more than 1 million km²
ABUNDANCE: sparse to common
STATUS: secure

(T Pescott)

Wandering Tattler

Tringa incana (in-kah'-nah: "hoary sandpiper")

The eastern Pacific representative of the Grey-tailed Tattler, the Wandering Tattler breeds in north-eastern Siberia, Alaska and far north-western Canada: it winters along the Pacific coast of North, Central and South America and on islands throughout the South Pacific Ocean. It is rare in Australia and virtually all records are from the coasts of Queensland and New South Wales, south to about Wollongong. It has also been seen on Norfolk and Lord Howe Islands. It has been recorded in all months but mainly between September and May.

Some uncertainty attaches to its Australian status because of the extreme difficulty in distinguishing it from the much more common grey-tailed Tattler. However, it has a different call (a ringing, excited trill of six to 10 notes) and tends to prefer surf-washed rocky reefs and platforms to the mudflats often frequented by Grey-tailed Tattlers. On its breeding grounds, it feeds mainly on aquatic insects and their larvae; otherwise it eats small crustaceans, molluscs and other invertebrates.

It breeds in dispersed pairs from May to August. The nest is a saucer of roots, twigs and other local vegetation, on the ground and often quite substantial. The clutch consists of four eggs, which hatch in 23 to 25 days. Both parents incubate and care for the young.

HABITAT: breeds at high altitudes on gravel flats of rivers; at other seasons exclusively coastal, favouring surf-washed rocks and reefs, coral cays
LENGTH: 26–28 cm
DISTRIBUTION: 10,000–30,000 km²
ABUNDANCE: very sparse to sparse
STATUS: probably secure

(H Shimura)

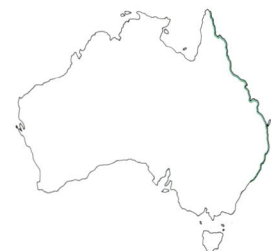

Greenshank

Tringa nebularia (neb'-ue-lah'-ree-ah: "clouded sandpiper")

(B Chudleigh)

The Greenshank breeds across northern Eurasia from Britain to Kamchatka and winters across Africa and southern Asia to New Guinea and New Zealand. In Australia it is one of the commonest and most widespread of migratory waders. Some remain all year, but it is most numerous from September to April.

Occurring usually in small parties and associating freely with other waders, it is gregarious in most activities but prefers to forage alone. It is alert, wary and excitable, and is often the first bird to give the alarm at the approach of an intruder, flying rapidly about the marsh uttering a loud, ringing "tyu-tyu-tyu". It forages over open mudflats or in shallow water, probing or pecking at the surface, often in quick dashing runs, and its varied diet includes insects, worms, molluscs, crustaceans, and small fishes.

It breeds in dispersed territorial pairs from May to August. Usually concealed in the shelter of a boulder or stump, the nest is a shallow scrape lined with grass and feathers. The clutch consists usually of four eggs, which hatch in 23 to 26 days. Cared for by both parents, the young fledge at 25 to 31 days.

HABITAT: breeds on tundra, moorland, peat bogs; at other seasons in most kinds of wetlands, coast and interior
LENGTH: 30–33 cm
DISTRIBUTION: more than 1 million km^2
ABUNDANCE: sparse to common
STATUS: secure

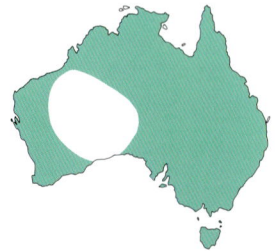

Marsh Sandpiper

Tringa stagnatilis (stag-nah'-til-is: "marshy sandpiper")

(B Chudleigh)

The Marsh Sandpiper breeds across central Eurasia from Bulgaria eastward to beyond Lake Baikal: it winters mainly in Africa and, in lesser numbers, across southern Asia to Indonesia and Australia, where it is widespread but generally uncommon, especially in the south. It occurs mainly from September to April, very few remaining through the austral winter.

Usually rather solitary, it sometimes occurs in flocks, and associates freely with other waders. Like the Greenshank (which it closely resembles in appearance except for its smaller size and needle-like bill) it is wary, active and excitable. It feeds mainly in shallow water, actively pursuing aquatic insects, molluscs and small crustaceans.

Breeding occurs from May to August, sometimes in loose colonies. The nest is a shallow scrape in the ground lined with grass, and four eggs constitute the clutch.

HABITAT: mainly fresh waters at all seasons; breeds on open river flats, water meadows, margins of swamps; at other seasons mainly in freshwater swamps, flooded fields, sewage farms, bore drains
LENGTH: 22–24 cm
DISTRIBUTION: more than 1 million km^2
ABUNDANCE: sparse to common
STATUS: secure

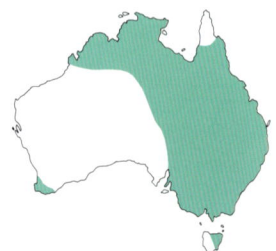

Terek Sandpiper

Tringa terek (te'-rek: "Terek [-River] sandpiper", in reference to the Terek River in the Georgian SSR)

The Terek Sandpiper breeds in Finland and across the northern Soviet Union almost to the Chukotskiy Peninsula: it winters across Africa and southern Asia to New Guinea, Australia and New Zealand. Though fairly common and widespread in Australia, it is almost entirely coastal; it is much scarcer in the west and south than in the north and east, and it is merely a vagrant in Tasmania. It also tends to show a marked preference for certain localities, and is not evenly distributed. Few birds remain through the austral winter, and it is most numerous from October to April.

One of the easiest of migratory waders to identify, it is a small tubby bird with a long, slightly upturned bill, short orange legs and (visible in flight) a conspicuous white trailing edge on the wings. It is markedly more gregarious than other *Tringa* sandpipers, though flocks tend to scatter when feeding. It forages on mud or sand, and pursues prey actively, feeding mainly on small crustaceans and molluscs.

It breeds in dispersed pairs from May to August. The nest is a shallow depression, sparsely lined with grass and leaf litter, on the ground in open or short vegetation. The clutch usually consists of four eggs, which hatch in about 23 to 24 days.

HABITAT: breeds on bogs, tussock grassland, margins of swamps; at other seasons mainly coastal— estuarine mudflats, coral reefs, mangrove swamps, beaches
LENGTH: 22–24 cm
DISTRIBUTION: 10,000–30,000 km^2
ABUNDANCE: sparse to common
STATUS: secure

(B Chudleigh)

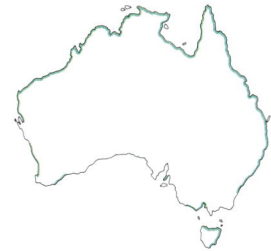

Redshank

Tringa totanus (toh-tah'-nus: "moorhen sandpiper")

(E & D Hosking)

One of the most widespread of migratory waders, the Redshank breeds across Eurasia from Iceland eastward to the Sea of Okhotsk and south to northern India and Tibet: it winters from the British Isles and the Mediterranean region to Indonesia and the Philippines. Its status in Australia is uncertain: although rarely reported, it is probably much more numerous and regular in occurrence, especially in the far north-west, than present data indicate.

Extremely noisy, wary and excitable, it somewhat resembles the Greenshank in general behaviour: it roosts, loafs and travels in flocks but prefers to forage alone. It feeds on aquatic insects, worms, molluscs and crustaceans.

HABITAT: breeds on tundra, moorland, and peat bogs; at other seasons mainly coastal: river flats, salt marshes, estuarine mudflats
LENGTH: 27–29 cm
DISTRIBUTION: 10,000–30,000 km^2
ABUNDANCE: rare
STATUS: probably secure

Genus Tryngites

(tring'-ee-tayz: "*Tringa*-like")

The characteristics of the genus are essentially those of the single species.

Buff-breasted Sandpiper

Tryngites subruficollis (sub-rue'-fee-kol'-is: "red-under-neck *Tringa*-like")

The Buff-breasted Sandpiper breeds in the high Arctic in Alaska and northern Canada and winters in central South America. In Australia it is a vagrant, but is recorded regularly in very small numbers, and with apparently increasing frequency.

It is strongly gregarious and associates freely with other waders, especially plovers. On its wintering grounds it occurs mainly on open prairies and grasslands, often far from water. It feeds mainly on insects and other arthropods.

Breeding occurs from June to August. Males congregate at communal display grounds, visited by itinerant females who, after mating, retire to raise their brood of young unaided. The nest is a shallow grass-lined depression in the ground, and usually four eggs constitute the clutch.

HABITAT: breeds on upland grass tundra in the high Arctic; at other seasons, short-grass prairie, dry or burnt-over grassland, ploughed or stubble fields, margins of lagoons and floodwaters
LENGTH: 18–20 cm
DISTRIBUTION: not applicable
ABUNDANCE: very rare
STATUS: probably secure

(M Wotton)

Family STERCORARIIDAE

(ster'-kor-ar-ee'-id-ee: "*Stercorarius*-family", after the genus of jaegers)

Skuas and jaegers resemble gulls in many respects, and are sometimes treated as a subfamily of the Laridae. They differ most strikingly in their mainly brown plumage and in their piratical habits: they steal much of their food from other seabirds, relentlessly pursuing them until they are forced to disgorge. About six species are grouped in two genera, both represented in Australia. One genus (*Stercorarius*) breeds in Arctic regions, the other (*Catharacta*) mainly in subantarctic regions; all skuas winter at sea.

In the past the terms "skua" and "jaeger" have been used almost interchangeably (though with a marked preference for the former in European usage, the latter in North America). The current trend is to restrict the term "skua" to species of *Catharacta*, "jaeger" to *Stercorarius*.

Genus Catharacta

(kath'-ar-ak'-tah: "bird of prey")

This genus consists of several extremely similar and closely related species whose interrelationships are a matter of debate. Currently four species are recognised; three breeding in subantarctic regions, the other at high latitudes in the Northern Hemisphere. Two species occur in Australian waters.

Skuas are large, powerful and aggressive seabirds. On their breeding grounds they prey on other seabirds, including the eggs and chicks of penguins. At sea in winter they feed on carrion and on fish stolen from other birds. The plumage is mainly brown, with conspicuous white flashes in the outer wing. The sexes are similar, young resemble adults, and there is little seasonal variation.

Southern Skua

Catharacta antarctica (ant-ark'-tik-ah: "southern bird of prey.")

The Southern (or Brown) Skua breeds in southern Argentina and on numerous subantarctic islands (including Heard and Macquarie Islands), dispersing northward in winter and, in Australia, commonly reaching the Houtman Abrolhos Archipelago in the west and Sandy Cape, Queensland, in the east. In general, it is most numerous in the south-west and in the Tasmanian region, numbers gradually declining eastward and northward.

It is virtually omnivorous in diet, on its breeding grounds preying heavily on nesting seabirds (especially prions and storm petrels) but largely shifting to carrion, fish and other marine life in winter.

Strongly territorial on its breeding grounds, it is mainly rather solitary at sea, although flocks may congregate at rich food sources. It often follows ships, harries other seabirds, and sometimes enters sheltered harbours and inlets.

Breeding occurs from September to January, in scattered single pairs or trios; the parents share nest building, incubation, and care of the young. The nest is a scrape in moss, lichen or gravel, sometimes sparsely lined with local vegetation. Two eggs form the clutch; incubation takes about 30 days and the young fledge at about 60 days.

HABITAT: grasslands and coasts of subantarctic islands when breeding; pelagic at other seasons
LENGTH: 61–66 cm
DISTRIBUTION: not applicable
ABUNDANCE: sparse
STATUS: secure

(B Hawkes)

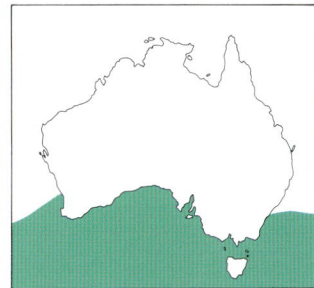

South Polar Skua

Catharacta maccormickii (mak-kor'-mik-ee-ee: "MacCormick's bird of prey")

This skua breeds on the shores of Antarctica. Adults are more or less sedentary but immature birds migrate to the northern Pacific, the western North Atlantic, and possibly the northern Indian Ocean. Its Australian status is uncertain owing to confusion with the Southern Skua, but it is reported with some frequency in the Tasmanian region and it has been recorded off Western Australia and New South Wales.

It is polymorphic, and some individuals are virtually impossible to distinguish from Southern Skuas in the field, but a pale grizzled area at the base of the relatively smaller and slenderer bill is often a valuable clue to identity.

Behaviour at sea and when nesting is similar to that of the Southern Skua. Breeding occurs from October to January. The clutch is two eggs, which hatch in 28 to 30 days; the young fledge at 49 to 59 days.

HABITAT: pelagic except when breeding; breeds on rocky or grassy, ice-free coasts
LENGTH: *c* 53 cm
DISTRIBUTION: not applicable
ABUNDANCE: very sparse
STATUS: probably secure

(PM & JL Sagar)

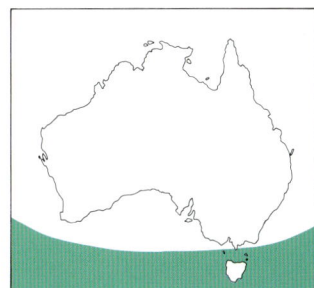

Genus *Stercorarius*

(ster'-kor-ar'-ee-us: "dung-associated [bird]")

The jaegers constitute a genus of three species which breed in the high Arctic and winter at sea, mainly in the Southern Hemisphere. All occur in Australian waters.

Females are slightly larger than males, but the sexes are otherwise similar. The plumage is basically black-capped, brown above, and near-white below, but polymorphism is pronounced in all three species. Light, dark and intermediate forms occur, and some individuals are extremely difficult to identify in the field. The two central tail feathers are extended into streamers in a manner unique to each species.

Long-tailed Jaeger

Stercorarius longicauda (lon'-jee-kaw'-dah: "long-tailed dung-bird")

The Long-tailed Jaeger breeds in the high Arctic. It is by far the most pelagic of the three jaegers in winter, spending this season far at sea in the southern Pacific and Atlantic Oceans. Its Australian status is uncertain, but fragmentary data suggests that the Tasman Sea is a significant wintering area; the Long-tailed Jaeger has been reported off south-eastern Australia with some frequency since about 1980.

It is smaller and more lightly built than the other two jaegers; the upperparts are smooth grey-brown, and the white in the wing is restricted to the outermost two or three primaries. The diagnostic long slender tail streamers extend up to 18 centimetres beyond the tip of the tail.

On its breeding grounds it feeds largely on lemmings, and may not breed at all in years when these rodents are rare. It resembles the other jaegers in behaviour but is somewhat less piratical at sea. It breeds from May to August. The nest is a shallow scrape in the ground. The two eggs take about 23 to 25 days to hatch and the young fledge at 24 to 26 days.

HABITAT: breeds on high Arctic tundra, often far from the sea; strongly pelagic at other seasons
LENGTH: 48–53 cm
DISTRIBUTION: not applicable
ABUNDANCE: rare
STATUS: secure

(W Lynch)

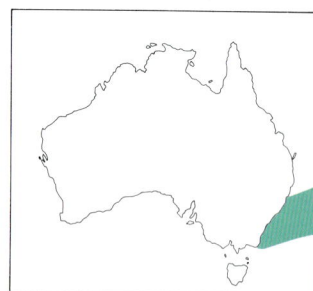

Arctic Jaeger

Stercorarius parasiticus (par-ah-sit'-ik-us: "parasitic dung-bird")

The breeding distribution of this jaeger is circumpolar on the fringes of the Arctic Ocean, south to Kamchatka, Scandinavia, northern Scotland and Iceland. It migrates southward after breeding, wintering at sea (though favouring inshore and coastal waters) mainly in the southern Atlantic and Pacific Oceans. It is common in Australian seas, especially in the south-east, and most numerous from October to April.

It is extremely variable in plumage, and some individuals are very difficult to identify. It is generally intermediate in size and build between the Pomarine and Long-tailed Jaegers, and its pointed tail streamers of medium length are diagnostic when present—though they are often broken off or moulting.

When nesting it feeds on small birds and mammals, insects, and berries, but at sea it feeds largely on fish stolen from other seabirds. It is not especially gregarious, but flocks sometimes gather at rich temporary food sources. Its flight is strong, dashing, and purposeful, with rather jerky wing-beats.

The breeding season is during June and July, and it occasionally nests in loose colonies. The nest is a shallow depression in the ground, sparsely lined with grass, moss or lichen. Most clutches consist of four eggs, which hatch in 25 to 28 days, the young fledging at 25 to 30 days.

HABITAT: breeds on coastal moorland and arctic tundra; pelagic at other seasons, mainly inshore waters
LENGTH: 41–46 cm
DISTRIBUTION: not applicable
ABUNDANCE: sparse
STATUS: secure

(*LE Löfgren*)

Pomarine Jaeger

Stercorarius pomarinus (pom'-ah-ree'-nus: "lidded-nose")

(*B Hawkes*)

and elongated central tail feathers when intact have a unique twist through 90° at the tip.

It is similar in habits to the Arctic Jaeger. It breeds in June and July. Two eggs constitute the usual clutch. Incubation takes about 26 days and the young fledge at 31 to 32 days.

HABITAT: breeds on low Arctic tundra, mainly near the coast; pelagic at other seasons, mainly coastal waters
LENGTH: 46–51 cm
DISTRIBUTION: not applicable
ABUNDANCE: sparse
STATUS: secure

The breeding distribution of the Pomarine Jaeger is circumpolar, mostly north of the Arctic Circle. It winters mostly south of the equator, especially off south-eastern Australia and western South America, and in the central Atlantic and the north-western Indian Ocean. It is common in Australian waters from October to April, especially in the region between Fraser Island and Melbourne, but numbers decline sharply westward, and it is uncommon in Western Australia.

Like the Arctic Jaeger, it is very variable in plumage and difficult to identify, but it is larger and much more powerfully built, and the blunt

Family LARIDAE

(lah'-rid-ee "*Larus*-family", after a genus of gulls)

This cosmopolitan family divides readily into three groups: gulls, terns and noddies (although some researchers remove the last two groups to the family Sternidae). As treated here, the family includes about 47 species of gulls, 42 terns, and five noddies. Members vary much in size but most are 30 to 50 centimetres long. They have relatively long and slender wings, and the typical plumage pattern is white with grey upperwings. Although there are numerous exceptions, the family as a whole is strongly associated with coastal environments. Many species breed in colonies, and both parents cooperate at all stages of the nesting cycle. The sexes are usually similar in appearance, but many species have a breeding plumage distinct from that worn at other times of year.

Although found on all continents, gulls are best represented in the Northern Hemisphere. Only three species breed in Australia, but four other species have occurred as vagrants. The group is remarkably homogeneous, most species falling into the single genus *Larus*. Most gulls have a relatively short, rounded tail and a distinctively shaped bill: moderately long and heavy, with a pronounced angle part-way along the lower ridge of the lower mandible. Gulls are common, conspicuous, noisy, aggressive and almost omnivorous. They are unusually adaptable, and many species have become successful commensals of humans.

Although superficially similar to gulls in many respects, terns are much less adaptable and differ in (usually) being substantially smaller, and slighter and more graceful in build. Most have black caps and deeply forked tails, and feed largely on fish obtained mainly by plunge-diving from dipping, hovering flight a few metres above the surface. Like gulls, terns form a remarkably homogeneous group, and most species fall into the genus *Sterna*. The genus *Chlidonias*, doubtfully distinct, differs in frequenting marshes and other fresh waters rather than the sea; in breeding plumage all species are black below, a characteristic otherwise unusual in terns.

Noddies are essentially tropical in distribution. The three species of the genus *Anous* are dark brown in plumage, with a white cap, but the Grey Noddy (genus *Procelsterna*) is almost entirely pale grey and the White Tern (genus *Gygis*) is almost entirely white. Noddies are gregarious and entirely marine, breeding mainly on isolated oceanic islands; several species differ from terns (among numerous other features) in building substantial nests, often in trees.

Genus Anous

(ah'-nows: "witless [-bird]")

The characteristics of the genus are described under the family heading.

Black Noddy

Anous minutus (min-ue'-tus: "small witless-bird")

The Black (or White-capped) Noddy is widespread but nowhere abundant in tropical and subtropical regions of the Atlantic and Pacific Oceans (it is replaced in the Indian Ocean by the very similar and closely related Lesser Noddy). In the Australian region it is a breeding resident at Norfolk Island and at many islands in the Coral Sea and along the Great Barrier Reef, south to the Capricorn and Bunker groups. It is frequently recorded at Lord Howe Island but it is not yet confirmed to have bred there. Vagrant individuals sometimes stray to inshore waters along the coast of Queensland and northern New South Wales, south to about Sydney. It differs little from other noddies in its habits.

One or sometimes two broods are raised per year, but there is no rigid breeding schedule. Nesting is usually in large dense colonies in groves of scrubby trees such as *Pisonia* or mangroves, rarely on the ground. The nest is a substantial structure of leaves, seaweed, and other local vegetation cemented with faeces. The clutch consists of one egg, which hatches in about 36 days. The chick fledges at about 52 days.

HABITAT: exclusively marine; breeds on oceanic islands, reefs, cays
LENGTH: 35–39 cm
DISTRIBUTION: 10,000–30,000 km²
ABUNDANCE: common
STATUS: secure

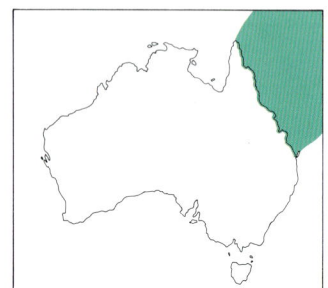

(K Atkinson)

Common Noddy

Anous stolidus (stol'-id-us: "stupid witless-bird")

(GA Hoye)

The Common (or Brown) Noddy has an extensive pantropical distribution in the Atlantic, Indian and Pacific Oceans. In the Australian region it breeds at Lord Howe and Norfolk Islands and at various islands off tropical northern Australia south to Lady Elliott in the east and Lacepede and (formerly) Bedout Islands in the west. It is most numerous at Lord Howe Island and on the Great Barrier Reef. It is generally sedentary and mainly pelagic, but individuals occasionally wander to inshore coastal waters south to about Sydney in the east and Esperance in the west.

It differs most notably from other noddies in its somewhat larger size and conspicuously two-toned brown upperwings. It is strongly gregarious and is usually encountered fishing over shallow inshore waters in large excited mobs or loafing, tight-packed, along the tideline on coral cays. It feeds largely on fish, captured mainly by dipping and snatching at the surface from low hovering flight. It seldom dives or submerges, though it frequently settles to rest on the water when not fishing.

The breeding schedule varies markedly from one colony to another, and sometimes follows a non-annual cycle; in Australian populations it is mainly from September to December. Nesting is conducted in large colonies, often mingled with Sooty Terns. The nest is a substantial stucture of leaves, seaweed, twigs, and debris, usually placed in a shrub but sometimes on the ground. The clutch consists of one egg, which hatches in 32 to 35 days.

HABITAT: exclusively marine; breeds on oceanic islands and coral cays
LENGTH: 40–45 cm
DISTRIBUTION: 10,000–30,000 km²
ABUNDANCE: common
STATUS: secure

Lesser Noddy

Anous tenuirostris (ten'-ue-ee-ros'-tris: "narrow-billed witless-bird")

(MJ Howard)

The Lesser Noddy breeds on scattered island groups along the eastern and western fringes of the tropical Indian Ocean. In Australian waters it nests only at the Houtman Abrolhos Islands, off Western Australia. It is mainly sedentary.

It closely resembles other noddies, differing from the Common Noddy mainly in being smaller and darker, with a slenderer bill and a slim, notched tail. Like other noddies it fishes mainly by day in dense excited mobs over shallow waters close to the colonies, straggling back at sunset to mass over the breeding islands. When not fishing it tends to loaf in flocks on sandy beaches.

In Australia it breeds in large dense colonies from August to January. Usually placed in the branches of a mangrove, the nest is a rough but substantial structure of seaweed and other local vegetation cemented together with faeces. One egg constitutes the clutch.

HABITAT: exclusively marine; breeds on oceanic islands, reefs, cays
LENGTH: 30–34 cm
DISTRIBUTION: less than 10,000 km²
ABUNDANCE: sparse
STATUS: probably secure

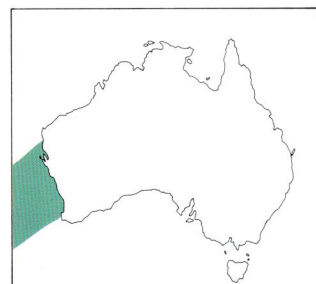

Genus *Chlidonias*

(klid-oh'-nee-as: "ornamented [-bird]")

The characteristics of the genus are discussed in the account of the family.

Whiskered Tern

Chlidonias hybridus (hie'-brid-us: "mongrel ornamented-bird")

The Whiskered Tern breeds in several widely scattered populations across southern Eurasia, Australia and south-eastern Africa and Madagascar. It occurs as a migrant, winter visitor or casual vagrant almost throughout the Old World. In Australia, breeding is largely restricted to the area south of 25°S. It is most numerous in the south-eastern interior, but widely distributed throughout the mainland when not breeding. It is a casual visitor to Tasmania. Substantial numbers of Australian birds migrate after breeding into Indonesia and south-eastern Asia, apparently mainly via the Northern Territory.

It is usually encountered in small groups, but occasionally congregates in flocks of thousands. It forages over fresh water, beating into the wind and periodically dipping to snatch prey from the surface; it seldom submerges. Its diet consists mainly of small fishes, amphibians, crustaceans, and aquatic insects and their larvae.

Breeding is usually in loose colonies, and occurs mainly from September to December, but is strongly influenced by flooding and other local conditions. The nest is a raft of vegetation in shallow water, usually moored to a tussock. Two or three eggs form the clutch. Incubation takes 18 to 20 days and the young fledge at about 23 days.

HABITAT: freshwater swamps, brackish and salt lakes, floodwaters, sewage farms, irrigated croplands
LENGTH: 25–26 cm
DISTRIBUTION: more than 1 million km²
ABUNDANCE: common
STATUS: secure

(*E & D Hosking*)

White-winged Tern

Chlidonias leucopterus (lue-kop'-te-rus: "white-winged ornamented-bird")

The White-winged Tern breeds in central and eastern Europe, central and western Russia, and eastern China, Mongolia and Manchuria, migrating southward after breeding to spend the winter in tropical Africa and from India across South-East Asia to New Guinea. In Australia it is a common non-breeding visitor in the tropical north; numbers generally decline sharply southward but it is subject to erratic influxes in the south (possibly related to tropical cyclones). It is usually most numerous in coastal regions.

In breeding plumage it is unmistakable; otherwise it closely resembles the Whiskered Tern in appearance and habits, though it is very much less likely to occur on salt or brackish waters.

Breeding is colonial, and occurs mainly from May to August. The nest is a rough mass of vegetation, floating or moored, and two to four eggs form the clutch. Incubation takes 18 to 22 days, and the young fledge at about 24 to 25 days.

HABITAT: freshwater swamps, floodwaters, sewage farms, irrigated croplands
LENGTH: 22–23 cm
DISTRIBUTION: 100,000–300,000 km²
ABUNDANCE: sparse to common
STATUS: secure

(*L. Jonsson*)

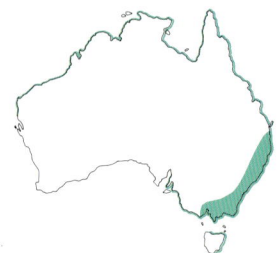

Genus *Gelochelidon*

(jel'-oh-kel'-id-on: possibly "laughing-swallow" but better treated as of unknown meaning)

The characteristics of the genus are those of the single species.

Gull-billed Tern

Gelochelidon nilotica (nee-lot'-ik-ah: "Nile *gelochelidon*")

The Gull-billed Tern occurs on all continents except Antarctica. In Australia it breeds mainly in the southern interior but disperses widely after nesting.

Large numbers winter in tropical Australia, and many apparently migrate to Indonesia and New Guinea.

(N Chaffer)

Often encountered alone or in pairs, it is not especially gregarious except when breeding. Nevertheless, it freely associates with other terns at high-tide roosts and in similar situations. Unlike other terns, it often forages over dry land (mudflats, grasslands and ploughed fields, for example) as well as over shallow water, and is attracted in large numbers to mouse or locust plagues. Its diet is extremely varied.

It breeds in colonies; in Australia usually from October to December. The nest is a shallow depression sparsely lined with grass or weeds. The normal clutch consists of two or three eggs; incubation takes 22 to 23 days, and the young fledge at 28 to 35 days.

HABITAT: freshwater swamps, brackish and salt lakes, coastal beaches and estuarine mudflats, floodwaters, sewage farms, irrigated croplands, flooded grasslands
LENGTH: 35–43 cm
DISTRIBUTION: more than 1 million km²
ABUNDANCE: common
STATUS: secure

Genus *Gygis*

(gie'-gis: "waterbird")

The characteristics of the genus are those of the single species.

White Tern

Gygis alba (al'-bah: "white waterbird")

The White Tern has a pantropical breeding distribution on oceanic islands in the Atlantic, Indian and Pacific Oceans; it disperses out to sea after breeding but its movements are largely unknown. In the Australian region it breeds only at

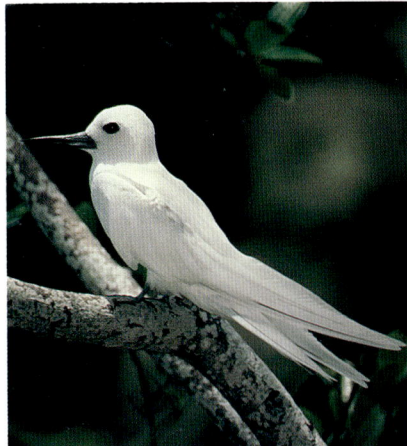
(CA Henley)

Norfolk and Lord Howe Islands but individuals wander with some frequency to inshore waters along the coast of Queensland and New South Wales, south to about Wollongong.

Only mildly gregarious, it seldom associates with other noddies or terns and usually forages alone far out at sea. It feeds mainly on small fishes. Its flight is distinctively airy and fluttering, with deep, hesitant wing beats. It is generally quiet and undemonstrative in behaviour; common calls include a subdued chatter and a soft, vibrant "tung".

In the Australian region, breeding occurs from October to March. The clutch consists of one egg, which is deposited directly onto the horizontal limb of a tree, being balanced precariously in minute crevices and

irregularities of the bark. Incubation takes about 28 days, and the chick fledges at about 60 to 75 days.

HABITAT: exclusively marine; breeds on oceanic islands
LENGTH: 28–33 cm
DISTRIBUTION: not applicable
ABUNDANCE: very sparse
STATUS: probably secure

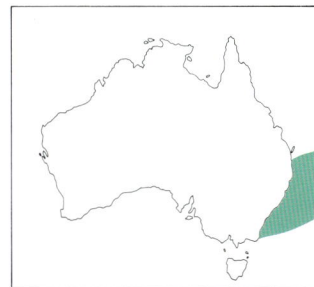

Genus Hydroprogne

(hie'-droh-prohg'-nay: "water-swallow", after mythical Greek Procne, who was transformed into a swallow)

The characteristics of the genus are those of the single species.

Caspian Tern

Hydroprogne caspia (kas'-pee-ah: "Caspian water-swallow")

Largest of all terns, this species occurs in North America, Eurasia, Africa and Australasia. In Australia it is generally uncommon but widespread. It is mainly sedentary.

It often joins flocks of other terns to roost or loaf, but it is among the least gregarious of terns. It usually forages alone, beating up and down over open water, often 15 metres or more above the surface, periodically hovering and plunging spectacularly. It feeds mainly on fish.

Breeding occurs in dispersed pairs or in dense colonies, and takes place mainly from October to December. The nest is a scrape in sand lined with a few wisps of grass. The usual clutch consists of two eggs, incubation takes 20 to 22 days, and the young fledge at about 30 to 35 days.

HABITAT: extensive wetlands, coast and interior; beaches, estuaries
LENGTH: 48–59 cm
DISTRIBUTION: more than 1 million km²
ABUNDANCE: common
STATUS: secure

(MJ Carter)

Genus *Larus*

(lar'-us: "gull")

The characteristics of the genus are discussed in the account of the family.

Kelp Gull

Larus dominicanus (dom-in'-ik-ah'-nus: "Dominican gull")

Often called the Dominican or Southern Black-backed Gull, the Kelp Gull occurs in subantarctic and cool temperate regions almost throughout the Southern Hemisphere, including many subantarctic islands (among them Macquarie and Heard Islands). In Australia it occurs mainly in the south-east but is commonly recorded west to Esperance in Western Australia, and north to Newcastle in New South Wales. The Kelp Gull was unknown in Australia before about 1940, and the existing local population was self-introduced from New Zealand; at present it breeds only in Tasmania and at a few sites in Victoria and New South Wales but it appears to be expanding vigor-ously, possibly to the detriment of the endemic Pacific Gull.

It is gregarious, noisy, aggressive, and conspicuous. It forages mainly along the shore but it freely follows ships and fishing vessels. It is virtually omnivorous.

It breeds from October to December, in dispersed pairs or small loose colonies, sometimes mingled with other gulls. The nest is a shallow depression copiously lined with seaweed and other local vegetation. Two or three eggs constitute the clutch, and incubation takes about 29 days. As in other gulls, both parents cooperate in nest construction, incubation and rearing the young.

HABITAT: seacoasts
LENGTH: 53–59 cm
DISTRIBUTION: 30,000–100,000 km²
ABUNDANCE: sparse
STATUS: secure

(D Watts)

Silver Gull

Larus novaehollandiae (noh'-vee-hol-an'-dee-ee: "New-Holland gull")

The Silver Gull occurs in Australia, New Zealand and New Caledonia. It is abundant and almost ubiquitous along the coast, especially in the south, but numbers decline northward and it is uncommon in the tropics. Inland it occurs at least

(B Chudleigh)

casually on all major river systems and the larger lakes. It is mainly sedentary, merely dispersing from its breeding islands to nearby coasts during winter, but some populations undertake longer movements. Its numbers are expanding rapidly near some major urban centres, especially in the south-east.

It is strongly gregarious, and is noisy, conspicuous and quarrelsome. Almost omnivorous, it is an abundant commensal of humans in many urban centres, scavenging at garbage dumps, beaches and city parks. Large flocks commonly attend fishing trawlers, but it seldom wanders far out of sight of land. At night the Silver Gull commutes to large communal roosts on sheltered waters.

It breeds mainly on offshore islands, usually from August to November, and in large, dense colonies. Sometimes two broods are

raised in succession. The nest is a shallow scrape lined with a few scraps of seaweed or other local vegetation. Two or three eggs form the clutch, incubated for about 24 days. Both parents are involved at all stages of the nesting cycle.

HABITAT: almost ubiquitous along coast; casual on extensive inland waters
LENGTH: 38–43 cm
DISTRIBUTION: more than 1 million km²
ABUNDANCE: abundant
STATUS: secure

Pacific Gull

Larus pacificus (pah-sif'-ik-us: "Pacific gull")

Found only in Australia, the Pacific Gull breeds in Tasmania and along the southern coast of Australia from Shark Bay in the west to Wilsons Promontory in the east. It is mainly sedentary but some individuals, especially immatures, occasionally wander farther north. There is some ground for concern that it may be adversely affected by competition from the recently expanding Kelp Gull.

Adults are relatively easily identi-

fied by their extremely deep bills and a narrow band of black towards the tip of the tail; immatures are difficult to distinguish from the young of the Kelp Gull.

The Pacific Gull is not especially gregarious but it often occurs in small flocks and associates freely with other gulls. It forages mainly along the tideline, feeding on fish, molluscs, crustaceans, eggs and fledgling seabirds, and carrion and human refuse.

Breeding occurs mainly from September to December, usually in dispersed pairs or small, loose colonies. The nest is a substantial bowl of seaweed and other local vegetation in a scrape in the ground. Two or three eggs form the clutch, and incubation takes about 26 to 28 days.

HABITAT: mainly coastal waters; seldom inland or far at sea; occasionally coastal swamps, sewage farms, rubbish tips, agricultural land
LENGTH: 58–66 cm
DISTRIBUTION: 30,000–100,000 km²
ABUNDANCE: common
STATUS: secure

(CA Henley)

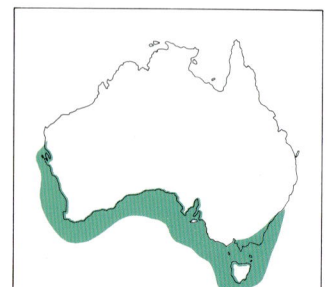

Genus *Procelsterna*

(proh'-sel-stern'-ah: "storm-tern")

The characteristics of the genus are those of the single species.

Grey Noddy

Procelsterna albivittata (al'-bee-vit-ah'-tah: "white-headbanded storm-tern")

The Grey Noddy is widespread in the tropical Pacific Ocean from Lord

(MJ Carter)

Howe and Norfolk Islands north to Hawaii and east to San Ambrose and San Felix Islands off Chile. It is mainly sedentary but individuals sometimes wander to inshore waters along the coast of Queensland and New South Wales, south to about Wollongong.

It differs little from other noddies in general behaviour except in being markedly less gregarious and demonstrative. It is often conspicuously tame and inquisitive. It feeds mainly on small fishes and crustaceans.

Breeding occurs from September to February in dispersed pairs or small loose colonies. No nest is built, the single egg being laid in a crevice on a cliff ledge or similar site. The incubation period is unknown, but the chick fledges at about 31 to 42 days.

HABITAT: exclusively marine; breeds on oceanic islands
LENGTH: 25–30 cm
DISTRIBUTION: not applicable
ABUNDANCE: rare
STATUS: probably secure

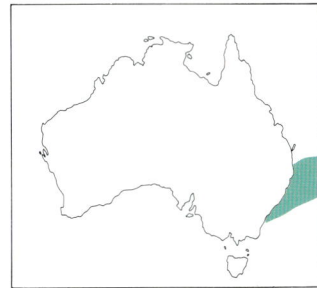

Genus *Sterna*

(ster'-nah: "tern")

The characteristics of the genus are described in the account of the family.

Little Tern

Sterna albifrons (al'-bee-fronz: "white-forehead tern")

The Little Tern has an extensive but scattered distribution in North America, Eurasia, western Africa and Australasia, although American populations may prove to belong to a different species. In Australia it is almost exclusively coastal in distribution, breeding from Tasmania to the Gulf of Carpentaria and dispersing in winter almost around the coast

(B Chudleigh)

except in the south-west. Native Australian populations are augmented in summer by migrants from southern China. It is extremely sensitive to human disturbance when breeding, and is rapidly declining in numbers and range.

It usually occurs in small flocks, associating freely with other terns at high-tide roosts. Active and excitable, it has a distinctive flight style: hovering and fluttering with quick, erratic, urgent strokes. It feeds mainly on small fishes.

The breeding schedule is variable, but is usually September to January in the south-east. It generally breeds in small colonies close to the tideline on open sand beaches, where it is extremely vulnerable to destruction by high tides, off-road vehicles and other natural and artificial hazards. The nest is a mere scrape in the sand,

sometimes rimmed with shell fragments. Two or three eggs form the clutch. Incubation takes 18 to 22 days, and the young fledge at about 19 to 20 days.

HABITAT: mainly coastal: open surf beaches, sheltered inlets, estuaries; occasionally lakes, sewage farms
LENGTH: 20–28 cm
DISTRIBUTION: more than 1 million km²
ABUNDANCE: very sparse
STATUS: endangered

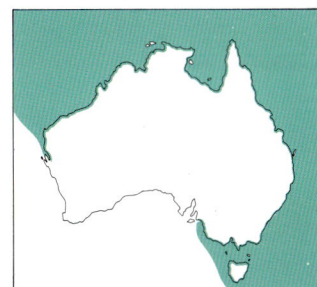

Bridled Tern

Sterna anaethetus (an-eeth'-et-us: "unburnt tern", referring to lighter pigmentation than that of the Sooty Tern)

The Bridled Tern has a scattered pantropical distribution in the Atlantic, Indian and Pacific Oceans. In Australia it breeds on a number of offshore islands along the north and west coasts, south to near Esperance in the west and the Bunker Group on the Great Barrier Reef in the east; since 1968 it has been recorded a number of times at Baudin Rocks, South Australia, where it has bred twice.

It closely resembles the Sooty Tern, but is slightly smaller and more slender, and the upperparts are dull brown rather than black; white on the forehead extends back as a prominent line over the eye. It also resembles the Sooty Tern in habits, but does not breed in huge colonies; it forages less at night, and settles freely on buoys, vessels, and floating driftwood. It feeds mainly on fish and small marine invertebrates.

The Bridled Tern nests in small, loose colonies, often mingled with other terns. In Australia breeding usually occurs from September to November, but the schedule varies markedly from colony to colony. Seldom far from the high tideline, and usually in the shelter of a boulder or driftwood, the nest is a mere scrape in the sand. The clutch usually consists of two eggs, which hatch in about 28 to 30 days. The young fledge at about 55 to 63 days.

HABITAT: exclusively marine; breeds on oceanic islands, offshore rocks, coral and sand cays
LENGTH: 30–32 cm
DISTRIBUTION: 30,000–100,000 km²
ABUNDANCE: common
STATUS: secure

(T & P Gardner)

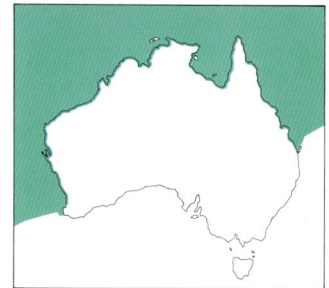

Lesser Crested Tern

Sterna bengalensis (ben'-gawl-en'-sis: "Bengal tern")

(T & P Gardner)

In Australia breeding occurs mainly from October to December. The clutch consists of one egg, which hatches in 21 to 26 days; the chick fledges at about 32 to 35 days.

HABITAT: exclusively coastal; beaches, sand cays, coral reefs, islands
LENGTH: 35–37 cm
DISTRIBUTION: 10,000–30,000 km²
ABUNDANCE: sparse to common
STATUS: probably secure

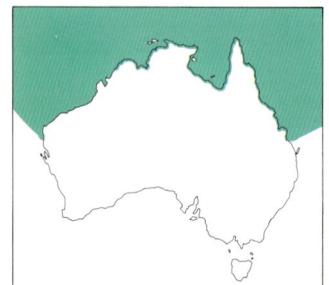

The Lesser Crested Tern closely resembles the Crested Tern but is much smaller and has a more extensive black cap and an orange (rather than yellow) bill. In Australia it is most numerous on the Great Barrier Reef, but it also breeds at Adele and Bedout Islands in Western Australia. Elsewhere it occurs in the Mediterranean Sea and from East Africa and Madagascar eastward along northern fringes of the Indian Ocean to Indonesia, New Guinea and the Solomon Islands.

Crested Tern

Sterna bergii (berg'-ee-ee: "Bergius's tern", after C. H. Bergius, Swedish biologist)

This is the commonest and most widespread of Australian terns, occurring along virtually the entire coastline. Elsewhere it has an extensive distribution around the fringes of the Indian Ocean and in the Pacific Ocean eastward through Micronesia and Polynesia.

It often forages alone, but freely congregates with other terns and gulls at high-tide roosts. The most common call is a loud, rasping "krrow", but it is generally a rather silent bird. It feeds mainly on fish.

In Australia breeding occurs mainly from October to January, and sometimes two broods are raised in succession. It usually nests in large colonies, often mingled with other terns. The nest is a mere scrape in sand, and the usual clutch consists of one egg (occasionally two), which hatches in 25 to 30 days. The chick fledges at 38 to 40 days.

HABITAT: almost exclusively coastal, seldom on inland waters; beaches, islands, lakes, inlets, harbours
LENGTH: 46–49 cm
DISTRIBUTION: 100,000–300,000 km²
ABUNDANCE: common
STATUS: secure

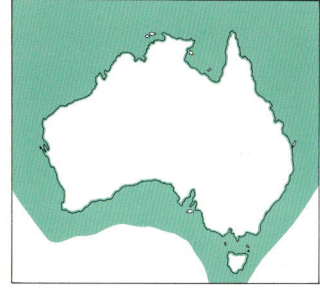

(KA Hindwood)

173

Roseate Tern

Sterna dougallii (due'-gal-ee-ee: "Dougall's tern", after a Scotsman, Dr MacDougall)

(G Chapman)

The Roseate Tern occurs along tropical and warm temperate coasts around the world. In Australia it is sparsely distributed in the tropics, south to the vicinity of Perth in the west and the southern end of the Great Barrier Reef in the east. It occurs as a casual vagrant further south, and it appears to be expanding steadily southward in the west. It seems to be mainly sedentary.

Except in its more tropical distribution, it closely resembles other members of the genus in habits and appearance. The Roseate Tern is, however, somewhat paler, slimmer and more graceful in build and has much longer tail streamers; when foraging it tends to hover less than other terns. It feeds mainly on small marine fishes, captured far out at sea.

The breeding schedule varies markedly with locality, but is mainly September to December on the Great Barrier Reef. Nesting is colonial, sometimes mingled with other terns, on isolated sand, coral or shingle cays or reefs. At some localities two broods are sometimes raised in succession. Usually two eggs form the clutch, incubation takes about 23 days, and the young fledge at about 27 to 30 days.

HABITAT: almost exclusively marine; ocean sand beaches and cays, coral reefs, offshore waters
LENGTH: 33–38 cm
DISTRIBUTION: 30,000–100,000 km^2
ABUNDANCE: sparse to common
STATUS: probably secure

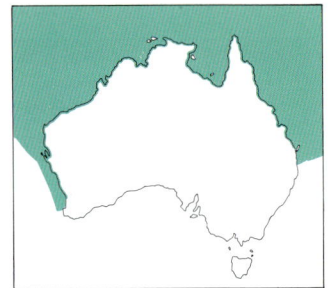

Sooty Tern

Sterna fuscata (fus-kah'-tah: "dusky tern")

(CN Challies)

The Sooty Tern is widespread in tropical and subtropical regions of the Atlantic, Indian and Pacific Oceans. In the Australian region it breeds at Lord Howe and Norfolk Islands; on islands from the Houtman Albrolhos to Ashmore Reef off the coast of Western Australia; in Torres Strait and the Coral Sea; and along the Great Barrier Reef south to Michaelmas Cay near Cairns. There are no known breeding colonies in the Northern Territory. It disperses far to sea after nesting and is absent from Norfolk Island from April to September, and from Lord Howe Island from April to August. Vagrants occur casually at any season along the Australian coastline south at least to Perth in the west and Sydney in the east.

The Sooty Tern often joins mobs of noddies over schools of fish close inshore near the breeding grounds, but usually forages alone, far at sea, often at night. Almost exclusively aerial when not breeding, it seldom settles on land or water. It feeds mainly on squid, fish and crustaceans.

The breeding schedule varies markedly with locality, not necessarily following an annual cycle, and nesting usually occurs in vast, dense colonies. The nest is a mere scrape in the ground. One egg constitutes the clutch, incubation takes about 26 days, and the chick fledges at about 30 days.

HABITAT: exclusively pelagic; breeds on oceanic islands
LENGTH: 33–36 cm
DISTRIBUTION: less than 10,000 km^2
ABUNDANCE: sparse
STATUS: secure

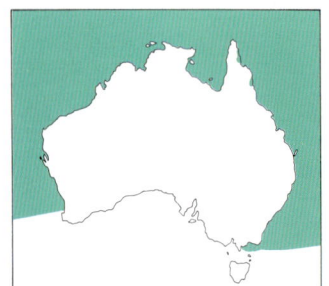

Common Tern

Sterna hirundo (hi-roon'-doh: "swallow tern")

The Common Tern breeds across much of northern North America and Eurasia and has an almost worldwide distribution at other seasons. In Australia it is widespread but almost exclusively coastal. It is much more numerous in the north and east than elsewhere, and is very rare in Tasmania and the south-west. It is normally present only from September to May, although individuals occasionally remain through the winter.

It is gregarious in all activities and at all seasons, and associates freely with other terns. It eats mainly small marine fish, obtained by plunge-diving.

Breeding occurs from May to August, usually in loose colonies. One to three eggs form the clutch, incubation takes 21 to 22 days, and the young fledge at about 25 to 26 days.

HABITAT: versatile when breeding, mainly coastal at other times: offshore waters, ocean beaches, estuaries; large lakes; occasionally freshwater swamps, brackish and salt lakes, floodwaters, sewage farms
LENGTH: 32–38 cm
DISTRIBUTION: 10,000–30,000 km²
ABUNDANCE: sparse
STATUS: probably secure

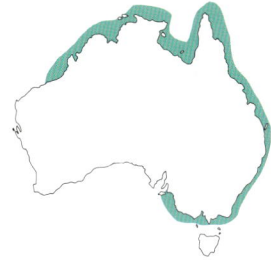

(B Hawkes)

Fairy Tern

Sterna nereis (ne-ray'-is: "sea-fairy tern")

The Fairy Tern occurs in New Zealand, New Caledonia, and Australia, where it is mainly confined to the south and west coasts, breeding from eastern Victoria and Tasmania west and north to the Lacepede Islands, Western Australia. It is mainly sedentary.

It is very similar to the Little Tern in appearance and habits, but during the breeding season the bill is a richer, deeper shade of yellow and the lores are white.

Breeding, which is conducted in colonies, occurs from September to January. The usual clutch consists of two eggs, which hatch in about 18 days. Its preference for offshore islands for nesting renders it somewhat less vulnerable to human disturbance than the Little Tern, and in consequence it seems less gravely threatened than that species.

HABITAT: almost exclusively coastal: beaches, islands, sheltered inlets; occasionally lakes, sewage farms
LENGTH: 22–27 cm
DISTRIBUTION: 30,000–100,000 km²
ABUNDANCE: common
STATUS: secure

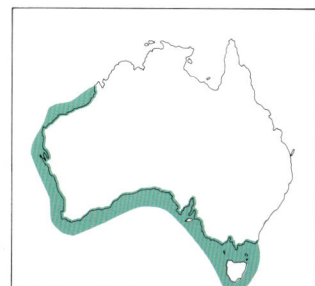

(G Moon)

175

Arctic Tern

Sterna paradisaea (pah'-rah-dee'-see-ah: "paradise tern")

This tern breeds in Arctic and subarctic regions and spends the winter months at sea, its distribution extending across all oceans south to the Antarctic pack-ice. It occurs as a migrant or casual visitor almost throughout the world and perhaps has the most extensive distribution of any bird. However, its chief migratory movements occur down the west coasts of South America and Africa and it is uncommon in the Australasian region. Its status is uncertain because of the extreme difficulty in distinguishing it from the Common Tern, but it is an apparently regular migrant in small numbers. Most records are from the south-west and south-east, from September to May. It usually occurs in small parties, and feeds mainly on small marine fishes.

Breeding occurs in loose colonies and extends from June to August. One to three eggs constitute the clutch, incubation takes 20 to 24 days, and the young fledge at 21 to 24 days.

HABITAT: breeds mainly near the coast on Arctic tundra, beaches, shingle banks, and grassy islands; pelagic at other seasons
LENGTH: 33–35 cm
DISTRIBUTION: not applicable
ABUNDANCE: very sparse
STATUS: probably secure

(E Hosking)

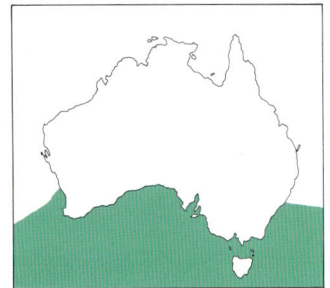

White-fronted Tern

Sterna striata (stree-ah'-tah: "striped tern")

The White-fronted Tern breeds in New Zealand, migrating across the Tasman Sea after breeding to winter in south-eastern Australia (mainly between Brisbane and Adelaide, and only casually further north and west), where it is most numerous from May to November. In Australia immature individuals greatly outnumber adults, and it appears that older birds are mainly sedentary. Since 1979 a few pairs have bred on islands in Bass Strait.

It closely resembles the other members of the genus in appearance and behaviour and, like the Roseate Tern, is almost exclusively marine. Only moderately gregarious, it frequently forages alone but mingles freely with other terns and gulls at high-tide roosts. It feeds mainly on small fishes, captured by plunge-diving from hovering flight a few metres above the surface.

Breeding occurs in colonies, usually on secluded sand or shingle beaches, and extends from October to February. One or two eggs constitute the clutch (but seldom is more than one young raised). Incubation takes about 23 days, and the young fledge at about 29 to 35 days.

HABITAT: exclusively marine beaches, reefs, offshore waters
LENGTH: 35–43 cm
DISTRIBUTION: 10,000–30,000 km²
ABUNDANCE: sparse
STATUS: probably secure

(B Chudleigh)

Black-naped Tern

Sterna sumatrana (soom'-ah-trah'-nah: "Sumatran tern")

The Black-naped Tern is widespread in the tropical Indian and western Pacific Oceans, north to Japan and east to Samoa. In Australia it occurs in the north and north-east from the vicinity of Darwin to Fraser Island and is most numerous on the Great Barrier Reef. It is sedentary.

It usually occurs in small groups rather than large flocks but it freely associates with other terns at high-tide roosts. It feeds mainly on small fishes, captured by plunge-diving in shallow waters of lagoons or just beyond the surf line along outer reefs.

Breeding occurs in small colonies, usually situated close to the high tideline on sand beaches on cays and small offshore islands; the season extends from October to December. The nest is a shallow scrape in sand, often lined with fragments of shell and coral; two eggs form the usual clutch.

HABITAT: exclusively tropical and marine: sand cays, coral reefs and lagoons
LENGTH: 30–32 cm
DISTRIBUTION: 10,000–30,000 km²
ABUNDANCE: common
STATUS: probably secure

(CA Webster)

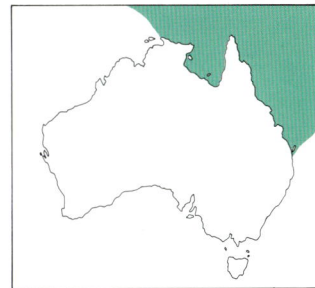

Order COLUMBIFORMES

(kol-um'-bee-for'-mayz; "*Columba*-order", after a genus of doves)

This order contains three families: Columbidae (pigeons and doves); Raphidae (Dodo); and Pezophapidae (Solitaire). Isolated on remote islands in the Indian Ocean and known only from a few specimens, subfossil fragments, and early travellers' tales, both the Dodo and Solitaire have been extinct for more than two centuries, and the characteristics of the order are now essentially those of the single surviving family, the Columbidae.

Until recently the order was traditionally regarded as also encompassing the Pteroclididae sand-grouse. Roughly pigeon-like in size and general appearance, these birds inhabit the steppes of central Asia and the deserts of northern Africa. They are short-legged and small-headed, have intricate plumage patterns, forage exclusively on the ground, and feed almost entirely on seeds. However, their resemblances to pigeons have been shown to be superficial, and the family is usually isolated in its own order, Pteroclidiformes.

Family COLUMBIDAE

(col-um'-bid-ee: "*Columba*-family", after a genus of doves)

About 255 species make up this rather homogeneous and virtually cosmopolitan family. It is well represented in Australia, which is inhabited by 22 native species, as well as three introduced species (Rock Dove, *Columba livia*; Spotted Turtledove, *Streptopelia chinensis;* and Laughing Turtledove, *S. senegalensis*), which are thriving. Ordinary English usage is inclined to separate "pigeons" (the larger species) from "doves" (the smaller species), but the distinction is ill-defined and lacks scientific significance.

Hard at the tip and relatively soft at the base, the columbid bill is distinctive in shape and structure; the nostrils are set in a naked, fleshy cere. Pigeons are otherwise unremarkable in structure and proportions, but the legs and tail are usually rather short and the head is rather small. They range in size from about that of a starling to nearly that of a hen turkey. Most species are soberly coloured in shades of grey, brown and rufous, but the fruit pigeons and fruitdoves are notable for their bright and intricate colour patterns. The sexes are usually similar, immatures somewhat resemble adults, and there is no seasonal variation.

In behaviour, pigeons range from exclusively arboreal to almost entirely terrestrial and they occupy a wide range of habitats from rainforest to desert. The diet consists largely of fruit or seeds, supplemented in some species by insects. Many species possess a muscular crop, and commonly gather grit to assist in the digestion of seeds, but most fruit pigeons and fruitdoves have a digestive system designed to slough off the fleshy part of fruits, leaving the inner seed to be voided intact (members of the largely African genus *Treron* constitute an important exception to this generalisation). Most pigeons are at least moderately gregarious, and many species breed in colonies.

The nests of pigeons are usually simple flat structures of sticks, often so flimsy that the eggs can be seen from beneath. Often the female builds, using material fetched by the male; both parents participate in incubation and care of the young. The eggs are plain white or yellowish. A distinctive pigeon trait is the habit of feeding the young on "pigeon's milk", a curd-like product of the lining of the crop.

Genus *Chalcophaps*

(kal'-koh-faps: "bronze-pigeon")

There are only two species in this genus of rather small, compact pigeons. One is widespread from India to the Philippines and eastern Australia; the other is confined to eastern Indonesia and New Guinea. The sexes are similar, and the plumage is metallic green above and dull rufous below. Quiet and unobtrusive in behaviour, they live in rainforest, nesting and roosting in trees but foraging on the ground, where they feed on seeds and small invertebrates.

Emerald Dove

Chalcophaps indica (ind'-ik-ah: "Indian bronze-pigeon")

The Emerald Dove has a wide distribution extending from India through South-East Asia to the Philippines, New Guinea and Vanuatu. In Australia it is common in the Kimberley and the Top End, and from Cape York along the east coast to the Victorian border; it also occurs on Lord Howe and Norfolk Islands, where it was possibly introduced. Historical data suggest that it formerly occurred in some areas west of the Great Dividing Range, but these environments have since been largely cleared for agriculture and the Emerald Dove is now essentially a bird of the coast and associated highlands. It is sedentary, and much less common in the south than in the north.

It is usually encountered alone or in pairs. It forages on the ground, but roosts and shelters in trees, and it also utters its low, mournful, monotonous territorial coos from perches in low branches. It is tame, and generally quiet and unobtrusive in behaviour. It feeds on the seeds and fruits of a wide variety of native and introduced plants.

The breeding season is variable and extended, and nests may be found in any month. The nest is usually a flimsy platform of twigs hidden in dense foliage several metres from the ground. Two eggs form the clutch. Incubation takes 14 to 16 days. Both parents incubate and rear the young.

HABITAT: mainly rainforest and similar habitats; also mangroves and melaleuca swamps, dense heathland, wet sclerophyll forest
LENGTH: 23–25 cm
DISTRIBUTION: 300,000–1 million km²
ABUNDANCE: common
STATUS: secure

(H & J Beste)

Genus *Columba*

(kol-um'-bah: "dove")

The ordinary and familiar domestic pigeon is typical of this genus of about 52 species found around the world, although a few tropical species show a trend towards occupying a rainforest niche comparable to that occupied by *Ducula* in Australasia. Most species are medium-sized pigeons that live on the ground, in trees or on cliffs, and feed largely on seeds or fruit. The sexes are usually similar, and the colour pattern generally involves shades of dull brown, wine red, smoke grey, slate grey and white. Like the domestic pigeon, many species have an area of iridescent pink or green on the side of the neck, and many have double black wingbars. Only one species occurs naturally in Australia, but another, the Feral Pigeon or Rock Dove (*Columba livia*), has been introduced and thrives in urban, suburban and farming areas.

White-headed Pigeon

Columba leucomela (lue'-koh-mel'-ah: "white-black crested-throat")

The White-headed Pigeon is widespread in coastal regions and associated highlands of eastern Australia from Cooktown in Queensland to about Eden in New South Wales, but its abundance declines abruptly south of Sydney. It is very common in north-eastern New South Wales but tends to be uncommon elsewhere. It is nomadic.

It somewhat resembles the Topknot Pigeon in many features of its general behaviour and the two species often occur in similar country. Of the two, the White-headed Pigeon is much the less gregarious: small groups are common but it seldom congregates in flocks exceeding about 50 individuals. It also differs in its tendency to frequent the middle and lower layers of the forest canopy rather than the upper levels, and it often gathers fallen fruit on the ground. It also descends to pools to drink. It is generally rather unobtrusive and wary. The vocal repertoire includes a distinctive double cooing note, quiet, slow and deep. Its diet includes a wide range of rainforest fruits and it is fond of the fruits of the introduced Camphor Laurel, *Cinnamomum camphora*.

Most breeding occurs from July to March, but the timing and extent of the nesting cycle are strongly influenced by local and seasonal conditions. The nest is a loose platform of sticks hidden in dense foliage in a tree, bush or vine tangle. A single egg is laid, which hatches in about 19 to 20 days; the chick fledges at about 20 days.

HABITAT: mainly rainforest and other dense forest, including relict fragments in cleared farming and dairy country; occasionally suburban parks and gardens
LENGTH: 38–41 cm
DISTRIBUTION: 100,000–300,000 km²
ABUNDANCE: very sparse to sparse
STATUS: vulnerable

(*MF Soper*)

Genus *Ducula*

(due'-kue-lah, "little-duke")

Often known as fruitpigeons or imperial pigeons, these birds constitute a genus of about 36 species that is very closely related to the fruitdoves (*Ptilinopus*), with an approximately similar distribution. Only one species occurs in Australia, but probably several others occur on the northern, seldom-visited islands in Torres Strait.

There are a number of exceptions but, in general, imperial pigeons differ from fruitdoves in colour pattern and in being very much larger, with proportionately much longer tails. In *Ducula* the colour pattern is simpler than in *Ptilinopus*, and the overall trend is to dark grey or iridescent green upperparts and plain heads and breasts, usually coloured in subtle shades of grey, fawn, pink or rufous. The sexes are usually similar.

All species are arboreal, inhabit rainforest or similar environments, and feed almost exclusively on fruit, digesting only the pulp and allowing the inner seed to be voided undamaged.

Torres Strait Pigeon

Ducula spilorrhoa (spee'-loh-roh'-ah: "spotted-tailed little-duke")

(*J Gray*)

Otherwise known as the Torresian Imperial-pigeon or Nutmeg Pigeon, this impressive bird occurs in northern and north-eastern Australia from the vicinity of Kunmunya in Western Australia to Mackay in Queensland, reaching its greatest abundance in the Great Barrier Reef region. It also occurs in New Guinea and the Bismarck Archipelago, and very closely related (possibly conspecific) populations also occur in Indonesia. In Australia the Torres Strait Pigeon remains common, but its numbers have been drastically reduced since European settlement by indiscriminate shooting at the nesting islands, an illegal practice only recently controlled and still not eradicated; there are now no significant breeding colonies south of Port Douglas. Populations in the Kimberley and the Top End are resident, but Queensland birds are strongly migratory, arriving in July or August to breed, and returning to New Guinea about February or March.

Virtually unmistakable, it is mainly white in plumage, with black tips to the wings and tail. It breeds mainly on small islands, and one of the characteristic sights among the islands of the Great Barrier Reef is of large flocks of the birds, commuting in low, fast and direct flight between the nesting islands and feeding grounds on the adjacent mainland, leaving early in the morning and returning at sunset. Once reaching the mainland, flocks disperse in small foraging parties, seldom penetrating far inland. The diet includes a very wide range of rainforest fruits (especially laurels), which are swallowed whole, the seeds being voided intact. Feeding birds are elusive and difficult to locate, usually remaining hidden in dense vegetation. The most common call is a loud, low and drawn-out double coo, with the emphasis on the second note.

Breeding occurs from August to February, and is normally (but not exclusively) conducted in dense colonies on small offshore islands. Sometimes several broods are raised in succession. The nest is a platform of sticks in a tree (often a mangrove), but the bird occasionally uses the abandoned nest of some other bird, or even nests on the ground. The clutch consists of one egg, which hatches in 26 to 28 days. Both parents incubate and help to rear the chick, which fledges at about 23 days.

HABITAT: in Australia, nests mainly on offshore islands; commutes to feed in rainforest and other dense forest
LENGTH: 38–44 cm
DISTRIBUTION: 300,000–1 million km²
ABUNDANCE: common
STATUS: vulnerable

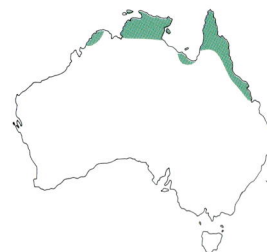

Genus *Geopelia*

(jee'-oh-pel'-ee-ah: "earth-dove")

This genus of three (possibly four) species occurs in Malaysia, Indonesia, New Guinea and Australia. They are closely related to the bronzewings (genus *Phaps*) but are generally smaller, lack the prominent face pattern that is characteristic of *Phaps*, and are unique in that the outermost primary is attenuated. The plumage is mainly fawn, dull brown and pale grey, and the orbital ring is naked, prominent, and brightly coloured. These doves live in open woodland, forage on the ground, feed mainly on seeds, and have distinctive calls, persistently uttered.

Diamond Dove

Geopelia cuneata (kue'-nay-ah'-tah: "wedge-shaped earth-dove")

One of the smallest of all pigeons, the Diamond Dove is common and widespread in arid and semiarid regions across northern and central Australia; is less numerous in the south, but it has been recorded in all mainland States. It is prone to erratic and irruptive dispersal, and some populations are migratory. In the arid interior it is often common around isolated bores and windmills.

It is sometimes confused with the Peaceful Dove, but the scattering of neat, small white spots on the otherwise plain grey upperparts, and the naked red orbital ring, are diagnostic (the Peaceful Dove is powder-blue around the eye, and has black-and-white bars around the neck). The sexes are similar.

It also resembles the Peaceful Dove in behaviour: like that species it is only mildly gregarious, occurring usually in pairs or small parties. It roosts and shelters in trees but forages on the ground, eating a wide variety of seeds, especially those of grasses and herbs.

It drinks daily, and often sunbathes on dusty ground. It is generally tame, quiet and unobtrusive, and seldom flies far. It calls are low and soft, with a distinctive mournful character.

Variable and extended, the breeding cycle is strongly influenced by local and seasonal conditions. The nest, built of twigs, grass and other vegetation, is very variable in structure and placement. Two eggs form the clutch.

Both parents build, incubate (for about 12 to 13 days), and care for the young, which fledge in about 12 to 14 days.

HABITAT: mainly arid and semiarid savannah woodland, close to water
LENGTH: 19–21 cm
DISTRIBUTION: more than 1 million km²
ABUNDANCE: common
STATUS: secure

(PD Munchenberg)

181

Bar-shouldered Dove

Geopelia humeralis (hue'-mer-ah'-lis: "shouldered earth-dove")

Commonest in the tropics and subtropics, the Bar-shouldered Dove occurs in New Guinea and in coastal and near-coastal regions of Australia from about Onslow in Western Australia to the Hunter River in New South Wales; casual occurrences extend south to about Bega. It is sedentary. Though generally common, it seems to be declining in some areas but extending its range in others, possibly in reaction to the spread of the introduced Spotted Turtledove, *Streptopelia chinensis*.

Except for its essentially tropical distribution and preference for humid coastal environments, it strongly resembles the Peaceful and Diamond Doves in general behaviour. It is usually encountered in pairs or small parties, foraging on the ground and roosting and sheltering in trees. It is loudly vocal, and has an extensive repertoire of calls. Its diet consists mainly of the seeds of grasses and sedges.

In the tropics the breeding season is closely linked with the height of the wet season; elsewhere breeding may occur at any time. The nest is a flimsy platform of twigs and rootlets, usually in tree foliage several metres from the ground. The clutch consists of two eggs, which hatch in 14 to 16 days; the young fledge at about 21 days.

HABITAT: varied: mainly eucalypt woodland with shrubby undergrowth; mangroves; pandanus groves and melaleuca swamps; also canefields, regrowth, heathland
LENGTH: 26–30 cm
DISTRIBUTION: more than 1 million km²
ABUNDANCE: common
STATUS: secure

(R Viljoen)

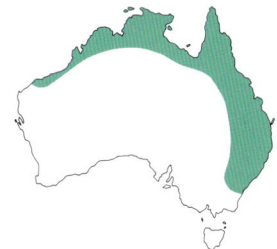

Peaceful Dove

Geopelia placida (plah'-sid-ah: "peaceful earth-dove")

The Peaceful Dove is absent from Tasmania and southern Western Australia but is otherwise widespread across Australia. It also occurs in southern New Guinea, and populations which some researchers consider to be conspecific occur in Malaysia and Indonesia. It has been introduced to the Philippines. Its occurrence is sparse and erratic in the arid interior of Australia but, in general, it is very common, and its distinctive, clear call, "doodle-doo", is a characteristic sound in most kinds of open woodland. It is mainly sedentary, or locally nomadic.

It sometimes congregates in large flocks near waterholes in times of drought, but it is ordinarily only mildly gregarious, usually being encountered in pairs or small parties. It roosts and loafs in trees but forages mainly on the ground. It is fond of sunbathing on bare dusty ground, reclining on its side with one wing raised and facing the sun. It is tame, quiet and undemonstrative in general behaviour. When disturbed it rises with whirring wings; the flight is low, swift, and direct, but seldom sustained. It feeds mainly on grass seeds and insects.

The breeding season is variable and extended: nests may be found in any month but activity declines to a minimum between June and August. Several broods may be raised in succession. The nest is a small platform of twigs, rootlets and grass placed in foliage usually 5 or 6 metres from the ground. Two eggs constitute the clutch. Incubation takes 13 to 14 days, the young fledge at 16 to 17 days, and both parents participate at all stages of the nesting cycle.

HABITAT: very flexible, but mainly open eucalypt woodland close to water; cultivated land; suburban parks and gardens
LENGTH: 17–23 cm
DISTRIBUTION: more than 1 million km²
ABUNDANCE: common
STATUS: secure

(M Seyfort)

Genus Geophaps

(jee'-oh-faps: "earth-pigeon")

Members of this exclusively Australian genus are very closely related to *Phaps*, but the iridescent patch on the wings is relatively obscure, they have a patch of naked, brightly coloured skin on the face, feed almost entirely on seeds, and are markedly more tropical in distribution, confined mainly to the arid interior. They are also markedly more gregarious, usually living in groups, and differ in several other aspects of behaviour and in their courtship displays. The Crested Pigeon *Ocyphaps lophotes* probably belongs in this genus.

Spinifex Pigeon

Geophaps plumifera (plume-if'-er-ah: "feather-bearing earth-pigeon", in reference to the crested head)

The Spinifex Pigeon is widespread and locally common across northern Australia from about Carnarvon in Western Australia to the Chillagoe and the Opalton area in Queensland, but its distribution is intricate and it is absent from large areas. It is mainly sedentary. A small cinnamon-coloured pigeon with a long slender crest, the species is unmistakable, but the plumage varies in detail from place to place: in particular, eastern birds have white bellies, western birds have reddish bellies.

It lives mainly in small coveys which split up into territorial pairs to breed. It is exclusively terrestrial, and seldom flies unless hard pressed. It drinks frequently during the day, and feeds mainly on seeds, especially those of *Triodia*; it also eats insects. Generally tame and unobtrusive in behaviour, it has a characteristic habit of perching motionless for extended periods on some conspicuous rock or similar vantage point.

Breeding activity tends to reach a peak between August and January, but its timing and extent is very variable and nests may be found in any month. The nest is a bare scrape in the ground, sometimes lined with a few wisps of grass, and usually in the shelter of a grass tussock or shrub. The clutch consists of two eggs, which hatch in 16 to 17 days. Both parents incubate and care for the young.

HABITAT: mainly spinifex grassland on rocky hills and outcrops
LENGTH: 20–23 cm
DISTRIBUTION: more than 1 million km²
ABUNDANCE: sparse
STATUS: secure

(D Green)

Squatter Pigeon

Geophaps scripta (skrip'-tah: "written earth-pigeon", in reference to the plumage pattern)

The Squatter Pigeon inhabits the interior of eastern Australia from central Cape York Peninsula westward to the vicinity of Normanton, Charleville and Long-reach in Queensland and south to about Dubbo, Tibooburra, and West Wyalong in New South Wales. It is much more common in the north

(P Klapste)

than in the south, and it is now infrequently recorded in New South Wales. It has perhaps declined since European settlement as a result of pastoral development, especially in the south. It is mainly sedentary.

There are two subspecies, divided approximately by the Burdekin River: birds living north of this point have the naked orbital skin red, those south have it blue.

Only mildly gregarious, it is usually encountered in pairs or in parties of about a dozen. It is entirely terrestrial, and prefers to walk or run rather than to fly, even when disturbed; if pressed, it flushes with a loud whirr of wings, but seldom flies far. It walks to water daily, and feeds almost entirely on seeds, especially those of various grasses. It is generally tame, quiet and unobtrusive.

The breeding season is variable and extensive, but nesting activity tends to peak during the dry season, from March to September. The nest

consists of a few bits of grass and leaves arranged in a shallow scrape in the ground, often sheltered by overhanging vegetation. The clutch consists of two eggs, which take about 17 days to hatch.

HABITAT: mainly dry eucalypt woodland, especially on lighter soils with bare or grassy clearings, and on gravelly ridges in sandy country
LENGTH: 27–32 cm
DISTRIBUTION: 300,000–1 million km²
ABUNDANCE: rare to sparse
STATUS: endangered

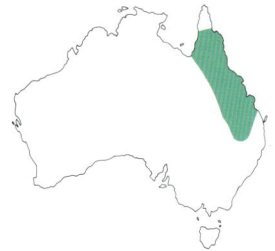

Partridge Pigeon

Geophaps smithii (smith'-ee-ee: "Smith's earth-pigeon", after J. E. Smith, English botanist)

The Partridge Pigeon is common and widespread but rather erratically distributed in northern and north-western Australia from King Sound, Western Australia, to the Macarthur River in the Northern Territory, including Melville Island. It is sedentary. Birds in the Kimberley

have yellow skin around the eye, while Top End birds have red skin.

It is the western relative of the Squatter Pigeon, and resembles that species in general appearance and behaviour. Exclusively terrestrial and only mildly gregarious, it is usually encountered in pairs or parties of up

to about a dozen birds. It prefers to walk or run rather than fly, and feeds largely on grass seeds.

Little is known of breeding, except that it occurs mainly during the dry season (March to October), the nest is merely a scrape in the ground, and two eggs form the clutch. Fledging takes about 10 to 12 days.

HABITAT: dry tropical eucalypt woodlands, especially with abundant leaf litter
LENGTH: 25–28 cm
DISTRIBUTION: 30,000–100,000 km²
ABUNDANCE: sparse
STATUS: endangered

(N Chaffer)

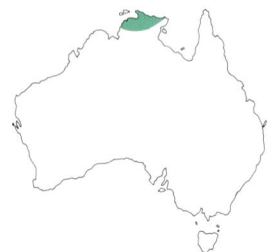

Genus *Leucosarcia*

(lue'-koh-sar'-kee-ah: "white-flesh")

The characteristics of the genus are essentially those of the single species.

Wonga Pigeon

Leucosarcia melanoleuca (mel'-ah-noh-lue'-kah: "black-white white-flesh")

(T & P Gardner)

The Wonga Pigeon is common in dense forests of eastern Australia, approximately between Rockhampton and Melbourne, mainly on the coast and associated highlands but locally also in the interior.

It is usually encountered alone or in pairs. It forages on the ground but roosts and shelters in trees. Usually frequenting dense undergrowth, it is wary, elusive and difficult to observe but, on quiet evenings, it often emerges to forage on open grassy areas at forest picnic grounds and similar sites, where it is often very tame. Its varied diet includes fruit and seeds, as well as insects and other small terrestrial invertebrates. Its territorial call is extremely distinctive, a monotonous series of loud, far-carrying, ventriloquial coos that may total 100 notes or more without pause.

Breeding occurs mainly from September to February. The nest is a substantial or frail platform of sticks placed in a tree fork or dense foliage 2 to 20 metres from the ground. Two eggs constitute the clutch. The incubation period is about 17 to 18 days and the young fledge at about 18 to 19 days.

HABITAT: temperate and subtropical rainforests and wet sclerophyll forests; also brigalow and vine-thickets
LENGTH: 42–45 cm
DISTRIBUTION: 300,000–1 million km²
ABUNDANCE: sparse to common
STATUS: probably secure

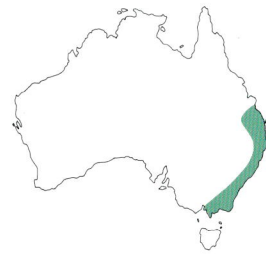

Genus *Lopholaimus*

(loh'-foh-lay'-mus: "crested-throat")

The characteristics of the genus are essentially those of the single species.

Topknot Pigeon

Lopholaimus antarcticus (ant-ark'-tik-us: "southern crested-throat")

(RH Lovell)

The numbers of the Topknot Pigeon have been substantially reduced by forest clearing but it remains common, especially in northern New South Wales and far northern Queensland. It is strongly nomadic.

The unusual crest produces a unique head shape, visible and identifiable at a considerable distance. The Topknot Pigeon is gregarious and strictly arboreal. Like other fruit pigeons it is agile and acrobatic when feeding, but spends much time loafing unobtrusively in the upper canopy. It seldom calls.

The diet includes a wide range of rainforest fruits and also that of the introduced Camphor Laurel, *Cinnamomum camphora*. Flocks fly far and high in search of fresh feeding grounds.

Most breeding occurs from June to January. The nest is a relatively substantial platform of sticks, well hidden in foliage at a considerable height in a tree. One egg constitutes the clutch. Incubation takes about 22 to 24 days, and the chick fledges at about 20 to 28 days. Both parents participate in incubation and in rearing the young.

HABITAT: tropical, subtropical and temperate rainforest
LENGTH: 40–46 cm
DISTRIBUTION: 30,000–100,000 km²
ABUNDANCE: sparse to common
STATUS: vulnerable

Genus *Macropygia*

(mak'-roh-pi'-jee-ah: "long-rump")

Confined to the Oriental and Australasian regions, this genus consists of about eight species with obscure affinities with other pigeons. One species occurs in Australia. Known as cuckoo doves, they are medium-sized pigeons with long, strongly graduated tails and a colour pattern that is deep brown, variably barred with black. They live in rainforests (especially at their margins or in clearings) and feed largely on fruit, foraging mainly in the shrub layer and lower levels of foliage. Species limits are difficult to define within the group.

Brown Pigeon

Macropygia phasianella (fah'-zee-ahn-el'-ah: "little-pheasant long-rump")

The Brown Pigeon (or Brown Cuckoo-dove) is widespread from Indonesia and the Philippines to New Guinea. In Australia it occurs, mainly in coastal regions and associated highlands, from Cape York Peninsula to the border between New South Wales and Victoria. It is sedentary and common but its abundance tends to decline southward.

It is not especially gregarious but it sometimes occurs in small parties. It feeds mainly in the lower levels of foliage, especially in shrubs and saplings, but it often descends to the ground to drink, bathe and take grit. It is acrobatic when feeding, but rather deliberate in most of its actions. It eats fruits and seeds, and in disturbed habitats shows a special fondness for fruits of the introduced weeds Lantana (*Lantana camara*) and the Tobacco Bush (*Solanum mauritianum*). Common calls include a clear "oo-wup".

Most breeding occurs from July to December, but nests may be found in any month. Constructed of sticks, the nest is very variable in form and placement. One egg constitutes the clutch (but two eggs have been recorded). Incubation takes 16 to 18 days, and the chick fledges at about 16 days.

HABITAT: mainly rainforest, especially margins, regrowth, roadsides and disturbed areas; rough and marginal farmland; occasionally suburban parks and gardens
LENGTH: 38–43 cm
DISTRIBUTION: 300,000–1 million km²
ABUNDANCE: common
STATUS: secure

(JD Waterhouse)

Genus Ocyphaps

(oh'-see-faps: "swift-pigeon")

The characteristics of the genus are essentially those of the single species.

Crested Pigeon

Ocyphaps lophotes (loh-foh'-tayz: "crested swift-pigeon")

Perhaps the most familiar of Australian pigeons, the Crested Pigeon is common almost throughout mainland Australia except for the Top End and Cape York Peninsula. It is vigorously expanding in range and number in many areas, especially in the arid interior (presumably associated with development of artificial water supplies intended for stock), and in urban and suburban areas along the east coast. It is mainly sedentary or locally nomadic.

It is usually encountered in small parties, but larger flocks sometimes congregate. It forages almost entirely on the ground but roosts and shelters in trees; it frequently perches conspicuously on roadside fences, utility wires and dead trees. In flight the wings produce a distinctive whistling note, and it has a characteristic habit of swinging the tail high over the back on landing at a perch. It feeds mainly on seeds, including spilled grain.

The breeding season is very variable and extended, and nesting may occur at any time. Often several broods are raised in succession (up to seven have been recorded). The nest is a rough and fragile platform of sticks placed in a dense (often prickly) bush several metres from the ground. Two eggs form the clutch. Incubation, by both parents, takes 18 to 20 days, and the young fledge at 20 to 25 days.

HABITAT: most kinds of dry, open woodland; also cultivated land; urban and suburban parks and gardens
LENGTH: 30–35 cm
DISTRIBUTION: more than 1 million km^2
ABUNDANCE: abundant
STATUS: secure

(GA Cumming)

Genus Petrophassa

(pet'-roh-fas'-ah: "rock-pigeon")

The characteristics of the genus are essentially those of the two very similar species.

White-quilled Rock Pigeon

Petrophassa albipennis (al'-bee-pen'-is: "white-feathered rock-pigeon")

The White-quilled Rock Pigeon is locally common but has a very restricted distribution in sandstone outcrops and similar rock formations in the Kimberley and the western Top End in north-western Australia. It is sedentary.

It seldom wanders more than a few hundred metres from rock escarpments, and it is usually encountered in pairs or small parties. Entirely terrestrial, it feeds mainly on the seeds of various grasses, herbs and shrubs. Generally tame and unobtrusive in behaviour, it often perches motionless for extended periods on a high rock or other vantage point.

Little is known of the breeding cycle, but nests have been found in most months. Two eggs are deposited in a substantial nest of sticks on a shady rock ledge. Both parents incubate (for 15 to 19 days) and care for the young, which fledge at about 15 days.

(A.L. Hertog)

HABITAT: tumbled boulders and cliffs at the base of sandstone escarpments and similar rock formations
LENGTH: 28–30 cm
DISTRIBUTION: 300,000–1 million km²
ABUNDANCE: common
STATUS: probably secure

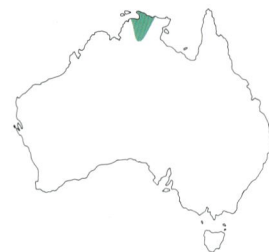

Chestnut-quilled Rock Pigeon

Petrophassa rufipennis (rue'-fee-pen'-is: "rufous-feathered rock-pigeon")

(H & J Beste)

Nests have been found in June, July and October. Two eggs form the clutch, deposited in a substantial stick nest on a shady rock ledge or crevice. Incubation takes 15 to 19 days, and the young fledge at about 21 days.

HABITAT: tumbled boulders and cliffs along sandstone escarpments and similar rock formations
LENGTH: 30–32 cm
DISTRIBUTION: 30,000–100,000 km²
ABUNDANCE: common
STATUS: probably secure

The Chestnut-quilled Rock Pigeon is locally common but is confined to sandstone outcrops in western Arnhem Land, northern Australia. It is sedentary.

Very similar to the White-quilled Rock Pigeon in appearance and behaviour, it differs in plumage mainly in having bright chestnut panels in the wing (usually obscured at rest) instead of white. Like its western relative it is exclusively terrestrial, is usually encountered in pairs or small parties, feeds mainly on the seeds of various grasses, herbs and shrubs, and seldom strays far from the tumbled boulders and cliffs of sandstone outcrops.

Genus *Phaps*

(faps: "pigeon")

Confined to Australia, this genus constitutes the core of a cluster of related pigeons conveniently referred to as bronzewings because of the patch of strongly iridescent and brightly coloured feathers on the inner wing, a feature common to almost all of the species. This cluster also includes the Australian genera *Ocyphaps, Geophaps* and *Petrophassa*, as well as *Henicophaps*, a genus of two species confined to rainforests of New Guinea and the Bismarck Archipelago.

The three species of *Phaps* tend to diverge ecologically, the Brush Bronzewing (*Phaps elegans*) inhabiting dense wet heathland, the Common Bronzewing (*Phaps chalcoptera*) living in open woodland, and the Flock Bronzewing (*Phaps histrionica*) occurring on arid grassy plains (the last-named is notable also for its extreme gregariousness). All three species are medium-sized pigeons unremarkable in general proportions. All have a well-marked colour pattern on the head, forage almost exclusively on the ground, and eat insects and seeds.

Common Bronzewing

Phaps chalcoptera (kal-kop'-te-rah: "bronze-wing pigeon")

(*G Rogerson*)

The Common Bronzewing occurs in most kinds of wooded country throughout Australia, avoiding only rainforest and dense wet sclerophyll forest in the humid east. It is mainly sedentary. It often occurs in flocks but is more usually encountered alone or in pairs. Mainly terrestrial, it forages on the ground and roosts and shelters under a bush, or in a low shrub. Active mostly in the early morning and evening, it is generally quiet, wary, and difficult to observe, but it is often seen at dusk, gathering grit at the margins of quiet country roads. It feeds mainly on seeds.

Most breeding occurs from August to January, but nests may be found in any month. The nest is a crude platform of sticks, usually within a few metres of the ground. The clutch consists of two eggs, which hatch in about 14 to 16 days. Both parents participate in incubation and in rearing the young.

HABITAT: very varied: most kinds of forest and scrub, especially eucalypt, Mulga and *Callitris* woodland
LENGTH: 30–35 cm
DISTRIBUTION: more than 1 million km²
ABUNDANCE: common
STATUS: secure

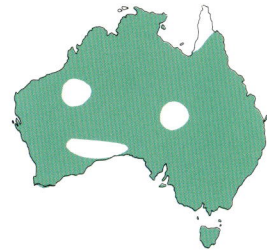

Brush Bronzewing

Phaps elegans (el'-eg-anz: "elegant pigeon")

(*M Seyfort*)

The Brush Bronzewing occurs in south-eastern Australia from about Fraser Island in Queensland to the vicinity of Ceduna in South Australia, and in Western Australia from about Dongarra to Eyre; it also occurs in Tasmania and on islands in Bass Strait, as well as on Kangaroo Island, the Abrolhos and the Recherche Archipelago. It is sedentary. It remains common, but historical records suggest a marked decline since European settlement.

It closely resembles the Common Bronzewing in appearance and behaviour, but it is more secretive, and more closely associated with moist, dense habitats. In the field it is most readily identified by the strong tinge of warm rufous at the nape and the fact that the upperparts are plain brown, not scalloped with buffy grey as in the Common Bronzewing. The Brush Bronzewing is rather solitary, wary, and almost entirely terrestrial. It feeds mainly on seeds. The call is a deep, low "ooom".

Breeding may occur at any time, but takes place especially from August to February. The nest is a scant platform of twigs, placed in or under a dense bush. The clutch consists of two eggs, which hatch in 15 to 18 days.

HABITAT: heathland; woodland and forest with dense understorey; wet sclerophyll forest; mallee
LENGTH: 28–30 cm
DISTRIBUTION: 300,000–1 million km²
ABUNDANCE: sparse
STATUS: probably secure

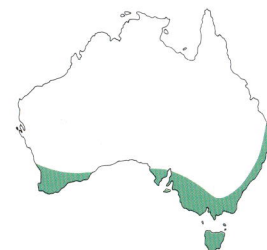

Flock Bronzewing

Phaps histrionica (his'-tree-on'-ik-ah: "acting pigeon")

From a core distribution centred on the Barkly Tableland in north-western Queensland, the Flock Bronzewing erupts erratically and unpredictably to occur almost anywhere in the arid interior of Australia from about Shark Bay across the Kimberley to western Queensland, northern South Australia and north-western New South Wales. The frequency of occurrence and the number of individuals involved in such movements decline sharply southward, and the species is seldom now recorded in New South Wales. Typically, the birds arrive in numbers to breed, then disappear, perhaps not to be recorded in the locality again for a decade or more. It formerly occurred in vast flocks, but it has been much reduced by pastoral development since European settlement.

The Flock Bronzewing is intensely gregarious and entirely terrestrial. it feeds mainly on the seeds of grasses. Breeding may occur at any time. The nest is a mere scrape in the ground, sparsely lined with grass. The clutch consists of two eggs, which hatch in about 16 days.

HABITAT: short-grass plains, especially Mitchell Grass and spinifex; also swales in sandhills
LENGTH: 28–30 cm
DISTRIBUTION: more than 1 million km?
ABUNDANCE: sparse
STATUS: probably secure

(V Serventy)

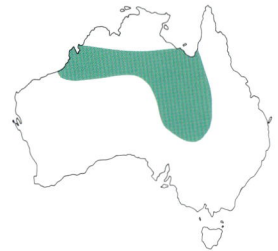

Genus Ptilinopus

(til-een'-oh-poos: "feathered-foot")

Nearly 50 species of fruitdoves are included in this genus, which is largely confined to south-eastern Asia (Malaysia, Indonesia and the Philippines), the Australasian region and the islands of the south-western Pacific, east to Tahiti and the Pitcairn Group. Many species are confined to small islands. Four species occur in Australia.

All are plump, small-headed, colourful and with very short tails. They are strongly arboreal, inhabit rainforest, and feed mainly on fruit. Most are nomadic, congregating in large numbers at fruiting trees and associating freely with other species of fruitdoves. Where several species occur together, they usually differ substantially in size as well as colour pattern.

The genus is closely related to *Ducula*, but in general its members are very much smaller (some species of *Ptilinopus* are among the smallest of all columbids) and more brightly coloured. The colour pattern is very variable, but the general trend is to plain green upperparts, with intricate patterns composed of very bright colours on the face and breast. Females usually somewhat resemble males but are very much duller.

Banded Fruitdove

Ptilinopus cinctus (sink'-tus: "banded feathered-foot")

The Banded Fruitdove occurs in eastern Indonesia and in a restricted distribution along the sandstone escarpments of western Arnhem Land in northern Australia, where it is sedentary. The only fruitdove in its environment, it is easily identified by its black breast band. The sexes are similar in appearance.

It is usually encountered alone or in small groups, but flocks of up to about 30 individuals sometimes congregate. It is strongly arboreal, feeding on a wide variety of rainforest fruits. It is especially active in the early morning and late afternoon, being inclined to perch idly in the crowns of trees at other times.

Very little is known of its breeding behaviour, but it seems to breed mainly from May to November. The nest is a frail platform of small sticks in the upper foliage of a tree or shrub. One egg constitutes the clutch.

(EE Zillmann)

HABITAT: monsoon forests growing along sandstone escarpments; occasionally adjacent eucalypt savannah woodland
LENGTH: 33–37 cm
DISTRIBUTION: 30,000–100,000 km²
ABUNDANCE: sparse
STATUS: vulnerable

Wompoo Fruitdove

Ptilinopus magnificus (mag-nif'-ik-us: "magnificent feathered-foot")

This large, impressive pigeon is common in New Guinea and in north-eastern Australia from Cape York peninsula south to the Illawarra district of New South Wales, though it is now much less common than formerly in New South Wales: it is rare south of Newcastle and extirpated in the Illawarra scrub. It is much less common in highlands than in coastal lowlands, and has been adversely affected by widespread clearing of forests. It is locally nomadic.

The Wompoo Fruitdove is easily identified by its large size and its mainly purple and yellow underparts; the sexes are similar in appearance, and immatures resemble adults except in being very much duller. Birds show a gradual decrease in average size progressing from south to north: birds in the extreme south are substantially larger than those in the far north.

Flocks sometimes congregate but it is usually encountered alone or in pairs. Rigidly arboreal, it favours the upper canopy of rainforest and is generally unobtrusive and difficult to locate, although, like other fruit-doves, it is acrobatic when feeding. It eats a wide variety of rainforest fruits, which are swallowed whole, the seeds being voided intact. It utters a variety of deep cooing notes, but its most distinctive call is a very deep, loud, far-carrying "g'wom-poo".

Most breeding occurs from July to December. The nest is an extremely scanty platform of twigs concealed high in the outer foliage of a rainforest tree. The clutch consists of one egg, which hatches in 12 to 14 days. Both parents participate in nest construction, incubation and rearing the chick.

HABITAT: almost exclusively rainforest
LENGTH: 38–50 cm
DISTRIBUTION: 100,000–300,000 km²
ABUNDANCE: sparse
STATUS: vulnerable

(EE Zillmann)

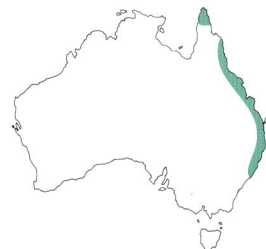

Rose-crowned Fruitdove

Ptilinopus regina (re-jee'-nah: "queen feathered-foot")

(J Gray)

This attractive, small, plump fruitdove occurs in Indonesia and in northern and eastern Australia in the Kimberley, the Top End, islands of Torres Strait and from Bamaga in Queensland to about Port Stephens in New South Wales. It is locally nomadic and perhaps migratory and is also prone to erratic and poorly understood movements southward with some frequency to about Sydney and, rarely, even as far as Tasmania. It is generally common, especially in the north. It has been recorded in New Guinea, but its status there is uncertain.

The adult female somewhat resembles the male but is very much duller and less conspicuously marked; the juvenile is mainly green, with a yellowish belly. Like the Superb Fruitdove, the Rose-crowned Fruitdove occasionally congregates in large flocks at fruiting trees, but is usually encountered alone or in pairs. Strictly arboreal, it is agile and acrobatic while feeding, but spends much of the day perched idly and unobtrusively in the upper canopy of trees, where its presence is difficult to detect. Its calls include a distinctive, lengthy series of short coos that gradually accelerates. It feeds on a wide variety of rainforest fruits.

Eastern birds breed mainly from August to February, but populations in the Kimberley and the Top End have a somewhat more protracted breeding season that starts a little later: nesting is uncommon during the latter part of the wet season. The nest is a small, frail platform of sticks hidden in the foliage of a tree or bush or in a vine tangle. The clutch consists of one egg, which hatches in 16 to 18 days. The young fledge at about 11 to 12 days.

HABITAT: rainforest, mangroves, adjacent eucalypt forest with rainforest shrubs, monsoon and riverine forest; also islands and beachside scrub
LENGTH: 20–23 cm
DISTRIBUTION: 100,000–300,000 km^2
ABUNDANCE: sparse
STATUS: vulnerable

Superb Fruitdove

Ptilinopus superbus (sue-per'-bus: "superb feathered-foot")

The Superb Fruitdove is widespread from Indonesia through New Guinea to the Solomon Islands and, in eastern Australia, from Cape York Peninsula generally south to the vicinity of Rockhampton, but casual occurrences extend commonly into New South Wales and, rarely, even to Tasmania. It tends to be most abundant in coastal lowland forests, but it is not uncommon at altitudes up to about 1000 metres above sea-level. It remains common in the north but is adversely affected by widespread clearing of forests. It is at least locally nomadic.

It occasionally congregates in large numbers at fruiting trees, but it is usually encountered alone or in pairs. The adult male is extremely colourful, but the female is more or less plain greenish in colour, with a white abdomen. Strongly arboreal, and generally unobtrusive in behaviour, its presence is usually detected by its call, a measured series of deep coos. It feeds on a wide variety of rainforest fruits, especially those of the many species of laurels.

Breeding occurs mainly from July to December. The nest is a small frail platform of sticks well hidden in foliage from a few to 30 metres or more up in a tree. The single egg hatches in about 14 days. Both sexes participate in nest construction, incubation (mainly the female by night, the male by day) and rearing of the chick, which reportedly fledges at about 7 days.

HABITAT: rainforest, including regrowth; also adjacent eucalypt forest, mangroves
LENGTH: 22–24 cm
DISTRIBUTION: 100,000–300,000 km²
ABUNDANCE: sparse to common
STATUS: vulnerable

(*T & P Gardner*)

Order PSITTACIFORMES

(sit'-ah-see-for'-mayz; "*Psittacus*-order", after a genus of parrots)

Few groups of birds are so sharply set apart from others as is this order, but few tangible, easily defined features reflect this isolation. Parrots, lorikeets and cockatoos all have blunt, strongly downcurved bills, zygodactylous feet (a condition in which the toes are so arranged that two point forwards and two point backwards), a naked fleshy area surrounding the nostrils, and tufts of powder down feathers.

None of these features (and others not mentioned) are unique among birds, but their combination is diagnostic. Only the pigeons and doves (Columbiformes) seem even remotely related.

These birds fall into three natural groups: cockatoos, lorikeets and "typical" parrots (all of which are represented in Australia), but few researchers agree on how this trichotomy should be reflected in their classification. Some place all psittaciform birds in a single family, Psittacidae, while others distinguish three families: Cacatuidae, Loriidae, and Psittacidae. The latter course is followed here.

Family CACATUIDAE

(kah'-kah-tue'-id-ee: "*Cacatua*-family", after a genus of cockatoos)

Cockatoos are large, gregarious parrot-like birds confined to the Australasian region. The family contains about 18 species, differing most conspicuously from typical parrots in possessing a large, mobile crest. Most species are either mainly black or mainly white in colour: the plumage lacks "Dyck-texture", a structural modification of the feathers that is largely responsible for the bright colours that characterise the typical parrots.

Genus Cacatua

(kah'-kah-tue'-ah: "cockatoo", from a similar Malay word)

Eleven species of white, conspicuously crested and relatively short-tailed cockatoos constitute the genus *Cacatua*. Five species occur in Australia, six elsewhere in Australasia. The Galah is sometimes isolated in its own genus, *Eolophus*.

The Australian species are all strongly gregarious, are very noisy and conspicuous, forage mainly on the ground, and (in general) inhabit open country rather than woodland. Sexual dimorphism is very slight, immatures resemble adults, and there is no seasonal variation.

Sulphur-crested Cockatoo

Cacatua galerita (gah'-ler-ee'-tah: "crested cockatoo")

The conspicuous sulphur-yellow crest (usually sleeked back but frequently raised in display, alarm or curiosity) renders this large white cockatoo unmistakable. Common and familiar, it has an extensive distribution across northern and eastern Australia, and is increasing in many south-eastern urban centres. Elsewhere it occurs in New Guinea and the Aru Islands, Indonesia, and has been introduced to New Zealand and Palau, and in Western Australia near Perth. It is mainly sedentary.

It is strongly gregarious: groups loaf in leafy trees during the midday heat, idly stripping leaves and branches, and gather in noisy conspicuous flocks at communal roosts. It often begs scraps at forest picnic areas. Its calls include an extraordinarily loud, harsh screech. The varied diet includes seeds, grain, nuts, fruit, bulbs, and insects and their larvae; most foraging is on the ground.

Breeding occurs mainly from May to September in the north, August to January in the south. The nest is a high tree cavity, usually in a eucalypt. Both parents prepare the nest cavity, incubate and care for the young. The clutch consists of one to three eggs, which hatch in 30 days; the young fledge at about 70 days, but remain with parents for an indefinite period thereafter.

HABITAT: forest and woodland, including wet sclerophyll forest and rainforest; also suburban parks and gardens, mangroves
LENGTH: 45–50 cm
DISTRIBUTION: more than 1 million km²
ABUNDANCE: common
STATUS: secure

(D Greig)

Pink Cockatoo

Cacatua leadbeateri (led'-beet-er-ee: "Leadbeater's cockatoo", after B. Leadbeater, British naturalist)

The Pink Cockatoo (or Major Mitchell's Cockatoo) has an

(PD Munchenberg)

extensive distribution across the arid interior of mainland Australia but it tends to be uncommon and local in occurrence, and rarely congregates in the large flocks typical of other "white" cockatoos in its genus. Although it occurs in most kinds of open country (wherever water and large nesting trees are available), it has in general been adversely affected by changes in land use since European settlement; marked declines have been noted in some areas.

Breeding pairs are generally sedentary, maintaining permanent territories at about 500 hectares; young birds form wandering parties. Most foraging is terrestrial, the diet consisting mainly of seeds, grain and tubers.

Breeding occurs mainly from July to January. One to three eggs form the clutch, laid in a tree cavity. They hatch in about 26 days and the young fledge at about 56 days. Both parents prepare the nest hollow, incubate and care for the young.

HABITAT: desert scrubs, open woodland, mallee, Mulga, callitris woodlands, and adjacent agricultural areas
LENGTH: 32–36 cm
DISTRIBUTION: more than 1 million km²
ABUNDANCE: sparse
STATUS: probably secure

Little Corella

Cacatua pastinator (pah'-stin-ah'-tor: "agricultural cockatoo")

The Little Corella is abundant and widespread in mainland Australia, especially in the far north and the arid interior, but a general decline since European settlement has been noted in the far south-west. In recent decades it has established isolated but rapidly expanding populations, almost certainly originating from aviary escapes, in the Hunter Valley, the Sydney region, and the Blue

(K Ireland)

Mountains of New South Wales, and elsewhere on the east coast. The Western Australian population *C. p. pastinator*, is distinguished by a very long upper mandible and is known as the Long-billed Corella. The species also occurs in southern New Guinea.

Nomadic and strongly gregarious, it often occurs in very large flocks. It forages mainly on the ground, eating bulbs and seeds, especially of grasses and cereal crops; it loafs in flocks in leafy trees during the midday heat. Late afternoon feeding is often followed by spectacular, wild and very noisy flights around the communal roost trees.

Breeding may occur at any time, but especially from June to October. The usual clutch is two eggs, laid in a tree cavity carpeted with wood fragments. The incubation period is about 25 days; both parents incubate and care for the young.

HABITAT: arid open country, usually near water
LENGTH: *c* 42 cm
DISTRIBUTION: more than 1 million km²
ABUNDANCE: common
STATUS: secure

Galah

Cacatua roseicapilla (roh-zay'-ee-kah-pil'-ah: "rosy-headed cockatoo")

Before European settlement the abundant and familiar Galah was a bird of the arid interior but, increasingly over the past few decades, its range has expanded to include most of mainland Australia and eastern Tasmania, where it was first recorded breeding in 1925. It is now common in several urban centres in the southeast, including Sydney.

Breeding adults are essentially resident, maintaining permanent territories, but young birds congregate in roving flocks. Active noisy and conspicuous, Galahs feed largely on seeds, gathered mainly on the ground. In some areas they feed almost exclusively on spilled grain. Highly mobile, Galahs may forage 15 kilometres or more from their communal roosts. They first breed when three or four years old.

The breeding season is very variable, but is mainly February to July in the north, July to December in the south. The nest is a tree cavity, and three or four eggs form the usual clutch. The incubation period is about 30 days, and the young fledge at about 56 days. Both parents incubate and care for the young.

HABITAT: open areas generally: semidesert, plains and open woodland; farmland, grainfields, golf courses, city parks
LENGTH: *c* 35 cm
DISTRIBUTION: more than 1 million km²
ABUNDANCE: abundant
STATUS: secure

(*D & V Blagden*)

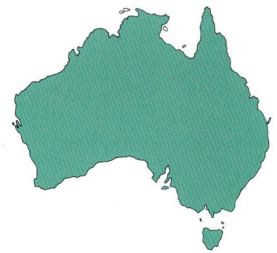

Slender-billed Corella

Cacatua tenuirostris (ten'-ue-ee-ros'-tris: "narrow-beaked cockatoo")

The Slender-billed Corella has a restricted distribution in southeastern Australia, extending from the Riverina south to the vicinity of Melbourne and west to the Coorong in South Australia. Its numbers and range declined drastically after European settlement but there has been some recent recovery, and it is now common in some areas. It is mainly sedentary. Its relationship with the Long-billed Corella of southwestern Australia has occasioned considerable debate, but currently the latter is considered a subspecies of the related Little Corella.

Little and Slender-billed Corellas are difficult to distinguish, but the latter has a variable reddish crescent on the breast, a naked eye ring of a different shape and a long pointed bill (not often evident in the field); the tremulous trisyllabic contact call is also distinctive.

The Slender-billed Corella is generally in pairs or small parties during the nesting season, but often congregates in large flocks at other times, especially when roosting and foraging. Foraging is almost entirely on the ground, and the diet consists primarily of the bulbs of native and introduced plants; it also eats waste, and spilled and sprouting cereal grains.

Most breeding occurs from July to November. The nest is a high cavity in (usually) a large River Red Gum. The clutch is one to three eggs, which hatch in about 24 days.

HABITAT: open country near woodland or River Red Gum forest: cereal crops, paddocks, native grasslands; seldom far from water
LENGTH: *c* 37 cm
DISTRIBUTION: 100,000–300,000 km²
ABUNDANCE: sparse
STATUS: vulnerable

(*LF Schick*)

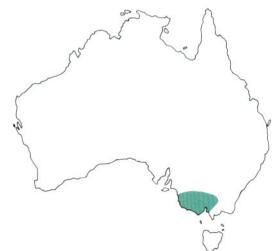

Genus *Callocephalon*

(kal'-oh-sef'-al-on: "beautiful-head")

The characteristics of the genus are those of the single species.

Gang-gang Cockatoo

Callocephalon fimbriatum (fim'-bree-ah'-tum: "fringed beautiful-head")

This cockatoo is restricted to south-eastern mainland Australia from the Portland district of Victoria (rarely west into South Australia) to the Blue Mountains of New South Wales (sporadically north to Barrington Tops); it has been introduced to Kangaroo Island where it is now seldom reported, and it is a vagrant in Tasmania. It has declined since European settlement and is now most common in the southern highlands.

Adult males are distinctive, with scarlet head and dusky grey body plumage narrowly scalloped in dull white. Females and immatures are dull grey but both sexes have a unique "feather-duster" crest.

It is mainly rather sedentary, but undertakes regular movement from higher altitudes in winter, at which time it is common in some highland cities, such as Canberra. Seldom encountered alone, it us usually in pairs or family parties, larger flocks often forming in winter. It is strongly arboreal but occasionally forages on the ground. Once located, a food source is usually exploited until exhausted, the birds spending much of the day feeding methodically and unobtrusively in the foliage. The tameness of feeding Gang-gang Cockatoos is well known: They can sometimes by touched by hand. They eat eucalypt seeds, fruit, and insects, and are especially fond of the berries of introduced hawthorn.

Breeding is mainly from September to January. Both parents prepare the nest-site (a high tree cavity, usually in a eucalypt near water), incubate and care for the young. Two eggs form the usual clutch, hatching in about 20 days; the young fledge at about 40 days but remain more or less dependent on their parents for four to six months.

HABITAT: tall eucalypt forest and adjacent woodland; suburban parks and gardens
LENGTH: *c* 35 cm
DISTRIBUTION: 100,000–300,000 km²
ABUNDANCE: sparse
STATUS: vulnerable

(J Christesen)

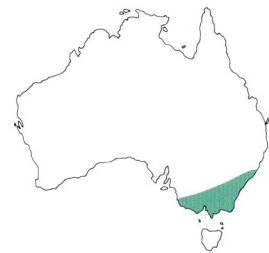

Genus *Calyptorhynchus*

(kah-lip'-toh-rink'-us: "covered-beak")

Four, perhaps five, species of very large, long-tailed, impressive cockatoos constitute this genus, which is restricted to Australia. The plumage is mainly black, with conspicuous panels of red, yellow or white in the tail. Sexual dimorphism is marked, and immatures generally resemble adult females.

The genus is characteristic of forest and woodland and only the Red-tailed Black Cockatoo is widespread. The relationship of the south-western "white-tailed" forms to the south-eastern "yellow-tailed" population is still debated: current consensus suggests an earlier trans-Nullarbor westward movement that resulted in the speciation of the White-tailed Black Cockatoo, followed by a more recent invasion resulting in a population that has not yet fully speciated (though also white-tailed) and is best regarded as a subspecies of the south-eastern Yellow-tailed Black Cockatoo.

White-tailed Black Cockatoo

Calyptorhynchus baudinii (boe-dan'-ee-ee: "Baudin's covered-beak", after N. Baudin, French navigator and naturalist)

The White-tailed Black Cockatoo is confined to dense forests of Marri

(L Pedler)

and Karri in far south-western Australia. It very closely resembles the western population of the Yellow-tailed Black Cockatoo and has only recently been separated from it as a distinct species, differing in subtle but distinct features of ecology and behaviour. It is locally migratory, strongly gregarious, and mainly arboreal; it feeds mainly on the seeds of Marri and Karri trees.

The breeding season extends from July to December. The female incubates alone, but both parents prepare the nest-site (a large tree cavity) and care for the young. Two eggs form the clutch; the incubation period is 28 to 30 days.

HABITAT: eucalypt forest and woodland, especially karri and marri
LENGTH: 50–60 cm
DISTRIBUTION: 100,000–300,000 km^2
ABUNDANCE: sparse
STATUS: vulnerable

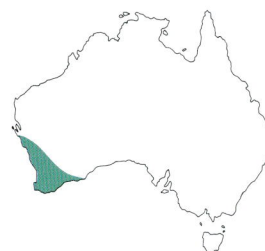

Yellow-tailed Black Cockatoo

Calyptorhynchus funereus (fue-nay'-ray-us: "funereal covered-beak")

More or less confined to humid regions of dense eucalypt forest, the Yellow-tailed Black Cockatoo (or Funereal Cockatoo) occurs in coastal regions and associated highlands from about Rockhampton in Queensland south to Tasmania and west to South Australia, and in south-western Australia from the Murchison River to Norseman.

It is a very large cockatoo, mainly dull black in plumage, with irregular yellow patches on the head and pale

(D & M Trounson)

yellow panels in the long tail. The south-western population (sub-species *C. f. latirostris*) is distinctive in having white panels in the tail: some authorities regard it as a distinct species. Females resemble males, and immatures resemble adults.

It is strongly arboreal in the east but western birds frequently forage on the ground. Once fledged, young birds remain with their parents for an indefinite period, and the family party is the basic social unit. Especially in winter, such parties frequently coalesce into large flocks to roost and forage. Wood-boring grubs and the seeds of eucalypts and exotic pines are prominent in the varied diet.

The breeding season is mainly from October to May in the east, July to December in the west. The female alone incubates and feeds the young for the first 20 days but the male contributes thereafter. The nest is a large, deep tree cavity carpeted with wood chips, generally at a great

height in a eucalypt. Two eggs, incubated for about 28 days, form the clutch, and the young fledge in about 70 days.

HABITAT: tall forests, pine plantations, heaths
LENGTH: 60–70 cm
DISTRIBUTION: more than 1 million km^2
ABUNDANCE: sparse
STATUS: probably secure

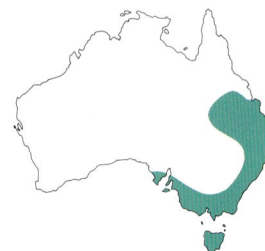

Glossy Black Cockatoo

Calyptorhynchus lathami (lay'-tham-ee: "Latham's covered-beak", after J. Latham, British ornithologist)

The Glossy Black Cockatoo is restricted to south-eastern Australia from the vicinity of Rockhampton in Queensland south to Victoria and westward (in isolated mountain ranges) to Cobar and Griffith in New South Wales; an isolated population occurs on Kangaroo Island in South Australia. It is uncommon and perhaps declining, especially in the south-western parts of its range and is now extinct in mainland South Australia.

Immatures and adult females are readily identified by irregular and variable yellow patches on the head. Adult males, because of the conspicuous red panels in the tail, are frequently misidentified as Red-tailed Black Cockatoos but calls, habitat and behaviour are very different.

The Glossy Black Cockatoo feeds almost exclusively on the seeds of casuarinas, only occasionally extending its diet to include insects and the seeds of eucalypts, angophoras, hakeas and acacias.

Strictly arboreal, it is usually encountered in family parties, feeding quietly and unobtrusively in groves of casuarinas, its presence often detectable only by the soft patter of fallen seed fragments on the leaf litter below and an occasional call—a quiet, wavering, grating squeal.

The Glossy Black Cockatoo breeds mainly from March to August. The female incubates and cares for the young alone, but is regularly fed by the male. The nest is a spacious tree cavity carpeted with wood chips and dust. One egg, which hatches in about 29 days, constitutes the usual clutch. The chick fledges in about 60 days but remains with its parents for an indefinite period thereafter.

HABITAT: casuarina forest and woodland

LENGTH: *c* 48 cm

DISTRIBUTION: 300,000–1 million km²

ABUNDANCE: very sparse

STATUS: vulnerable

(D & M Trounson)

199

Red-tailed Black Cockatoo

Calyptorhynchus magnificus (mag-nif'-ik-us: "magnificent covered-beak")

This large, gregarious and conspicuous cockatoo is found throughout mainland Australia but its numbers and distribution have declined drastically in the south-east since European settlement: it is now most common in the tropical north. It is mainly nomadic, but migratory movements have been detected in some areas.

The adult male is entirely black except for panels of brilliant red in the long tail; the female is mainly dusky black, but yellow edges to the feathers of the underparts produce a scalloped effect and the head and back are liberally speckled with yellow; her tail has panels of yellow or dusky orange with obscure black bars. Immatures resemble adult females, young males reaching adult plumage gradually over about four years.

Its varied diet includes seeds, fruits, bulbs, and insects. It is mainly arboreal, but in many areas regularly forages on the ground. Calls are loud, varied and harsh.

The breeding season varies with locality: July to January in the south-east, March to September in the north, and July to October in the west. The female incubates alone, but both parents feed the young. The usual clutch is one egg, which hatches in about 30 days; the chick fledges in about 90 days but usually remains dependent on its parents for a further three months or so.

(D & M Trounson)

HABITAT: mainly open woodland and riparian forest; also mallee, cultivated lands, savannah, Mulga, rainforest
LENGTH: 55–66 cm
DISTRIBUTION: more than 1 million km²
ABUNDANCE: common
STATUS: probably secure

Genus *Nymphicus*

(nim'-fik-us: "nymph")

The characteristics of the genus are those of the single species. Often listed among the typical parrots, its true affinities with the cockatoos have only recently been demonstrated.

Cockatiel

Nymphicus hollandicus (hol-and'-ik-us: "[New] Holland nymph")

Avoiding only Cape York Peninsula and the humid south-east, the Cockatiel (or Quarrion) is common over much of the Australian mainland. It has been recorded in Tasmania but these reports almost certainly involve escaped aviary birds; escaped birds also frequently occur in south-eastern towns and cities. It is strongly migratory, being rare south of about 30°S in winter, but movements are also strongly influenced by rainfall.

The adult female resembles the male but is very much duller, and her crest is grey rather than yellow; immatures resemble adult females. In any plumage, the slim silhouette, overall grey colour, and conspicuous white wing patches are distinctive.

The Cockatiel usually occurs in pairs or small parties, but often travels in large flocks which, perched conspicuously on telephone wires or dead roadside trees, are a common sight in outback Australia. It feeds mainly on the ground, on seeds, cereal crops, fruits and berries.

The timing of breeding is strongly influenced by rainfall and other local conditions but is usually from July to December in the interior south-east. Both parents prepare the nest hollow, incubate, and care for the young. The clutch is usually five eggs (varying from two to eight), which hatch in about 19 days; the young fledge at about 30 days.

(M Seyfort)

HABITAT: arid and semiarid open country, usually near water
LENGTH: c 32 cm
DISTRIBUTION: more than 1 million km²
ABUNDANCE: common
STATUS: secure

Genus Prosciger

(proh-bos'-i-jer: "nose-bearer")

The characteristics of the genus are those of the single species.

Palm Cockatoo

Prosciger aterrimus (ah-te'-rim-us: "very-black nose-bearer")

The sedentary Palm Cockatoo is common in its extremely limited Australian distribution in rainforests at the northern end of Cape York Peninsula: it also occurs in New Guinea and the Aru Islands. Its huge bill, great size, and bare red facial skin are diagnostic. Females are slightly smaller than males but the sexes are otherwise similar.

It is often alone but is usually encountered in pairs or small parties. Active and conspicuous, it spends much time perched high in tall trees or flying over the forest canopy; it seldom descends to the ground. It eats fruit, seeds, and leaf buds. Social activities include groups of up to seven birds congregating at "meeting" trees for joint displays in the early morning before dispersing to forage; males use short sticks and similar objects to drum on tree stubs in courtship display.

Most breeding occurs from August to January. The nest is a large cavity in a (usually) dead tree, the floor covered with a thick layer of chewed twigs and leaves. Both sexes prepare the nesting hollow and care for the young, but only the female incubates. The clutch consists of one egg, which hatches in about 35 days; the fledging period is about 42 days but the chick often remains dependent on its parents for a further two or three months.

HABITAT: chiefly rainforest; occasionally eucalypt and palm woodland; forest edges
LENGTH: *c* 56 cm
DISTRIBUTION: 30,000–100,000 km²
ABUNDANCE: common
STATUS: vulnerable

(D & M Trounson)

Family LORIIDAE

(lor'-ee-id-ee: "*Lorius*-family", after a genus of lories)

Parrots of this family, known as lories or lorikeets, are strictly arboreal, strongly gregarious, and usually brightly coloured with relatively tight, glossy plumage. Nectar and pollen are prominent in the diet, and the family shows several corresponding adaptations, especially a tongue tipped with elongate papillae and a weak, non-muscular gizzard. Lorikeets are usually strongly nomadic, and a common element of behaviour is their habit of moving from one source of food to another in conspicuous, high-flying, screeching flocks. When foraging, they clamber around in foliage and flowers with considerable agility. Flight is very swift and direct.

Males resemble females, immatures resemble adults, and there is no seasonal variation.

Some 55 species constitute the family, which is essentially restricted to Australasia, reaching its greatest diversity in New Guinea.

Genus *Glossopsitta*

(glos'-op-sit'-ah: "tongued-parrot")

Three small species constitute this genus, which is restricted to southern and eastern Australia.

Musk Lorikeet

Glossopsitta concinna (kon-sin'-ah: "neat tongued-parrot")

The Musk Lorikeet is widespread in south-eastern Australia (including Tasmania) but is most common in Victoria: its occurrence is increasingly rare and erratic north of about Brisbane, Queensland. There are some grounds for suspecting that its distribution has changed since European settlement, especially in the north. It is strongly nomadic.

It is most readily identified by its red ear coverts, red-tipped bill and green underwing coverts.

It is strongly gregarious, and is perhaps most frequently encountered in dense flocks at groves of flowering eucalypts. These are often scenes of almost frenetic activity, the birds feeding noisily with much chattering and excitement, small groups constantly dashing to and fro. The diet is mainly nectar, supplemented with pollen, fruit and seeds.

Breeding occurs from August to January. Only the female incubates but both parents feed the young. Two eggs constitute the clutch, laid in a tree cavity and hatching in 24 days.

HABITAT: eucalypt forests and woodlands; avoids rainforest and wet sclerophyll forest
LENGTH: *c* 22 cm
DISTRIBUTION: 300,000–1 million km²
ABUNDANCE: sparse
STATUS: vulnerable

(D & M Trounson)

Purple-crowned Lorikeet

Glossopsitta porphyrocephala (por-fie'-ro-sef'-ah-lah: "purple-headed tongued-parrot")

(R Garstone)

The distribution of the strongly nomadic Purple-crowned Lorikeet extends across southern Australia from the vicinity of Perth to the Mallacoota region of Victoria: it has not been recorded on the Nullarbor Plain and it is rare east of Melbourne. It does not occur in Tasmania but vagrants have occurred in New South Wales and Queensland. It is often common in suburban parks and gardens.

It is strongly gregarious, active, noisy and conspicuous. Foraging birds are often extremely tame, ignoring a human observer even at very close range. Small, mainly green, with a deep purple crown and red underwing coverts, it is easily identified. Females are much duller than males but otherwise similar. The diet is mainly nectar, supplemented with pollen, fruit and seeds.

Breeding (often in loose colonies) occurs from August to December.

Both parents feed the young but only the female incubates. Three or four eggs (usually four) are laid in a tree cavity; incubation takes 22 days.

HABITAT: open eucalypt woodland; heathland; mallee and banksia scrub; jarrah and karri forests in the west; suburban parks and gardens
LENGTH: *c* 16 cm
DISTRIBUTION: 300,000–1 million km²
ABUNDANCE: sparse
STATUS: vulnerable

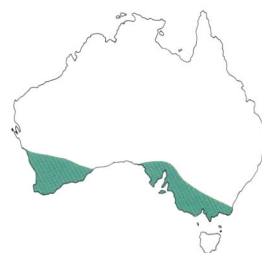

Little Lorikeet

Glossopsitta pusilla (pue-sil'-ah: "very-small tongued-parrot")

The Little Lorikeet is abundant and widespread across much of eastern Australia from the vicinity of Cairns in Queensland to western Victoria and (formerly) the Mount Lofty Ranges of South Australia. It has been recorded as a rare vagrant in Tasmania.

It differs most conspicuously from other small lorikeets with which it might be confused by its red face (not ear coverts as in the Musk Lorikeet), black bill and (in flight) green underwing coverts.

Strictly arboreal, it is gregarious in all activities except breeding; it is usually encountered in small parties, but large dense flocks often congregate at groves of profusely flowering eucalypts. The diet is mainly nectar, supplemented with pollen, fruit and seeds.

The breeding season extends from August to January. Both parents prepare the nesting hollow and share in caring for the young, but only the female incubates. The nest is a tree cavity (usually in a eucalypt) carpeted with wood dust. Three to five eggs form the clutch, which hatches in 22 days.

HABITAT: dry open woodlands and forests; also heath and banksia scrub; riverine woodland
LENGTH: *c* 15 cm
DISTRIBUTION: more than 1 million km²
ABUNDANCE: sparse
STATUS: probably secure

(D & M Trounson)

Genus *Psitteuteles*

(sit'-ue-tay'-layz: "worthless-parrot")

With one included species, this genus is restricted to Australia; it is sometimes merged with *Trichoglossus*. Its characters are those of the species.

Varied Lorikeet

Psitteuteles versicolor (ver'-see-kol'-or: "varicoloured worthless-parrot")

This small lorikeet is abundant across the tropical lowlands of northern Australia from about Broome to Cape York: it is rare east of the Gulf of Carpentaria.

The female is somewhat duller than the male but the sexes are otherwise similar. As the only small lorikeet in the tropical north, it is generally unmistakable, but its white "goggles" are a striking feature.

Nomadic and strongly gregarious, it is sometimes unobtrusive when foraging in a flowering tree but is otherwise active and conspicuous. It feeds on nectar, pollen and fruit.

Nesting usually occurs from April to August. Both parents feed the young but only the female incubates. The two to four eggs forming the clutch are laid in a tree hollow, which is often carpeted with wood dust or chewed leaf pieces. The eggs take 22 days to hatch, the young fledging in about 40 days.

HABITAT: eucalypt and melaleuca woodland
LENGTH: *c* 19 cm
DISTRIBUTION: more than 1 million km²
ABUNDANCE: sparse
STATUS: probably secure

Genus *Trichoglossus*

(trik'-oh-glos'-us: "hair-tongue")

This genus comprises a group of some nine species of medium-sized lorikeets with long, strongly graduated tails. Two species occur in Australia, the others in New Guinea.

Scaly-breasted Lorikeet

Trichoglossus chlorolepidotus (klor'-oh-lep'-id-oh'-tus: "green-yellow-scaled hair-tongue")

Found only in Australia, the Scaly-breasted Lorikeet is common in coastal areas and associated highlands from about Cooktown in Queensland to the vicinity of Wollongong in New South Wales; an introduced population occurs near Melbourne in Victoria. Nomadic flocks frequently penetrate inland to the western slopes of the Great Dividing Range, where they usually favour riverine woodland.

It is not an especially colourful bird, the head and upperparts being bright green and the underparts yellow, scaled with green: in these respects it differs from the gaudy Rainbow Lorikeet but the two species are otherwise very similar in size, general appearance, calls and behaviour. They often occur in mixed flocks, and—at any distance too great to permit discrimination of colour—are difficult to distinguish.

The diet is mainly nectar, but the Scaly-breasted Lorikeet also eats pollen, seeds, fruit, and ripening grain.

The breeding season extends from May to December. The female apparently incubates alone, although the male may roost with her at night; both parents feed the young. The two eggs forming the usual clutch are laid in a tree cavity, frequently at a considerable height. The incubation period is 25 days, and the young fledge after 40 to 50 days.

HABITAT: most types of moist coastal forests, including rainforest and melaleuca and banksia scrub
LENGTH: *c* 23 cm
DISTRIBUTION: more than 1 million km²
ABUNDANCE: common
STATUS: probably secure

Rainbow Lorikeet

Trichoglossus haematodus (hee'-mah-toh'-dus: "blood-red hair-tongue")

The Rainbow Lorikeet is an abundant species throughout the coastal lowlands of northern and eastern Australia, but it has apparently declined in the south since European settlement and it is now uncommon south of about Sydney in New South Wales. An introduced population appears to be well established in Perth in Western Australia. Elsewhere it is widespread from Indonesia to Vanuatu and New Caledonia.

Brilliantly marked with a deep blue head, bright green back and yellow and scarlet underparts, the Rainbow Lorikeet is well known and unmistakable. The sexes are similar, there is no seasonal variation, and immatures closely resemble adults except in being somewhat duller. The population in the Top End is distinguished by a conspicuous red band across the nape, and this form is sometimes separated as a distinct species (the Red-collared Lorikeet).

It eats mainly nectar, but the diet is commonly supplemented by pollen, fruit, seeds, and occasionally insects. Active, belligerent, and noisy, it is strongly gregarious, and its screeching, high-flying or roosting flocks are a conspicuous feature of many eastern cities. It comes readily to garden feeders.

Breeding occurs mainly from August to January, at least in the south-east. Both parents care for the young but apparently only the female incubates. The nest is in a tree cavity, usually at a considerable height. The clutch consists of two or three eggs (usually two), which hatch in 26 days; the young fledge in about 50 to 55 days.

HABITAT: wet sclerophyll forest, rainforest, mangroves and other coastal forests; suburbs and urban areas with trees
LENGTH: *c* 30 cm
DISTRIBUTION: more than 1 million km²
ABUNDANCE: abundant
STATUS: secure

(JD Waterhouse)

Family PSITTACIDAE

(sit-ah'-sid-ee: "*Psittacus*-family", after a genus of parrots)

Doubtfully distinct from the cockatoos (Cacatuidae) and the lorikeets (Loriidae), the parrots constitute a family of about 260 species grouped in at least 60 genera; 31 species occur in Australia. Parrots occur on all continents except Antarctica but few extend into the northern hemisphere. The group is so large and diverse that there are exceptions to every generalization, but parrots are usually sedentary, gregarious, granivorous or frugivorous medium-sized arboreal birds that are best represented in tropical forests and often bright green in colour. Many Australian species, however, are multicoloured, inhabit the arid interior, and feed mainly on the ground.

In all species the upper bill articulates with the skull and, strongly curved, overlaps with the lower bill. The musculature of the jaw is very highly developed, and there is a muscular gizzard and a well-developed crop. The nostrils are set in a prominent naked cere. The family lacks members with movable crests.

Many parrots mate for life. Breeding occurs mainly in dispersed territorial pairs, but there are some colonial species. The clutch of plain white eggs is almost invariably deposited in a tree cavity and, usually, the female incubates unaided and broods the helpless young for the first few days while her mate feeds her at or near the nest. When the chicks are a few days old the male joins in their care. Once fledged, the young frequently remain with their parents until the following breeding season.

Genus *Alisterus*

(al'-is-ter-us: "Alister[-bird]", after Alister Mathews, son of ornithologist G. M. Mathews)

This is a genus of three very closely related species with a superficial resemblance in size, appearance and behaviour to some rosellas of the genus *Platycercus*. However, they differ in many anatomical details and lack the scalloped upperparts and contrasting cheek patches characteristic of the rosellas. They inhabit forests and feed on fruit, seeds, buds and insects. Males usually differ in appearance from females. Only one species occurs in Australia (the other two inhabit Indonesia and New Guinea).

King Parrot

Alisterus scapularis (scah'-pue-lah'-ris: "shouldered Alister-bird")

(G Little)

The King Parrot is common in forests of the coast and associated highlands in eastern Australia from about Cooktown in Queensland south to the vicinity of Melbourne. It is mainly sedentary or locally nomadic. The adult female resembles the male except that her head and upper breast are dull green, not bright red.

It roosts communally, but it is usually encountered in pairs or small parties. It spends much time in the foliage of trees, but frequently descends to forage on open grassy ground.

Active, noisy and conspicuous, as it is generally wary but often comes to garden bird feeders. It feeds mainly on seeds, nuts, berries and buds.

Breeding occurs mainly from July to January. The nest is a cavity high in a tree, usually a large eucalypt. Three to five eggs constitute the clutch. Only the female incubates, but she is fed at the nest by the male, who also helps feed the young once they are about half grown. The fledging period is about 35 days.

HABITAT: forest (including rainforest) and woodland; suburban parks and gardens

LENGTH: 41–43 cm

DISTRIBUTION: 300,000–1 million km²

ABUNDANCE: common

STATUS: probably secure

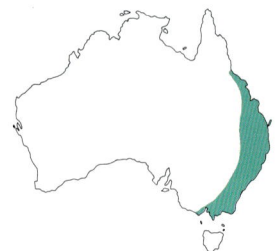

Genus Aprosmictus

(ah'-proh-smik'-tus: "solitary-bird")

The characteristics of the genus are those of the single species.

Red-winged Parrot

Aprosmictus erythropterus (e'-rith-rop'-te-rus: "red-winged solitary-bird")

The Red-winged Parrot is nomadic, and has an extensive but rather patchy distribution across northern Australia from the vicinity of Broome in Western Australia eastward to Cape York Peninsula and southward into northern New South Wales. It also occurs in southern New Guinea.

The only Australian parrot that is mainly green with large red wing patches, the Red-winged Parrot is unmistakable, but females are very much duller than males.

It lives in pairs or small family parties, sometimes congregating in larger flocks after breeding. Wary and rather difficult to approach, it spends most of its time in the foliage of trees, feeding on seeds, nectar, pollen and blossoms, and usually descending to the ground only to drink. Its flight is distinctive: wavering and deeply undulating, with hesitant wing beats.

Breeding occurs mainly from August to February in the south, May to October in the far north. The nest is a deep cavity, usually high in a tall eucalypt near water. The clutch consists of three to five eggs, which hatch in about 18 days. The young fledge and leave the nest at about 35 days, but remain with their parents indefinitely.

HABITAT: most kinds of temperate and tropical open woodland
LENGTH: 31–33 cm
DISTRIBUTION: more than 1 million km²
ABUNDANCE: common
STATUS: secure

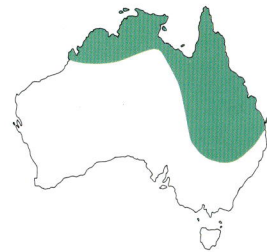

(D & M Trounson)

207

Genus *Barnardius*

(bar-nar'-dee-us: after E. Barnard, British ornithologist)

The characteristics of the genus are essentially those of the single species.

Ringneck

Barnardius barnardi (bar-nar'-dee: "Barnard's Barnard-bird", after E. Barnard, British ornithologist)

(GK Taylor)

The Ringneck is an abundant, widespread and familiar species across most of mainland Australia west of the Great Dividing Range. There are several populations, differing in many details of appearance but broadly in that western birds have black heads, south-eastern birds green, while birds inhabiting the Cloncurry–Mount Isa region of north-western Queensland are smaller and relatively pallid overall. All have a conspicuous band of yellow across the nape. Originating from escaped cage-birds, there are now growing feral populations in some south-eastern urban centres, such as Sydney.

In general behaviour the Ringneck closely resembles the rosellas, except in its preference for relatively arid environments. It is usually encountered in pairs or small parties.

It forages mainly on the ground, but roosts and shelters in trees. Generally tame, active and conspicuous, it drinks at dawn and dusk, forages during the early morning and late afternoon, and loafs through the midday heat in the shady crowns of trees. It feeds mainly on seeds, supplemented by flowers, buds, nectar and insects.

The breeding season varies with locality and with seasonal conditions, but is generally August to February in the south, February to July in the north. Sometimes two broods are raised in succession. The clutch of usually five or six eggs is deposited in a tree cavity, most often 2 to 5 metres from the ground in a eucalypt. Incubation takes about 19 to 20 days, and the young fledge at about 35 days. The female alone incubates but both parents participate in rearing the young.

HABITAT: most kinds of open and semiarid woodland and scrub; especially riparian woodland in the north, mallee in the east, and eucalypt forest (including coastal marri and karri forests) in the west; also roadside timber, farmland, orchards

LENGTH: 36–38 cm

DISTRIBUTION: more than 1 million km^2

ABUNDANCE: common

STATUS: secure

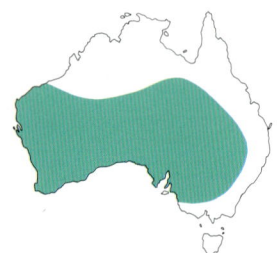

Genus *Eclectus*

(ek-lek'-tus: "chosen-[bird]")

The characteristics of the genus are essentially those of the single species.

Eclectus Parrot

Eclectus roratus (ror-ah'-tus: "bedewed chosen-bird")

In Australia the Eclectus Parrot is confined to the rainforests of eastern Cape York Peninsula, from the Pascoe River to Massey Creek, but elsewhere it is widespread, from the Moluccas through New Guinea to the Solomon Islands. It is sedentary, and generally common.

A large, conspicuous and impressive species, the Eclectus Parrot is notable for the striking difference in appearance between the sexes: the adult male is mainly bright green, with red flanks and underwing coverts, whereas the female is mainly deep red, with clear blue underparts. Immatures are much duller but somewhat resemble adults of their sex.

It is usually encountered alone or in pairs, or occasionally in small parties. It roosts communally. Strictly arboreal, it spends most of its time in the upper foliage. When travelling, it flies high above the canopy. The varied diet includes seeds, nuts and fruits. Its calls include a harsh, double screech.

Breeding occurs mainly from August to January. The nest is a deep cavity high in a tree. The clutch consists of two eggs, which hatch in about 26 days. The chicks fledge at about 60 days. Only the female incubates but both parents care for the young, sometimes assisted by other birds.

HABITAT: rainforest and adjacent open forest
LENGTH: *c* 42 cm
DISTRIBUTION: 10,000–30,000 km^2
ABUNDANCE: sparse
STATUS: vulnerable

(H & J Beste)

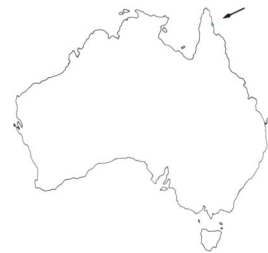

Genus *Geoffroyus*

(jef-roy'-us: "Geoffroy [-bird]", after E. Geoffroy Saint-Hilaire, French zoologist)

Colourful and conspicuous birds inhabiting mainly lowland rainforest and feeding largely on fruit, these parrots are stocky and medium-sized, with relatively short tails. The sexes differ in appearance, and several anatomical details indicate an affinity with *Eclectus*. There are three species: two in New Guinea (one of which also occurs at the tip of Cape York Peninsula) and another in the Bismarck Archipelago and the Solomon Islands.

Red-cheeked Parrot

Geoffroyus geoffroyi (jef-roy'-ee: "Geoffroy's Geoffroy-bird", after E. Geoffroy Saint-Hilaire, French zoologist)

The Red-cheeked Parrot is widespread in Indonesia and New Guinea but in Australia it occurs only in the rainforests of north-eastern Cape York Peninsula, between the Pascoe and Rocky Rivers, where it is common. It is sedentary.

It sometimes congregates in small parties, but it is usually encountered alone or in pairs. Roosting is communal. It seldom descends from the upper canopy of forest, where it forages for seeds, fruits, nuts and berries. Its varied calls include loud double honks and screeches. When travelling, it often flies high above the canopy.

Little is known of the breeding cycle, but nests have been found from August to November. The clutch consists of three eggs, laid in a cavity high in a tree. The female incubates, but both parents rear the young.

HABITAT: rainforest; also mangroves and adjacent open forest
LENGTH: *c* 22 cm
DISTRIBUTION: 10,000–30,000 km²
ABUNDANCE: rare
STATUS: endangered

(WS Peckover)

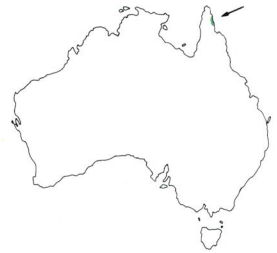

Genus *Geopsittacus*

(jee'-oh-sit'-ah-kus: "earth-parrot")

The characteristics of the genus are essentially those of the single species.

Night Parrot

Geopsittacus occidentalis (ok'-sid-ent-ah'-lis: "western earth-parrot")

Late in 1990, a dead Night Parrot was found by the side of the road near Boulia, northern Queensland. The first specimen to have been collected since 1912, it provided undeniable physical evidence in support of sporadic reports of sightings of the species over the intervening period.

Historical records show that the Night Parrot once had an extensive distribution across the arid interior of the continent. Although its range has contracted and its abundance has diminished, it is reasonable to suppose that it is surviving, however patchily, over a considerable area with spinifex (*Triodia*) and samphire-like (*Sclerolaena*) vegetation.

Being nocturnal, almost entirely terrestrial, and secretive, the Night Parrot is unlikely to be seen unless intensively searched for.

Almost nothing is known of its biology but there is some slight evidence that it may feed mostly on *Triodia* seeds.

HABITAT: spinifex plains; samphire flats around salt lakes and along large watercourses
LENGTH: *c* 23 cm
DISTRIBUTION: 100,000–300,000 km²
ABUNDANCE: rare
STATUS: endangered

Genus *Lathamus*

(lay'-tham-us: "Latham [-bird]", after J. Latham, British ornithologist)

The characteristics of the genus are essentially those of the single species.

Swift Parrot

Lathamus discolor (dis'-kol-or: "varicoloured Latham-bird")

The Swift Parrot is a common species, which breeds only in Tasmania but migrates across Bass Strait to spend the winter in the south-eastern mainland States, mainly in Victoria but casually and erratically north to about Mackay in Queensland and west to Adelaide in South Australia.

It is strongly gregarious in all activities, but is usually encountered in small parties rather than large flocks. It freely descends to the ground to drink, but is otherwise entirely arboreal, foraging mainly in the canopy and favouring the crowns and topmost foliage. It feeds on nectar and pollen, a diet sometimes supplemented with insects and their larvae. Aptly named, it commutes between flowering trees in low, direct and rocketing flight. In much of its behaviour it resembles lorikeets (with which it often associates) rather more than other parrots, being noisy, active and conspicuous. The sexes are similar.

Breeding occurs from October to December, and is often loosely colonial. The nest is a cavity, usually high in a eucalypt, and three to five eggs constitute the clutch. Only the female incubates but both parents participate in rearing the young, which fledge at about 42 days.

HABITAT: most kinds of forest and woodland; suburban parks and gardens
LENGTH: 23–26 cm
DISTRIBUTION: 300,000–1 million km²
ABUNDANCE: sparse
STATUS: vulnerable

(*LF Schick*)

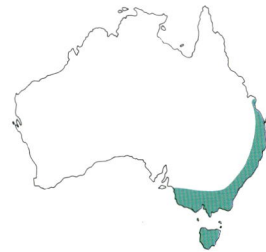

Genus *Melopsittacus*

(mel'-op-sit-ah-kus: "song-parrot")

The characteristics of the genus are essentially those of the single species.

Budgerigar

Melopsittacus undulatus (un'-due-lah'-tus: "wave [-marked] song-parrot")

The Budgerigar occurs virtually throughout mainland Australia west of the Great Dividing Range, avoiding only the far south-west and the Top End. It is often extremely abundant, but it is also highly mobile, its erratic and extensive movements tending to result in a concentration in the north in winter, the south in summer.

Because of its abundance in captivity, escaped pets and cage birds also occur frequently in urban areas, especially in the south-east.

Intensely gregarious, it usually occurs in flocks of up to 100 birds, but it often congregates in vast hordes. It feeds mainly on the ground, shelters and roosts in trees, and regularly visits waterholes to drink, especially at midday. It is active, tame, noisy and conspicuous, and flocks often indulge in swift, agile, and well-coordinated aerial maneouvres. It feeds mainly on the seeds of grasses, especially those of species of *Triodia* and *Astrebla*. Its calls include a variety of chattering notes, as well as a distinctive warbling note uttered usually in flight.

Breeding may occur at any time of year in immediate response to local rainfall, but tends to reach peaks of activity between August and January in the south, and June and September in the north. Breeding is often loosely colonial, and several broods may be raised in succession. The nest is a cavity in almost any available tree, fallen log or fence post. The clutch consists usually of four to six eggs, which hatch in about 18 days. The female alone incubates, but both parents participate in rearing the young, which fledge at about 35 days.

HABITAT: most kinds of arid open country, but seldom far from water
LENGTH: *c* 18 cm
DISTRIBUTION: more than 1 million km^2
ABUNDANCE: abundant
STATUS: secure

(LF Schick)

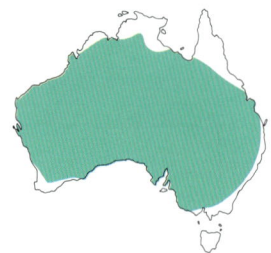

Genus *Neophema*

(nee'-oh-fem'-ah: "new-voice")

The seven species in this genus all inhabit southern Australia and are mainly characteristic of open, well-grassed woodland and open country. They are small, slim-bodied and long-tailed, forage mainly on the ground and feed almost exclusively on the seeds of various grasses and herbs. They are gregarious, and the distribution of some species overlaps extensively with others in the genus; they often occur in mixed flocks, especially in winter. Few are sedentary and several are strongly migratory. With two exceptions, sexual dimorphism is slight but young birds are much plainer than adults.

Bourke's Parrot

Neophema bourkii (ber'-kee-ee: "Bourke's new voice" after Sir Richard Bourke, a Governor of New South Wales)

Bourke's Parrot is widespread and generally common across the arid interior of southern Australia from western New South Wales westward

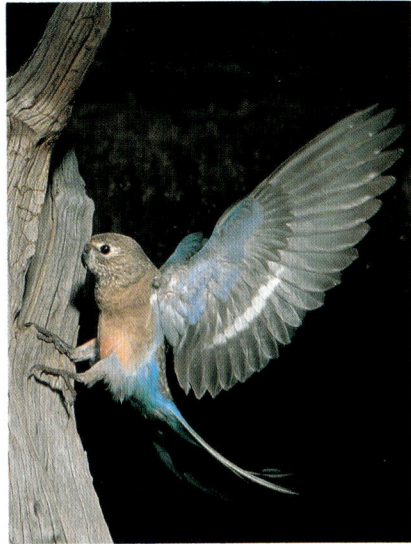

(B & B Wells)

to North West Cape and Shark Bay, Western Australia. It is mainly sedentary or locally nomadic. Distinctively clad in pastel shades of brown, pink and blue, it is easily recognised. The sexes are approximately similar in appearance.

Bourke's Parrot roosts communally and sometimes occurs in large flocks, but it is usually encountered in pairs or small parties. It congregates at waterholes to drink at dawn and dusk, often while it is still dark. Quiet, unobtrusive and crepuscular in behaviour, it feeds mainly on seeds, especially those of various grasses and of species of *Acacia* and *Cassia*.

Most breeding occurs from August to November. The clutch of three to six eggs is deposited in a tree cavity

and incubated by the female for about 18 days. Both parents participate in rearing the young, which fledge in about 28 days.

HABITAT: mainly arid Mulga scrub
LENGTH: *c* 19 cm
DISTRIBUTION: more than 1 million km²
ABUNDANCE: very sparse
STATUS: vulnerable

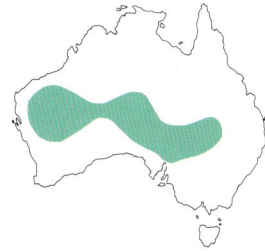

Orange-bellied Parrot

Neophema chrysogaster (krie'-soh-gas'-ter: "golden-bellied new-voice")

The total population of the Orange-bellied Parrot is probably fewer than 500 individuals, and it must be regarded as among the rarest and most gravely threatened of all

(D Watts)

Australian birds. It breeds only in remote areas of south-western Tasmania, migrating across Bass Strait to spend the winter on coastal flats around Port Phillip Bay and (formerly) westwards along the Victorian coast to Lake Alexandrina in South Australia.

It is difficult to distinguish from several other *Neophema* parrots, but the overall colour impression is distinctly bright green rather than yellowish or olive as in others of its genus, and there is a patch of orange on the belly. The sexes differ little in appearance.

On its breeding ground it is usually encountered in pairs or family parties, but small flocks congregate in winter. Closely resembling other *Neophema* species in behaviour, it forages almost exclusively on the ground, feeding mainly on the seeds of grasses, herbs and salt-loving plants.

Breeding occurs from November to February. The clutch of usually

four to six eggs is deposited in a tree cavity, usually in a eucalypt. The female incubates alone for 20 to 21 days, and both parents rear the young to fledging at about 30 to 35 days.

HABITAT: tidal and samphire flats, grasslands, salt marshes in winter; when breeding, moorland and open woodland
LENGTH: 20–21 cm
DISTRIBUTION: 10,000–30,000 km²
ABUNDANCE: rare
STATUS: endangered

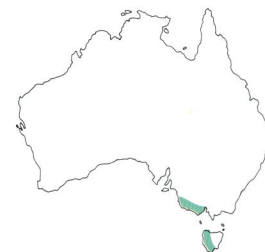

Blue-winged Parrot

Neophema chrysostoma (krie'-soh-stoh'-mah: "golden-mouthed new-voice")

(*L Robinson*)

The Blue-winged Parrot breeds in Tasmania, Victoria, and the South-East district of South Australia but, after breeding, many individuals migrate northwards to spend the winter on the saltbush and bluebush plains of South Australia and western New South Wales. It is generally common.

Large flocks often congregate in winter, but during the breeding season it occurs mainly in pairs or small parties. Generally rather quiet and inconspicuous in behaviour, it forages almost entirely on the ground, but roosts and shelters in trees. It feeds mainly on the seeds of various grasses.

Most breeding occurs between October and February, and only one brood is raised in a season. The clutch of four to six eggs is deposited in a tree cavity and incubated by the female for 20 days. Both parents cooperate in raising the young, which fledge at about 30 days.

HABITAT: breeds in open woodland and heath, migrating to saltbush and bluebush steppe during winter
LENGTH: 20–21 cm
DISTRIBUTION: 300,000–1 million km²
ABUNDANCE: sparse to common
STATUS: probably secure

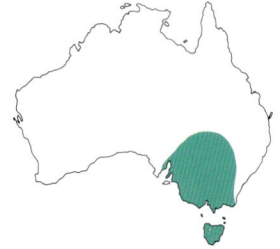

Elegant Parrot

Neophema elegans (el'-e-ganz: "elegant new-voice")

(*LF Schick*)

The Elegant Parrot occurs in south-eastern Australia from far south-western New South Wales to the Mount Lofty Ranges and Kangaroo Island in South Australia, and in south-western Western Australia from about Esperance to Perth. Partly nomadic, it is uncommon at the eastern fringes of its distribution but is expanding vigorously in the south-west.

Like other *Neophema* parrots, it is tame, quiet and unobtrusive in behaviour, foraging almost entirely on the ground and feeding largely on grass seeds. Large flocks may congregate in winter but otherwise it is usually encountered in pairs or small parties. It sometimes associates with Blue-winged Parrots; the two species are very similar but the Elegant Parrot differs in that the blue panel on the wings is distinctly two-toned, and it has more extensive yellow about the face.

Its flight is high, swift and direct, and its calls include a distinctive, terse "zizt-zizt", usually uttered in flight.

Breeding occurs mainly from August to January. Four or five eggs constitute the normal clutch, laid in a small tree cavity near the ground and incubated for about 18 days. Reared by both parents, the young fledge at about 30 to 35 days.

HABITAT: heathland and open country, open woodland, cropland and semiarid shrubland
LENGTH: *c* 22 cm
DISTRIBUTION: 300,000–1 million km²
ABUNDANCE: sparse to common
STATUS: probably secure

Rock Parrot

Neophema petrophila　(pet-rof'-il-ah: "rock-loving new-voice")

(D & V Blagden)

Seldom seen far from the seacoast, the Rock Parrot inhabits southern Australia from the vicinity of Kingston to Ceduna in South Australia and from Cape Arid National Park to Geraldton in Western Australia. It is partly migratory and locally common, though it has recently declined markedly at the eastern fringes of its range.

Rarely congregating in large flocks, it is usually encountered in small parties. It often perches on bushes, but it is otherwise almost entirely terrestrial, feeding on the seeds and fruits of grasses, shrubs and various beachside plants. Like other *Neophema* species, it is quiet, unobtrusive and undemonstrative in general behaviour.

Breeding occurs mainly from August to December, almost invariably on small offshore islands. The normal clutch is four or five eggs, deposited in a rock crevice or ledge, frequently sheltered behind a hanging veil of succulent shrubs. Incubation takes about 18 days, and the young fledge at about 30 days but remain loosely associated with their parents throughout the subsequent winter.

HABITAT: breeds on offshore islands; in winter, tidal flats, salt marshes, sand dunes, rocky foreshores
LENGTH: *c* 22 cm
DISTRIBUTION: 100,000–300,000 km^2
ABUNDANCE: sparse to common
STATUS: probably secure

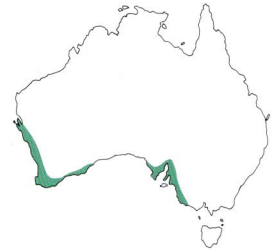

Turquoise Parrot

Neophema pulchella　(pool-kel'-ah: "pretty new-voice")

The Turquoise Parrot is widespread but erratically distributed in south-eastern Australia from south-eastern Queensland to northern Victoria, mainly west of the Great Dividing Range but locally extending to the east coast. For reasons that are still unclear, it suffered a catastrophic decline between 1900 and 1920 and became extremely rare. It gradually recovered through the 1940s and 1950s until it has now apparently regained much of its former status.

it is mainly sedentary or locally nomadic.

The adult male is immediately recognisable by the presence of a patch of dull red at the shoulder, but in the female this feature is absent, and she is easily confused with the females of several other *Neophema* parrots.

The Turquoise Parrot lives in pairs or small family parties and, like other *Neophema* parrots, forages almost exclusively on the ground, though roosting and sheltering in trees. It feeds on the seeds of various native grasses and of some shrubs and herbaceous plants. It drinks daily at dawn. In behaviour it is generally quiet, confiding and undemonstrative.

It breeds mainly from August to December, and several broods may be raised in succession. The clutch of four or five eggs is deposited in a cavity in a tree. Incubation takes about 18 days, and the young fledge at about 30 days.

HABITAT: mainly open eucalypt woodland, especially with a grassy understorey and rocky outcrops
LENGTH: *c* 20 cm
DISTRIBUTION: 100,000–300,000 km^2
ABUNDANCE: sparse
STATUS: probably secure

(D Schick)

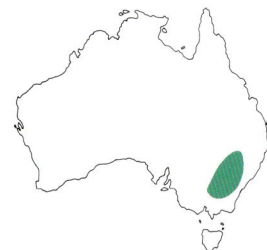

215

Scarlet-chested Parrot

Neophema splendida (splen'-did-ah: "bright new-voice")

The beautiful Scarlet-chested Parrot is widely but erratically distributed across the arid interior of southern Australia from far western New South Wales to the vicinity of Kalgoorlie, Western Australia. Its movements are inadequately understood: sometimes locally common, it seems subject to local irruptions but is normally rarely reported.

The sexes are similar, except that the female is much duller than the male, and has a dull green, not scarlet, breast.

Usually encountered in pairs or small flocks, it closely resembles other *Neophema* parrots in behaviour: it is tame and unobtrusive, forages almost entirely on the ground, and feeds largely on the seeds of grasses (especially *Triodia*).

Breeding occurs mainly from August to December, and is often loosely colonial. The clutch of three to six eggs is deposited in a cavity in a tree, and is incubated by the female for about 18 days. Reared by both parents, the young fledge at about 30 days.

HABITAT: arid mallee and acacia scrub, especially with *Triodia* ground cover and a sandy substrate
LENGTH: *c* 20 cm
DISTRIBUTION: more than 1 million km^2
ABUNDANCE: rare
STATUS: possibly endangered

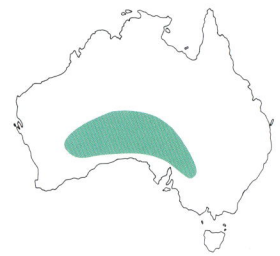

(*LF Schick*)

Genus Northiella

(nor'-thee-el'-ah: "little North[-bird]", after A. J. North, Australian ornithologist)

The characteristics of the genus are essentially those of the single species.

Bluebonnet

Northiella haematogaster (hee'-mah-toh-gas'-ter: "blood-bellied North-bird")

Mainly pale brown with an indigo face and yellow underparts, this unusually coloured parrot is locally common but erratically distributed across much of interior southeastern and southern Australia. Birds of the western population occupying the Nullarbor Plain are distinctly smaller than eastern birds and differ in several plumage details, and are sometimes regarded as a distinct species. Bluebonnets are generally sedentary or locally nomadic.

The Bluebonnet is believed to mate for life, and usually occurs in pairs or small family parties, only occasionally in large flocks. Like other arid zone parrots, it feeds mainly on the ground but roosts and shelters in trees, and drinks mainly at dawn and dusk. Its varied diet consists largely of the seeds of various grasses and herbaceous plants but also includes nectar, flowers and fruit. It has a very distinctive sharp double clucking call, uttered chiefly in flight.

Breeding is strongly influenced by local rainfall but tends to reach a peak of activity between August and January. The clutch of four to nine eggs is deposited in a tree cavity, often very small and usually within 5 or 6 metres of the ground. Incubation takes about 19 days and the young fledge at about 42 days.

HABITAT: arid and semiarid scrub and cropland; Mulga scrub, mallee, dry open woodland
LENGTH: 28–34 cm
DISTRIBUTION: more than 1 million km^2
ABUNDANCE: sparse to common
STATUS: probably secure

(*M Seyfort*)

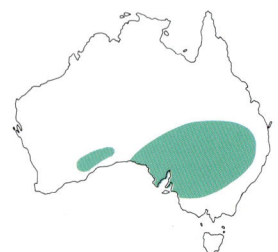

Genus *Pezoporus*

(payz'-oh-por'-us: "pedestrian[-bird]")

The characteristics of the genus are essentially those of the single species.

Ground Parrot

Pezoporus wallicus (wahl'-ik-us: "[New-South-]Wales pedestrian-bird")

The Ground Parrot is most numerous in Tasmania but it also occurs elsewhere in south-eastern Australia, mainly along the coast and associated highlands, from Fraser Island in Queensland to the Adelaide Plains in South Australia. A population in south-western Australia is seldom reported and its status is uncertain. It is highly specialised and extremely sensitive to habitat modification, being dependent on a precise fire regime; its distribution is accordingly much fragmented and its numbers have been drastically reduced since European settlement. It has been almost extirpated in many areas. It is mainly sedentary, but there is a pronounced dispersal after breeding.

The Ground Parrot is entirely terrestrial and largely nocturnal, roosting singly by day in dense vegetation. Its diet includes green shoots and the seeds of various grasses and herbs. Secretive and elusive, it is usually encountered only when inadvertently flushed, and its presence is best detected by its distinctive calls: a series of three or four mellow, clear, and high-pitched whistled notes, uttered from dense cover at dawn and dusk.

Adults are thought to mate for life and defend permanent territories, but the role of the sexes in the nesting cycle is uncertain. Breeding occurs mainly from September to December. Well hidden, the nest is a shallow saucer of plant stems and leaves on or near the ground. The clutch consists of three or four eggs, which hatch in about 21 days. The young leave the nest at about 20 days and fledge at about 30 days.

HABITAT: low heath; swampy sedgeland; buttongrass plains; occasionally estuarine flats, pastures
LENGTH: 29–30 cm
DISTRIBUTION: 30,000–100,000 km²
ABUNDANCE: very sparse
STATUS: vulnerable

(J Gray)

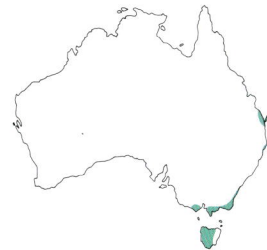

Genus *Platycercus*

(plat'-ee-ser'-kus: "flat-tail")

The rosellas are among the most common, familiar and widespread of Australian parrots. Seeds constitute the chief component of the varied diet, and all feed primarily on the ground. Rosellas are medium-sized parrots with long, graduated tails and long, pointed wings. In all species the back is black, but each feather is broadly rimmed with some other colour (green, yellow or red) that varies with the species, and all have patches of colour on the lower cheek that contrast with the remainder of the plumage. The various populations fall into three groups on the basis of the colour of the cheeks (white, yellow or blue), but species limits within the rosellas remain a matter of debate (some authorities recognizing seven species). In all but the Western Rosella the sexes are virtually identical in appearance.

Green Rosella

Platycercus caledonicus (kal'-ed-on'-ik-us: "[New-]Caledonian flat-tail", a name given in error to a Tasmanian specimen)

The Green Rosella is confined to Tasmania and the islands of Bass Strait, where it is very common. It is mainly sedentary or locally nomadic. It differs strikingly in appearance from the Crimson Rosella of mainland Australia, being mainly green in colour instead of red, but the two forms are very close in general behaviour, and some researchers prefer to combine the two in a single species.

Breeding occurs mainly from November to February. The clutch of four or five eggs is deposited in a cavity, usually high in a tall eucalypt, and hatches in about 20 to 21 days. The young fledge at about 35 days.

(*D Watts*)

HABITAT: most kinds of wooded country; heath, farmland, suburban parks and gardens
LENGTH: *c* 37 cm
DISTRIBUTION: 30,000–100,000 km²
ABUNDANCE: common
STATUS: secure

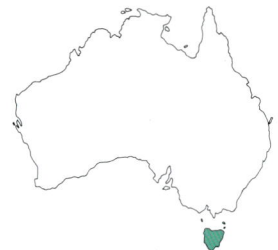

Crimson Rosella

Platycercus elegans (el'-eg-anz: "elegant flat-tail")

The abundant and familiar Crimson Rosella is widespread in eastern and south-eastern Australia from the vicinity of Cairns to south-eastern South Australia. It has been introduced to Norfolk Island and to New Zealand. It is mainly sedentary, and occurs in several populations, formerly regarded as species, that are conveniently referred to as the "red", "yellow" and "orange" groups: the bird of south-eastern forests is mainly crimson, but similar birds occurring in riverine forests of the Murray–Darling Basin have the red areas replaced with yellow, and a population (widely known as the Adelaide Rosella) centred in the Mount Lofty and Flinders Ranges of South Australia is considered to represent a hybrid swarm between the "red" and "yellow" forms, and is essentially orange in colour. The complex may well be extended to include the Green Rosella of Tasmania, and the name "Blue-cheeked" Rosella has been proposed to cover the entire assemblage.

It is usually encountered in pairs or small parties, but immatures sometimes congregate in large flocks. It feeds on the ground or in trees, and its diet includes a wide range of seeds (especially of eucalypts), fruits and blossoms, as well as insects and their larvae. Although often rather quiet and unobtrusive when feeding, it is generally active, noisy and conspicuous. It often visits forest picnic areas and similar sites, where it frequently becomes so tame as to permit feeding by hand.

Most breeding occurs from September to January. The usual clutch consists of five eggs, deposited in a cavity, usually high in a eucalypt. Incubation (by the female) takes 19 to 21 days, and the young fledge at about 35 days, though they normally continue in the company of their parents for a further month or more.

(*EE Zillmann*)

HABITAT: eucalypt forest, woodland and adjoining habitats
LENGTH: *c* 36 cm
DISTRIBUTION: 300,000–1 million km²
ABUNDANCE: common
STATUS: secure

White-cheeked Rosella

Platycercus eximius (ex-im'-ee-us: "excellent flat-tail")

(IP Rowles)

in a tree.

Breeding occurs from August to January in the south, from June to September in the north, and sometimes two broods are raised in succession. The clutch consists of usually five eggs, deposited in a tree hollow. The female incubates but both parents participate in rearing the young. The eggs hatch in about 19 days and the young fledge at about 35 days, but they often remain with their parents until the following breeding season.

HABITAT: mainly open eucalypt woodland; also melaleucas, heath, savannah woodland, shrubs, cropland, suburban parks and gardens
LENGTH: 28–30 cm
DISTRIBUTION: more than 1 million km²
ABUNDANCE: common
STATUS: secure

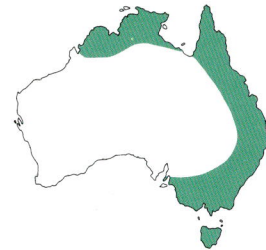

This rosella is abundant and widespread in eastern Australia from Cape York to Tasmania, and in the Top End from the Kimberley to the Queensland border, but it is much less numerous in the tropics than in the south-east; it has been introduced to New Zealand. Generally sedentary or locally nomadic, it occurs in three populations differing in appearance, and formerly regarded as distinct species (called Eastern, Pale-headed and Northern Rosellas, respectively): south-eastern birds have red heads and yellow underparts; Queensland birds have pale yellow heads and blue underparts;

and Top End birds have black heads and pale yellow underparts. All differ from other rosellas in having white cheeks.

It is believed to mate for life, and occurs in pairs or family parties which congregate into larger flocks during winter. It is unobtrusive when feeding, but is generally active, vocal and conspicuous. It forages on the ground or in trees, feeding on seeds, fruits, nectar, blossoms and insects. Its vocabulary is extensive and includes a clear, high-pitched ringing call uttered in flight, soft conversational chattering notes, and a soft, clear piping call uttered from a perch

Western Rosella

Platycercus icterotis (ik'-ter-oh'-tis: "yellow-eared flat-tail")

(G Rogerson)

December. The clutch of usually five eggs is deposited in a tree cavity and hatches in about 19 days. The young fledge at about 35 days.

HABITAT: most kinds of open forest, woodland; heath, farmland, orchards
LENGTH: *c* 26 cm
DISTRIBUTION: 300,000–1 million km²
ABUNDANCE: common
STATUS: probably secure

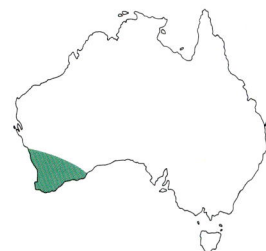

This rosella is confined to far southwestern Australia. It is sedentary. It differs from other rosellas in being substantially smaller, in having yellow cheeks, and in that adult females are readily distinguishable from

males in being very much duller.

Like other rosellas, it lives in pairs or small parties that may congregate in flocks during winter. It feeds on seeds, nuts, fruits and insects.

Breeding occurs from August to

Genus *Polytelis*

(pol'-ee-tay'-lis: "magnificent [-bird]")

This genus consists of three species confined to Australia and shows a number of features linking *Alisterus* and *Aprosmictus* with *Platycercus* and *Barnardius*. All members are medium-sized parrots with very long tails, in which the central pair of feathers greatly exceed the others in length. The bill is small and slender. Sexual dimorphism is pronounced in two species, very slight in the third. These parrots inhabit arid and semiarid woodlands and scrub, forage in the trees or on the ground, and feed on seeds, fruit, buds and flowers.

Princess Parrot

Polytelis alexandrae (al-ex-and'-ree: "[Princess-]Alexandra's magnificent-bird")

One of the most elusive and least known of Australian parrots, the Princess (or Alexandra's) Parrot is widely but erratically distributed across interior northern Australia. It is strongly nomadic.

It is usually encountered in small parties. Tame and unobtrusive in general behaviour, it forages mainly on the ground, feeding largely on the seeds of grasses and shrubs.

Breeding may occur at any time, and is often loosely colonial. The nest is a cavity in a tree, and four to six eggs constitute the clutch. Incubation takes about 21 days, and the young fledge at about 42 days.

HABITAT: arid woodland and scrub, especially Mulga; spinifex
LENGTH: *c* 40 cm
DISTRIBUTION: 300,000–1 million km²
ABUNDANCE: rare
STATUS: endangered

(D & M Trounson)

Regent Parrot

Polytelis anthopeplus (an'-thoh-pep'-lus: "flower-robed magnificent-bird")

The Regent Parrot has a very restricted distribution in south-eastern Australia but it is widespread and common in the south-west. Both populations have, however, declined markedly since European settlement, and the south-eastern population is probably threatened.

In general behaviour it strongly resembles the Superb Parrot, but it is nomadic, strongly favours mallee for foraging, and feeds mainly on the ground. Its diet includes seeds, fruits, nuts, blossoms and leaf buds, as well as spilled cereal grains.

Breeding occurs mainly from August to December. The nest is a deep cavity, usually high in a River Red Gum (in the east) or a Wandoo (in the west) near water. The clutch usually consists of four eggs, which hatch in about 21 days. The male brings food to the female while she incubates; the young fledge at about 40 days but remain with their parents for a month or more thereafter.

HABITAT: in the east, mainly River Red Gum forest and woodland with adjacent mallee; in the west, especially Wandoo and Salmon Gum forest, agricultural land, suburban parks and gardens
LENGTH: 39–41 cm
DISTRIBUTION: 300,000–1 million km²
ABUNDANCE: sparse to common
STATUS: vulnerable

(PD Munchenberg)

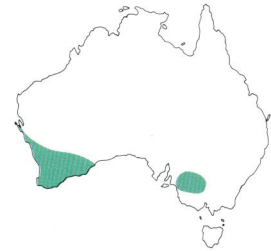

Superb Parrot

Polytelis swainsonii (swane'-sun-ee-ee: "Swainson's magnificent-bird", after W. Swainson, English ornithologist)

(IP Rowles)

Virtually restricted to interior New South Wales, the Superb Parrot is confined as a breeding bird to River Red Gum forests along the Murray and Edwards Rivers near Deniliquin and Mathoura; along the Murrumbidgee River from Wagga Wagga to about Darlington Point; and along the Lachlan River to near Cowra. Many birds disperse northwards after breeding to the vicinity of Narrabri, Gunnedah, and Coonabarabran. It remains fairly common, but it has declined markedly since European settlement, a decline that appears to have accelerated since about 1970 and now offers some cause for concern.

The adult male is bright green, with a yellow head and a band of scarlet across the breast; the female is similar but the head is green, and she lacks red on the breast.

The Superb Parrot lives in small flocks which disperse in pairs to breed. Foraging in trees or on the ground, it is often active in the early morning and in the late afternoon, but on hot days it spends most of the time idling unobtrusively in the shady crowns of trees. Its varied diet includes seeds, nectar, blossoms, fruits and insects; it is also fond of spilled cereal grains, gathered around silos and along the margins of country roads. Its calls have a distinctive husky quality.

Breeding occurs from September to December. The nest is a deep cavity high in a large River Red Gum near water. The clutch consists of three to six eggs, which hatch in about 20 days. The female alone incubates but both parents feed the young, which fledge at about 35 to 45 days.

HABITAT: forest and woodlands (especially River Red Gum); box or mixed box and White Cypress Pine woodlands; cereal crops
LENGTH: *c* 40 cm
DISTRIBUTION: 100,000–300,000 km²
ABUNDANCE: sparse to common
STATUS: vulnerable

Genus *Psephotus*

(sef-oh'-tus: "jewelled [-bird]")

The members of this genus bear a superficial resemblance to the rosellas in general appearance and behaviour but they are smaller and slimmer, lack well-marked cheek patches, have uniformly coloured backs and less robust bills, and in general inhabit more arid environments. Males differ substantially from females in appearance. Exclusively Australian in distribution, the genus has five member species, including one that has almost certainly become extinct since European settlement.

Golden-shouldered Parrot

Psephotus chrysopterygius (krie'-sop-te-ri'-jee-us: "golden-winged jewelled-bird")

Now very rare and severely threatened, the Golden-shouldered Parrot is confined to the south-western part of Cape York Peninsula, though it has not been recorded during this century from the southern part of its range. It is locally nomadic, tending to disperse away from its breeding grounds after nesting.

The adult male is multicoloured and unmistakable, but the female is essentially plain light green, washed with bronze on the crown and nape, and pale blue on the face and belly.

It is usually encountered in pairs or small family parties. It roosts and shelters in trees but forages almost entirely on the ground, its diet being largely restricted to the seeds of various grasses, only occasionally supplemented by small fruits and insects. Easily overlooked, it is quiet and undemonstrative in most of its actions.

Breeding is sometimes prolonged but occurs mainly from April to June. The nest is a chamber at the end of a tunnel drilled by the birds into a terrestrial termite mound. Three to six eggs constitute the clutch. The male feeds his mate at the nest while she incubates (for 19 to 20 days) and assists in rearing the young, which fledge at about 35 days.

(H & J Beste)

HABITAT: mainly open savannah woodland studded with large termite mounds
LENGTH: *c* 26 cm
DISTRIBUTION: 30,000–100,000 km²
ABUNDANCE: rare
STATUS: endangered

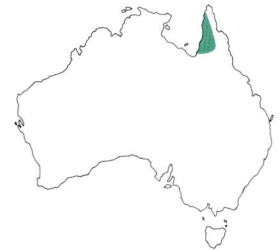

Hooded Parrot

Psephotus dissimilis (dis-sim'-il-is: "different jewelled-bird")

The Hooded Parrot is confined to northern parts of the Northern Territory, where it is generally uncommon and possibly declining. It is perhaps the same species as the Golden-shouldered Parrot of Cape York Peninsula, but the Hooded Parrot is the larger bird, and it differs also in lacking the yellow frontal band that is characteristic of the Golden-shouldered Parrot and in several other plumage details. The female differs from the female Golden-shouldered Parrot in having red undertail coverts.

The two forms are also very similar in behaviour, except that the Hooded Parrot is more gregarious, often occurring in flocks of up to about 30 birds. It feeds almost exclusively on grass seeds.

Breeding occurs mainly from August to December. The clutch of three to six eggs is deposited in a chamber at the end of a tunnel drilled by the breeding birds in a termite mound. Incubation takes about 19 to 20 days, and the young fledge at about 35 days.

(G Chapman)

HABITAT: mainly open savannah woodland with abundant termite mounds
LENGTH: *c* 26 cm
DISTRIBUTION: 100,000–300,000 km²
ABUNDANCE: sparse
STATUS: possibly endangered

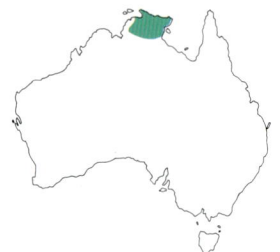

Red-rumped Parrot

Psephotus haematonotus (hee'-mah-toh-noh'-tus: "blood-backed jewelled-bird")

The Red-rumped Parrot is abundant and widespread across most of south-eastern Australia from the vicinity of Brisbane in Queensland to Kangaroo Island and Yorke Peninsula in South Australia. It has profited from some changes in land use by humans and is now common in many eastern coastal districts where it was formerly rare. It is mainly sedentary or locally nomadic.

Small, plump, long-tailed, and mainly bright green with a red rump, the adult male is unmistakable, but the female is among the drabbest of Australian parrots, being dull olive and lacking any conspicuous markings.

It is gregarious, roosting communally, foraging in small parties and sometimes congregating in large flocks. It roosts and shelters in trees but forages mainly on the ground, gathering seeds from a variety of grasses and herbaceous plants. It is generally active, tame and conspicuous, and its mannerisms have a distinctively brisk, bustling quality. It drinks at dawn and dusk, and intermittently through the day. Its calls include a trilled, warbling note, and an abrupt "seweet-seet", usually uttered in flight.

Breeding occurs from August to January, and is often loosely colonial. The clutch ranges from four to seven eggs, but five is usual; they are deposited in a cavity in a tree or sometimes in a fence post, fallen log or similar site. The male feeds his mate while she incubates, and also participates in rearing the young. Incubation takes 19 to 20 days and the young fledge at about 28 days.

HABITAT: farmland; roadside scrub; urban and suburban parks and gardens; lightly wooded grasslands; riparian woodland in the arid zone; seldom far from water
LENGTH: *c* 27 cm
DISTRIBUTION: more than 1 million km²
ABUNDANCE: abundant
STATUS: secure

(*T Howard*)

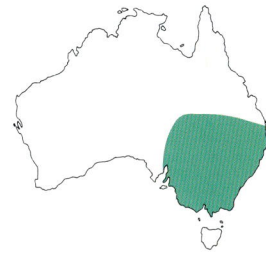

Paradise Parrot

Psephotus pulcherrimus (pool-ke'-rim-us: "very-pretty jewelled-bird")

The Paradise Parrot has not been reported with certainty since 1927, and is now almost certainly extinct. It formerly occurred from central interior Queensland southward into northern New South Wales. Almost nothing is known of its movements.

It was usually encountered in pairs or small parties, foraging almost exclusively on the ground and feeding on the seeds of various grasses. It was reportedly tame, and most active in the early morning and late afternoon.

Little is known of breeding: the few recorded nests span the months of September to May. The clutch of three to five eggs was deposited in a chamber at the end of a tunnel in a termite mound, or occasionally in the bank of a dry creek-bed or in a cavity in a tree.

(*Australian Museum*)

HABITAT: arid savannah woodland with an abundance of termite mounds
LENGTH: *c* 27 cm
DISTRIBUTION: unknown
ABUNDANCE: unknown
STATUS: possibly extinct

Mulga Parrot

Psephotus varius (vah'-ree-us: "varied jewelled-bird")

The Mulga Parrot is widespread but somewhat erratically distributed across interior mainland Australia from central New South Wales westwards. It is locally common but usually rather rare, and mainly sedentary or locally nomadic.

The male is bright and multi-coloured, mainly green with bright yellow patches on the forehead and wings, but the female is plain dull olive, with dull red and blue markings on the wing.

It is usually encountered in pairs or small parties, only rarely in flocks, but it otherwise strongly resembles the Red-rumped Parrot in general behaviour. It feeds on seeds, especially those of grasses and herbaceous plants. It is usually quiet and unobtrusive, and easily overlooked.

Breeding is strongly influenced by local rainfall but occurs mostly from August to December. The clutch consists usually of five eggs, which are deposited in a tree cavity and hatch in about 19 days. The female alone incubates, but both parents rear the young, which fledge at about 28 days.

HABITAT: arid woodland, especially Mulga, but also mallee, cypress pine and box woodlands
LENGTH: *c* 28 cm
DISTRIBUTION: more than 1 million km²
ABUNDANCE: sparse to common
STATUS: probably secure

(*D & V Blagden*)

Genus *Psittaculirostris*

(sit-ak'-ue-lee-ros'-tris: "*Psittacula*-beak", after a genus of parrots)

Most of the five species in this genus are confined to New Guinea. Strongly arboreal and restricted to rainforest, they are stocky, small to medium-sized parrots with very short tails, and differ from other parrots in a number of anatomical details. The bill is relatively large and broad, and there is a prominent notch in the upper mandible. All are brightly coloured, with intricate plumage patterns, and feed largely on fruit, especially figs. In most species, males differ in appearance from females. Together with one other New Guinea species, the single Australian species is sometimes placed in the genus *Opopsitta*.

Double-eyed Figparrot

Psittaculirostris diophthalma (die'-of-thal'-mah: "double-eyed *Psittacula*-beak")

The Double-eyed Figparrot is widespread in Indonesia and New Guinea, and in Australia occurs in north-eastern Cape York Peninsula, and approximately between Cairns and Cardwell, also from about Maryborough southward into northern New South Wales. It occurs in highland forests but is mainly a bird of coastal lowlands. It is nomadic. The two northern populations remain common, but widespread habitat destruction has caused a drastic decline in the southernmost population, which is now very seldom reported.

The three Australian populations are very distinct, differing mainly in size and in details of the head pattern: all are very small, plump, short-tailed parrots, mainly green but marked with red and blue on the head.

Large numbers sometimes congregate at fruiting trees but it usually occurs in pairs or small parties. It roosts communally, dispersing at dawn to forage. Its diet consists mainly of the seeds of various fruits, especially native figs, supplemented by nectar and insects. It is generally quiet and unobtrusive when feeding, but agile and active, and it often flies high above the canopy. Its calls include a distinctive high-pitched "zeet, zeet", often uttered in flight.

Breeding occurs mainly from August to December. The nest is a tree cavity usually 10 to 20 metres from the ground. The clutch consists of two eggs, which hatch in about 21 days. Only the female incubates and broods the young, but the male fetches food. The young fledge at about 50 to 55 days.

HABITAT: rainforest, open forest and mangroves, especially with abundant fruiting or flowering trees
LENGTH: *c* 13 cm
DISTRIBUTION: 30,000–100,000 km^2
ABUNDANCE: rare
STATUS: possibly endangered

(H & J Beste)

225

Genus *Purpureicephalus*

(per'-pue-ray'-ee-sef'-ah-lus: "purple-head")

The characteristics of the genus are essentially those of the single species.

Red-capped Parrot

Purpureicephalus spurius (spue'-ree-us: "false purple-head")

This striking parrot is generally common but is confined to south-western Australia, approximately from Perth to Esperance and mainly in coastal lowlands.

Adults apparently mate for life and occupy permanent territories, but immatures join wandering flocks until reaching maturity. The Red-capped Parrot forages in trees or on the ground. It is often quiet and unobtrusive when feeding, but is generally active, conspicuous and tame. The usual flight call is a distinctive, grating "checkacheck". It eats a wide range of foods, including nectar, insects and fallen fruit, but its bill is uniquely adapted to deal with the seeds of one species of tree, the Marri *(Eucalyptus calophylla)*, with which it has a close ecological relationship.

Breeding occurs mainly from August to November, and only one brood is raised in a season. The nest is a cavity in a tree, and the clutch usually consists of five eggs, which hatch in about 20 days. The female alone incubates and feeds the young until they are about 12 to 14 days old, after which time the male joins in their care; they fledge at about 35 days.

HABITAT: mainly eucalypt forest and woodland; also orchards, suburban parks and gardens
LENGTH: 33–37 cm
DISTRIBUTION: 100,000–300,000 km²
ABUNDANCE: common
STATUS: probably secure

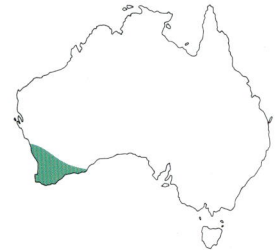

(K Stepnell)

Order CUCULIFORMES

(kue-kue'-lee-for'-mayz: "*Cuculus*-order", after a genus of cuckoos)

The order Cuculiformes is usually construed as consisting of two families: Musophagidae and Cuculidae. The Cuculidae are almost cosmopolitan, but the Musophagidae are an entirely African group of about 20 colourful, arboreal, fruit-eating species which inhabit rainforest. The latter family has been associated with the gallinaceous birds, and even made a separate order, but its links with the cuckoos are reinforced by the results of studies on their egg-white proteins.

Family CUCULIDAE

(kue-kue'-lid-ee: "*Cuculus*-family", after a genus of cuckoos)

About 130 species, grouped in 34 genera, constitute the cosmopolitan family Cuculidae. Cuckoos are notable for their parasitic breeding behaviour, laying their eggs in the nests of other birds and abandoning them to be raised by the unwitting foster-parents. However, this behaviour is not universal: more than half of all cuckoo species build nests and incubate and rear their own young in the conventional manner.

The family is diverse and difficult to categorise concisely. Most members are slender, long-tailed birds of medium size; the plumage is loose, and the toes are arranged so that two are directed forwards and two back. Most species live in woodland and feed on insects (especially hairy caterpillars), but some eat fruit.

Genus Centropus
(sen'-troh-poos: "spur-foot")

The coucals form a group of about 27 species distributed from Africa to the Solomon Islands; one species occurs in Australia. They are mainly secretive birds of dense undergrowth, with limited powers of flight. They do not practise brood-parasitism.

Pheasant Coucal
Centropus phasianinus (fah'-zee-ah-nee'-nus: "pheasant-like spur-foot")

Widely known in country areas as the "water-bottle bird" because of its distinctive bubbling call, the Pheasant Coucal is common and widespread in coastal and near-coastal regions of northern and eastern Australia, as well as in Indonesia and New Guinea. A large bird, it has a tail that is longer than the head and body. Outside the breeding season, the wings and tail have a mottled brown pattern resembling that of the Ringneck Pheasant.

It is mainly sedentary, living in territorial pairs when breeding, but may wander locally at other times. It is secretive and usually keeps to dense undergrowth, but it frequently emerges to perch in low bushes or on fence posts and similar sites, especially after showers of rain. It seldom flies, but runs nimbly. Its varied diet includes insects, eggs, small reptiles, mammals and nestling birds.

It may nest twice during a breeding season that extends mainly from October to March. The nest is a bowl of green leafy twigs built on a platform of trampled grass, over which surrounding vegetation is woven to form a canopy. The clutch ranges from two to five eggs, which hatch in about 15 days. The young fledge at about 12 to 15 days. Both parents build the nest, incubate and care for the young.

HABITAT: woodland with a dense ground cover of grass; salt marshes; abandoned pasture; canefields
LENGTH: *c* 65 cm
DISTRIBUTION: more than 1 million km²
ABUNDANCE: sparse to common
STATUS: probably secure

(P Green)

227

Genus *Chrysococcyx*

(krie'-soh-kok'-ix: "golden-cuckoo")

About 13 species found in Africa, southern Asia and Australasia comprise this genus, characterised by generally small size, glossy green upperparts, and closely barred underparts.

Horsfield's Bronze-cuckoo

Chrysococcyx basalis (bah-sah'-lis: "based golden-cuckoo")

(J Handel)

Almost ubiquitous, this is perhaps Australia's most numerous and widespread cuckoo. Most birds leave Tasmania in winter, and the species is at least partly migratory elsewhere in southern Australia, some birds reaching Indonesia and New Guinea in winter. It is solitary, quiet and unobtrusive, and feeds mainly on insects and their larvae, especially caterpillars. The common call is a series of mournful descending whistles. Its hosts include various species of the families Maluridae and Acanthizidae, especially fairywrens and scrubwrens.

HABITAT: most kinds of woodland
LENGTH: 15–17 cm
DISTRIBUTION: more than 1 million km^2
ABUNDANCE: sparse to common
STATUS: secure

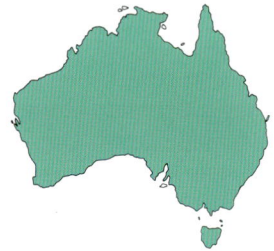

Shining Bronze-cuckoo

Chrysococcyx lucidus (lue'-sid-us: "shining golden-cuckoo")

The Shining Bronze-cuckoo closely resembles Horsfield's Bronze-cuckoo in appearance and habits but, in Australia, its distribution is limited to areas of relatively high rainfall in the east and south; elsewhere it occurs widely from the Philippines and Indonesia to New Zealand and islands of the south-western Pacific, including Norfolk

and Lord Howe Islands. Australian populations are migratory and most birds winter in Indonesia or New Guinea. The Shining Bronze-cuckoo is more richly coloured than Horsfield's Bronze-cuckoo, and lacks the dusky cheek-patch characteristic of that species.

Most breeding occurs from August to January. As is the case with most

other cuckoos, only one egg is laid in the nest of each host, but up to 16 nests may be parasitised in a season. The egg hatches in about 14 to 17 days, and the chick fledges at about 19 days. The most common hosts include various species of the families Maluridae and Acanthizidae, especially fairywrens and scrubwrens.

HABITAT: mainly eucalypt forest and woodland, mainly in regions with an average annual rainfall exceeding 380 mm
LENGTH: 17–18 cm
DISTRIBUTION: more than 1 million km^2
ABUNDANCE: sparse to common
STATUS: secure

(RH Green)

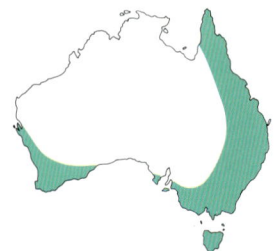

Little Bronze-cuckoo

Chrysococcyx minutillus (min'-ue-til'-us: "very-small golden-cuckoo")

The three Australian populations of this tropical cuckoo are representatives of a cluster of related forms widespread in Indonesia and New Guinea: species limits are controversial and uncertain. Here they are tentatively grouped in a single species, the Australian forms being characterised in the field by possession of a bright red ring around the eye. Northern birds are sedentary but the eastern population is migratory. Members of the genus *Gerygone* constitute the most frequent hosts.

HABITAT: mainly rainforest and its margins; melaleuca swamps; mangroves; suburban parks and gardens
LENGTH: 14–16 cm
DISTRIBUTION: 100,000–300,000 km²
ABUNDANCE: rare to very sparse
STATUS: probably secure

(*GB Baker*)

Black-eared Cuckoo

Chrysococcyx osculans (os'-kue-lanz: "kissing golden-cuckoo")

The Black-eared Cuckoo occurs across Australia, but it tends to avoid the coastal areas of high rainfall in the east and south. Almost all breeding occurs south of about 26°S. Its movements are imperfectly understood, but most southern populations appear to be migratory, some birds reaching Indonesia and New Guinea in winter. Like other cuckoos, it is generally solitary, quiet, and unobtrusive in behaviour, although active, noisy chases occur during courtship. It eats insects and their larvae.

Breeding occurs mainly from August to January. The usual hosts are the Speckled Warbler and the Redthroat, both of which build domed nests and lay distinctive deep reddish eggs, which are closely matched by those of the Black-eared Cuckoo.

HABITAT: mainly arid scrublands; open eucalypt woodland; mallee; spinifex
LENGTH: 19–21 cm
DISTRIBUTION: more than 1 million km²
ABUNDANCE: sparse
STATUS: probably secure

(*L Pedler*)

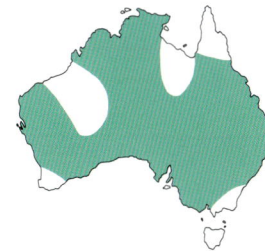

Genus *Cuculus*

(kue'-kue-lus: "cuckoo")

About 20 species of cuckoos comprise this genus, which is widespread in the Old World. Five species occur in Australia, but generic limits are controversial and three of these (the Fan-tailed, Chestnut-breasted, and Brush Cuckoos) together with several non-Australian species are sometimes isolated in the genus *Cacomantis*. Most species are medium-sized and have a sober plumage pattern and restrained colours, with the outer tail feathers intricately notched in black and white.

229

Chestnut-breasted Cuckoo

Cuculus castaneiventris (kah'-stah-nay'-ee-ven'-tris: "chestnut-bellied cuckoo")

The Chestnut-breasted Cuckoo is common in New Guinea but its distribution in Australia is restricted to northern Cape York Peninsula south to the Daintree region. It closely resembles the Fan-tailed Cuckoo in appearance and behaviour but very little is known of its biology. It mostly parasitises scrubwrens.

HABITAT: mainly rainforest
LENGTH: 23–25 cm
DISTRIBUTION: 30,000–100,000 km²
ABUNDANCE: rare to sparse
STATUS: vulnerable

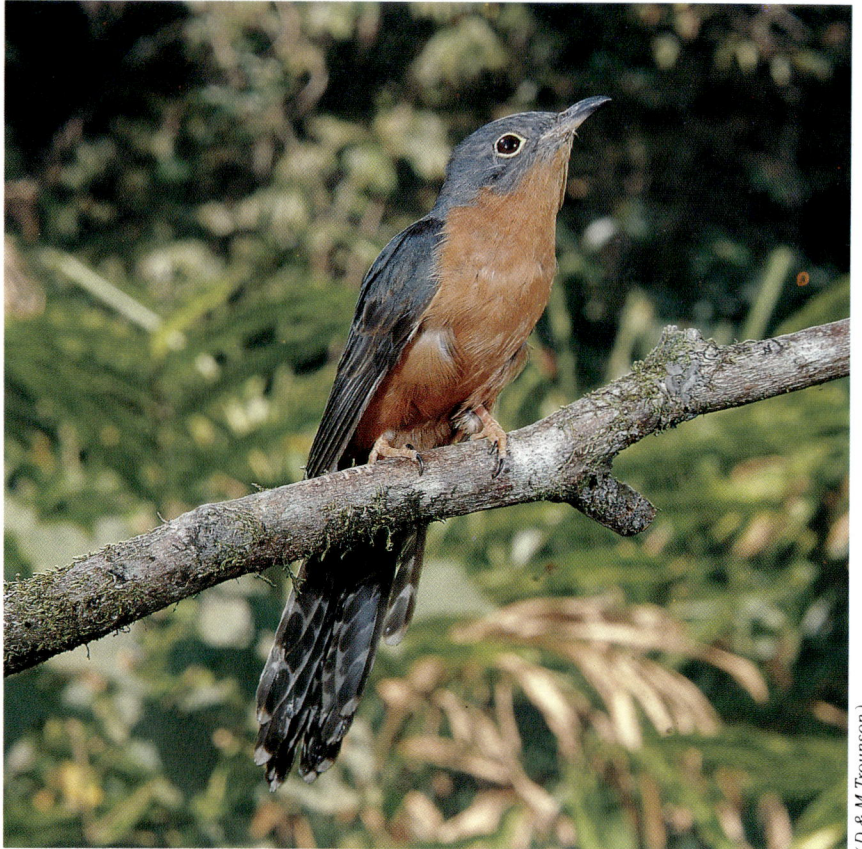

(*D & M Trounson*)

Pallid Cuckoo

Cuculus pallidus (pal'-id-us: "pallid cuckoo")

The Pallid Cuckoo occurs virtually throughout Australia. South of about 26°S it is mainly migratory, adults and most young birds moving northwards to spend the winter in northern Australia. Some birds annually reach Indonesia and New Guinea.

It is noisy when breeding but otherwise quiet, unobtrusive and rather solitary. It feeds mainly on large insects and their larvae, especially caterpillars. Prey is usually captured on the ground or in low foliage by pouncing on it from an exposed vantage perch, often a dead branch.

Breeding occurs from September to January. The Pallid Cuckoo selects as hosts those birds that build open, cup-shaped nests, especially Yellow-faced, White-plumed and Singing Honeyeaters.

HABITAT: open woodland of almost all kinds, rarely rainforest
LENGTH: 28–32 cm
DISTRIBUTION: more than 1 million km²
ABUNDANCE: abundant
STATUS: secure

(*D & M Trounson*)

Fan-tailed Cuckoo

Cuculus pyrrhophanus (pie'-roh-fah'-nus: "fire-bright cuckoo")

The distinctive plaintive, descending trill of this common cuckoo is one of the characteristic sounds of eastern and southern Australian forests, but the bird itself, like other cuckoos, is rather solitary and unobtrusive in behaviour. It frequents the lower levels of foliage, generally within a few metres of the ground, though it often selects high exposed perches from which to sing. It feeds on insects and their larvae, exhibiting a special fondness for hairy caterpillars. Elsewhere the Fan-tailed Cuckoo occurs in New Guinea and islands of the south-western Pacific.

Breeding occurs mainly from August to January. Most hosts are small songbirds of the families Maluridae and Acanthizidae, which build domed nests: the Brown Thornbill is a particularly frequent host. The clutch size is uncertain. The egg hatches in about 14 to 15 days and the chick fledges at about 16 to 17 days.

HABITAT: rainforest; eucalypt forest and woodland; mallee; heath; occasionally suburban parks and gardens
LENGTH: 24–27 cm
DISTRIBUTION: more than 1 million km²
ABUNDANCE: common
STATUS: secure

(*JD McNaughton*)

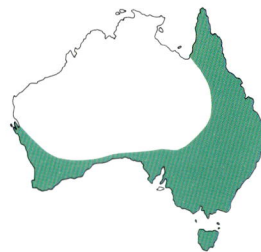

Oriental Cuckoo

Cuculus saturatus (sat'-ue-rah'-tus: "richly-coloured cuckoo")

The Oriental Cuckoo breeds in the Soviet Union, Mongolia, China, Japan and northern India, and migrates southward for the winter. The limits of its wintering quarters are uncertain but it is known to winter commonly in South-East Asia and Indonesia, less commonly in New Guinea. In Australia it occurs, mainly from November to May, in the Kimberley and the Top End, and from the islands of Torres Strait south along the east coast to southern New South Wales, although its occurrence south of the tropics is very rare and erratic. Vagrants have occurred in New Zealand.

Adults are grey above and boldly barred below, but the immature birds frequently seen in Australia are obscurely mottled brown and nondescript. Although small groups occasionally congregate, especially just before the northward migration, it is usually solitary, shy, quiet and elusive. It feeds mainly on large insects and their larvae.

HABITAT: rainforest, monsoon forest, eucalypt forest, melaleuca swamps, mangroves
LENGTH: 30–33 cm
DISTRIBUTION: 300,000–1 million km²
ABUNDANCE: rare
STATUS: secure

(*EE Zillmann*)

Brush Cuckoo

Cuculus variolosus (var'-ee-oh-loh'-sus: "spotted cuckoo")

The Brush Cuckoo inhabits coastal and subcoastal Australia from Victoria north to Cape York and westward to the Kimberley, and it is also widespread from South-East Asia eastward to the Solomon Islands. It is sedentary over much of its range, but birds in south-eastern Australia are migratory, wintering mainly in New Guinea.

Like other cuckoos it is noisy when breeding but otherwise quiet, unobtrusive and rather solitary. It feeds mainly on large insects and their larvae (especially caterpillars), captured mainly in foliage.

In the tropics breeding may occur at any time; in the south-east it usually takes place from October to February. Open, cup-shaped nests are selected, at least in the south, and the most common hosts include Grey Fantails, Leaden Flycatchers and Brown-backed Honeyeaters. Eggs resemble those of the host, and hatch in about 12 to 14 days; the chick fledges at about 15 to 17 days.

(EE Zillmann)

HABITAT: dense eucalypt forest, melaleuca swamps, rainforest, mangroves
LENGTH: 22–24 cm
DISTRIBUTION: more than 1 million km²
ABUNDANCE: sparse
STATUS: secure

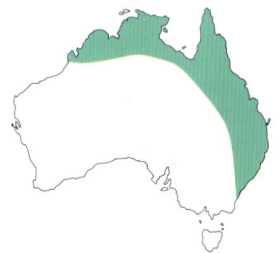

Genus *Eudynamis*

(ue'-die-nah'-mis: "good-power" [?], possibly meaningless)

The characteristics of the genus are essentially those of the Common Koel.

Common Koel

Eudynamis scolopacea (skol'-oh-pah'-say-ah: "snipe-like *eudynamis*")

The persistent, loud, and shrill "coo-ee" calls of the Common Koel are a familiar feature of the suburban environment to many eastern city dwellers, announcing the arrival of the birds from their wintering grounds in New Guinea. Sometimes beginning long before dawn, these vocal performances often accelerate to frantic crescendos. The Common Koel arrives in late September or early October and leaves in March, much less obtrusively than it arrived. It is also common in the tropical north, and ranges widely elsewhere from India to the Solomon Islands.

The adult male is mainly black in colour, with a bright red eye; the female is brown and intricately barred. Notwithstanding its strident calls, it is secretive in behaviour, seldom leaving the dense foliage of the upper canopy of trees where it is difficult to observe. It feeds mainly on fruit.

Breeding occurs mainly from September to March. A variety of bird species are parasitised, especially figbirds, friarbirds, wattlebirds, and others that build open nests in the upper canopy of trees. The egg hatches in about 13 to 14 days; the chick may eject the host chicks from the nest or starve them by monopolising the food supply brought by the foster-parents.

(R Viljoen)

HABITAT: rainforest and its margins; eucalypt forest; riparian woodland; urban parks and gardens
LENGTH: 39–46 cm
DISTRIBUTION: more than 1 million km²
ABUNDANCE: common
STATUS: secure

Genus Scythrops

(skith'-rops; "sullen-face")

The characteristics of the genus are essentially those of the single species.

Channel-billed Cuckoo

Scythrops novaehollandiae (noh'-vee-hol-and'-ee-ee: "New-Holland sullen-face")

The Channel-billed Cuckoo occurs across northern Australia and southward along the east coast almost to the Victorian border, but is rare south of about Moruya, New South Wales, and may be less numerous in the north than in the east. It is strongly migratory, leaving for Indonesia and New Guinea in March and returning in September or October. It is generally solitary, but small parties commonly congregate after breeding. It feeds mainly on fruit—especially figs—and large insects.

Breeding has been recorded from August to December, but its distribution may be limited in the south-east by the fact that most potential hosts have already begun breeding before its usual time of arrival in late September. Common hosts include currawongs, magpies and ravens; not infrequently two or more cuckoo's eggs may be laid in the same nest, and the cuckoo chick does not eject the host chicks, which usually starve and are removed by the foster-parents.

HABITAT: rainforest and its margins; eucalypt forest; riparian woodland; urban parks and gardens
LENGTH: 59–65 cm
DISTRIBUTION: more than 1 million km^2
ABUNDANCE: sparse to common
STATUS: secure

(A Davies)

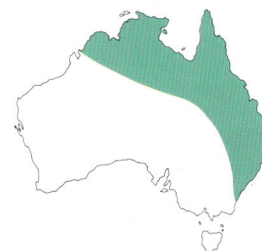

Order STRIGIFORMES

(stri'-jee-for-mayz: "*Strix*-order", after a genus of owls)

Owls have soft fluffy plumage, short tails, large heads, taloned feet, and large, forward-facing eyes; they resemble no other birds. Most are nocturnal and all are carnivorous. In many respects they resemble the raptors—the diurnal birds of prey in the order Falconiformes—but they are more closely related to the Caprimulgiformes, which are, however, essentially insectivorous.

Although owls are very distinct, relationships within the order are far from clear. Two families, Tytonidae and Strigidae, are generally recognised but there are few trenchant distinctions, and many researchers favour merging the two.

Family TYTONIDAE

(tie-ton'-id-ee: "*Tyto*-family", after a genus of owls)

Eight species in two genera constitute the family Tytonidae, or barn owls: the genus *Phodilus* occurs in Africa and India, but *Tyto* is almost cosmopolitan. The family is best represented in the Australasian region where five (perhaps six) species occur.

Varying relatively little in size, typical barn owns have long slender legs and conspicuously heart-shaped facial discs. All are nocturnal or mainly so. The eyes are dark, never yellow. There is pronounced asymmetry in the size and shape of the ear openings in the skull and hearing is extremely acute: it has been shown in laboratory experiments that a Barn Owl is capable of locating and catching a mouse in absolute darkness.

The female is much bigger than the male but the sexes are similar in appearance.

Genus Tyto

(tie'-toh: "owl")

The characteristics of the genus are essentially those of the family.

Barn Owl

Tyto alba (al'-bah: "white owl")

This species is among the most widespread of birds, being found in North and South America, Africa and Eurasia, and on some islands of the south-west Pacific, as well as across the whole of Australia. It is more common in the east and south-west than in the arid interior, but it is rare in Tasmania and perhaps does not breed there.

It requires open country over which to hunt and therefore normally avoids dense eucalypt woodland and rainforest, but it occurs in most other kinds of lightly wooded country and is often common in towns and cities. It feeds largely on mice, especially the House Mouse, *Mus musculus*, and (as is the case with most predators) numbers and movements are strongly influenced by fluctuations in abundance of these small rodents: mouse plagues are followed by an abrupt rise in the density of breeding Barn Owls, followed by large-scale movements (all recent Tasmanian records followed mainland mouse plagues) and widespread deaths of starving owls (one such event occurred in the Riverina in 1986, when many owls were found dead on roads).

Elsewhere in the world Barn Owls often hunt by day but Australian individuals seem more rigidly nocturnal and the bird is seldom seen despite its abundance. It lives alone or in pairs and seldom calls. It roosts in caves, sheds and barns, tree hollows, or densely foliaged trees.

Breeding is very flexible in timing and the pair bond is temporary. Most nests are in tree cavities. The female incubates the clutch of three to seven eggs, which hatch in about 33 to 35 days. Young chicks are fed by the female, later assisted by the male; they fledge in 60 to 70 days.

HABITAT: mainly temperate and tropical woodland and scrub; croplands and farming country; towns and cities
LENGTH: 30–40 cm
DISTRIBUTION: more than 1 million km²
ABUNDANCE: sparse to common
STATUS: secure

(G Threlfo)

Grass Owl

Tyto capensis (kay-pen'-sis: "Cape owl")

The Grass Owl is unusual among Australian birds in living in two distinct populations: these occur in different habitats and show corresponding differences in behaviour. It inhabits heathland and extensive grasslands on the flood plains of large rivers along the east coast from Cape York to the Manning River in New South Wales, where its populations, although small and scarce, seem relatively permanent and stable. It also occurs in the grasslands of the Barkly Tableland and the channel country of western Queensland. Here its ecology is closely linked with fluctuations in its major prey, the Long-haired Rat, *Rattus villosissimus*; its numbers are variable and its movements erratic. In the inevitable crash following periodic plagues of rats, starving Grass Owls disperse in large numbers, and vagrants may reach as far afield as Melbourne or Darwin.

Elsewhere the Grass Owl occurs in southern Africa and from India east to Fiji (but not in New Guinea). Uncertainty exists as to whether African birds are specifically distinct from eastern populations, and the Australian form is often listed as the Eastern Grass Owl, *Tyto longimembris*.

Aptly named, the Grass Owl shelters by day in dense grass tussocks, in squats reached by hidden runways in the grass. By night it hunts methodically, quartering surrounding grassland at a height of several metres, locating prey by sight and sound. It is readily distinguished from other *Tyto* owls in flight by the fact that the colours of the upperwing (tawny brown and slate) are distributed in bold slabs; the upperwings of other owls, although intricately mottled, offer a much more even impression of colour.

Nesting may occur at any time, but especially from March to June. The nest is a hidden platform of grass, and the usual clutch consists of three to eight eggs. Only the female incubates, but the male brings her food at night and roosts with her during the day.

HABITAT: grassland, coastal heath, lignum swamps
LENGTH: 32–38 cm
DISTRIBUTION: 300,000–1 million km²
ABUNDANCE: sparse
STATUS: probably secure

(A Wilson)

Masked Owl

Tyto novaehollandiae (noh'-vee-holl-an'-dee-ee: "New-Holland owl")

The Masked Owl is a sedentary inhabitant of eucalypt forest and woodland in southern New Guinea and Australia. Secretive, relatively silent and strictly nocturnal, its presence is difficult to detect and some details of its distribution remain uncertain. It apparently occurs nowhere more than 300 kilometres from the coast, and is common in Tasmania.

The female is much bigger than the male and has relatively larger feet. Each sex is thought to take a different range of prey animals but, in general, the Masked Owl feeds mainly on terrestrial mammals up to about the size of a rabbit. Unlike the Barn Owl, it apparently mates for life and maintains permanent territories, but the two species resemble each other in many other respects and are deceptively similar in appearance.

Nesting may occur at any time. The male prepares the nest cavity in a tree but only the female incubates the two or three eggs, which hatch in about 35 days; she is fed at the nest by the male. The young fledge at 70 to 84 days. The same tree cavity is often used for nesting for several years in succession.

(E Hosking)

HABITAT: eucalypt forest and woodland
LENGTH: (females) 40–50 cm
DISTRIBUTION: more than 1 million km²
ABUNDANCE: sparse to common
STATUS: probably secure

Sooty Owl

Tyto tenebricosa (ten'-eb-rik-oh'-sah: "dark owl")

Apart from its distribution in New Guinea, the Sooty Owl occurs in two disjunct Australian populations: in the Cairns–Atherton region of north-eastern Queensland; and from near Maryborough in Queensland south to Melbourne. The northern population is sometimes separated as a distinct species (the Lesser Sooty Owl, *Tyto multipunctata*).

Like the Masked Owl, the Sooty Owl is a sedentary species which apparently mates for life and maintains permanent territories. It tends to hunt from perches rather than from flight, waiting quietly at the edge of clearings and periodically flying out to pounce on prey on or near the ground. Flexible in diet, it eats small mammals, roosting birds, frogs, small reptiles and, occasionally, insects. It is notable for its extraordinary territorial call, a long descending whistled scream that sounds like a falling bomb; it also utters a variety of twittering, churring and rasping notes.

Nesting may occur at any time. The nest is in a large, high tree cavity, and usually two eggs are laid (usually about 4 days apart). The incubation period is about 42 days and the young fledge at about 80 to 85 days. Only the female incubates, fed by her mate at the nest.

(H & J Beste)

HABITAT: rainforest, dense eucalypt forest
LENGTH: (females) 45–50 cm
DISTRIBUTION: 100,000–300,000 km²
ABUNDANCE: sparse
STATUS: probably secure

Family STRIGIDAE

(stri'-jid-ee: "*Strix*-family", after a genus of owls)

The family Strigidae comprises about 135 species of owls found on all continents except Antarctica. All are carnivorous and most are nocturnal. The characteristics of the family are essentially those of the order, its members differing from the Tytonidae largely in having less highly modified ears and facial disks, and tending to depend rather more on sight than on sound in their hunting. Females are usually substantially larger than males, but the sexes are otherwise similar in appearance. All Australian species belong to a single genus, *Ninox*.

Genus Ninox

(nin'-ox: meaning unclear but probably intended as "night-bird")

The genus *Ninox* consists of about 16 to 18 species largely restricted to the Australasian and eastern Oriental regions; four species occur in Australia. Members of the genus differ from other owls largely in their relatively unmodified facial discs. All are nocturnal, inhabit woodland and forest, and have yellow eyes. Hearing seems relatively poor (compared to other owls) and hunting is mainly by sight.

Barking Owl

Ninox connivens (kon-ee'-venz: "winking night-bird")

Characteristic of open eucalypt woodland, the Barking Owl occurs across mainland Australia but is rare or absent over much of the arid interior. It also occurs in southern New Guinea. It is sedentary, living in pairs that maintain permanent territories. It roosts by day in trees, and at night preys upon birds (up to the size of a Tawny Frogmouth), mammals, and nocturnal insects. The introduced Rabbit is an important prey animal, at least in southern Australia.

Although generally nocturnal in habits, the Barking Owl often calls and sometimes hunts by day, particularly in cool or cloudy weather. The territorial call is a brisk disyllabic barking note uttered in series by both sexes throughout the year but especially in autumn; it also utters a loud, tremulous scream.

Breeding occurs mainly from July to November. The nest is a bed of wood debris in a roomy tree cavity, selected and prepared by the male. The female lays and incubates a clutch of two or three eggs, which hatch in 34 to 38 days. The male feeds his mate at the nest, and later brings food for the young, which fledge at about 42 days.

HABITAT: mainly savannah woodland, open eucalypt woodland and riverine forest
LENGTH: *c* 40 cm
DISTRIBUTION: more than 1 million km²
ABUNDANCE: rare to sparse
STATUS: possibly endangered

(*EE Zillmann*)

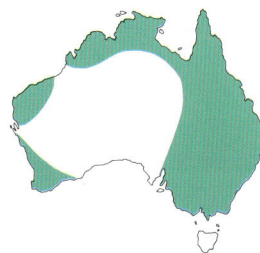

Boobook Owl

Ninox novaeseelandiae (noh'-vee-see-land'-ee-ee: "New-Zealand night-bird")

The Boobook Owl, most numerous and widespread Australian member of the Strigidae, occurs in virtually all kinds of wooded country throughout the continent as well as in eastern Indonesia, New Guinea

(E Hosking)

and New Zealand. It also occurs on Norfolk Island, and has been introduced to Lord Howe Island. In most areas it seems sedentary or locally nomadic, but young birds disperse widely and there is some evidence of migration in south-eastern Australia.

The Boobook Owl preys on small birds and mammals such as the House Mouse, *Mus musculus*, but many insects are also eaten, especially moths and nocturnal beetles. The distinctive territorial call, uttered by both sexes throughout the year, is probably the most familiar nocturnal bush sound to most Australians: it is a loud, deliberate "boo-book", the second note pitched somewhat lower than the first. The bird is often called the mopoke, after this call.

Breeding occurs mainly from August to January. For nesting, the male selects a tree cavity and prepares in this a bed of twigs, dead

leaves and wood debris, on which the female lays and incubates two to four eggs. The incubation period is about 30 days. The male feeds his mate at the nest, and later brings food for the young, which fledge at about 40 days.

HABITAT: wooded country of all kinds, including urban parks and gardens
LENGTH: 25–35 cm
DISTRIBUTION: more than 1 million km^2
ABUNDANCE: common
STATUS: secure

Rufous Owl

Ninox rufa (rue'-fah: "rufous night-bird")

In Australia this elusive and secretive owl is found in the Kimberley and the Top End, and in north-eastern Queensland from Cape York to the vicinity of Rockhampton; it also

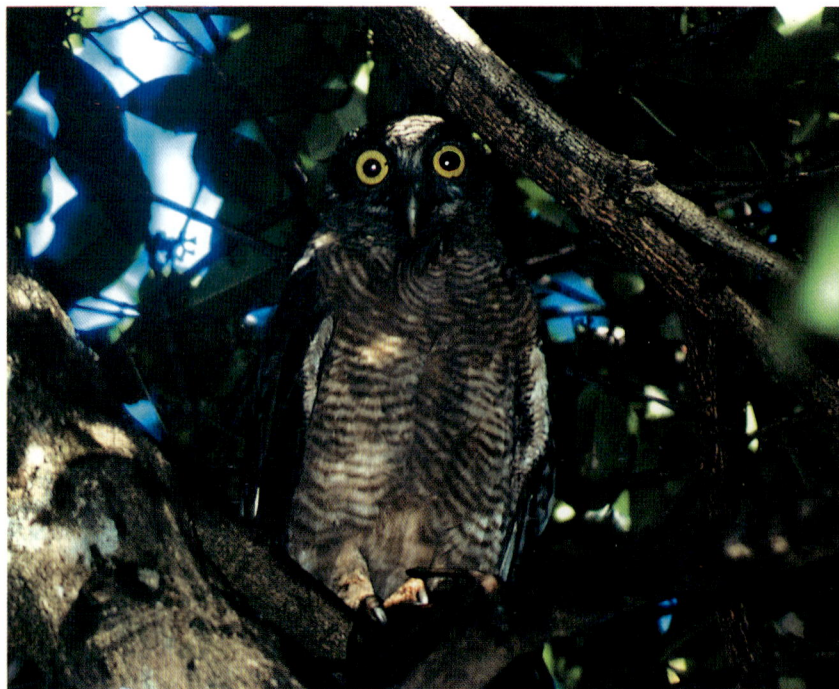

(EE Zillmann)

occurs in New Guinea. It seems sedentary but little is known of its movements, habits, or life history. At night it preys upon birds, arboreal mammals, frogs and large insects. It

roosts by day in dense cover.

The breeding season extends from June to November. The nest is in a large, high tree cavity, carpeted with dead leaves and wood debris, and used repeatedly. The usual clutch consists of two eggs, which hatch in about 37 days. Only the female incubates.

HABITAT: rainforest; riverine woodland; occasionally melaleuca swamps and mangroves
LENGTH: 45–55 cm
DISTRIBUTION: 300,000–1 million km^2
ABUNDANCE: rare to very sparse
STATUS: possibly endangered

Powerful Owl

Ninox strenua (stren'-ue-ah: "powerful night-bird")

The Powerful Owl is confined to south-eastern Australia (except Tasmania) from the vicinity of Gladstone in Queensland to the Otway Ranges in Victoria. It is characteristically a bird of densely forested gullies on the coastal slopes of the Great Dividing Range but it also occurs sporadically on the western slopes and in some outlying mountain ranges such as the Warrumbungles in New South Wales. It lives in pairs (apparently mated for life) that occupy permanent territories of about 1000 hectares. At night it preys on birds and arboreal mammals; by day it roosts in leafy trees, selecting a perch that is shaded from above but which commands a wide view. Pairs roost together or apart, but maintain several roosts used in rotation, shifting every few days.

The most important prey species include the Greater Glider *Petauroides volans*, the Common Ringtail Possum *Pseudocheirus peregrinus*, and the Sugar Glider *Petaurus breviceps*. Prey is torn apart, the head usually being eaten first; the tail and hindquarters are often held carefully under one foot at roost through the following day, then consumed before the owl leaves to hunt in the evening.

The territorial call, uttered by both sexes throughout the year, is a loud, deliberate "whoo-hoo".

Breeding occurs mainly from May to October. The nest is in a large tree cavity, usually high in a very large eucalypt. Pepared by the male, the cavity is carpeted with dead leaves and wood debris. The female alone incubates the two eggs (laid four nights apart) until they hatch in about 37 days; she is fed near the nest by the male, who also hunts for the young. Fledging takes about 60 days.

HABITAT: mainly dense eucalypt forest
LENGTH: 60–66 cm
DISTRIBUTION: 100,000–300,000 km²
ABUNDANCE: very sparse
STATUS: possibly endangered

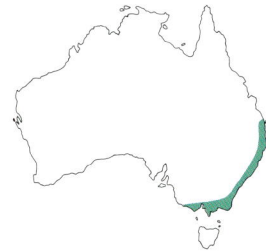

(P. Klapste)

Order CAPRIMULGIFORMES

(kap'-ree-mul'-gee-for'-mayz: "*Caprimulgus*-order", after a genus of nightjars)

The order Caprimulgiformes comprises two suborders: Steatornithes and Caprimulgi. The former contains a single family (Steatornithidae) and species, the remarkable Oilbird of Trinidad, Venezuela and elsewhere in northern South America, which feeds largely on fruit and uses a sonar-like technique to manoeuvre in its dark communal nesting caves.

The Caprimulgi consists of several families of nocturnal birds linked by their highly cryptic plumage and insectivorous diets. They somewhat resemble owls (order Strigiformes) but have (by comparison) feeble legs and feet that lack talons, broad blunt bills, and much less well-developed stereoscopic vision and hearing. Widespread, but best represented in the tropics, the four included families are: the Caprimulgidae (nightjars); Podargidae (frogmouths); Aegothelidae (owlet-nightjars); and Nyctibiidae (potoos). The last family is exclusively American in distribution, but all of the others are represented in Australia. The Aegothelidae and Podargidae are virtually restricted to the Australasian region.

Family CAPRIMULGIDAE

(kap'-ree-mul'-gid-ee: "*Caprimulgus*-family", after a genus of nightjars)

The nightjars form a group of some 79 species arranged in two subfamilies: Chordeilinae and Caprimulginae. The former is exclusively American while the latter group is virtually cosmopolitan but best represented in the tropics. By far the largest genus (about 45 species) is *Caprimulgus*, which may be regarded as broadly representative of the family as a whole; three species occur in Australia.

239

Nightjars are notable for their extraordinarily cryptic plumage; all are intricately and subtly mottled and vermiculated in shades of grey, black, brown and rufous that effectively hide them from predators at their daytime roosts on the ground. They are seldom observed by day except when accidentally flushed, when they rise from almost underfoot to flutter a hundred metres or so before settling again in the leaf litter. Males usually differ from females only in detail, and immatures resemble adults.

Nightjars have short but broad bills surrounded by strong bristles; the skull is flattened, the eyes are large, the legs short, and the toes small. All species feed by night on large flying insects such as moths and beetles, captured in flight.

Genus *Caprimulgus*

(kap'-ree-mul'-gus: "goat-sucker", in reference to an ancient belief that these birds suckled milk from goats)

The characteristics of the genus are discussed in the account of the family.

Spotted Nightjar

Caprimulgus argus (ar'-gus: "Argus [i.e., many-eyed] goat-sucker", referring to eye-like spotting of the plumage)

The Spotted Nightjar occurs throughout Australia except Tasmania and the eastern coastal strip.

Its movements are inadequately understood, but some southern populations seem migratory and the species has been recorded in winter in Indonesia and New Guinea. In some southern districts it is sedentary or locally nomadic. It lives alone or in pairs but travelling birds may form loose flocks.

A common species, it is fond of resting on quiet gravel roads where the ruby-red reflection of oncoming vehicular headlights from its eyes renders it conspicuous. It is readily distinguished from other nightjars by the large white panel in the outer wing, and the absence of white in the tail. Like other nightjars, it roosts by day on the ground and feeds at night on large flying insects captured in flight.

It breeds from September to January. The single egg is laid among leaf litter on the ground, where it is incubated for about 20 days. The chick fledges at about 28 days. Both parents incubate and care for the young.

HABITAT: sparse arid woodland; stony and sandy rises
LENGTH: 29–31 cm
DISTRIBUTION: more than 1 million km²
ABUNDANCE: sparse to common
STATUS: probably secure

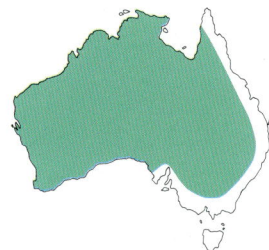

(PD Munchenberg)

Large-tailed Nightjar

Caprimulgus macrurus (mak-rue'-rus: "long-tailed goat-sucker")

(*WS Peckover*)

This tropical nightjar is widely distributed from India through South-East Asia to New Guinea; in Australia it occurs in the Top End and along the north-east coast south to Maryborough and Fraser Island in Queensland. It is sedentary, living in pairs that maintain permanent territories, roosting alone in the leaf litter by day and foraging over ranges of about 50 hectares for flying insects by night.

It is easily distinguished from other nightjars by its relatively unmarked wings and the large patches of white in the outer tail. The persistent call is very distinctive, a monotonous "chop, chop" suggesting a distant axe cutting wood; the call is uttered (usually by the male) from a lengthwise perch on a low horizontal limb or branch, or sometimes a fence post.

Breeding occurs from August to January. Two eggs, laid on the ground, constitute the usual clutch. Incubation takes about 21 to 22 days and the young fledge at about 25 days. Both parents care for the young but incubation is mainly by the female.

HABITAT: rainforest margins and clearings
LENGTH: 27–29 cm
DISTRIBUTION: 300,000–1 million km²
ABUNDANCE: common
STATUS: probably secure

White-throated Nightjar

Caprimulgus mystacalis (mis'-tah-kah'-lis: "moustached goat-sucker")

This nightjar occurs in New Guinea, the Solomon Islands, New Caledonia, and eastern Australia, mainly on the coast and associated highlands, south to the vicinity of Melbourne. It has not been recorded in Tasmania. North of about Townsville in Queensland, it is mainly sedentary; in the south it is migratory, being absent from Victoria and New South Wales in winter; some individuals migrate across Torres Strait to New Guinea.

Favouring rocky ridges for roosting and nesting, it lives in pairs that maintain nesting territories of about one hectare but forage over larger ranges (100 hectares or more) that may overlap with those of neighbouring pairs. Individuals apparently mate for life, returning to the same nesting territory year after year. The diet consists mainly of large flying insects caught on the wing in the course of regular patrols along forest margins, along streams, or in clearings. The distinctive call is a deliberate, low but penetrating, mellow chuckle. The White-throated Nightjar frequently visits camp sites to feed on moths attracted by the light.

Nesting takes place from September to February. No nest is built; the single egg is laid in leaf litter on the ground, where it is incubated for 22 to 28 days. The chick fledges at about 21 days. Both parents incubate and care for the young.

HABITAT: temperate and tropical eucalypt woodland
LENGTH: 32–36 cm
DISTRIBUTION: 300,000–1 million km²
ABUNDANCE: common
STATUS: secure

(*K Ireland*)

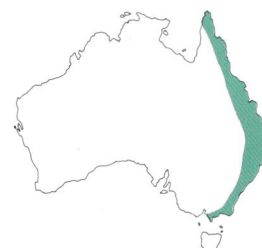

241

Family PODARGIDAE

(pod-arg'-id-ee: "*Podargus*-family", after a genus of frogmouths)

Twelve species in two genera constitute the family Podargidae, closely related to the nightjars (Caprimulgidae) but with several distinctive features. They resemble the nightjars in being nocturnal and insectivorous, and in the extreme development of cryptic plumage patterns. They differ in numerous anatomical features and, behaviourally, in preying on ground-living rather than flying insects, and in roosting in trees, not on the ground. Frogmouths usually hunt by pouncing on crawling prey from a low perch.

The larger genus, *Batrachostomus*, contains nine species restricted to India and South-East Asia; *Podargus* consists of three species widespread in Australasia (but not in New Zealand). Members of *Batrachostomus* are somewhat smaller in size (on average); they possess a small oil gland (*Podargus* has none); are more agile in flight; and build different nests.

Podargus is notable for its distinctive roosting behaviour, shared by all species. A sleeping frogmouth adopts a rigid upright stance along a tree limb or branch, the tail pressed against the bark and the bill pointed obliquely upwards. In this position, because of its greyish, cryptic plumage, it bears a remarkable resemblance to a broken-off stub.

Genus Podargus

(pod-arg'-us: possibly "trap-foot")

The characteristics of the genus are discussed in the account of the family.

Marbled Frogmouth

Podargus ocellatus (oh'-sel-ah'-tus: "eye [-spotted] trap-foot")

This rare and little-known species is distributed in two widely separated tracts of rainforest in eastern Australia: from the vicinity of Grafton in New South Wales to near Gympie in Queensland, and on Cape York Peninsula. Elsewhere it occurs in New Guinea and the Solomon Islands. Southern birds are substantially larger than northern birds.

The Marbled Frogmouth is sedentary, and is thought to live in pairs that maintain permanent territories. It feeds on small animals, such as beetles, spiders and frogs, snatched from the ground or tree trunks. The call is a low, pulsating, disyllabic booming note.

Breeding occurs from August to December. Usually only one egg is laid, in a shallow nest of interwoven vine tendrils and sticks in a horizontal fork or an epiphyte up to 15 metres from the ground. Both parents build, incubate and care for the young.

HABITAT: rainforest
LENGTH: 40–45 cm
DISTRIBUTION: 30,000–100,000 km²
ABUNDANCE: rare to sparse
STATUS: possibly endangered

(*EE Zillmann*)

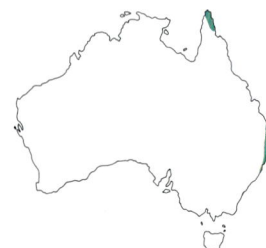

Papuan Frogmouth

Podargus papuensis (pah'-pue-en'-sis: "Papuan trap-foot")

The Papuan Frogmouth occurs in New Guinea and in north-eastern Australia from Cape York south to the vicinity of Townsville. It is sedentary, living in pairs that maintain permanent feeding territories of about 20 to 30 hectares. Defended breeding territories are apparently somewhat smaller and, under some circumstances, nests may be little more than 100 metres apart. It is similar to other frogmouths in habits but tends to hunt somewhat larger prey.

Breeding occurs from August to January (especially October and November). One egg (rarely two) is laid in a shallow saucer of sticks in a horizontal fork up to 20 metres from the ground. Both parents build, incubate and care for the young.

HABITAT: rainforest margins; eucalypt woodland; mangroves
LENGTH: 50–60 cm
DISTRIBUTION: 100,000–300,000 km²
ABUNDANCE: common
STATUS: secure

(WS Peckover)

Tawny Frogmouth

Podargus strigoides (strig-oy'-dayz: "owl-like trap-foot")

The Tawny Frogmouth is probably the best-known of Australian nocturnal birds, especially because of its habit of feeding along country roads. It usually hunts by pouncing from a low perch, such as a fence post or road sign, onto small terrestrial animals crossing bare open ground. Quiet roads provide ideal conditions for this technique, but the Tawny Frogmouth is not especially agile, either on the ground or on the wing, and many are killed by vehicles. The chief prey are such arthropods as centipedes, spiders, scorpions and cockroaches, but it also eats frogs, mice and—occasionally—small birds.

It occurs almost throughout Australia (and southern New Guinea), wherever there are trees for roosting and open areas in which to hunt, but it does not generally inhabit rainforest. It is often common in urban parks and gardens. It is sedentary, and seems to live in pairs that maintain permanent territories some 40 to 80 hectares in extent. From place to place it varies markedly in size and colour (ranging from grey to dull rufous) but, on average, females are somewhat smaller, more rufous in plumage, and less heavily mottled than males.

Territories are marked and maintained by a persistent series of 10 to 50 low booming notes. Breeding occurs from August to December. The nest is a flimsy platform of sticks, placed in a horizontal tree fork up to 15 metres from the ground. Usually two eggs form the clutch. Incubation takes about 30 days, and the young fledge at 25 to 35 days. Both parents build, incubate and care for the young.

HABITAT: most kinds of open woodland, especially eucalypt
LENGTH: 35–53 cm
DISTRIBUTION: more than 1 million km²
ABUNDANCE: common
STATUS: secure

(PD Munchenberg)

Family AEGOTHELIDAE

(ee'-goh-thay'-lid-ee: "*Aegotheles*-family", after a genus of owlet-nightjars)

Small, fluffy, and with engagingly large dark eyes, owlet-nightjars are closely related to frogmouths but differ in general appearance and in several anatomical features, including the absence of powder-down tufts.

Eight species constitute the family, which is essentially Australasian in distribution, although one representative occurs in eastern Indonesia. The group is best represented in New Guinea, which has five species; Australia has one.

The plumage is strongly cryptic. The sexes are similar, immatures resemble adults, and there is no seasonal variation.

Genus *Aegotheles*

(ee'-goh-thay'-layz: "goat-sucker", referring to an old European belief that nightjars suckled milk from female goats)

The characteristics of the single genus are those of the family.

Australian Owlet-nightjar

Aegotheles cristatus (kris-tah'-tus: "crested goat-suckler")

The Australian Owlet-nightjar occurs in southern New Guinea and throughout Australia. Tolerant of most kinds of wooded country, it is probably Australia's most widespread nocturnal bird. Adults are apparently sedentary.

It roosts by day in a variety of sheltered sites but prefers tree cavities some 2 to 10 metres from the ground. At night it forages mainly by pouncing on terrestrial insects from a low perch, but it occasionally hawks for moths on the wing. The distinctive call—a loud rattling churr—is uttered by both sexes throughout the year. Adults mate for life but roost separately.

Breeding occurs between August and December. The nest is usually a tree cavity up to 5 metres from the ground. The clutch is usually three eggs (but varies from two to five), which hatch in 27 to 29 days. The young are brooded constantly for the first 4 or 5 days, then by day for a further week; they fledge at about 25 days but remain with their parents for several months thereafter. Both parents incubate and care for the young.

HABITAT: forest and woodland
LENGTH: 21–25 cm
DISTRIBUTION: more than 1 million km^2
ABUNDANCE: common
STATUS: secure

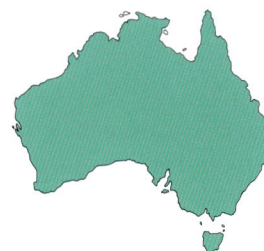

(E Hosking)

Order APODIFORMES

(ay-poh'-dee-for'-mayz: "*Apus*-order", after a genus of swifts)

Two groups of birds constitute this order: the hummingbirds, which form the suborder Trochili, and the swifts, suborder Apodi, consisting of two families: Apodidae (swifts and swiftlets) and Hemiprocnidae (crested or tree swifts).

In all species the feet and legs are very small, and effectively useless for locomotion: the wings are narrow and rigid, with a very short inner section (humerus), and they differ from other birds in certain features of the structure of the skull. There are ten primaries.

Remarkable in several respects, the hummingbirds are an exclusively American group of more than 300 species. The majority are extremely small: one species, the Bee Hummingbird of Cuba, weighs only about two grams and is the smallest of all birds. The metabolic rate of hummingbirds far exceeds that of other birds. The pectoral girdle is uniquely arranged to permit full rotation of the wing along the long axis, resulting in extraordinary aerial manoeuvrability: hummingbirds can fly backwards or in any other direction, or hover motionless in one spot, with equal facility. Wing-beat rates range from about 20 to 78 strokes per second, producing a distinctive humming note from which the group receives its English name.

Like hummingbirds, the swifts are highly evolved to an aerial existence. There are a number of other similarities but also some profound differences. The structure of the bill is strikingly different (long and slender in hummingbirds, very short and broad in swifts). Swifts depend on aerial insects for food, hummingbirds largely on nectar. Swifts are cosmopolitan, found on all continents except Antarctica (the crested swifts comprising a group of four or five species found only from India to the Solomon Islands). Only the Apodidae are represented in Australia.

All members of the order lay plain white, rather elongated eggs, and the young are hatched naked and helpless. It is usual among the Apodi for both parents to be involved at all stages of the nesting cycle whereas, among hummingbirds, the male seldom plays any active part in raising the young. Sexual dimorphism is rare among swifts, but usual among hummingbirds.

Family APODIDAE

(ah-poh'-did-ee: "*Apus*-family", after a genus of swifts)

The swifts make up a well-defined family of worldwide distribution with some 83 member species. Their possible relationship with hummingbirds constitutes one of the perennial debates in avian taxonomy; otherwise they have no close relatives except the crested swifts (Hemiprocnidae), a group of three of four species inhabiting South-East Asia, Indonesia and New Guinea.

Swifts spend most of their time in flight, foraging for aerial insects. The wing is unusual in structure, the inner wing (humerus) being extremely short, the outer wing (carpal) correspondingly long, and the whole very narrow, rigid and shallow in camber. Flight consists mainly of high-speed glides interspersed with brief bursts of wing-flapping. The feet are very small, and most species use them for little more than clinging to vertical surfaces.

The sexes are alike, immatures generally resemble adults, and there is no seasonal variation.

Genus Apus

(ah'-poos: "footless [-bird]")

Best represented in Africa, this genus contains 15 species, several of which extend across Europe and Asia. One species, strongly migratory, breeds across northern Asia and winters mainly in Australia; another (the Little or House Swift) has occurred as a vagrant.

Perhaps the most aerial of birds, species of *Apus* spend virtually their entire lives in the air except when breeding: they eat, drink, sleep and even mate on the wing. In addition to several other distinctive anatomical features, the toes are uniquely opposed in a manner that results in a laterally inward (right–left) grasp (not front-to-back as in other birds).

Fork-tailed Swift

Apus pacificus (pah-sif'-ik-us: "Pacific footless-bird")

Somewhat smaller than the Spine-tailed Swift, the Fork-tailed Swift also has a very different flight silhouette, characterised by a long slender tail and back-curved narrow wings; the fork in the tail is seldom obvious.

It breeds across northern Asia and, in winter, migrates southward through Indonesia to Australia. Although widespread across the continent, it is uncommon along the east coast and is most numerous in the interior, especially in regions where the average annual rainfall is between 130 and 220 millimetres. It is rare in Tasmania.

It often moves in loose flocks, which may be enormous, but it also frequently occurs in small groups, and often associates with Spine-tailed Swifts where the two species occur together. There are only scattered reports of perch-roosting, and it seems to habitually spend the night on the wing, circling at great height. Like those of the Spine-tailed Swift, flocks tend to move ahead of advancing weather fronts, where changes in air pressure tend to concentrate the aerial insects on which it feeds.

Nests are constructed in clusters; they are made of grass and other plant materials cemented with saliva against a rock face or the side of a building. Two or three eggs form the normal clutch; both parents incubate and care for the young.

HABITAT: aerial
LENGTH: 16–18 cm
DISTRIBUTION: more than 1 million km²
ABUNDANCE: sparse
STATUS: secure

(*J Paton*)

245

Genus *Collocalia*

(kol'-oh-kah'-lee-ah: "glue-nester")

About 20 species constitute this genus, restricted to South-East Asia, the Australasian region, and various islands in the Indian and south-western Pacific Oceans. Its members are small, obscurely patterned, extremely similar, and unusual among birds in their ability to echolocate. Most breed in dark caves, in which they navigate by uttering streams of high-pitched clicking notes, locating and avoiding obstacles by analysis of the returning echoes. Not all members have this ability, and some researchers further divide the group—largely on the presence or absence of the faculty—into two genera: *Aerodramus* and *Collocalia*. The single resident Australian species (two others have occurred as vagrants) is a member of the former group.

Grey Swiftlet

Collocalia spodiopygia (spoh'-dee-oh-pi'-jee-ah: "ash-grey-rumped glue-nester")

The Grey (or White-rumped) Swiftlet is widespread in the New Guinea region and on some islands of the south-west Pacific, as well as in north-eastern Australia from Cape York to the vicinity of Rockhampton; vagrants sometimes occur as far south as north-eastern New South Wales. It is sedentary, and mainly restricted to the coast and associated highlands. Although it is common and widespread, only six Australian breeding sites have so far been located.

Gregarious in all activities, it usually forages in large, loose flocks. When not roosting, it spends virtually all of its time on the wing, gathering insects—mainly flies and bugs—over all kinds of country from rainforest to open paddocks; it may range 30 kilometres or more from the breeding colony. Most foraging is above the canopy but within about 50 metres of the ground.

Breeding within the colonies is not synchronised, and pairs at any stage of the breeding cycle may occupy adjacent nests; several broods may be raised in succession. Most nesting occurs from July to February. The nest is tiny, deep and bracket-shaped, constructed of plant debris and feathers, cemented with saliva against the wall of a cave. The single egg takes about 22 days to hatch; the chick fledges at 40 to 50 days. Both parents construct the nest, incubate and care for the chick.

HABITAT: aerial
LENGTH: 11–11.5 cm
DISTRIBUTION: 100,000–300,000 km²
ABUNDANCE: sparse to common
STATUS: vulnerable

(T & P Gardner)

Genus *Hirundapus*

(hi-roon'-dah-poos: "footless-swallow")

This genus has four species restricted to eastern and south-eastern Asia; one species is strongly migratory, wintering in Australia. Its members have somewhat paddle-shaped wings and short stumpy tails in which the shafts of the rectrices extend beyond the vanes in stiff, sharp spines, a feature thought to be associated with clinging to rock faces, where most species nest.

Spine-tailed Swift

Hirundapus caudacutus (kaw'-dah-kue'-tus: "sharp-tailed footless-swallow")

Also called White-throated Needle-tail, this species may fly faster than any other bird, speeds of more than 100 kilometres per hour having been claimed for it. It breeds in Japan and across much of the Soviet Union; in winter it migrates southward to New Guinea and Australia, where it is largely restricted to the eastern part of the continent. Only occasional stragglers reach Western Australia and the Northern Territory; some reach Tasmania.

Like other swifts, it feeds, drinks and—apparently—even sleeps and mates on the wing. It is usually seen in large, loose flocks, their passing often associated with thunderstorms and weather fronts. It forages at almost any height from canopy level to 100 metres or more above the ground, over all kinds of country. Travelling flocks are a common sight over eastern seaboard cities from October to April.

Unlike many swifts, the Spine-tailed Swift does not use saliva in nest construction: the five to seven eggs constituting the clutch are laid on a carpet of wood fragments in a tree cavity.

HABITAT: aerial
LENGTH: 20–22 cm
DISTRIBUTION: more than 1 million km²
ABUNDANCE: sparse
STATUS: secure

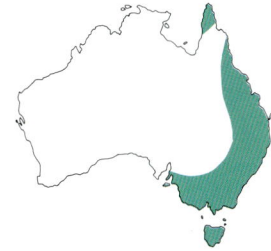

(AF D'ombrain)

Order CORACIIFORMES

(ko'-rah-see'-ee-for'-mayz: "*Coracius*-order", after a genus of rollers)

This order encompasses several hundred species of birds distributed in ten families, forming a heterogenous group so diverse that it is difficult to characterise concisely. A long bill, brightly coloured plumage, and syndactylous feet (a condition in which two forward-pointed toes are joined along part of their length) are common features, but most of the trenchant characters indicating mutual relationship are anatomical or physiological.

Best represented in tropical Africa, the order as a whole is virtually cosmopolitan, but several families are severely restricted in distribution, and only the kingfishers occur on all continents (except Antarctica). Three families (Alcedinidae, Meropidae, and Coraciidae) are represented in Australia, the last two by only a single species each.

Most members of the order are at least partly arboreal, foraging on the ground, in mid-air or in water, and almost all feed mainly either on large flying insects or on small vertebrates, especially fish and reptiles. Almost all breed in tunnels in earth or sand banks or in cavities in trees, laying round, plain white eggs. The young are hatched blind, naked and helpless.

Family ALCEDINIDAE

(al'-say-din'-id-ee: "*Alcedo*-family", after a genus of kingfishers)

The cosmopolitan family Alcedinidae (the kingfishers) consists of about 85 species generally grouped in three subfamilies (Daceloninae, Alcedininae, and Cerylinae); generic limits are controversial. The last-mentioned subfamily is unrepresented in Australia, where 10 species of the first two subfamilies occur. All are large-headed birds, with long strong bills and small weak, syndactylous, feet. Despite their name, most kingfishers are woodland birds that feed largely on insects and small reptiles, preying on fish only occasionally or not at all. Most are brightly coloured (the kookaburras constituting a notable exception), and green or blue upperparts are a widespread element of plumage pattern. The sexes are usually similar or nearly so, and immatures resemble adults except in being duller.

Genus Ceyx

(sayx: "kingfisher", after mythical Greek king Ceyx, who was changed into a bird)

These small kingfishers are the sole Australian representatives of the alcedinine kingfishers, a group widespread in the Old World. The eight or so species frequent rivers, streams and lakes and feed largely on small fishes. The genus has an extensive distribution in South-East Asia and the New Guinea region; two species occur in Australia.

Azure Kingfisher

Ceyx azurea (ah'-zue-ray'-ah: "blue kingfisher")

Largely a sedentary and solitary species, the Azure Kingfisher occurs in Indonesia and New Guinea, and in northern and eastern Australia from the Kimberley to Cape York Peninsula and southward to Tasmania, mostly near the coast. Established pairs maintain permanent linear territories incorporating stretches of up to about 500 metres of river or stream.

Although not restricted to such environments, the Azure Kingfisher is characteristic of slow-moving streams overhung by trees and bushes. It perches quietly in the shadows, on stubs and overhanging branches within about a metre of the water, intently scanning the surface for prey and diving at intervals with a brief splash and a quick return to its perch. Prey is swallowed headfirst. The Azure Kingfisher feeds on small fishes, frogs, insects and small crustaceans such as crabs and yabbies. Its flight is low, direct and rapid. The call is a very thin, high-pitched squeal.

Breeding occurs from September to January in the south, where two broods are usually raised in succession; in the tropics it breeds in the wet season. The nest is a chamber at the end of a tunnel dug in a stream bank. Both sexes incubate the clutch of four to seven plain white eggs, which hatch in about 22 days. The young fledge at about 25 days.

HABITAT: wetlands, including quiet rivers and streams, lagoons, lake shores, ornamental pools and bore drains; mangroves
LENGTH: *c* 18 cm
DISTRIBUTION: more than 1 million km²
ABUNDANCE: sparse
STATUS: secure

(*G Chapman*)

Little Kingfisher

Ceyx pusilla (pue-sil'-ah: "very-small kingfisher")

The Little Kingfisher occurs in New Guinea and in tropical Australia from Joseph Bonaparte Gulf in the Northern Territory to the vicinity of Mackay in Queensland. Solitary and sedentary, it closely resembles the Azure Kingfisher in habits and behaviour, but is more secretive and difficult to observe. It breeds in the wet season, but little is known of its nesting habits; four to six eggs form the clutch.

HABITAT: mangroves; rainforest pools and creeks
LENGTH: *c* 12 cm
DISTRIBUTION: 100,000–300,000 km²
ABUNDANCE: sparse
STATUS: secure

(*G Chapman*)

Genus *Dacelo*

(dah-say'-loh: "Kingfisher", anagram of *Alcedo*, a genus of kingfishers)

The kookaburras are large to very large kingfishers that live communally and lack the brilliant and colourful plumage characteristic of other members of the family. They feed mainly on small reptiles and are not especially closely associated with water; the habits and behaviour of the well-known Laughing Kookaburra are characteristic of the group. There are four species, two of which occur in Australia.

Blue-winged Kookaburra

Dacelo leachii (lee'-chee-ee: "Leach's Kingfisher", after W. E. Leach, British zoologist)

The Blue-winged Kookaburra inhabits southern New Guinea and northern Australia, where it occurs from southern Queensland in the

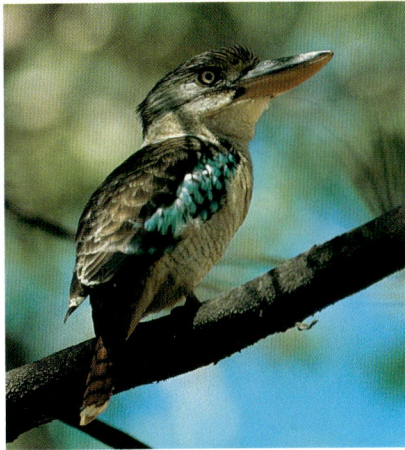

(G Chapman)

east to the vicinity of Shark Bay in the west. It is to some extent the tropical representative of the Laughing Kookaburra, but the two coexist over much of Queensland. The two species are extremely similar in ecology, habits and behaviour and their territories are mutually exclusive. The Blue-winged Kookaburra prefers somewhat wetter environments, has a somewhat larger group size (up to 12 birds), and has less ritualised territorial behaviour.

Breeding occurs from September to January. Most nests are cavities in arboreal termite mounds, or in baobab trees. The clutch usually consists of two eggs, but may vary from one to four. Incubation takes

about 24 days, and the young fledge at about 35 days.

HABITAT: swampy riparian forest and woodland; melaleuca swamps; eucalypt forest; canefields
LENGTH: 38–40 cm
DISTRIBUTION: more than 1 million km²
ABUNDANCE: sparse to common
STATUS: secure

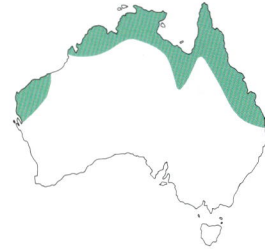

Laughing Kookaburra

Dacelo novaeguineae (noh'-vee-gin'-ay-ee: "New-Guinean Kingfisher")

The Laughing Kookaburra occurs throughout almost all of eastern Australia from Cape York Peninsula in Queensland to Eyre Peninsula in South Australia; it has been

(IP Rowles)

introduced to Tasmania, south-western Western Australia, and New Zealand, where it is now widespread. It is sedentary.

It lives in groups, consisting of a dominant pair and several auxiliaries, that together maintain permanent territories, announcing and defending these by means of loud, rollicking calls (the famous "laughter") uttered in chorus, especially at dawn and dusk. The diet consists mainly of reptiles, frogs, crustaceans and large insects; prey is usually captured on the ground by pouncing on it from a low perch.

Breeding occurs from September to January. The nest is in a large cavity, generally in a tree or arboreal termite mound. Usually two eggs form the clutch, incubation takes about 24 days, and the young fledge at about 35 days. Other birds may assist the female during incubation, and all group members share in the care and feeding of the young.

HABITAT: mainly eucalypt forest and woodland; also partly cleared agricultural land; suburban parks and gardens
LENGTH: *c* 43 cm
DISTRIBUTION: more than 1 million km²
ABUNDANCE: abundant
STATUS: secure

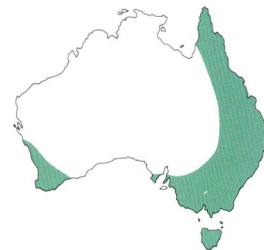

Genus *Halcyon*

(hal'-see-on: "kingfisher")

The genus *Halcyon* comprises about 40 species widespread in the Old World from Africa to Polynesia; generic limits are debatable and Australian species (together with certain others) are sometimes isolated in the genus *Todiramphus*. *Halcyon* kingfishers are medium-sized woodland species which feed largely on insects and small reptiles; their occurrence is not necessarily associated with water and they do not depend on fish. A common feature of plumage pattern consists of a collar of white or near-white extending around the nape. Four species occur in Australia.

Collared Kingfisher

Halcyon chloris (klo'-ris: "yellow-green kingfisher")

The Collared (or Mangrove) Kingfisher has an extensive distribution from the Arabian Peninsula across southern Asia to

(J Purnell)

islands of the south-western Pacific, including northern and north-eastern Australia. In Australia it inhabits mangroves almost exclusively, but elsewhere it occupies a range of (mainly coastal) habitats including dune scrub and coconut plantations. It seems to be largely sedentary, except in the south of its range where it is migratory. The Collared Kingfisher is somewhat larger, darker, and heavier-billed than the Sacred Kingfisher but the two species are otherwise extremely similar in appearance and difficult to distinguish; they are also closely similar in habits and behaviour. The Collared Kingfisher feeds mainly on insects, small reptiles and crustaceans.

The breeding season begins in September. Arboreal termitaria are favoured for nesting, but the birds

sometimes use tree cavities or earthen banks. Two or three eggs form the clutch, incubated by both parents.

HABITAT: mangroves
LENGTH: *c* 27 cm
DISTRIBUTION: 30,000–100,000 km²
ABUNDANCE: sparse to common
STATUS: secure

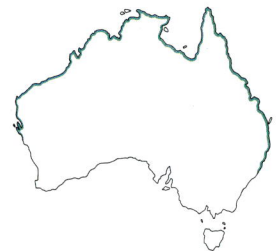

Forest Kingfisher

Halcyon macleayii (mak-lay'-ee-ee: "Macleay's kingfisher", after W. J. Macleay, Australian naturalist)

(EE Zillmann)

The Forest Kingfisher occurs in New Guinea and in northern and north-eastern Australia south to the Manning River in New South Wales. In the east it is strongly migratory and many Australian birds spend the winter in New Guinea. A common and conspicuous species, it lives in pairs and feeds mainly on large terrestrial insects and small reptiles, captured in typical kingfisher fashion by pouncing from a low perch. Often confused with the Sacred Kingfisher, it is identifiable by the diagnostic white flash in the upperwing.

Most breeding occurs from September to December. Nests are usually in tunnels excavated in arboreal termite mounds, but tree cavities or earth banks are

sometimes used; four to six eggs constitute the clutch, incubated by both sexes.

HABITAT: mainly eucalypt forest and woodland
LENGTH: *c* 20 cm
DISTRIBUTION: more than 1 million km²
ABUNDANCE: sparse
STATUS: secure

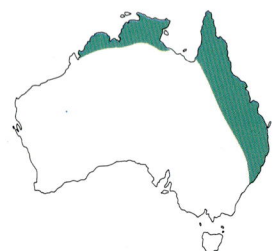

250

Red-backed Kingfisher

Halcyon pyrrhopygia (pie'-roh-pi'-jee-ah: "flame-colour-rumped kingfisher")

Most desert-adapted of the Australian kingfishers, the Red-backed Kingfisher is common over much of mainland Australia, avoiding only regions of relatively high rainfall in the south and east. It is largely migratory in the south, nomadic in the north. It is mainly solitary but sometimes occurs in loose groups when breeding. Like other kingfishers it feeds largely on insects and small reptiles, captured on the ground.

Most breeding occurs in October and November. The nest is a chamber at the end of a tunnel, usually drilled in the bank of a dry watercourse, but a termite mound is sometimes used. The usual clutch consists of four eggs. Incubation and fledging both take about 20 days, and both parents incubate and care for the young.

HABITAT: semiarid and arid eucalypt woodland; acacia scrub; tussock grassland with scattered trees
LENGTH: *c* 22 cm
DISTRIBUTION: more than 1 million km²
ABUNDANCE: sparse to common
STATUS: secure

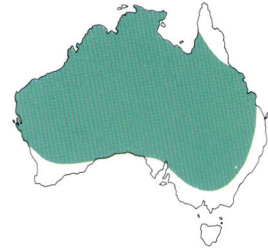

(AJ Olney)

Sacred Kingfisher

Halcyon sancta (sank'-tah: "sacred kingfisher")

This is perhaps the most familiar of the smaller Australian kingfishers, well known by its distinctive call—a deliberate, staccato, "ek-ek-ek-ek".

It is generally common throughout most of mainland Australia but is rare or absent in the most arid regions of the interior. It is a rare visitor to Tasmania. Other populations exist on Norfolk and Lord Howe Islands and it is also widespread from Indonesia, through New Guinea, to New Zealand and islands of the south-western Pacific. In Australia it is sedentary in the north, partly migratory in the south. It lives in pairs and feeds mostly on insects and small reptiles.

Breeding occurs from September to March. The nest is a chamber at the end of a tunnel, usually drilled into an arboreal termite nest but sometimes in an earth bank or even a fence post. Three to six eggs constitute the clutch. Incubation takes about 17 days, and the young fledge at about 28 days. Both parents incubate and care for the young.

HABITAT: most kinds of open temperate and tropical woodland; suburban parks and gardens; mangroves
LENGTH: *c* 21 cm
DISTRIBUTION: more than 1 million km²
ABUNDANCE: common
STATUS: secure

(P Rowles)

251

Genus *Syma*

(sie'-mah: "sea-nymph")

This genus consists of two species that are essentially Papuan in distribution; one of them occurs at the tip of Cape York Peninsula. Members of the genus closely resemble *Halcyon* kingfishers, but their colour pattern is strikingly different; the sexes differ in appearance; and the bill is yellow, not black. They also differ in inhabiting the lower levels of rainforest.

Yellow-billed Kingfisher

Syma torotoro (tor'-oh-tor'-oh: supposed Aboriginal name of the species)

The Yellow-billed Kingfisher is widespread in the New Guinea region but in Australia is restricted to the northern part of Cape York Peninsula. It seems mainly sedentary and, except when breeding, is a quiet, solitary and unobtrusive inhabitant of the lower levels of rainforest, feeding mainly on insects and small reptiles. The sexes differ in appearance: the crown of the male is entirely orange whereas the female has a black cap.

Arboreal termitaria are favoured for nesting. Breeding has been recorded from November to January, and three or four eggs constitute the clutch; little is otherwise known of breeding behaviour.

HABITAT: rainforest
LENGTH: *c* 20 cm
DISTRIBUTION: 30,000–100,000 km²
ABUNDANCE: sparse to common
STATUS: probably secure

(*EE Zillmann*)

Genus *Tanysiptera*

(tah'-nee-sip'-te-rah: "spread-feather")

Like *Syma*, this genus is essentially Papuan in distribution, but one species migrates annually to breed in north-eastern Australia. The eight species closely resemble *Halcyon* kingfishers in appearance, general behaviour and habits, but the plumage pattern is more colourful, the bill is red not black, and all species have extremely long attenuated tail streamers. All live in rainforest.

Buff-breasted Paradise-kingfisher

Tanysiptera sylvia (sil'-vee-ah: "forest spread-feather")

This spectacular kingfisher breeds in north-eastern Australia from Cape York south to about Townsville, and migrates to New Guinea to spend the winter, returning (usually abruptly) in early November. It is restricted to rainforest, where it lives in pairs frequenting the lower foliage levels, but it tolerates small relict patches and can reach very high densities (about one pair per hectare according to one study).

It is extremely active, noisy and conspicuous at the start of the breeding season but otherwise rather quiet and unobtrusive.

Breeding begins in November and most chicks hatch in early January.

The nest is usually a chamber in a termite mound at the base of a tree; tunnelling takes several weeks and the chamber is usually neglected for a week before laying begins. The clutch consists of three or four eggs; the incubation period is unrecorded, but the young fledge at about 24 days. Both parents prepare the nest, incubate, and care for the young.

HABITAT: rainforest
LENGTH: *c* 19 cm, plus tail streamers up to 13 cm
DISTRIBUTION: 100,000–300,000 km²
ABUNDANCE: sparse to common
STATUS: secure

(H & J Beste)

Family MEROPIDAE

(me-roh'-pid-ee: "*Merops*-family", after a genus of bee-eaters)

The family Meropidae, or bee-eaters, consists of about 24 species, typically resident in open country in tropical and warm temperate regions across the Old World. Most species are found in Africa but several extend into Europe and across southern Asia; two species reach New Guinea and one is widespread in Australia. A few aberrant tropical forms are placed in monotypic genera, but the majority are grouped in the genus *Merops*.

Among the most attractive of birds, typical bee-eaters are slim, graceful, colourful, active, and strongly aerial; they feed mainly on flying insects (especially bees and wasps) and usually occur in flocks. Most prey is captured during sallies from a high, exposed perch. Breeding is often colonial.

The sexes are similar, immatures resemble adults, and there is no seasonal variation.

Genus Merops

(me'-rops: classical Greek name for a species of bee-eater)

The characteristics of the genus are essentially those of the family.

Rainbow Bee-eater

Merops ornatus (or-nah'-tus: "ornate bee-eater")

The Rainbow Bee-eater breeds across much of mainland Australia and (locally) in southern New Guinea. It does not occur in Tasmania. Southern birds are strongly migratory and, in late summer, form flocks of several hundred which begin to move northwards, gradually at first. Knowledge of their wintering grounds is fragmentary but large numbers are known to cross Torres Strait: the species occurs throughout New Guinea in winter, some birds reaching Micronesia and southern Japan.

The Rainbow Bee-eater is gregarious in all activities, but usually occurs in small parties rather than large flocks (except when migrating). The diet consists mainly of dragonflies, wasps, swarming termites, and other flying insects of similar size; virtually all food is captured in the course of elegant, convoluted sallies in midair, the bird returning to a high, exposed perch to dismember prey.

Breeding, which extends from November to January in the south, is communal, typically involving about 30 to 40 birds. The birds help one another in tunnel excavation, incubation and care of the young, though most such supplementary assistance is provided by young, unmated males. The usual clutch is four or five eggs, laid in a chamber at the end of a one metre tunnel excavated in an earth bank or similar site.

The eggs hatch in about 21 to 25 days and the young fledge at about 28 to 30 days.

HABITAT: temperate to tropical woodland; savannah; forest edges; farmland
LENGTH: 21–24 cm
DISTRIBUTION: more than 1 million km^2
ABUNDANCE: common
STATUS: secure

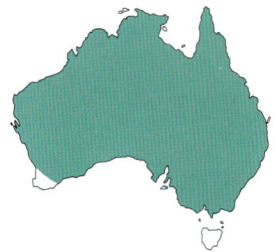

(MJ Seyfort)

254

Family CORACIIDAE

(koh'-rah'-sid-ee: "*Coracias*-family", after a genus of rollers)

Strongly represented in Africa but with several species extending northward into Europe and eastward across Asia to the Australasian region, the family Coraciidae, or rollers, consists of 11 species arranged in two genera: *Coracias* and *Eurystomus*. The latter has three species, one of which is a common summer migrant in northern and eastern Australia.

Species of *Coracias* have crow-like beaks and feed on large terrestrial insects, pouncing on them from low perches. Species of *Eurystomus* have broad bills and feed on large flying insects captured in sallies from high perches in treetops; they are markedly crepuscular.

Rollers are far from agile except in flight, and spend much of the day inactive on exposed perches, waiting and watching for prey. They usually occur in monogamous pairs. Few species are conspicuously patterned, and plumage colours are generally rich but muted. The sexes are similar, there is no seasonal variation, and immatures resemble adults.

Genus Eurystomus

(ue'-ree-stoh'-mus: "broad-mouth")

The characteristics of the genus are as discussed in the account of the family.

Dollarbird

Eurystomus orientalis (o'-ree-en-tah'-lis: "eastern broad-mouth")

In Australia the Dollarbird occurs in the north and east from about Broome to Melbourne, but there are a few records of vagrants in South Australia and a specimen was taken in Tasmania in 1918. Elsewhere the species is widespread from India to China, the Philippines and the Solomon Islands. The Australian population is migratory, spending the winter in New Guinea: departure is in March or April and, in the south-east, most birds are back on their breeding territories in October.

The Dollarbird is usually seen in pairs or (towards the close of the breeding season) in family parties; it occasionally forms flocks. It is a conspicuous bird, spending much of its time on high perches, especially on the edge of clearings or along rivers or streams. It favours dead trees and usually selects the tallest available; its persistent harsh rattling call (uttered by both sexes) is a characteristic sound of eastern woodlands in summer. It feeds mainly on large flying insects, such as cicadas, beetles and moths, normally captured in flight, but it occasionally descends to the ground to take grasshoppers. The flight is very buoyant, with deep hesitant wing-beats: complex aerobatics are used in the courtship display.

Breeding extends from October to January. The usual clutch is four eggs, laid in a tree cavity, often at a considerable height. Both parents care for the young and (probably) incubate.

HABITAT: temperate to tropical eucalypt woodland: rainforest margins and regrowth; the banks of rivers and streams
LENGTH: 26–30 cm
DISTRIBUTION: more than 1 million km^2
ABUNDANCE: common
STATUS: secure

(J Warham)

Order PASSERIFORMES

(pass'-e-ree-for'-mayz: "*Passer*-order", after a genus of sparrows)

Of a grand total of some 9000 bird species, about two-thirds belong to the order Passeriformes: they are referred to collectively as passerines. They occur on all continents except Antarctica and, in fact, dominate all but the most specialised and aberrant environments around the world. Only polar regions and a few small remote islands are entirely without them. None are marine, and only a very few obtain most of their food in fresh water.

The group is so large and diverse that common characters are hard to find but, in general, most are small to very small (only the ravens and lyrebirds could be described as large); very few are carnivorous; most breed in single pairs; parental care is highly developed and involves both parents; the young are hatched naked and helpless; and most have a very highly developed use of song as a territorial and courtship device. However, exceptions can be found for all of these features. The structure of the foot is characteristic: never webbed, it has three forward-pointing toes and one (non-reversible) pointing backward; all four are at the same level. The arrangement is ideally suited to the firm grasp of slender twigs and stems, and the group is often referred to as the "perching birds". All species have either nine or 10 distinct primaries, and most have 12 rectrices.

The taxonomy of the Passeriformes is an extremely intricate and uncertain field; it is also the focus of vigorous current research. Few families exhibit distinctive, unambiguous characters, and their boundaries are debatable: of Australian families, the relationships of the Pittidae, Menuridae, and Atrichornithidae are especially enigmatic.

In general, the passerines may be divided into two groups: the Deutero-oscines and the Oscines, although a few researchers favour isolating the former into a distinct order (Tyranniformes), in the belief that this group has a separate origin descending from the Coraciiformes. The Deutero-oscines are mainly American in distribution, but some researchers include the Pittidae within them.

In most classifications of the order some importance is attached to the number of primaries: the so-called "nine-primaried" passerines form a cluster of families, together involving about 1000 species, that seem closely related on other features. Like the Deutero-oscines, this group is also largely American in distribution but includes two families (Emberizidae and Fringillidae) with a wide range across the Northern Hemisphere: several of its members have been introduced to Australia.

In accordance with the foregoing outline, the vast majority of Australian passerines belong to the "10-primaried" Oscines. The larks (Alaudidae) and swallows (Hirundinidae) are very distinct (largely African and Asian in distribution, they are poorly represented in Australia and may be relatively recent immigrants), but several major lines of radiation may be distinguished in the remainder. Thus the thrushes, warblers, flycatchers and babblers (Turdidae, Sylviidae, Muscicapidae and Timaliidae) seem to be closely interrelated, and the same might be said of the monarchs, fantails and whistlers (Monarchidae and Pachycephalidae); the fairywrens, thornbills, scrubwrens and gerygones (Maluridae and Acanthizidae); and the honeyeaters, sunbirds, flowerpeckers, pardalotes, and Australian chats (Meliphagidae, Nectariniidae, Dicaeidae, Pardalotidae and Ephthianuridae). Similarly, the crows, birds of paradise, orioles, woodswallows, currawongs and several other groups (Corvidae, Paradisaeidae, Oriolidae, Artamidae, Cracticidae and others) also appear to have had a common origin, possibly within the Australasian region.

The above outline can at least plead traditional usage in its support, but much recent work has significantly modified it at several points, and other research is still too recent and rapidly evolving to allow useful synthesis. Meanwhile, the arrangement used here must be regarded as makeshift and temporary at best.

Family PITTIDAE

(pit'-id-ee: "*Pitta*-family", after the genus of pittas)

Grouped in a single genus, about 26 species of terrestrial rainforest birds constitute the family Pittidae. Their relationships are obscure. One species inhabits Africa but the group reaches its highest diversity in South-East Asia; three species occur in Australia (a fourth has occurred once as a vagrant).

Pittas have a characteristic silhouette: stocky, long-legged, extremely short-tailed, and with an alert, upright stance. Persistent tail-flicking is a characteristic mannerism. They roost in trees but forage on the forest floor, eating small ground invertebrates. Most species are brightly coloured and intricately patterned, because of which they are sometimes known as "jewel-thrushes". In the Australian species the sexes are nearly similar, there is no seasonal variation, and immatures resemble adults except in being duller.

Genus Pitta

(pit'-ah: "pitta", the common and specific names being derived from Tamil *ponunki pitta*, a jay-like bird)

The characteristics of the genus are those of the family.

Red-bellied Pitta

Pitta erythrogaster (e-rith'-roh-gas'-ter: "red-bellied pitta")

In Australia the Red-bellied (or Blue-breasted) Pitta occurs only in the monsoon rainforests of Cape York Peninsula (south to the McIlwraith Range), but it is also common in New Guinea and from the Philippines to Indonesia and the Bismarck Archipelago. The Australian population is strongly migratory, leaving for New Guinea in March or April and usually returning in November, sometimes as early as October.

Generally quiet, unobtrusive and rather solitary, it roosts in trees and forages on the ground, feeding on insects, worms, snails and other small invertebrates. The distinctive call is a low, penetrating, mournful whistle, often delivered from a perch up to 25 metres from the ground.

It nests from November to January. The nest is domed, with a side entrance, and constructed of roots, vine tendrils, sticks and moss, on or near the ground, often in a vine tangle or on top of a stump. Two to four eggs form the clutch.

HABITAT: mainly tropical monsoon rainforest
LENGTH: 16–18 cm
DISTRIBUTION: 30,000–100,000 km²
ABUNDANCE: sparse to common
STATUS: probably secure

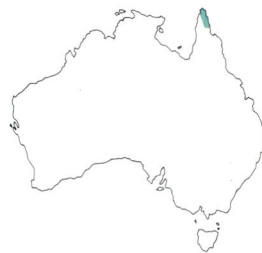

Rainbow Pitta

Pitta iris (ie'-ris: "rainbow pitta")

This pitta, found only in Australia, occurs in the Top End (including Groote Eylandt and Melville Island) and in the Kimberley, and is apparently sedentary. Usually encountered alone or in pairs, it forages on the forest floor for insects, worms, snails and other small invertebrates. Little is known of its habits.

It breeds at the onset of the summer monsoon, from November to March; the nest is a dome of sticks, roots, moss, and other plant material, situated on or near the ground. Three or four eggs form the clutch.

HABITAT: tropical vine-thickets; sometimes eucalypt woodland, mangroves, and streamside vegetation.
LENGTH: 16–18 cm
DISTRIBUTION: 100,000–300,000 km²
ABUNDANCE: sparse
STATUS: secure

(R Drummond)

257

Noisy Pitta

Pitta versicolor (ver'-see-kol-or: "varicoloured pitta")

The Noisy Pitta occurs in New Guinea and in eastern Australia from Cape York and the islands of Torres Strait to the Gloucester Tops region of New South Wales, but with notable gaps in distribution between the Cape York, Atherton Tableland, Clarke Range and southernmost populations. It has been recorded as a vagrant south to the Illawarra region. It mainly inhabits rainforest but sometimes occurs in adjacent eucalypt woodland and in mangroves; it seems most numerous in highlands (above an altitude of about 500 metres) and some researchers suspect that it migrates to coastal lowlands in winter.

It is almost exclusively terrestrial, but roosts in trees and often selects a perch up to 10 metres from the ground from which to deliver its loud, penetrating territorial whistle: "walk-to-work!". Except for these calls, it is rather quiet and unobtrusive in behaviour, though not especially shy.

It generally forages alone. Some fruit is eaten, but it feeds mainly on insects and other ground invertebrates: it sometimes uses a boulder as an anvil against which to crack open the shells of snails, leaving characteristic heaps of shell fragments.

Most nesting occurs from October to January. Usually secluded among the buttresses of a forest tree, or on a dead stump, the nest is a loose, bulky dome of sticks, roots, moss and ferns, sometimes bound with mud and lined with fragments of rotten wood; it often has a "doormat" of mammal dung at the entrance. Three to five eggs form the clutch, which is incubated by both sexes.

HABITAT: mainly warm temperate to tropical rainforest; sometimes eucalypt woodland, mangroves
LENGTH: 17–20 cm
DISTRIBUTION: 300,000–1 million km²
ABUNDANCE: common
STATUS: secure

(J Purnell)

Family ATRICHORNITHIDAE

(ah'-trik-orn-ith'-id-ee: "*Atrichornis*-family", after the genus of scrub-birds)

Restricted to Australia, this distinctive family has only two species in a single genus. Certain unusual features of the syrinx (voice-box) indicate a relationship with the lyrebirds (Menuridae) but otherwise the family has no close relatives. The furcula (wishbone) is cartilaginous, and scrub-birds are feeble and reluctant fliers, spending virtually all of their time in dense cover on the ground. They are sedentary and have loud penetrating songs.

Males, which take little active part in the nesting cycle, are markedly larger than females but the sexes are otherwise similar. Immatures resemble adults, and there is no seasonal variation.

Genus *Atrichornis*

(ah'-trik-or'-nis: "bristle-less-bird")

The characteristics of the genus are those of the family.

Noisy Scrub-bird

Atrichornis clamosus (klah-moh'-sus: "noisy bristle-less-bird")

One of Australia's most critically endangered birds, the Noisy Scrub-bird now occurs only in the vicinity of Two Peoples Bay in Western Australia, where (formerly considered extinct) it was rediscovered in 1961. In 1983 the total population was estimated at 137 individuals but, under careful and vigorous management, its numbers seem to be gradually increasing. Extremely rigid habitat requirements are a factor in its scarcity and, even in the last century, it was known only from a few scattered localities around the south-west: Albany, Augusta, Torbay, Margaret River, Drakesbrook.

It seldom leaves dense cover, where it is quick, agile, and extremely difficult to observe. It rarely flies. The diet consists of insects and other small invertebrates. The species is sedentary, and males permanently occupy dispersed territories of about six hectares, which they advertise with extraordinarily loud songs. Males take no part in the nesting cycle, and females occasionally nest beyond the boundaries of a territory.

Breeding occurs from May to October. Well hidden in a tussock or similar site, the bulky domed nest is constructed of rushes with some leaves and twigs; it has a low side entrance, often with a woven entrance ramp or platform; the nest cavity is lined with a cardboard-like substance. The single egg hatches in 36 to 38 days; the chick fledges at about 25 days.

HABITAT: dense thickets in temperate wet coastal heath and eucalypt woodland
LENGTH: 21–23 cm
DISTRIBUTION: less than 10,000 km²
ABUNDANCE: sparse
STATUS: endangered

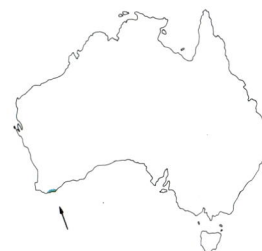

Rufous Scrub-bird

Atrichornis rufescens (rue-fes'-enz: "reddish bristle-less-bird")

Extremely difficult to observe, the Rufous Scrub-bird spends almost all of its time on the ground under dense cover, where it is quick and agile, rummaging through (and sometimes under) deep leaf litter. It feeds on insects, worms, other invertebrates, and seeds.

Its habitat requirements are extremely rigid, and its distribution is restricted to scattered pockets in highland forest from Mount Mistake in Queensland south to Barrington Tops in New South Wales. Males permanently defend territories of one to two hectares, advertising these (during the breeding season) by means of extraordinarily loud, penetrating songs that frequently include mimicry of other birds. Territories are widely dispersed, and prime habitat contains, on average, about four territories per square kilometre. Females maintain their own territories, which only partly overlap those of males. The total population is probably little over 1000 individuals.

Breeding occurs from September to December. The nest is domed, with a side entrance, constructed of rushes and grass, and lined with a hard cardboard-like substance. Two eggs constitute the clutch.

HABITAT: temperate and cool temperate highland forest, especially *Nothofagus*, with dense undergrowth and abundant leaf litter
LENGTH: 16–18 cm
DISTRIBUTION: 10,000–30,000 km²
ABUNDANCE: sparse to common
STATUS: possibly endangered

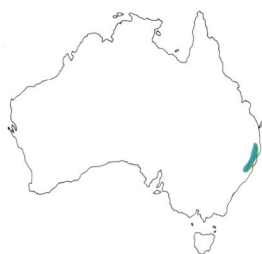

259

Family MENURIDAE

(men-ue'-rid-ee: "*Menura*-family", after the genus of lyrebirds)

Restricted to Australia, this family contains only two species in a single genus. Its relationships are obscure but it seems to be most nearly related to the Atrichornithidae. Lyrebirds are notable for their courtship displays and extraordinary song, which is audible for a kilometre or more and incorporates much mimicry. The name refers to the highly modified tail, which consists of two outer lyrate feathers, two median feathers and 12 extremely long filamentous feathers.

The sexes differ markedly in size (the female being much smaller) and in development of the tail, but are otherwise rather similar. There is no seasonal variation and immatures resemble adults except in tail structure; males take several years to attain full plumage and sexual maturity.

Genus Menura

(men-ue'-rah: probably "mighty-tail")

The characteristics of the genus are those of the family.

Albert's Lyrebird

Menura alberti (al'-ber-tee: "[Prince] Albert's mighty-tail", after the husband of Queen Victoria)

Albert's Lyrebird has an extremely restricted distribution extending from the Mistake Mountains of Queensland south to the Nightcap Range in New South Wales. It is sedentary.

Like the Superb Lyrebird, it roosts in trees but is otherwise terrestrial; it feeds on insects and other ground-living invertebrates obtained by raking through leaf litter. It is also similar to the Superb Lyrebird in much of its behaviour.

Unlike the Superb Lyrebird, the male does not construct display mounds: instead, it partly clears arenas on the forest floor. It also often sings and displays on fallen logs and other prominent low perches. The tail plumes are comparatively less developed, and the display performance is less spectacular than that of the Superb Lyrebird—but still remarkable.

Breeding occurs in June and July. The nest is a bulky dome of sticks, roots, moss and ferns lined with finer plant material and feathers, on or near the ground. The clutch consists of one egg, which hatches in about 42 days.

HABITAT: warm temperate wet sclerophyll forest; rainforest
LENGTH: *c* 85 cm
DISTRIBUTION: 30,000–100,000 km²
ABUNDANCE: sparse
STATUS: possibly endangered

(*W Lawler*)

Superb Lyrebird

Menura novaehollandiae (noh'-vee-hol-and'-ee-ee: "New-Holland mighty-tail")

The Superb Lyrebird inhabits dense rainforests and wet sclerophyll forest on the coastal plain and associated highlands of south-eastern Australia from the Dandenong Ranges in Victoria to the vicinity of Brisbane; it has been introduced into Tasmania, where it thrives. It is sedentary.

It roosts in trees but is otherwise terrestrial, roving over the forest floor, raking leaf litter aside in wide sideways sweeps of its powerful feet and feeding on insects and other small invertebrates thus exposed. It seldom flies except when going to roost. Although generally rather solitary, it frequently forms loose temporary foraging parties after breeding. It is wary but not necessarily shy and, where undisturbed, it may become extremely tame.

In courtship, males scrape together shallow mounds of earth at various points within their territories of about two to three hectares, visiting these daily during the season and using them as a stage for perhaps the most extraordinary performance among birds. Spreading and vibrating the tail and throwing it forward over the head, it prances and stamps on the mound while uttering a long, rambling, complex and extremely varied song that includes much mimicry of natural and artificial sounds. The notes are remarkably loud, rich and full.

Females, which maintain their own foraging territories overlapping those of the males, respond to this display by approaching the mound, where copulation takes place. No pair bond is formed and the male takes no part in nesting. He may mate with several females.

Breeding is mainly restricted to June and July but may extend from May to October. The nest is a bulky dome of sticks, moss, and ferns lined with finer plant material and feathers; generally on the ground or within three metres of it, it is usually constructed under the lip of an overhanging bank, between boulders, or against a tree stump. The single egg takes about 42 days to hatch, and the chick fledges at about 42 days.

HABITAT: temperate wet sclerophyll forest; rainforest
LENGTH: 80–100 cm
DISTRIBUTION: 100,000–300,000 km²
ABUNDANCE: sparse to common
STATUS: possibly endangered

(*P Klapste*)

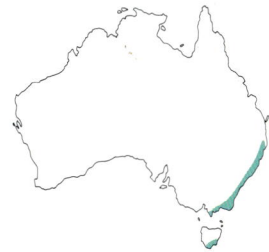

Family ALAUDIDAE

(ah-law'-did-ee: "*Alauda*-family", after a genus of larks)

One species of lark occurs in North America but the family is otherwise restricted to the Old World. The family contains about 78 species distributed in about 15 genera; most species occur in Africa and about one-half are restricted to that continent, but one species extends to Australia (another, the Skylark, has been introduced from Europe).

Almost all species are small, cryptically coloured birds of grassland, steppe and desert; none is arboreal. With few exceptions, the sexes are similar, and immatures resemble adults. They superficially resemble pipits (family Motacillidae) and frequently share the same habitats, but are stockier in silhouette, have generally shorter, deeper bills and do not bob or wag their tails.

Genus *Mirafra*

(mi-raf'-rah: "wonderful-African [-bird])

This genus incorporates about 25 species of larks, most of which occur in Africa, only one species extending to Australia, where it is widespread. These birds inhabit open, grassy plains, have streaked backs, are only slightly crested, and have short, rather deep bills. They feed on seeds and insects.

Singing Bushlark

Mirafra javanica (jah-vah'-nik-ah: "Javan wonderful-African-bird")

This lark occurs from Africa across southern Asia to Australia, where it is widespread in the north and east; it does not occur in Tasmania. It seems to be nomadic under some circumstances but there is little evidence for widespread movement. A bird of open, grassy country and crops, it vigorously colonises newly cleared farmland, and appears to have increased in numbers since European settlement. Regional populations vary in colour in an intricate manner that seems linked with variations in the soil types on which they live.

Though sometimes encountered alone, bushlarks generally occur in loose groups (of a few to several hundred) that wander and nest together. They feed on the ground on seeds and insects.

The male advertises his territory by hovering at some height above the ground, or (occasionally) by singing from a perch such as a fence post or low bush. Breeding is strongly influenced by rainfall but is usually from September to January in the south-east. The typical nest is a domed structure of grass and rootlets in the shelter of a tussock. Two to four eggs are laid.

HABITAT: cool temperate to tropical open grasslands; crops, pastures, farmland
LENGTH: *c* 13 cm
DISTRIBUTION: more than 1 million km^2
ABUNDANCE: sparse
STATUS: secure

(*MJ Seyfort*)

Family MOTACILLIDAE

(moh'-tah-sil'-id-ee: "*Motacilla*-family", after a genus of wagtails)

Largely Old World in distribution, this family consists of about 54 species in three genera. The longclaws (*Macronyx*) are exclusively African but the other two genera (*Motacilla* and *Anthus*) are more widespread (*Anthus* is virtually cosmopolitan) and are both represented in Australia, the former only as an uncommon winter visitor.

All are insectivorous and terrestrial birds of open country and are characterised, in addition to several internal anatomical features, by an elongated claw on the hind toe.

Genus *Anthus*

(an'-thus: Greek name of a small bird of uncertain identity)

The 35 species of pipits resemble wagtails (genus *Motacilla*) in many respects, but are generally less elegant in outline and more soberly coloured, commonly plain brown and streaked. Most are confusingly similar and difficult to identify. Best represented in Africa and Asia, the genus has at least one or two representatives in most parts of the world. Australia has one species.

Females are very similar to males, the young resemble adults, and there is no seasonal variation.

Richard's Pipit

Anthus novaeseelandiae (noh'-vee-see-land'-ee-ee: "New-Zealand *anthus*")

Richard's Pipit is found almost wherever there is open country throughout Australia; it also occurs across the Old World from Africa to New Zealand. It is not migratory but some populations undertake local seasonal movements and individuals tend to wander in winter, sometimes extensively.

If alarmed, a Richard's Pipit will fly to the top of a fence post, a low branch or some other conspicuous perch, but otherwise it spends virtually all of its time on the ground, running nimbly about, persistently wagging its tail up and down, and snatching insects from the surface and from low vegetation. It prefers bare, open areas with little cover, and is often common on golf courses, ocean beaches and other exposed environments. It is usually in pairs but may form flocks in winter.

The breeding season is August to December. The nest is a shallow cup of grass built into a shallow depression in the ground. Two to four eggs are laid, which hatch in about 13 to 14 days; the young fledge in 13 to 14 days. Most incubation is by the female (the male standing guard nearby) but both parents feed the young.

HABITAT: most kinds of temperate to tropical open country: alpine tussocks, gibber plains, pasture, clearings, beaches, golf courses and playing fields
LENGTH: 16–18 cm
DISTRIBUTION: more than 1 million km^2
ABUNDANCE: common
STATUS: secure

(B & B Wells)

Genus Motacilla

(moh'-tah-sil'-ah: Latin name for wagtail)

Wagtails are small, slender, elegant birds with extremely long tails, which they constantly wag up and down in a distinctive manner. There are 11 species in the genus. Common plumage colours are yellow, grey and black. Most species breed in Eurasia, migrating southward in winter. Three species (perhaps four) have been recorded in Australia, one regularly.

Yellow Wagtail

Motacilla flava (flah'-vah: "yellow wagtail")

A familiar farmland bird in Europe, the Yellow Wagtail has a breeding distribution that extends across northern Asia to Japan. In winter it migrates southward to Africa, India and New Guinea. Largely undetected in Australia before about 1970, it is now known to be a regular winter visitor in small numbers in the tropical north, especially the Kimberley and the Top End.

It is normally gregarious in its winter quarters but flocks are seldom recorded in Australia. Strictly terrestrial, it especially favours the boggy margins of streams and ponds in open country, where it feeds on small insects.

HABITAT: most kinds of tropical open country, including pastures, croplands, airfields and urban grasslands, usually near water
LENGTH: 17–19 cm
DISTRIBUTION: 30,000–100,000 km²
ABUNDANCE: rare
STATUS: secure

(*J Fennell*)

Family HIRUNDINIDAE

(hi'-roon-din'-id-ee: "*Hirundo*-family", after a genus of swallows)

The swallows and martins constitute a virtually cosmopolitan group and are among the most familiar of small songbirds to most people. They spend much of their time in flight, usually low over open land, or resting in twittering groups on telephone wires and similar perches; several species are common in urban parks and gardens. The various species are highly specialised to exploit small flying insects as a food resource, employing a range of behavioural and physiological adaptations to minimise the energy expended in capturing their food. Most species are small (typically weighing about 20 grams, and seldom exceeding 16 centimetres in total length), gregarious, and migratory (at least in temperate regions).

The family contains about 74 species, usually placed in 17 genera, but the group is unusually uniform and there is no consensus on generic boundaries. The family is best represented in tropical Africa, where about half of the species occur. Australia has five species: one in the endemic genus *Cheramoeca*, two martins and two swallows, one of which is a migrant from the Northern Hemisphere.

Most species are glossy blue-black above, white or pale reddish below. The sexes are similar, immatures resemble adults, and seasonal variation is absent or negligible. The term "swallow" is often applied to those species that have long, deeply forked tails, whereas "martins" typically have short, square tails and white rumps. These distinctions (among others) have been used as generic characters; thus the Australian swallows are placed in the genus *Hirundo* and the martins in the genus *Cecropis*, but the differences seem superficial and many researchers place all in the genus *Hirundo*.

Genus Cecropis

(sek-roh'-pis: Latinisation of Kekrops, mythical Greek king said to have founded Athens)

The characters of the genus are discussed in the account of the family.

Fairy Martin

Cecropis ariel (ah'-ree-el: "fairy *cecropis*")

The Fairy Martin occurs throughout mainland Australia but is very rare in Tasmania and seldom breeds north of 14°S: vagrants have occured in New Guinea and New Zealand. It is the most clearly migratory of Australian swallows, especially in the south-east where it is uncommon in winter. It is also the most gregarious, and seldom breeds except in colonies, sometimes quite large. It roosts communally and freely associates with other swallows and martins.

It closely resembles the Tree Martin in appearance, but is somewhat smaller and more graceful in build and has a pale rufous, rather than steely blue, crown. It feeds on small flying insects, captured in flight.

Breeding occurs mainly from August to January. The distinctive nest is a bottle-shaped structure of compacted pellets of mud, plastered against a vertical or overhanging surface; nests are normally crammed close together, and the underside of a bridge or a roadside culvert is a typical site. The same site is utilised year after year. The usual clutch consists of four or five dull white, faintly speckled eggs, which hatch in about 15 days. Both parents incubate.

HABITAT: cool temperate to tropical open country, especially near water
LENGTH: 11–12 cm
DISTRIBUTION: more than 1 million km²
ABUNDANCE: common
STATUS: secure

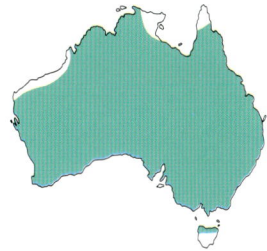

(MJ Seyfort)

Tree Martin

Cecropis nigricans (nig'-rik-ans: "blackish *cecropis*")

The Tree Martin occurs throughout Australia but virtually all breeding takes place south of 20°S; elsewhere the species occurs from southern Indonesia east to the Solomon Islands and New Caledonia. In Australia its distribution is strongly associated with forest and woodland (especially where large mature trees provide abundant cavities for roosting and nesting) and it is probably the most numerous and widespread Australian swallow. Some southern populations are migratory but the exodus is seldom complete.

It differs little from the Welcome Swallow in general habits and behaviour. Like that species, it breeds in single pairs or small groups, but it is otherwise strongly gregarious: roosting flocks in winter may be vast and spectacular. The Tree Martin associates freely with other swallows and martins. It feeds on small flying insects captured in steady flight.

Breeding occurs from July to January. Nests are usually placed in tree cavities, but caves and buildings are sometimes used; the cavity is usually padded with grass and other plant fragments or a few feathers, and its shape may be modified with mud. The same nest is often used year after year. Three to five eggs form the clutch; the incubation period is 15 to 16 days.

HABITAT: open cool temperate to tropical woodland and open country
LENGTH: 12–13 cm
DISTRIBUTION: more than 1 million km²
ABUNDANCE: common
STATUS: secure

(MJ Seyfort)

Genus *Cheramoeca*

(ke'-ram-eek'-ah: "hole-nester")

This genus contains only one species, which is confined to Australia; its characters are those of the species.

White-backed Swallow

Cheramoeca leucosternum (lue'-koh-stern'-um: "white-breasted hole-nester")

Found only in Australia, the White-backed Swallow occupies most of the arid southern interior of the continent but its occurrence is very rare and erratic north of about 20°S; its range has expanded eastward since European settlement and it is now locally common in south-east coastal regions where previously unknown. It is sedentary or locally nomadic.

It seldom forms large flocks, most colonies containing up to about 80 birds, but often far fewer. Since it nests in tunnels excavated in sand banks and similar sites, its distribution is closely linked to inland watercourses, gullies, eroded land and other geographical features providing access to such sites; soil erosion associated with land clearing and poor land management may be a factor in its eastward spread. Like all swallows, it feeds on flying insects captured in almost constant flight, generally over open country. It periodically rests in groups on roadside telephone wires or other high exposed perches. It roosts communally (unusually among birds, it has been found torpid in burrows). It breeds in small colonies.

Breeding occurs from August to December. The nest is a chamber at the end of a burrow about 1 metre in length, the chamber sparsely lined with grass and other plant fragments. The usual clutch consists of four to six pure white, glossy eggs, which hatch in about 14 days. Both parents incubate.

HABITAT: arid and semiarid temperate open country, especially near watercourses
LENGTH: 14–15 cm
DISTRIBUTION: more than 1 million km²
ABUNDANCE: sparse
STATUS: secure

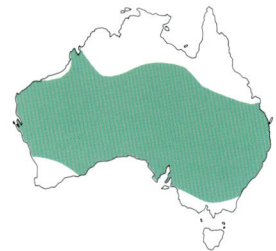

(M J Seyfort)

Genus *Hirundo*

(hi-roon'-doh: "swallow")

The characteristics of this genus are discussed in the account of the family.

Welcome Swallow

Hirundo neoxena (nay'-ox-ay'-nah: "new-foreign swallow")

(G Rogerson)

The Welcome Swallow occurs throughout Australia (including Tasmania) but its population reaches maximum density in the south and south-east; its occurrence is local and erratic in the arid interior and it does not breed in the Kimberley, the Top End or Cape York Peninsula. Some southern populations are migratory. Although formerly endemic in Australia, it reached New Guinea and New Zealand during the 20th century, and now breeds commonly in these regions.

It breeds in single pairs or small groups but is otherwise strongly gregarious: wintering or travelling flocks may number tens of thousands of individuals but are usually very much smaller. It feeds on small flying insects captured in steady flight— usually low, graceful and erratic— over pastures, playing fields and other open country. It is common in many towns and cities.

The breeding season extends from August to December and several broods are often raised in succession. The nest is a cup-shaped structure of mud, grass and plant fragments plastered to the wall of a building, the underside of a bridge or other structure, or a tree cavity or cave. Both parents incubate. The clutch consists of three to six (usually four) eggs, which hatch in about 15 days. The young fledge at about 28 days.

HABITAT: cool temperate to subtropical open country, especially near water
LENGTH: 14–16 cm
DISTRIBUTION: more than 1 million km²
ABUNDANCE: common
STATUS: secure

Barn Swallow

Hirundo rustica (rus'-tik-ah: "rural swallow")

With a breeding distribution extending across much of the Northern Hemisphere, this species is among the most widespread of all swallows. In winter it migrates south to Central America, southern Africa, India and South-East Asia, but it was not recorded in Australia until 1960 (except for a specimen taken in Torres Strait in 1860), when birds were seen neary Derby in Western Australia. Since then it has been recorded with increasing frequency, and it is now regarded as a common winter visitor to tropical Australia. Isolated occurrences have been recorded as far south as Perth, Adelaide and Sydney.

In Australia it frequently associates with flocks of Welcome Swallows and may be difficult to detect (the most conspicuous difference in appearance being possession of a narrow dark breast band). It is not clear whether the southward spread of records reflects a genuine shift in winter quarters or is an artefact associated with increased observer activity in the tropical north, but a similar southward trend has been noted in Africa. The Barn Swallow closely resembles the Welcome Swallow in habitat and general behaviour.

HABITAT: warm temperate to tropical open country, especially near water
LENGTH: 14–16 cm
DISTRIBUTION: 100,000–300,000 km²
ABUNDANCE: sparse
STATUS: secure

(P Klapste)

Family CAMPEPHAGIDAE

(kam'-pe-fah'-jid-ee: "*Campephagus*-family", after a genus of cuckooshrikes)

The family Campephagidae comprises about 70 species of songbirds found across the Old World, mainly in the tropics. Two clusters of genera are significant. The minivets (genus *Pericrocotus*), which are mainly black and red in plumage, are common and conspicuous members of the local avifauna in Oriental forests. The cuckooshrikes, comprising about eight other genera, are widespread in Africa, southern Asia and Australia (where seven species, grouped in two or three genera, occur). Apart from the Ground Cuckooshrike, they are strongly arboreal, gregarious birds of forest and woodland. The family is essentially insectivorous but some species feed largely on fruit. The plumage is generally grey, black and white in various combinations; the sexes are usually similar. A common feature rests in the feathers of the rump, which are unusually dense, partly erectile, and equipped with sharp rigid shafts.

Genus Coracina

(kor'-ah-see'-nah: "little-raven")

Cuckooshrikes are mainly soft grey in colour and have a distinctive suavity and elegance. They are strongly arboreal (except for the Ground Cuckooshrike) and gregarious. The diet comprises fruit or insects. They build characteristically meagre nests, and both sexes are involved in nesting, their respective involvement varying among the species. Vocalisations are simple but persistent. The group consists of about 40 species, best represented in Indonesia and New Guinea; five occur in Australia.

Barred Cuckooshrike

Coracina lineata (lin'-ay-ah'-tah: "lined little-raven")

The Barred (or Yellow-eyed) Cuckooshrike is narrowly and boldly barred on the underparts, a feature that distinguishes it from all other Australian cuckooshrikes. It inhabits New Guinea and eastern Australia from Cape York to the Manning River in New South Wales. Except when breeding, it is gregarious and strongly nomadic; roosting is communal. It is arboreal, and feeds mainly on fruit (swallowed whole), especially figs.

Breeding occurs from October to January. The nest is a small saucer of twigs and cobweb placed in a high horizontal tree fork; two eggs form the clutch.

HABITAT: mainly warm temperate to tropical rainforest and its margins
LENGTH: *c* 25 cm
DISTRIBUTION: 100,000–300,000 km²
ABUNDANCE: sparse
STATUS: probably secure

(EJ Whitbourn)

Ground Cuckooshrike

Coracina maxima (max'-im-ah: "largest little-raven")

The largest Australian member of the genus, the Ground Cuckooshrike also differs from the other species in being strongly terrestrial. It is widespread across Australia, mainly west of the Great Dividing Range. It is nomadic and social, and feeds largely on grasshoppers, beetles and similar ground-living insects.

Breeding is strongly influenced by local rainfall, but generally occurs from August to December. Built of twigs, grass and bark fragments bound with cobweb, the nest resembles that of other cuckooshrikes but is bulkier and looser in construction than most; it is placed on a horizontal bough or tree fork several metres from the ground. Two or three eggs form the clutch, incubated by both parents; all group members assist in feeding the young.

(JD Waterhouse)

HABITAT: temperate to subtropical open short-grass plains; savannah woodland; shrub steppe
LENGTH: 34–36 cm
DISTRIBUTION: more than 1 million k$^m2$
ABUNDANCE: sparse
STATUS: probably secure

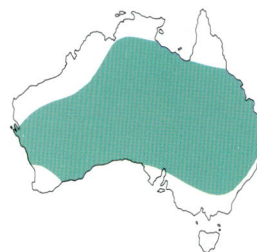

Black-faced Cuckooshrike

Coracina novaehollandiae (noh'-vee-hol-an'-dee-ee: "New-Holland little-raven")

(N Chaffer)

The Black-faced Cuckooshrike is perhaps the most widely recorded and ubiquitous Australian songbird, common wherever there are a few trees. It also occurs in New Guinea and a very closely related form (sometimes considered conspecific) extends westward across Indonesia through South-East Asia to India. It is partly migratory and locally nomadic but is seldom entirely absent from any area. It lives in pairs or small parties (but may flock in winter) and eats mainly insects, their larvae and, sometimes, fruit and seeds. It is strongly arboreal, favouring exposed perches and outer foliage, and seldom descends to the ground. Most prey is pounced on from hovering flight or from an exposed perch.

Black-faced Cuckooshrikes fussily shuffle their wings on landing, a persistent and distinctive mannerism. Slender, smooth and suave, adults are distinctive in appearance; immatures are similar but the facial black is limited to the lores and ear-coverts, thus offering some risk of confusion with the White-bellied Cuckooshrike. The sexes are similar.

Breeding occurs from August to December. On arrival from their winter quarters, birds establish territories several hectares in extent by means of conspicuous swooping display flights from tree to tree, frequently involving several birds. Both sexes cooperate in nest construction, incubation and care of the young. The nest is very small, a saucer of twigs and bark fragments copiously bound with cobweb, usually placed on a horizontal limb a considerable distance from the ground. Two or three eggs form the clutch, and the young fledge at about 25 days.

HABITAT: cool temperate to tropical, mostly open woodland; also scrub, heath, urban parks and gardens
LENGTH: *c* 33 cm
DISTRIBUTION: more than 1 million km^2
ABUNDANCE: abundant
STATUS: secure

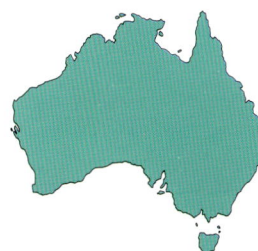

269

White-bellied Cuckooshrike

Coracina papuensis (pah'-pue-en'-sis: "Papuan little-raven")

The White-bellied Cuckooshrike closely resembles the Black-faced Cuckooshrike in almost every aspect of appearance, ecology and behaviour (it even shuffles its wings on landing). However, facial black is limited to the lores and ear coverts; it perhaps favours somewhat denser (but not necessarily taller) forest and woodland; and it is thought to focus on somewhat smaller insect prey. Its distribution is northern and eastern Australia, and from Indonesia through New Guinea to the Solomon Islands. Breeding occurs from August to March; a clutch of two or three eggs is incubated by both sexes.

HABITAT: mostly tropical to temperate open woodland; also scrub, heath, urban parks and gardens (especially in the tropics)
LENGTH: 27–28 cm
DISTRIBUTION: more than 1 million km²
ABUNDANCE: sparse to common
STATUS: probably secure

(E McNamara)

Cicadabird

Coracina tenuirostris (ten'-ue-ee-ros'-tris: "narrow-beaked little-raven")

Aptly named for its ringing, buzzy, cicada-like call, the Cicadabird is widespread in the New Guinea region and in northern and eastern Australia (mostly coastal and near-coastal) from the Kimberley to the vicinity of Melbourne, Victoria; in recent decades it seems to have declined markedly in much of Victoria. In the south-east it is strongly migratory. Rather solitary, it spends most of its time high in the canopy of dense forest, and is difficult to observe. It feeds mainly on insects.

The sexes differ in plumage, the male being almost entirely dark grey in colour, the female largely dull brown above, faintly and obscurely barred below.

Breeding occurs mainly from November to January. The nest, constructed largely by the male, is a small saucer of twigs, bark fragments and cobweb placed in a high horizontal tree fork. Only the female incubates the single egg, but both sexes feed the young; the incubation period is about 22 days, and fledging takes 27 to 28 days.

HABITAT: temperate to tropical rainforest; eucalypt and melaleuca forest; mangroves
LENGTH: *c* 26 cm
DISTRIBUTION: 300,000–1 million km²
ABUNDANCE: sparse to common
STATUS: probably secure

(J Purnell)

Genus *Lalage*

(lal'-ah-jay: "prattler", but also referring to an attractive girl immortalised in a Latin ode by Horace)

Aptly known as trillers, members of this genus are widespread in the Australasian region. They closely resemble cuckooshrikes except in having a plumage pattern which tends to be black and white rather than grey and black. There are about nine species, two of which occur in Australia (another, now extinct, occurred on Norfolk Island).

Varied Triller

Lalage leucomela (lue'-koh-mel'-ah: "white-black prattler")

The Varied Triller is an unobtrusive inhabitant of the upper canopy in heavy forest, where established pairs apparently maintain permanent territories. It is widespread in the New Guinea region and in northern and eastern Australia, south to the mid-north coast of New South Wales. It feeds mainly on fruit and large insects; foraging is methodical, and insects are gleaned rather than pounced upon.

Most breeding occurs from September to December. Extremely small, the nest is a shallow cup of twigs, cobweb and bark fragments placed in a high horizontal fork; only one egg is laid.

HABITAT: mainly warm temperate to tropical rainforest
LENGTH: *c* 19 cm
DISTRIBUTION: 300,000–1 million km²
ABUNDANCE: sparse to common
STATUS: probably secure

(K Ireland)

White-winged Triller

Lalage sueurii (sue-er'-ee-ee: "Le Sueur's prattler", after C. A. Le Sueur, French zoologist)

(IL Morgan)

The White-winged Triller occurs in Indonesia and New Guinea and is common throughout mainland Australia, although it does not often breed north of the Tropic of Capricorn. It is a casual vagrant to Tasmania, but has been known to breed there. It is strongly migratory in the south. It migrates and winters in small flocks but, on its breeding grounds, it lives in pairs (several nesting pairs frequently congregating in loose colonies). Common and conspicuous in general behaviour, it feeds mainly on insects, gleaned on the ground or in trees.

Unusually among Australian songbirds, the male has a pied nuptial plumage distinct from the non-breeding feminine plumage worn in its winter quarters. Females, immatures and winter males are dull brown above, with grey rumps and white-edged wing feathers; breeding males are crisply black above and white below.

Breeding extends from September to March. Males mark out their territories with flamboyant hovering display flights and persistent trilling songs from exposed perches. The nest is constructed of grass and bark fragments bound with cobweb and placed in a horizontal fork several metres from the ground. Two or three eggs form the clutch. Both parents cooperate in nest-site selection, construction, incubation and care of the young. Incubation takes 14 days, and the young fledge about 15 days after hatching.

HABITAT: temperate to tropical, mainly open eucalypt woodland, mallee, acacia scrub
LENGTH: *c* 17 cm
DISTRIBUTION: more than 1 million km²
ABUNDANCE: sparse to common
STATUS: secure

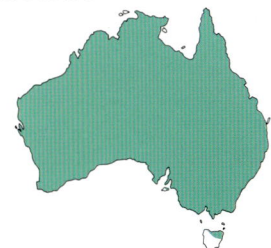

Family TURDIDAE

(ter'-did-ee: "*Turdus*-family", after a genus of thrushes)

About 315 species constitute this cosmopolitan family. In Africa, where they are best represented, chats are small, relatively brightly coloured and conspicuously patterned, and inhabit open country; members of the family are otherwise generally medium-sized, soberly coloured birds of forest and woodland. Males usually resemble females, and the young have a characteristic spotted plumage. Most species have attractive, melodious songs. They forage mostly on the ground, and they eat worms, snails and insects, often supplemented with fruit.

With nearly 70 species, *Turdus* is the largest and most widespread genus, and may be considered representative of the family as a whole. No species occurs naturally in Australia, but two (the familiar Blackbird *Turdus merula* and the Song Thrush *T. philomelos* of Europe) have been introduced. The genus *Zoothera* (about 33 species) closely resembles *Turdus*, but its members are sedentary (many species of *Turdus* are strongly migratory), inhabit denser woodland (often rainforest), have white bases to the primaries, and (with the exception of two species in western North America) are largely restricted to the Old World tropics: two species occur in Australia.

Genus Zoothera

(zoh'-oh-thee'-rah: "animal-hunter")

The characteristics of the genus are discussed in the account of the family.

Russet-tailed Thrush

Zoothera heinei (hie'-nee: "Heine's animal-hunter", after F. Heine, German ornithologist)

Only recently distinguished from the Bassian Thrush, the Russet-tailed Thrush occurs, mainly above 250 metres altitude, from the Gloucester Tops region of New South Wales north to the Atherton Tableland in Queensland. It also occurs in New Guinea. It is sedentary.

Rather solitary, terrestrial and unobtrusive, it closely resembles the Bassian Thrush in behaviour and appearance but has a somewhat shorter tail, a stronger rufous tinge on the rump and tail, and different songs and calls: the typical song consists of a loud, drawn-out, disyllabic whistle. It eats worms, insects and other small invertebrates.

HABITAT: temperate and highland tropical rainforest and wet eucalypt forest
LENGTH: 25–28 cm
DISTRIBUTION: 100,000–300,000 km²
ABUNDANCE: sparse
STATUS: probably secure

(*N Chaffer*)

Bassian Thrush

Zoothera lunulata (lue'-nue-lah'-tah: "crescentic animal-hunter")

Also called Ground, Scaly, White's, or Golden Mountain Thrush, the Bassian Thrush occurs in coastal regions and associated highlands of eastern Australia from the Mount Lofty Ranges in South Australia to the Atherton Tableland in Queensland, as well as Tasmania and islands in Bass Strait. It is sedentary.

Since it closely resembles the related Russet-tailed Thrush, field identification is very difficult, but its songs and calls are distinctive: the song is a melodious, subdued series of whistles and trills, usually slurred upwards. It often sings from treetops, mainly at dawn and dusk.

It is generally solitary, but is sometimes encountered in pairs or family parties. It feeds mainly on the ground, using the bill to probe in leaf litter, briskly flicking debris aside. Earthworms provide most of the diet, but it also eats insects, other small invertebrates, and some fruit. Rather furtive and unobtrusive in general behaviour, it spends most of its time in dense cover but often emerges to forage on adjacent lawns and roadside verges.

Breeding occurs from July to December, and sometimes two broods are raised in succession. Placed on a stump or in a tree fork up to 15 metres from the ground, the nest is a bulky bowl of bark strips, grass and leaves, decorated outside with moss and lined with rootlets. Two or three eggs form the clutch.

HABITAT: mainly temperate and cool temperate or highland rainforest and wet eucalypt forest; disperses in non-breeding season to more open woodland
LENGTH: 25–28 cm
DISTRIBUTION: 300,000–1 million km²
ABUNDANCE: sparse to common
STATUS: probably secure

(A Selby)

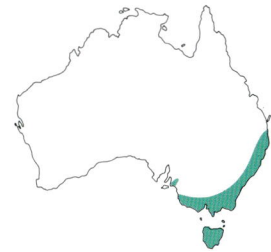

Family ORTHONYCHIDAE

(or'-thoh-nik'-id-ee: "*Orthonyx*-family", after the genus of logrunners)

Almost exclusively Australasian in distribution, the family Orthonychidae consists of about 21 species, grouped into seven genera. Three genera, comprising 10 species, inhabit Australia. The group is almost certainly artificial, but a better understanding is hampered by a notable lack of detailed information on the constituent species: almost nothing is known of several Papuan representatives. All members are essentially sedentary and terrestrial, most inhabit rainforest or other dense and complex plant communities (the quailthrushes *Cinclosoma* are a conspicuous exception), and most are distinctively broad-tailed, short-billed, and strong-legged, with soft, fluffy plumage.

Genus Cinclosoma

(sin'-kloh-soh'-mah: "bird-body")

The genus *Cinclosoma* is almost exclusively Australian, but one quailthrush is found only in New Guinea. The New Guinean species inhabits rainforest but the others are more characteristic of dry eucalypt woodland and arid scrublands. All are extremely quiet and unobtrusive in behaviour but have the distinctive characteristic of flushing, when disturbed, with a noisy, quail-like whirr of wings. Quailthrushes are sedentary, terrestrial and mainly insectivorous.

Chestnut-breasted Quailthrush

Cinclosoma castaneothorax (kah'-stan-ay'-oh-thor'-ax: "chestnut-breasted bird-body")

Very closely related to the Cinnamon Quailthrush, the Chestnut-breasted Quailthrush occupies two widely separated regions in the eastern and western interior of Australia. The adult male is distinguishable from other quailthrushes by a prominent horseshoe mark of black on the breast. It is locally nomadic, lives in small family groups, and closely resembles other quailthrushes in behaviour.

HABITAT: temperate Mulga and other arid woodland scrubs on stony ground
LENGTH: 21–25 cm
DISTRIBUTION: more than 1 million km^2
ABUNDANCE: sparse
STATUS: probably secure

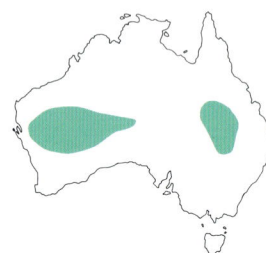

(*J Gray*)

Chestnut Quailthrush

Cinclosoma castanotum (kah'-stoh-noh'-tum: "chestnut-backed bird-body")

In many respects, the Chestnut Quailthrush is the arid-zone representative of the Spotted Quailthrush, closely resembling it in general behaviour. It occurs in the southern interior of Australia from central New South Wales westward to the coast of Western Australia.

Mainly sedentary and fairly common, it is somewhat local in distribution. It is unobtrusive, elusive and shy; its contact call is a thin, sibilant, extremely high-pitched whistle.

Breeding occurs mainly from July to December but is influenced by local rainfall; sometimes two or three broods are raised in succession. The nest consists of a shallow depression in the ground lined with grass; two eggs form the clutch.

HABITAT: temperate arid and semiarid scrublands, especially mallee and Mulga scrub, and almost always on substrates of sand
LENGTH: 23–26 cm
DISTRIBUTION: more than 1 million km^2
ABUNDANCE: sparse
STATUS: probably secure

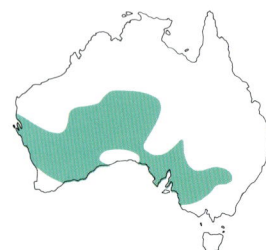

(*G Chapman*)

Cinnamon Quailthrush

Cinclosoma cinnamomeum (sin'-ah-moh-may'-um: "cinnamon-coloured bird-body")

Smallest of the quailthrushes, this species occurs in two distinct and widely separated populations, sometimes regarded as species: one on the Nullarbor Plain and the other in the basin of Lake Eyre and its vicinity. Males are easily dis-tinguished from the Chestnut Quail-thrush by the absence of chestnut on the breast, but females are very difficult to identify.

It resembles other quailthrushes in general behaviour, but little is known of its biology.

HABITAT: warm temperate low arid scrub on stony hillsides
LENGTH: 19–22 cm
DISTRIBUTION: 300,000–1 million km²
ABUNDANCE: sparse to common
STATUS: probably secure

(L Robinson)

Spotted Quailthrush

Cinclosoma punctatum (punk-tah'-tum: "fine-spotted bird-body")

The Spotted Quailthrush inhabits eastern Australia from the vicinity of Rockhampton in Queensland to extreme south-eastern South Australia. It is generally uncommon, locally distributed, and now very rare at the western extremity of its range.

It is sedentary and territorial, living in pairs or, sometimes, small family groups. It forages slowly and deliberately on the ground, where it is very shy and unobtrusive.

It is usually encountered when inadvertently flushed, whereupon it rockets away in low, direct flight on a quail-like burst of whirring wings. Birds maintain contact with each other by means of a thin, soft, high-pitched whistle. The varied diet includes seeds, insects, and sometimes small vertebrates such as lizards.

Breeding occurs mainly from August to December and one to three broods may be raised in succession. The female builds the nest and incubates the eggs but both parents feed the young. The nest is a cup of dry grass, leaves and bark fragments placed in a shallow depression in the ground, usually at the base of a tree, shrub, rock, or grass tussock. The usual clutch consists of two eggs; the young fledge at about 19 days.

HABITAT: mainly temperate dry sclerophyll woodland with abundant leaf litter, especially on rocky hillsides
LENGTH: 26–28 cm
DISTRIBUTION: 300,000–1 million km²
ABUNDANCE: sparse
STATUS: probably secure

(NW Longmore)

Genus Orthonyx

(orth-on'-ix: "straight-claw")

This genus consists of two species, both found in Australia. A striking zoogeographical anomaly rests in the fact that the Logrunner occurs in New Guinea and the southern block of eastern Australian rainforests but is absent from much of Queensland, while the Chowchilla of highland rainforests of north-eastern Queensland is endemic. Both species are sedentary, terrestrial, gregarious, and insectivorous, and are notable for their loud, penetrating calls uttered in chorus, their exceptionally robust legs and feet, and the fact that the quills of the tail feathers end in stiff, sharp emergent spines.

Chowchilla

Orthonyx spaldingii (spawl'-ding-ee-ee: "Spalding's straight-claw", after E. Spalding, Australian collector and taxidermist)

The Chowchilla (or Northern Logrunner) is restricted to the highlands of north-eastern Queensland from near Cooktown south to Paluma, generally above an altitude of 450 metres but locally at lower altitudes in regions of especially high rainfall.

The sexes are approximately similar but the male has a white throat whereas that of the female is rufous tan.

The Chowchilla closely resembles the Logrunner in general behaviour, being common, sedentary, territorial and terrestrial, and living in small groups; it is also notable for its ringing dawn choruses, which have a rhythmic, chanting quality, and frequently involve mimicry of other birds. At night it roosts in trees.

Little is known of its breeding, but nests may be found at any time of year (mainly from April to August). The clutch consists of one egg, and the nest resembles that of the Logrunner.

HABITAT: tropical rainforest
LENGTH: 28–30 cm
DISTRIBUTION: 30,000–100,000 km²
ABUNDANCE: common
STATUS: possibly endangered

(*L. Robinson*)

Logrunner

Orthonyx temminckii (tem'-ink-ee-ee: "Temminck's straight-claw", after C. J. Temminck, Dutch ornithologist)

Also known as the Spine-tailed Logrunner, Southern Logrunner or Spine-tailed Chowchilla, the Logrunner occurs in coastal regions and associated highlands of eastern Australia from the Bunya Mountains of south-eastern Queensland south to the Illawarra district of New South Wales. Despite extensive forest clearance, it remains common in the north, but numbers decline southward and it is now extremely rare in the Illawarra district. It also occurs in the highlands of New Guinea.

The sexes are approximately similar, except that the male has a white throat whereas in the female the throat is rufous tan.

The Logrunner lives in small parties which maintain permanent territories. It is sedentary, strictly terrestrial, and not especially shy or difficult to observe, but it is made most conspicuous by its loud choruses of rich, rapid, penetrating notes, uttered especially at dawn. It feeds on insects and other small invertebrates living in the soil and leaf litter, obtained by vigorously scratching the leaf litter aside with wide sweeps of its long, powerful feet. It seldom flies.

Most breeding occurs from May to August, two broods often being raised in succession. The nest is an elaborate domed structure of sticks and debris, with a side entrance overhung by moss and opening onto a platform of sticks, usually placed against a tree trunk on or near the ground. Construction, incubation and care of the young is largely by the female. Two plain white eggs constitute the clutch; incubation takes about 24 days and the young fledge at about 16 to 18 days.

HABITAT: warm temperate rainforest
LENGTH: 18–21 cm
DISTRIBUTION: 100,000–300,000 km^2
ABUNDANCE: sparse to common
STATUS: possibly endangered

(E & N Taylor)

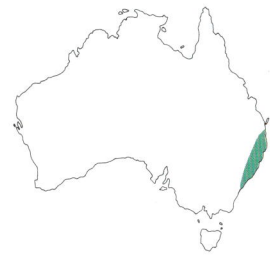

Genus *Psophodes*

(soh-foh'-dayz: "noisemaker")

The genus *Psophodes* consists of four species found only in Australia (although the enigmatic *Androphobus* of New Guinea may well prove to constitute a fifth species). Two species of whipbirds inhabit dense plant communities in southern and eastern Australia, while two wedgebills inhabit scrublands of the arid interior. Whipbirds live in pairs occupying permanent territories; are shy and furtive in behaviour; forage mainly on the ground or near it; and utter songs in antiphonal duets. Wedgebills are communal and were only recently distinguished as two species, mainly on ecological and behavioural grounds: they differ only slightly in appearance.

Chirruping Wedgebill

Psophodes cristatus (kris-tah'-tus: "crested noisemaker")

The Chirruping Wedgebill inhabits the arid interior of south-eastern Australia, west to Lake Eyre and Lake Torrens in South Australia. It is generally common but somewhat local in distribution. It appears to be mainly sedentary and often lives in small communal groups, occupying territories that are maintained throughout the year. It spends most time in lower cover, where it is unobtrusive but not especially hard to observe. The diet consists of seeds and insects.

The Chirruping Wedgebill is distinguishable from the Chiming Wedgebill only by its distribution and calls; by the faint streaking on the breast; and by its slightly longer tail. The sexes are similar and immatures resemble adults. The persistent song, uttered from a conspicuous perch on a high twig or the top of a bush, is an antiphonal duet: the male's part consists of two soft, sweet notes followed by two shrill notes, the female's response being a brisk, "chir-up".

Breeding which is strongly influenced by local rainfall, occurs from March to May and from August to November: several broods may be raised in succession. The nest is a cup of grass, bark and twigs lined with fine grass and placed in a low fork, a mistletoe clump or dense vegetation within about 3 metres of the ground; two or three eggs form the clutch.

HABITAT: low temperate shrublands, including bluebush and emubush; acacia scrub, lignum and canegrass associations
LENGTH: 18–22 cm
DISTRIBUTION: 300,000–1 million km²
ABUNDANCE: common
STATUS: probably secure

(WJ Labbett)

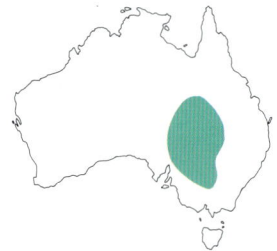

Western Whipbird

Psophodes nigrogularis (nig'-roh-gue-lar'-is: "black-throated noisemaker")

(LA Moore)

Reduced by fires and clearing to scattered and isolated pockets of distribution, the Western Whipbird is restricted to parts of the Murray Mallee districts of South Australia and Victoria; Eyre Peninsula, Yorke Peninsula, and Kangaroo Island in South Australia; and the coastal mallee of southern Western Australia. It is generally rare to very rare.

It lives in pairs that maintain partly overlapping and erratically defended territories up to 10 hectares in extent. Foraging almost entirely in dense cover on or near the ground, it is extraordinarily shy and furtive in behaviour, its presence normally detectable only by its calls. Uttered always from dense cover, and most freely during winter, the territorial song is an antiphonal duet initiated by the male with a series of creaky whistles, to which the female responds promptly with a phrase sometimes transliterated as "pick-it-up". The varied diet includes insects and other invertebrates, small reptiles, and seeds.

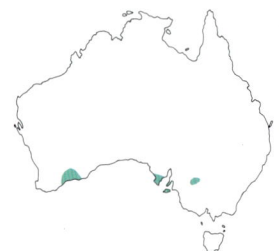

One brood is raised each year, normally between July and November. In contrast to the Eastern Whipbird's nesting behaviour, both parents build, incubate and rear the young. Constructed over about 7 to 14 days, the nest is a substantial bowl of twigs and bark strips, lined with fine grass and placed in or under a low shrub, usually within one metre of the ground. Two eggs form the clutch. Incubation takes 21 days, and the young are able to fly at about 14 days but leave the nest several days earlier.

HABITAT: temperate mallee with a dense shrub understorey, broombush and associated heathlands
LENGTH: 22–24 cm
DISTRIBUTION: 30,000–100,000 km²
ABUNDANCE: rare to very sparse
STATUS: possibly endangered

Chiming Wedgebill
Psophodes occidentalis (ok'-sid-en-tah'-lis: "western noisemaker")

(G Chapman)

The Chiming Wedgebill inhabits central Australia from the Simpson Desert westward to the coast of Western Australia. It is common but somewhat locally distributed and locally nomadic. In appearance, habits and behaviour it closely resembles the Chirruping Wedgebill of eastern Australia, but it favours slightly different habitat, is markedly more shy and elusive, and does not sing in duet.

The persistent, attractive, musical song is most often transliterated as "sweet-Kitty-Lintof", with the emphasis on the last note.

Breeding occurs from February to May and from August to November, but is influenced by local rainfall; several broods are sometimes raised in succession. Two or three eggs form the clutch, and incubation takes about 17 days.

HABITAT: warm temperate arid scrublands with acacia (especially Mulga), broombush, mallee and spinifex along watercourses; especially favours areas with abundant mistletoe
LENGTH: 19–23 cm
DISTRIBUTION: more than 1 million km²
ABUNDANCE: common
STATUS: probably secure

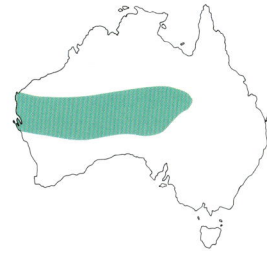

Eastern Whipbird
Psophodes olivaceus (ol'-iv-ah'-say-us: "olive noisemaker")

(E Hosking)

The Eastern Whipbird inhabits eastern Australia from near Rockhampton in Queensland south to the vicinity of Melbourne in Victoria; isolated populations also occur on the Atherton Tableland and the Clarke Range in north-eastern Queensland.

It is sedentary, living in pairs that maintain permanent territories up to 10 hectares in extent. The Eastern Whipbird spends much of its time in dense undergrowth, where it is extremely difficult to observe, but it occasionally forages at some height in bushes and trees. Although furtive in behaviour, it is inquisitive and can sometimes be coaxed into the open by a cautious observer. Insects constitute the chief food.

The sexes are similar but immatures lack the white cheek patch characteristic of the adult. The famous call is an antiphonal duet, in which the male produces a long whistle, rising to crescendo and climaxing in an explosive whipcrack note, to which the female responds by uttering a brisk "awee awee" or "chew chew". Whipbirds also utter various chuckling and scolding notes when foraging.

The breeding season extends from July to January, during which period two broods are often raised. The nest is a loose cup of sticks and bark fragments, lined with grass and rootlets, and hidden in dense undergrowth or in a low shrub up to two metres from the ground. Two or three eggs form the clutch. Incubation takes 18 days, and the young leave the nest at 11 to 12 days, several days before they are able to fly. The female alone incubates and broods the young but the male feeds her at the nest and also assists in feeding the young.

HABITAT: mainly temperate to subtropical rainforest and dense, lush eucalypt forest; sometimes wet heath, suburban parks and gardens
LENGTH: 26–31 cm
DISTRIBUTION: 300,000–1 million km²
ABUNDANCE: common
STATUS: probably secure

Family TIMALIIDAE

(tim'-ah-lee'-id-ee: "*Timalia*-family", after a genus of babblers)

About 277 species, mainly characteristic of the Old World tropics, constitute the family Timaliidae. Closely related to the Muscicapidae (flycatchers) and Turdidae (thrushes), the group is extremely diverse, and concise definition is difficult. They occupy a wide range of environments from desert to jungle and rainforest. Many of the approximately 42 genera are difficult to allocate to family, and some researchers prefer to combine all three families, together with Sylviidae (warblers), into a single large family, Muscicapidae.

Within this amorphous assemblage, typical babblers differ most conspicuously in being sedentary, in living in permanent close-knit groups or communes, and in lacking a distinct juvenile plumage. They are generally rather stocky in build compared to thrushes and flycatchers, with relatively large bills and long, stout powerful legs. The sexes are usually similar in appearance. Most lack tuneful songs, but the vocabulary is characteristically wide and varied, consisting of chuckles, whistles and yelping notes. Most species eat insects and other small invertebrates, captured on the ground. Babblers are noisy, active, playful and conspicuous.

The only Australian genus, *Pomatostomus*, conforms broadly with this synopsis. All four species have highly structured social behaviour. They build domed nests, and also build nests specifically for communal roosting (several are usually constructed within the group's territory). All live in permanent groups of up to about 20 individuals, inhabit arid or semiarid scrubland, and feed mainly on the ground.

These characteristics notwithstanding, evidence is accumulating that *Pomatostomus* may not be closely related to other babblers, and some researchers have urged that the genus be placed in a separate family.

Genus Pomatostomus

(poh'-mah-toh-stoh'-mus: "covered-mouth")

The characteristics of the genus are discussed in the account of the family.

Hall's Babbler

Pomatostomus halli (haw'-lee: "Hall's covered-mouth", after Major H. W. Hall, Australian philanthropist)

Sometimes called the Black-bellied Babbler, this species is closely related to the White-browed Babbler and largely replaces it in arid mulga scrubs of the central eastern interior of Australia, where it is common but somewhat patchy in distribution. It remained undiscovered until 1963. Like other babblers, it lives in close-knit communes of five to 25 birds (occasionally in pairs) occupying permanent territories up to about 20 hectares. It is noisy, active and conspicuous in behaviour, feeds mostly on the ground, and eats insects and spiders.

Easily confused with the White-browed Babbler, it may be distinguished by its broad white eyebrow and blackish lower breast and belly, sharply cut off from the white breast. Its calls are somewhat less varied than those of other babblers.

It breeds from June to October, or at any time after suitable rainfall. All group members participate but only the prime female incubates and broods, being fed by others while on the nest. Two eggs constitute the usual clutch. The nest is a rough domed structure of sticks and twigs in a small tree or bush up to 10 metres from the ground.

HABITAT: arid acacia scrubs, especially Mulga, and especially those with a grassy understorey
LENGTH: 23–25 cm
DISTRIBUTION: 100,000–300,000 km²
ABUNDANCE: sparse to common
STATUS: probably secure

(DD Dow)

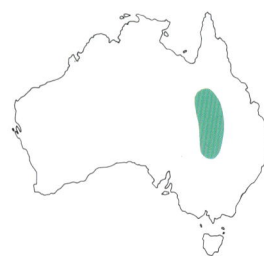

Chestnut-crowned Babbler

Pomatostomus ruficeps (rue'-fee-seps: "red-headed covered-mouth")

This attractive babbler generally inhabits more arid environments than those occupied by other babblers with which its range overlaps in the south-eastern interior of mainland Australia. It is sedentary, living in troops of about 12 to 20 individuals that maintain permanent territories. Like other babblers, it is noisy, active, conspicuous and playful, but its behaviour has not been systematically studied. It forages mainly on the ground, where it eats insects and spiders, as well as some seeds. Its double white wing bar is a useful identification character in the field.

Breeding usually occurs from July to November, and sometimes two broods are raised in succession. The nest is the typical domed babbler structure of sticks and twigs placed in a tree fork up to 10 metres from the ground; three to five eggs form the normal clutch.

HABITAT: arid temperate scrubs including mallee, Mulga, belah; saltbush plains; also lignum and coolabah woodland
LENGTH: *c* 22 cm
DISTRIBUTION: 300,000–1 million km²
ABUNDANCE: sparse to common
STATUS: probably secure

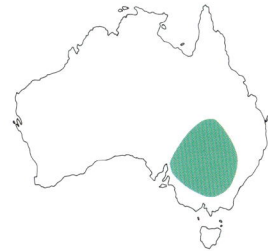

(WJ Labbett)

White-browed Babbler

Pomatostomus superciliosus (sue'-per-sil'-ee-oh'-sus: "eyebrowed covered-mouth")

The White-browed Babbler is common and widespread across southern Australia, north to southern Queensland in the east and Shark Bay in the west; it does not occur in Tasmania nor, in general, east of the Great Dividing Range. It is sedentary or locally nomadic. Its relatively small size, dark crown, white breast and narrow white eyebrow are useful characters in field identification.

Noisy, active and conspicuous in behaviour, it resembles the Grey-crowned Babbler in general habits, living in compact groups, foraging mostly on the ground, and feeding on insects and spiders.

Breeding extends from June to December, and two broods may be raised in succession. All group members participate (but only the dominant female incubates), and occasionally two or more females lay in the same nest. The nest is a large dome of sticks and twigs lined with softer materials and placed in a small tree or bush up to 6 metres from the ground. Two or three eggs constitute the normal clutch. The incubation period is about 16 days.

HABITAT: arid temperate scrubby woodland, especially eucalypt and acacia
LENGTH: 20–22 cm
DISTRIBUTION: more than 1 million km²
ABUNDANCE: common
STATUS: secure

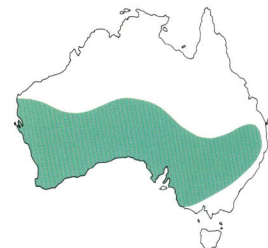

(G & J Little)

Grey-crowned Babbler

Pomatostomus temporalis (tem'-por-ah'-lis: "temple [-marked] covered-mouth")

The Grey-crowned Babbler occurs in southern New Guinea and is common and widespread over most of mainland Australia except for the south and south-west. The sexes are similar, and immatures differ most conspicuously in having brown (not yellow) eyes. North-western birds are markedly darker and more rufous than eastern birds, and the best field marks are the pale greyish crown and yellow eyes.

The Grey-crowned Babbler is sedentary, living in communal groups of four to 12 individuals that maintain permanent territories, usually of about 12 to 20 hectares. Members of a group remain in close contact through most activities, even sleeping together in a roosting nest constructed especially for the purpose. Squabbles and vigorous chases involving neighbouring groups are common but fighting is rare; groups huddle tightly together when alarmed. Most foraging is on the ground, the birds covering the area in noisy active mobs in follow-the-leader fashion, probing soft soil, overturning small stones and similar objects, rummaging through leaf litter, and exploring the rough bark of fallen branches and limbs. The diet consists mainly of insects and spiders, but reptiles and seeds are also eaten. Calls include a variety of chuckles and harsh notes, and an antiphonal duet between male and female.

Most breeding occurs from June to February, and two broods may be raised in succession. All group members participate. The nest is a bulky dome of sticks and twigs, lined with grass and fur and placed conspicuously in a tree fork up to 12 metres from the ground. Two to four eggs constitute the clutch; incubation takes 18 to 23 days and the young fledge in about 21 days.

HABITAT: tropical and temperate open eucalypt forest and woodland
LENGTH: 26–29 cm
DISTRIBUTION: more than 1 million km^2
ABUNDANCE: common
STATUS: secure

(B & B Wells)

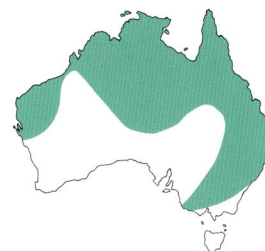

Family EOPSALTRIIDAE

(ee'-op-sahl'-tree-id-ee: "*Eopsaltria*-family", after a genus of robins)

This is a family of about 45 species, only a few of which have distributions extending beyond Australia and New Guinea. Conveniently referred to as Australo-Papuan robins, and traditionally placed within the essentially Northern Hemisphere Muscicapidae, these birds have relatively recently been recognised to be quite distinct from this family. The nature of their relationships with other Australian insectivorous forest birds, especially the whistlers (family Pachycephalidae) and monarchs (family Monarchidae), remains much less certain. They are only tentatively assigned to family rank.

They are generally quiet and unobtrusive inhabitants of the lower storeys of dense forest. Most share a common foraging technique: selecting perches a metre or two above the ground (often clinging sideways to the lower trunk of a tree), scanning the ground beneath, and periodically pouncing on crawling insects, emerging earthworms, and other small invertebrates. The legs are relatively long and the wings pointed, and they are agile in flight, in foliage and on the ground. The sexes are usually similar (the genus *Petroica* is a conspicuous exception) and patches of yellow, red, orange or pink are common elements of the colour pattern, although several species are black and white. Usually, the female builds the nest and incubates unaided but both parents rear the young.

Genus *Drymodes*

(drie-moh'-dayz: "woods-dweller")

Confined to Australia and the New Guinea region, the scrub-robins constitute a genus of two species with a vexed taxonomic history. One species inhabits tropical rainforest and the other inhabits arid scrublands of the Australian interior. They have many thrush-like characteristics and were long associated with that family (the Turdidae). They have also been tentatively linked with several other groups of birds, but recent biochemical studies strongly indicate a relationship with the Australo-Papuan robins.

Southern Scrub-robin

Drymodes brunneopygia (broon'-ay-oh-pi'-jee-ah: "brown-rumped woods-dweller")

The Southern Scrub-robin has an extensive distribution across southern Australia from the vicinity of Mount Hope in central New South Wales westward to Shark Bay, but its occurrence is somewhat patchy and it is generally uncommon. It is sedentary.

It lives in pairs and forages mainly on the ground, usually in dense cover, flicking leaf litter aside with its bill and feeding on insects and other small invertebrates. Often tame and inquisitive, it is generally quiet and unobtrusive in behaviour. Its calls include a musical, attractive song, "chip chip par-eee", delivered either from an exposed perch or from dense cover, and a thin, slow whistle; it also directs harsh chattering scolding notes at an intruder. It has a distinctive habit of cocking its tail, either slowly or with an abrupt flick, then lowering it gradually.

It breeds in dispersed territorial pairs from July to December, raising only one brood per season. The nest is a cup of grass, twigs and bark fragments placed in a depression in the ground near the base of a tree, shrub or other sheltering object. One egg constitutes the clutch.

HABITAT: mallee and other semiarid scrub, especially broombush; also heath, coastal melaleuca thickets
LENGTH: 21–23 cm
DISTRIBUTION: 300,000–1 million km²
ABUNDANCE: sparse
STATUS: probably secure

(L Robinson)

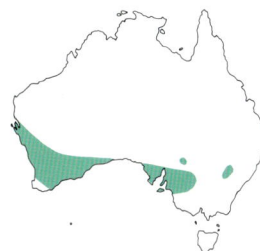

Northern Scrub-robin

Drymodes superciliaris (sue'-per-sil'-ee-ah'-ris: "eyebrowed woods-dweller")

The Northern Scrub-robin inhabits the Aru Islands and New Guinea as well as the north-eastern part of Cape

(D & M Trounson)

York Peninsula and the Roper River region of the Northern Territory. It is sedentary and common on Cape York Peninsula but the population along the Roper River has not been recorded since its discovery in 1910.

It is generally encountered alone or in pairs. Quiet, unobtrusive and furtive, it forages on the ground under cover, feeding on insects, land snails and other small invertebrates. It has a distinctive habit of raising and lowering the tail slowly. Its calls include a long, thin whistle.

Breeding occurs from October to January. The nest is a saucer-shaped depression in the ground in the shelter of a tree or bush, lined with

twigs and bark fragments. Two eggs constitute the clutch.

HABITAT: rainforest and vine-thickets
LENGTH: 21–22 cm
DISTRIBUTION: 10,000–30,000 km²
ABUNDANCE: sparse
STATUS: probably secure

Genus *Eopsaltria*

(ee'-op-sahl'-tree-ah: "dawn-harpist")

This genus of birds is very similar to *Petroica* robins in many respects, but most species are substantially larger; in general they frequent dense, humid forests rather than open woodland; most have some element of yellow instead of red in the plumage (especially on the breast); and the sexes are similar in appearance. The juvenile plumage is reddish and strongly mottled. Communal breeding has been reported in several species and is known to be usual in one. Largely confined to Australia and New Guinea (one species occurs in New Caledonia), the group consists of five or six species, four of which occur in Australia.

Eastern Yellow Robin

Eopsaltria australis (os-trah'-lis: "southern dawn-harpist")

The Eastern Yellow Robin does not occur in Tasmania but it is otherwise widespread and common in forests and woodland almost throughout south-eastern Australia from extreme south-eastern South Australia north to the Atherton Tableland in Queensland, and extending well into the interior. It is mainly sedentary, but there is a marked dispersal from high altitudes in winter.

It lives in permanent territorial pairs and is usually encountered alone. It closely resembles other robins in general behaviour, being mainly arboreal but favouring the lower levels of foliage and hunting by pouncing from a low perch onto insects and other small invertebrates on the ground. It has a distinctive habit of clinging sideways on the lower trunks of trees. It is generally quiet and unobtrusive in behaviour, but tame and confiding with humans. One of the earliest of woodland birds to wake in the morning, it utters a distinctive "chop, chop" note long before dawn and sometimes after dark; its song is a lengthy series of piping notes and it has a variety of harsh scolding calls.

Breeding occurs from July to January, and up to three broods are sometimes raised in succession. The female builds and incubates but both parents participate in rearing the young, usually assisted by other birds. The nest is a cup made of bark fragments and grass bound with cobweb and lined with grass, bark or dry leaves, and usually placed in an upright fork of a small tree seldom more than about 5 metres from the ground. Two or three eggs constitute the clutch. Incubation takes about 15 days and the young fledge at about 10 to 14 days.

HABITAT: rainforest, eucalypt forest and woodland, mallee and acacia scrub, urban parks and gardens
LENGTH: 15–17 cm
DISTRIBUTION: more than 1 million km²
ABUNDANCE: common
STATUS: secure

(*R Slater*)

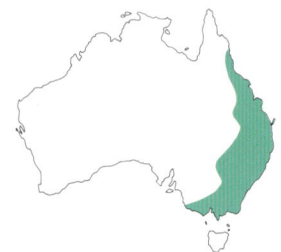

White-breasted Robin

Eopsaltria georgiana (jor'-jee-ah'-nah: "[King-] George [-Sound] dawn-harpist")

The White-breasted Robin is confined to coastal regions of far south-western Australia from north of Geraldton to about Albany. It is common and sedentary, and lives in territorial pairs that, like other robins, hunt by pouncing from an elevated perch on insects and other small invertebrates on the ground. It prefers the shelter of undergrowth and thickets. Its calls include a simple whistled song, a single repeated "zhit", and a sharp, abrupt double alarm call, "chit chit".

Two successive broods are sometimes raised during a breeding season extending from July to January. The female builds and incubates unaided but both parents participate in rearing the young, sometimes assisted by other birds. Usually well hidden in dense foliage, the nest is a loose cup of grass and rootlets bound with spider web and placed in an upright tree fork usually within 2 metres of the ground. Incubation takes about 14 days.

HABITAT: wet eucalypt forest, dense acacia and melaleuca thickets, heathland
LENGTH: 15–16 cm
DISTRIBUTION: 30,000–100,000 km²
ABUNDANCE: common
STATUS: vulnerable

(*B & B Wells*)

Western Yellow Robin

Eopsaltria griseogularis (griz'-ay-oh-gue-lar'-is: "grey-throated dawn-harpist")

The Western Yellow Robin is confined to south-western Australia, extending eastward along the coast to the Eyre Peninsula. It is sedentary, common and widespread, but it has declined markedly in the vicinity of Perth and the Darling Ranges since about 1930.

It is closely related to the Eastern Yellow Robin (some researchers argue that the two forms should be combined in a single species) and resembles the eastern bird in general behaviour and appearance.

One or two broods are raised during a breeding season extending from July to December. The female builds and incubates unaided but both parents rear the young, often assisted by other birds. The nest resembles that of the Eastern Yellow Robin and the usual clutch consists of two eggs. Incubation takes about 15 days.

HABITAT: most types of temperate eucalypt forest and woodland
LENGTH: 15–16 cm
DISTRIBUTION: 300,000–1 million km²
ABUNDANCE: common
STATUS: probably secure

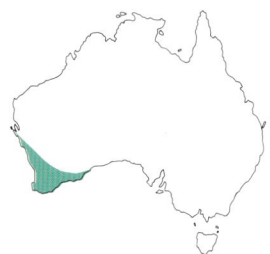

(*B & B Wells*)

Mangrove Robin

Eopsaltria pulverulenta (pul'-ve-rue-len'-tah: "powdery dawn-harpist")

The Mangrove Robin occurs in New Guinea and the Aru Islands, and it is erratically distributed but locally common along the coast of northern Australia from about Townsville to Exmouth Gulf. It is sedentary and restricted almost exclusively to mangroves.

It lives in territorial pairs but usually forages alone, pouncing on prey on the ground from a perch low in a mangrove. It feeds on insects, small crustaceans and other invertebrates. It is extremely quiet, elusive and inconspicuous in behaviour, but often tame and confiding. The song is a clear whistled phrase; it also has a soft, long, mournful note and a harsh "chak".

Breeding is variable but occurs mainly from September to February; sometimes several broods are raised in succession. The nest is a compact cup of bark fragments and fibres bound with cobweb, lined with fur and placed several metres from the ground in the fork of a mangrove.

(*G Chapman*)

HABITAT: mangroves
LENGTH: 15–17 cm
DISTRIBUTION: 30,000–100,000 km²
ABUNDANCE: sparse
STATUS: probably secure

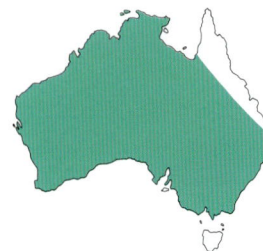

Genus Melanodryas

(mel'-ah-noh-drie'-as: "black-woodnymph")

This genus consists of only two very closely related species inhabiting southern Australia. In one species the female differs strikingly from the male, thus recalling *Petroica*. They lack yellow (or any other bright colour) in the plumage but otherwise strongly resemble *Eopsaltria* robins in much of their behaviour, and in several details of nest construction, appearance of the eggs, and juvenile plumage.

Hooded Robin

Melanodryas cucullata (kue'-kul-ah'-tah: "hooded black-woodnymph")

Mainly sedentary, the Hooded Robin is common and widespread over much of mainland Australia; although mainly found west of the Great Dividing Range, it is not uncommon in some eastern coastal districts.

Adult males are strikingly black and white; females are similar in pattern but duller, and the areas of black are replaced with dull brown.

It occurs alone, in pairs or in small parties; in winter it sometimes joins wandering mixed feeding parties of other birds. Mainly terrestrial, it forages on open ground by pouncing on its prey from a low perch such as a fence post or a stump, feeding on insects and other small invertebrates. It is usually silent, but active and conspicuous in behaviour.

Breeding is influenced by local rainfall but occurs usually from July to November. Sometimes two broods are raised in succession. The female builds and incubates but both parents participate in rearing the young, sometimes assisted by other birds. The nest is a cup of grass and bark placed in a tree fork or shallow cavity usually within a few metres of the ground. Two or three eggs constitute the clutch. Incubation takes 15 to 18 days, and the young fledge at about 12 to 14 days.

(*WJ Labbett*)

HABITAT: mainly dry open temperate woodland, including mallee, Mulga, and scrub
LENGTH: 14–17 cm
DISTRIBUTION: more than 1 million km²
ABUNDANCE: sparse to common
STATUS: probably secure

Dusky Robin

Melanodryas vittata (vit-ah'-tah: "banded black-woodnymph")

(D Watts)

A large, drab robin, this species is confined to Tasmania and the islands of Bass Strait, where it is common and widespread, breeding in most kinds of wooded country up to about 1200 metres altitude. It is closely related to the Hooded Robin of the Australian mainland and resembles that species closely in behaviour. The sexes are similar, both resembling the female plumage of the Hooded Robin.

It breeds in dispersed territorial pairs but often congregates in small, loose wandering flocks in winter. Like the Hooded Robin, it feeds mainly on the ground, pouncing on insects and other arthropods from a low perch. Tame, confiding and inquisitive, it has a variety of low mournful calls but is generally silent.

Breeding occurs from July to December, and sometimes two or three broods are raised in succession. The female builds and incubates, but both parents rear the young. Placed in almost any kind of situation within a few metres of the ground, the nest is a cup of grass and bark fragments lined with fur. The usual clutch is three eggs, which hatch in about 14 days.

HABITAT: flexible: occurs in most kinds of wooded country, but shows a special fondness for ecotones, forest margins and fire-damaged clearings; also coastal heath, sedgeland, woodlots and rough farmland
LENGTH: *c* 16 cm
DISTRIBUTION: 30,000–100,000 km²
ABUNDANCE: common
STATUS: probably secure

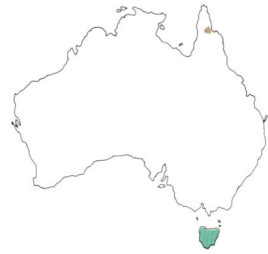

Genus Microeca

(mie'-kroh-eek'-ah: "small-house", in reference to the nest)

This genus differs most notably from other robins in its development of aerial flycatching techniques: *Microeca* robins tend to sally out from a low perch to catch flying insects rather than pouncing onto crawling insects on the ground. They have relatively shorter legs than other robins, their plumage is relatively plain, and the conspicuous wing bars that are characteristic of most other robins are absent. Their vocalisations are more complex and musical, the young are spotted, not mottled, and they build distinctive small cup-shaped nests placed on horizontal branches. There are about six species, three of which occur in Australia.

Lemon-bellied Flycatcher

Microeca flavigaster (flah'-vee-gas'-ter: "yellow-bellied small-house")

The Lemon-bellied Flycatcher is common and widespread in New Guinea and in northern Australia from the Kimberley to Cape York and

(G Chapman)

south to the vicinity of Townsville, Queensland. It is sedentary.

Usually encountered alone, in pairs or in small family parties, it is largely arboreal, hawking for insects in outer foliage or pouncing on prey on the ground from a low perch. It is usually quiet and unobtrusive in behaviour. The song, often delivered during a display flight, is a rich and varied phrase of seven or eight notes, and other calls include an animated, rhythmic "chauncey chauncey chew".

Sometimes two successive broods are raised during a breeding season extending from August to February. One egg constitutes the clutch, deposited in a small shallow cup-shaped nest of grass and bark fragments bound with cobweb and

placed in a horizontal limb up to 10 or 12 metres from the ground.

HABITAT: mainly savannah woodland, melaleuca swamps, riverine woodland and scrub; occasionally mangroves
LENGTH: 12–13 cm
DISTRIBUTION: 300,000–1 million km²
ABUNDANCE: common
STATUS: probably secure

287

Yellow-footed Flycatcher

Microeca griseoceps (griz'-ay-oh-seps: "grey-headed small-house")

The little-known Yellow-footed Fly-catcher is widespread in New Guinea but in Australia it is restricted to forests in the north-eastern part of Cape York Peninsula, where it is apparently sedentary.

It is usually encountered alone, in pairs or small family parties, foraging in the lower canopy and outer foliage, hawking actively after insects with much fluttering of wings. It is generally quiet and inconspicuous in behaviour but its calls include a loud clear whistle and a low piping note.

Very few nests have ever been found, and the timing and duration of the breeding cycle are uncertain. Two eggs form the clutch, and the nest is a small neat cup of rootlets, bound with cobweb, decorated with pieces of bark and lichen, and placed on a horizontal branch several metres from the ground.

(D & M Trounson)

HABITAT: rainforest and its margins; adjacent eucalypt forest
LENGTH: 11–13 cm
DISTRIBUTION: 10,000–30,000 km^2
ABUNDANCE: rare
STATUS: endangered

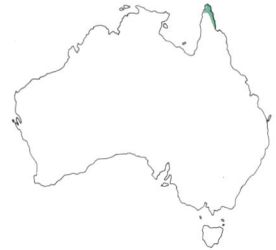

Jacky-winter

Microeca leucophaea (lue'-koh-fee'-ah: "white-grey small-house")

The Jacky-winter is a common, familiar and widespread bird almost throughout mainland Australia except for parts of the far north, frequenting almost any kind of open, lightly wooded country, but it is declining in some urban areas in the south-east. It is generally sedentary but some populations are partly migratory. It also occurs in southern New Guinea.

A plain bird, it is best identified by its lack of white wing bars and its black tail with conspicuous white outermost feathers. While foraging it is active and conspicuous, selecting any available perch in open spaces and periodically flying out to snatch flying insects or to pounce on small invertebrates on the ground. On returning to its perch it habitually flirts and swivels its tail. It is often solitary but it usually occurs in pairs or small family parties. Its song is a distinctive, persistent and vivacious "peter-peter-peter".

Breeding usually occurs from July to December, but in the arid interior it is heavily influenced by local rainfall. Sometimes two broods are raised in succession. The nest is a small shallow cup of grass and rootlets, bound with cobweb and placed on a horizontal branch at almost any height. Two or three eggs form the clutch. Incubation takes 16 to 17 days, and the young fledge at about 14 to 17 days.

HABITAT: most kinds of open woodland; partly cleared paddocks, farmland, golf courses, urban parks and gardens
LENGTH: 12–14 cm
DISTRIBUTION: more than 1 million km^2
ABUNDANCE: common
STATUS: probably secure

(GD Anderson)

Genus *Petroica*

(pet-roy'-kah: "rock-dweller")

This genus of about 12 species differs from other robins most strikingly in its pronounced sexual dimorphism: adult males have bold colour patterns often involving a bright red, pink or flame-orange breast and a conspicuous spot of white (bright red in one species) on the forehead, but in most species the female is dull brown and very plain. Most robins are sedentary inhabitants of rainforests in Australia and New Guinea, but the genus *Petroica* is atypical in being widespread in New Zealand and islands of the south-western Pacific as well as in Australia and New Guinea, in its penetration of arid and open environments, and in including a number of populations that are migratory or partly so.

Red-capped Robin

Petroica goodenovii (good'-en-ov'-ee-ee: "Goodenough's rock-dweller", after S. Goodenough, British biologist)

The Red-capped Robin has an extensive distribution across southern Australia (breeding north to about 21°S latitude), mostly inland of the Great Dividing Range. It is generally sedentary in the east, but pronounced seasonal movements take place in the western and southern interior, involving mainly immature birds; at this season wandering individuals reach the Pilbara, the Kimberley and northern Queensland.

It is very closely related to the Scarlet Robin—so much so that, where their ranges overlap, the two species defend mutually exclusive territories. They are distinguished most notably by their habitat preferences and distribution— relatively humid southern forests in the case of the Scarlet Robin, arid interior scrublands in the case of the Red-capped Robin—and the two resemble each other very closely in appearance and general behaviour.

Breeding occurs from August to December, and sometimes two broods are raised in succession. The female builds and incubates unaided but the male brings food for her and the chicks. The nest is a cup of grass and bark fragments placed in a tree fork within a few metres of the ground. Two to four eggs constitute the clutch.

HABITAT: mainly semiarid temperate woodland and scrub, especially mallee and Mulga
LENGTH: 11–12 cm
DISTRIBUTION: more than 1 million km²
ABUNDANCE: common
STATUS: secure

(MJ Seyfort)

289

Scarlet Robin

Petroica multicolor (mul'-tee-kol'-or: "many-coloured rock-dweller")

(MF Soper)

The Scarlet Robin is common in south-western Australia from about Esperance nearly to Geraldton, and in the south-east from Eyre Peninsula to Tasmania and north to south-eastern Queensland. It is mainly sedentary, but immature birds disperse widely in autumn. Elsewhere it occurs on Norfolk Island and many other islands of the south-western Pacific, westwards to Samoa.

The adult male has black upperparts and a black throat, a bright red breast, and a white wing bar, but females and immatures are dull grey-brown above and pale below and are difficult to distinguish from other robins.

It lives in pairs and often forages alone (though it may join wandering mixed foraging flocks of other birds in winter). It is arboreal, favouring undergrowth and lower levels of foliage, but it forages mainly on clear open ground, pouncing on its prey from an elevated perch. It feeds on insects and other small invertebrates. It is quiet, tame and confiding. Its calls include an animated whistling trill and soft ticking notes.

Two or three broods may be raised in succession during a breeding season extending from August to January. The female builds and incubates while the male feeds her, and both parents rear the young. The nest is a rough cup of grass and bark fragments, bound with cobweb and decorated with lichen and placed in a tree fork, a hollow stump or some other small cavity, usually within a few metres of the ground. Three eggs form the clutch, and incubation and fledging both take about 15 days.

HABITAT: eucalypt forests and woodland, dispersing in winter to more open habitats including suburban parks and gardens
LENGTH: 12–13 cm
DISTRIBUTION: 300,000–1 million km²
ABUNDANCE: common
STATUS: secure

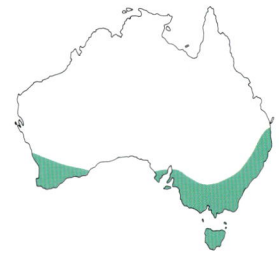

Flame Robin

Petroica phoenicea (fee-nis'-ay-ah: "purple-red rock-dweller")

The Flame Robin is very common throughout much of south-eastern Australia, especially the highlands of Tasmania and the Australian Alps. It is not sedentary, and in winter there is a general movement of birds from the high country to adjacent lowlands, a pronounced migration across Bass Strait, and also a westward dispersal extending to extreme south-eastern South Australia and north into the Riverina.

With grey upperparts and brilliant flame-orange underparts, the adult male is unmistakable, but the female is generally dull brown above and paler below, with pale buff wingbars, and is difficult to distinguish from several other species.

It breeds in dispersed territorial pairs but often congregates in loose flocks at other seasons. It feeds on insects and other small invertebrates, selecting a perch on a fence post, fallen limb or similar site and pouncing on its prey on the ground.

It is generally active and conspicuous in behaviour and has a cheerful animated song, but its calls are soft and it is not especially vocal.

Breeding occurs from August to January, and sometimes two broods are raised in succession. The female builds the nest (taking about 14 days) and incubates while her mate feeds her at the nest, and both parents feed the young. The nest is a bulky cup of bark and grass, bound with spider web, lined with fur or plant fibre, decorated with lichen, and placed in a cavity of a tree or rock, behind bark, or among root tangles, usually close to the ground. Three or four eggs form the clutch. Incubation takes 14 days and the young fledge at about 14 to 18 days.

(G Chapman)

HABITAT: eucalypt forest and woodland, dispersing in winter to more open habitats; farmland, urban parks and gardens
LENGTH: 12–14 cm
DISTRIBUTION: 300,000–1 million km²
ABUNDANCE: common
STATUS: secure

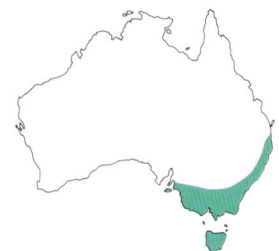

Pink Robin

Petroica rodinogaster (roh'-din-oh-gas'-ter: "rosy-bellied rock-dweller")

The Pink Robin is most common and widespread in Tasmania but it also breeds in highland forests in southern and central Victoria and (perhaps) the Australian Capital Territory and extreme south-eastern New South Wales. It is apparently largely sedentary, but it disperses from higher altitudes in winter, and has been recorded as a vagrant north to Sydney and west to Adelaide.

The adult male closely resembles the adult male Rose Robin, but is much darker grey above, has the breast a different shade of pink, and has no white in the tail. The female is distinguished from the female Rose Robin by her double tan wing bars.

The Pink Robin lives in pairs and is usually encountered alone. It is mainly arboreal, but prefers lower levels of foliage and undergrowth, and often forages on the ground. it feeds on insects and forages actively, but it is generally quiet and unobtrusive in behaviour, with soft calls, including one that sounds like a small twig snapping.

It often raises two broods in succession during a breeding season extending from October to January. The nest is a small deep cup of moss and plant fibre, bound with spider web, lined with grass and fur, often decorated on the outside with lichen, and placed in a horizontal tree fork usually within 5 metres of the ground. Three or four eggs constitute the clutch.

HABITAT: rainforest and wet eucalypt forest, dispersing into drier more open habitats in winter
LENGTH: 11–13 cm
DISTRIBUTION: 100,000–300,000 km²
ABUNDANCE: common
STATUS: vulnerable

(G Chapman)

Rose Robin

Petroica rosea (roh-zay'-ah: "rose-coloured rock-dweller")

(MJ Seyfort)

The Rose Robin does not occur in Tasmania but it is locally common in the coastal and highland forests of Victoria and eastern New South Wales. Partly migratory, it deserts higher altitudes in winter, descending to coastal lowlands and dispersing northward to about Rockhampton, Queensland. It has been recorded several times in south-eastern South Australia.

The adult male is plain dark grey above, with a pink breast and white outer tail feathers, but the female is mainly dull brown, with paler underparts and pallid wing bars.

It lives in pairs and usually forages alone. Largely arboreal, it prefers the middle to upper levels of foliage. It is very active when foraging, catching insects and other small arthropods with much fluttering of the wings, but it is generally quiet and unobtrusive in behaviour. It has a soft sweet trilling song, and also utters an abrupt dry "tick", resembling the snapping of a dry twig.

Breeding occurs from September to January, and several broods may be raised in succession. The female builds the nest (in about 7 to 14 days) and incubates but both parents rear the young. The nest is a small deep cup of moss and fibre lined with fur or plant down, decorated externally with lichen, and placed in a horizontal tree fork up to 20 metres from the ground. The clutch consists of two or three eggs, which hatch in 12 to 14 days.

HABITAT: breeds in rainforest and wet eucalypt forest, especially with acacias; sometimes drier, more open woodland in winter
LENGTH: 11–12 cm
DISTRIBUTION: 300,000–1 million km²
ABUNDANCE: sparse
STATUS: vulnerable

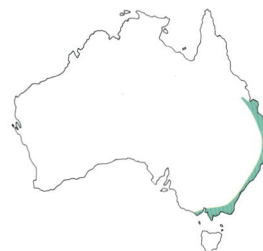

Genus *Poecilodryas*

(pee'-sil-oh-drie'-as: "mottled-woodnymph")

The limits of this small genus are doubtful, there being several New Guinea species of uncertain affinities, which may or may not belong within it. Its members are rather large robins which inhabit dense forests, have intricate plumage patterns, and share a number of behavioural characteristics, including that of habitually raising their tails. The sexes are similar, and the young are rufous and unmarked.

Grey-headed Robin

Poecilodryas albispecularis (ahl'-bee-spek'-ue-lar'-is: "white-mirrored mottled-woodnymph")

The Grey-headed Robin occurs in New Guinea and in north-eastern Queensland from Mount Amos south to Mount Spec, mainly in the highlands but locally or at some seasons extending down to an altitude of about 200 metres. It is sedentary and common.

(AIG Halton)

It lives in pairs and usually forages alone. Quiet and undemonstrative in general behaviour, it is tame and inquisitive and often approaches humans for handouts at forest picnic areas and similar sites. Like other robins it hunts mainly by pouncing from a low perch (often clinging sideways to the lower trunk of a tree) onto prey on the ground. It feeds on insects, worms, molluscs and occasionally small lizards. Its song is a whistled phrase of one long note followed by three shorter, lower notes, repeated monotonously; it also utters various chattering notes.

Sometimes two successive broods are raised during a breeding season extending from August to January. A single egg constitutes the clutch, laid in a cup-shaped nest built of twigs, rootlets, bark fragments and leaf skeletons, decorated outside with moss and placed in an upright tree fork or in a lawyer-vine, usually within about 3 metres of the ground.

HABITAT: rainforest
LENGTH: 18–20 cm
DISTRIBUTION: 10,000–30,000 km²
ABUNDANCE: common
STATUS: probably secure

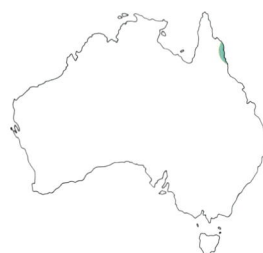

White-browed Robin

Poecilodryas superciliosa (sue'-per-sil'-ee-oh'-sah: "eyebrowed mottled-woodnymph")

Strongly associated with riverine vegetation, the White-browed Robin has an extensive but patchy and disrupted distribution across northern Australia from about Derby in Western Australia to Rockhampton in Queensland. It is sedentary and generally common, but it has declined markedly since European settlement at the western limits of its range, and is now very rare or absent in some major north-western river systems such as the Ord and Fitzroy Rivers.

There are two well-marked subspecies, separated approximately at 140°E longitude: *P. s. superciliosa* to the east and *P. s. cerviniventris* to the west. Birds of the western population (often known as the Buff-sided Robin) are markedly larger than eastern birds, have rufous buff flanks and undertail coverts and a darker crown and face, and differ to some extent in behaviour and habitat preference as well.

It is usually encountered alone or in pairs, favouring the lower levels of foliage and foraging mainly on the ground, pouncing on prey from an elevated perch. It feeds on insects, earthworms, and other small invertebrates. It is quiet and inconspicuous in general behaviour, but very tame and inquisitive, responding readily to an observer's mimicry of its calls. Sometimes nervous and fidgety, it often cocks its tail and flicks its wings. The song is a piping whistle, repeated three or four times, and other calls include a thin drawn-out whistle and various chattering notes.

Breeding occurs from August to February, and sometimes two broods are raised in succession. The female builds and incubates alone, but both parents participate in rearing the young. Two eggs constitute the clutch. The nest is a cup of bark rootlets, bound with cobweb and placed in a tree fork usually within about 5 metres of the ground. Incubation takes about 14 days.

HABITAT: rainforest and adjacent eucalypt woodland, vine scrubs, melaleuca swamps, pandanus swamps, bamboo thickets, mangroves, riverine vegetation
LENGTH: 14–18 cm
DISTRIBUTION: 300,000–1 million km²
ABUNDANCE: sparse to common
STATUS: probably secure

(AJ Dominelli)

Genus *Tregellasia*

(tre-gel'-az-ee'-ah: "Tregellas [-bird]", after T.H. Tregellas, Victorian naturalist)

This small genus is characteristic of dense tropical rainforests in New Guinea and north-eastern Australia. The two species closely resemble *Eopsaltria* robins in their predominantly yellow plumage and in various elements of behaviour, but they differ strikingly in their juvenile plumages, which are rusty red in colour (as in *Eopsaltria*) but are plain, not mottled.

Pale-yellow Robin

Tregellasia capito (kap'-it-oh: "large-headed Tregellas-bird")

The Pale-yellow Robin is common in rainforests of the coast and associated highlands of eastern Australia from Mount Amos to near Townsville and from Caloola southward to the Barrington Tops region of New South Wales. It is sedentary.

A very quiet and inconspicuous bird, it lives in pairs and, like other robins, is usually encountered alone, frequenting the understorey and lower levels of foliage and foraging mainly on the ground, pouncing on insects and other small invertebrates from a low perch. Its calls include a series of soft mournful whistles.

Breeding occurs from July to January, and two broods are sometimes raised in succession. The clutch consists of two eggs, laid in a neat cup-shaped nest constructed of dry leaves, bark and plant fibre, bound with spider web, decorated on the outside with bark, lichen and moss, and placed in a tree fork of a sapling or in a tangle of lawyer-vine.

HABITAT: rainforest (especially areas with abundant lawyer-vine) and adjacent dense eucalypt forest
LENGTH: 12–13 cm
DISTRIBUTION: 30,000–100,000 km²
ABUNDANCE: sparse to common
STATUS: probably secure

(W Lawler)

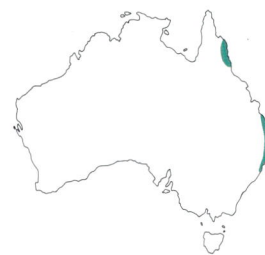

White-faced Robin

Tregellasia leucops (lue'-kops: "white-faced Tregellas-bird")

This little-known species is widespread in New Guinea but, in Australia, it is restricted to forests at the north-eastern part of Cape York Peninsula, southwards to the Rocky River. It is sedentary and rather solitary, though it sometimes joins mixed feeding parties of other birds. In general behaviour it closely resembles the Pale-yellow Robin, which it replaces in tropical forests.

Breeding occurs from September to January. Two eggs form the clutch, and the nest is a cup of bark and other plant fibre, bound with spider web and decorated with moss, lichen and bark, and placed in a tree fork or a tangle of lawyer-vine within a few metres of the ground.

HABITAT: mainly tropical lowland rainforest and vine scrubs
LENGTH: 12–13 cm
DISTRIBUTION: 10,000–30,000 km²
ABUNDANCE: common
STATUS: probably secure

(D & M Trounson)

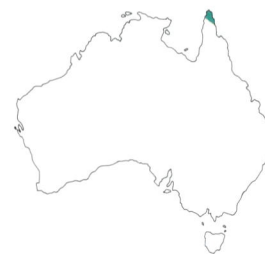

Family PACHYCEPHALIDAE

(pak'-ee-sef-ah'-lid-ee: "*Pachycephalus*-family", after a genus of whistlers)

Whistlers and their relatives are widespread from India to Samoa, Tonga, and other islands in the south-western Pacific but the group is best represented in New Guinea and Australia. The family consists of about 40 species, grouped mainly in three closely-related genera (*Pachycephala, Colluricincla,* and *Pitohui*), with about six divergent genera each containing a single species.

With 28 to 30 species, *Pachycephala* is the largest and most widespread genus. Whistlers are common birds of forest and woodland, where they forage for insects on the ground or in foliage. In most species the male is rather brightly coloured, the female plain and dull: the typical whistler colour pattern consists of a black cap; a white throat; yellow, white or rufous underparts; and a bold band of black across the breast. Whistlers are notable for their persistent, vivacious whistled songs, but in behaviour they are generally rather quiet, sedate and inconspicuous.

Members of the genus *Colluricincla* (shrike-thrushes) closely resemble whistlers in appearance and behaviour, and often coexist with them. However, most species are somewhat larger, their songs and calls are somewhat lower-pitched and mellower, and the plumage pattern is plain and dull, females differing little from males. *Pitohui* is an uniquely Papuan genus of about seven species, closely resembling *Colluricincla* but more strongly terrestrial and often communal.

Several species in New Guinea and Australia lack especially close relatives and are placed in monotypic genera: in Australia these include the Crested Bellbird (*Oreoica*) and the Shriketit (*Falcunculus*). *Oreoica* is the sole arid-zone inhabitant of the family, while *Falcunculus* is strongly adapted to foraging for insects on a substrate of bark.

Genus Colluricincla

(kol'-ue-ree-sink'-lah: "shrike-bird")

The characteristics of the genus are discussed in the account of the family.

Bower's Shrike-thrush

Colluricincla boweri (bow'-er-ee: "Bower's shrike-bird", after T. H. Bowyer-Bower, British naturalist)

Bower's Shrike-thrush is a common but little-studied species found only in the highland rainforests (mainly above 400 metres altitude) of north-eastern Queensland from Mount Amos south to Mount Spec.

It is solitary, territorial and largely sedentary, but there is some dispersal to nearby coastal lowlands in winter.

The song is rich, full and varied, but Bower's Shrike-thrush is among the quietest and least conspicuous of its family. It forages deliberately in the foliage of trees, favouring the lower and middle levels. Its streaked underparts are distinctive.

Breeding occurs from October to January. The nest is a cup of twigs, leaves and bark fragments placed several metres from the ground in a tree fork or vine tangle. Two eggs form the clutch.

HABITAT: tropical rainforest
LENGTH: 19–21 cm
DISTRIBUTION: 10,000–30,000 km²
ABUNDANCE: common
STATUS: vulnerable

(J Purnell)

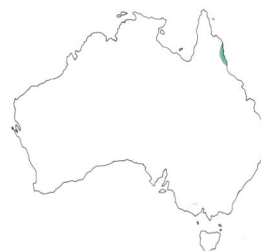

Grey Shrike-thrush

Colluricincla harmonica (har-mon'-ik-ah: "harmonious shrike-bird")

(*D & V Blagden*)

The Grey Shrike-thrush occurs almost wherever there are trees throughout Australia, including Tasmania and most offshore islands, excepting only the most arid and sterile parts of the Simpson and Great Sandy Deserts. It also occurs in southern New Guinea.

It is sedentary, generally solitary, and territorial at all seasons, except in more rigorous climates in the south-east, where it often wanders in winter, sometimes joining roving flocks of other birds. It is quiet, slow and methodical in foraging, searching leaf litter on the ground, fallen logs, limbs and trunks of trees, and the foliage of shrubs and trees at almost any height. The varied diet includes insects, spiders and other invertebrates and, occasionally, lizards, nestling birds and small mammals. Its song is strikingly loud, pure, rich and varied; it is most vocal when breeding.

Most breeding occurs from July to February, and several broods may be raised in succession. The nest is a cup of bark fragments, twigs and grass in a very wide variety of sites, but usually within a few metres of the ground. The usual clutch consists of three eggs, which hatch in about 17 to 18 days. Both parents share in nest construction, incubation and care of the young.

HABITAT: most kinds of temperate to tropical wooded country, including rainforest, sclerophyll forest, mallee, Mulga, and urban parks and gardens
LENGTH: 22–25 cm
DISTRIBUTION: more than 1 million km²
ABUNDANCE: common
STATUS: secure

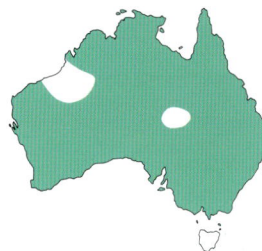

Little Shrike-thrush

Colluricincla megarhyncha (meg'-ah-rink'-ah: "great-beaked shrike-bird")

The Little Shrike-thrush inhabits forests in New Guinea and in coastal regions and associated highlands in northern and eastern Australia from the Kimberley to the Barrington Tops region of New South Wales. Generally fairly common, it is sedentary and rather solitary: established pairs appear to maintain permanent territories. Rather quiet, deliberate and inconspicuous, it somewhat resembles a whistler in general behaviour, but, for a forest bird, it shows unusual latitude in its feeding behaviour, foraging indiscriminately in foliage or on limbs and branches at all levels from the ground to the upper canopy. It eats insects, seeds and fruit. It calls include a clear melodious song and harsh chattering notes.

Breeding occurs from September to February. The nest is a deep cup of twigs, bark fragments and dry leaves bound with cobweb and lined with fine plant material, usually placed within about 5 metres of the ground in an upright tree fork, in dense foliage or in a vine tangle; two or three eggs form the clutch.

HABITAT: warm temperate to tropical rainforest; mangroves; melaleuca swamps; coastal and riparian woodland
LENGTH: 17–19 cm
DISTRIBUTION: 300,000–1 million km²
ABUNDANCE: sparse to common
STATUS: probably secure

(*MJ Seyfort*)

Sandstone Shrike-thrush

Colluricincla woodwardi (wood'-war-dee: "Woodward's shrike-bird", after B. H. Woodward, sometime director of the WA Museum)

The rich, clear and liquid song phrases of this large shrike-thrush are an evocative component of the remote sandstone gorges and escarpments of the Kimberley, Arnhem Land and north-western Queensland. The Sandstone Shrike-thrush is found nowhere else. It sings from prominent high perches on boulders or cliff tops, and spends much of its time foraging on open rock faces, catching insects and other small invertebrates.

Breeding occurs from November to January. The nest is a cup of spinifex rootlets and fine grass placed on a rock ledge or in a rock crevice. Two or three eggs form the clutch.

(*G Chapman*)

HABITAT: tropical sandstone escarpments and gorges
LENGTH: 25–27 cm
DISTRIBUTION: 300,000–1 million km²
ABUNDANCE: sparse to common
STATUS: probably secure

Genus *Falcunculus*

(fal-kun'-kue-lus: "little-falcon")

The characteristics of the genus are those of the single species.

Shriketit

Falcunculus frontatus (fron-tah'-tus: "foreheaded little-falcon")

The Shriketit occurs in three widely separated populations: in the Kimberley and Top End; in the far south-west; and on the east coast from the Atherton Tableland in Queensland to the Mount Lofty Ranges in South Australia. It is generally common except in the north, where it is rare and local. The three populations differ slightly in plumage, but the throat of the male is always black, that of the female dull green.

It is sedentary, and established pairs maintain apparently permanent territories. It usually keeps to the treetops, digging insects and other small invertebrates from crevices in the trunk and limbs, prising away loose bark and rummaging noisily in pendent tangles of shed bark: sometimes it gleans in foliage. It is solitary, quiet and methodical. The most common call is a high, mellow, mournful whistle.

Breeding occurs from August to January. The female builds and incubates alone, assisted only casually by the male, but both parents feed the young, sometimes helped by other birds. The nest is a very deep narrow cup of bark fibres bound with cobweb, usually placed in a high three-pronged fork. Two or three eggs constitute the clutch. Incubation takes 16 to 19 days, and the young fledge at about 15 to 17 days.

(*M Wright*)

HABITAT: mainly temperate eucalypt forest and woodland
LENGTH: 16–19 cm
DISTRIBUTION: more than 1 million km²
ABUNDANCE: sparse to common
STATUS: probably secure

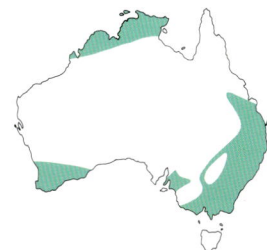

Genus Oreoica

(or'-ay-oy'-kah: "mountain-dweller")

The characteristics of the genus are those of the single species.

Crested Bellbird

Oreoica gutturalis (gut'-er-ah'-lis: "throated mountain-dweller")

The Crested Bellbird is widespread across the southern interior of mainland Australia. It is generally sedentary, and established pairs maintain apparently permanent territories. The male has a small crest and a distinctive black and white facial pattern, but females are plain and mousy brown. The unmistakable and evocative call is a mellow, bell-like, far-carrying phrase that sounds somewhat like "pan pan panella". The male sings from an elevated perch, but otherwise the birds spend most time on the ground, feeding on insects and seeds.

One or sometimes two broods may be raised during a breeding season extending mainly from August to January. The nest is a deep cup of bark strips, short twigs, leaves and grass, lined with fine dry vegetation, and placed in a tree fork, on top of a dead stump, or in a bush up to 3 metres from the ground. An intriguing (and unexplained) peculiarity is that a number of paralysed caterpillars are often arranged carefully around the rim. The clutch varies from two to four eggs.

HABITAT: mainly temperate semiarid scrublands
LENGTH: 21–23 cm
DISTRIBUTION: more than 1 million km²
ABUNDANCE: sparse to common
STATUS: probably secure

(G Chapman)

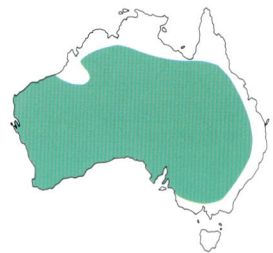

Genus *Pachycephala*
(pak'-ee-sef'-ah-lah: "thick-head")

The characteristics of the genus are discussed in the account of the family.

Gilbert's Whistler
Pachycephala inornata (in'-orn-ah'-tah: "plain thick-head")

It is not always easy to distinguish Gilbert's Whistler from the Red-lored Whistler in the field, (especially since they sometimes behave similarly and may occur together) but Gilbert's Whistler has black lores and no dull red on the forehead. Further, although Gilbert's Whistler occurs in the broombush or whipstick mallee required by the Red-lored Whistler, it also occupies a much wider range of habitats, showing a special fondness for lignum. It is locally common across interior southern Australia from the Lachlan River in New South Wales westward to the vicinity of Northam in Western Australia.

Like other whistlers, it has a rich and varied repertoire of songs and calls, but is rather quiet, sedate and unobtrusive in general behaviour. It lives in pairs that together defend apparently permanent territories. It forages mainly on the ground, feeding largely on insects.

Breeding extends from September to December. A single brood is raised, and the parents cooperate in building, incubation and care of the young. The nest is a cup of bark strips and dry grass bound with cobweb and lined with finer material. It is usually placed 2 to 3 metres from the ground in an upright fork, on a dead stump or, occasionally, in the disused nest of some other bird. The clutch consists of two or three eggs.

HABITAT: temperate mallee, Mulga and other drier woodlands; lignum and melaleuca thickets
LENGTH: 19–21 cm
DISTRIBUTION: 300,000–1 million km²
ABUNDANCE: very sparse to sparse
STATUS: vulnerable

(R Miller)

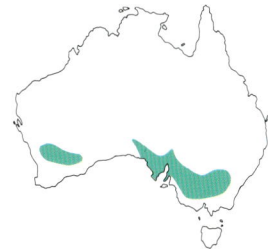

White-breasted Whistler
Pachycephala lanioides (lah'-nee-oy'-dayz: "shrike-like thick-head")

The White-breasted Whistler is locally common in mangroves along the northern coast of Australia (including offshore islands) from Shark Bay in Western Australia to the vicinity of Karumba in Queensland. Sedentary, sedate, and rather solitary, it is in every way a typical whistler except for its rigid habitat requirements: it is seldom found far from mangroves. It frequently forages on mud exposed at low tide, and crustaceans form a significant component of its diet.

Breeding may occur at any time, but especially from March to November. The female performs most building, incubation and care of the young, assisted only casually by the male. The nest is usually placed in an upright mangrove fork just above the reach of the highest tides; one or two eggs form the clutch.

HABITAT: subtropical and tropical mangroves
LENGTH: 18–20 cm
DISTRIBUTION: 30,000–100,000 km²
ABUNDANCE: sparse
STATUS: probably secure

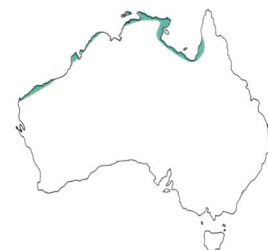

(G Chapman)

Mangrove Golden Whistler

Pachycephala melanura (mel'-ahn-ue'-rah: "black-tailed thick-head")

The Mangrove Whistler is fairly common in coastal mangroves and on some offshore islands around northern Australia from about Carnarvon in Western Australia to Repulse Bay in Queensland. It also occurs in New Guinea and the Bismarck Archipelago.

The adult male differs in appearance from the adult male Golden Whistler only in subtle features of plumage: the colours are richer, the collar slightly broader, and the tail somewhat shorter. Females differ from female Golden Whistlers in having a distinctly stronger tinge of yellow on the underparts. The two species are also nearly alike in general behaviour, but differ markedly in ecology.

Breeding occurs from October to December, but few details of the nesting cycle are known; two or three eggs form the usual clutch.

HABITAT: subtropical and tropical mangroves and adjacent waterside vegetation
LENGTH: 15–17 cm
DISTRIBUTION: 30,000–100,000 km²
ABUNDANCE: common
STATUS: secure

(G Chapman)

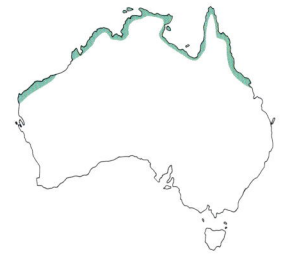

Olive Whistler

Pachycephala olivacea (ol'-iv-ah'-say-ah: "olive thick-head")

The Olive Whistler occurs in dense highland forests of south-eastern Australia from the vicinity of Toowomba in Queensland to extreme south-eastern South Australia, including Tasmania and the islands of Bass Strait. It is generally uncommon and local in the northern part of its range, much more common and widespread in Tasmania and the Australian Alps, but very scarce or absent in a broad intervening zone corresponding roughly with the Blue Mountains of New South Wales. The two populations, which are thus effectively isolated from each other, have very different songs and occupy subtly different habitats.

The Olive Whistler is mainly sedentary and solitary. It is quiet, deliberate and unobtrusive in behaviour but its song is loud, elaborate and attractive. It feeds on insects and berries, sought mainly in dense thickets, in the lower foliage of trees, or on the ground.

Breeding occurs from September to January, and only one brood is raised. The nest is a cup of twigs, bark and grass, lined with fine grass and rootlets, and placed in a low tree fork, a shrub or a grass tussock; two or three eggs form the clutch.

HABITAT: mainly cool temperate beech forest and other dense, wet forests, including alpine thickets; coastal scrubs; heaths
LENGTH: 19–22 cm
DISTRIBUTION: 100,000–300,000 km²
ABUNDANCE: sparse to common
STATUS: probably secure

(M Wright)

Golden Whistler

Pachycephala pectoralis (pek'-tor-ah'-lis: "breasted thick-head")

The adult male Golden Whistler can be confused only with the Mangrove Golden Whistler but females are very plain, dull brown: they lack conspicuous plumage feathers and are often misidentified. The song is notably persistent, loud and attractive, encompassing a diverse repertoire of musical whistled phrases that often end with a distinctive, emphatic flourish. Both members of a mated pair sing, and both defend their joint breeding territory with equal vigour. However, at other seasons they are, like all whistlers, rather sedate, silent and unobtrusive birds.

They inhabit most kinds of dense woodland and forest, showing a distinct preference for the shrub layer and lower levels of canopy foliage. The varied diet includes small fruits and insects and other small invertebrates.

Strongly migratory in the south-east but mainly sedentary elsewhere, the Golden Whistler is widespread in eastern and southern Australia from Cooktown in Queensland to Shark Bay and the far south-west of the continent. Elsewhere it occurs in Indonesia and New Guinea and many islands of the south-western Pacific, including Norfolk and Lord Howe Islands.

It breeds from September to January, raising only one brood per season. Nests are usually placed in a vertical fork up to 6 metres from the ground in a shrub or sapling. The clutch consists usually of two or three eggs, which hatch in 14 to 17 days. The chicks take 10 to 13 days to fledge.

HABITAT: cool temperate to tropical rainforest, eucalypt forest, woodland, riverine vegetation, mallee, brigalow and other denser dry scrubs, occasionally mangroves, urban parks and gardens
LENGTH: *c* 16 cm
DISTRIBUTION: more than 1 million km^2
ABUNDANCE: common
STATUS: secure

(G Weber)

Rufous Whistler

Pachycephala rufiventris (rue'-fee-ven'-tris: "rufous-bellied thick-head")

The adult male Rufous Whistler is unmistakable, but the female is dull and plain, her finely streaked pale buff underparts offering the best clue to identity.

The song is unusually rich, loud, persistent and varied even in a genus notable for its vocalisations: a strongly characteristic phrase is a ringing, emphatic "eee-chong", frequently interspersed among other whistled phrases. The Rufous Whistler is also notable for its willingness to be provoked into song by any loud, sudden noise, such as a thunderclap or rifle shot. Both sexes sing, and defend joint breeding territories with equal vigour. Like other whistlers it is otherwise generally rather quiet, deliberate and unobtrusive in behaviour, frequenting the upper foliage of trees where it feeds mainly on insects and their larvae.

It is widespread across mainland Australia. It also occurs in New Caledonia, and related forms, perhaps conspecific but of controversial status, inhabit New Guinea and Indonesia. It is one of the most common and ubiquitous birds of open eucalypt woodlands but tends to avoid denser, wetter forest in regions of relatively high rainfall, where its place is taken by the Golden Whistler: only rarely do the ranges of the two species overlap. It is strongly migratory in the southeast, but variously sedentary, nomadic or partly migratory (depending on locality) elsewhere.

Breeding occurs mainly from September to February, and often two broods are raised in succession. The female builds the nest alone but both parents incubate and feed the young. The nest is a bowl of twigs and dry grass bound with cobweb and lined with fine grass and rootlets and usually placed in a vertical tree fork up to 10 metres from the ground. Two or three eggs form the clutch. Incubation takes about 13 to 15 days, and the young fledge at about 10 to 15 days.

HABITAT: temperate to tropical open eucalypt forest and woodland; mallee, Mulga, and other open scrublands
LENGTH: 16–18 cm
DISTRIBUTION: more than 1 million km²
ABUNDANCE: common
STATUS: secure

(FG Craven)

Red-lored Whistler

Pachycephala rufogularis (rue'-foh-gue-lar'-is: "red-necked thick-head")

The Red-lored Whistler inhabits interior south-eastern Australia, mainly in the Ninety Mile and Big Deserts on the border between Victoria and South Australia, but extending locally eastward into mid-western New South Wales. It is generally rare and elusive, with extremely rigid habitat requirements.

It is sedentary, solitary and, except for its persistent and distinctive call, very quiet and secretive in behaviour. It eats mainly insects, sought on the ground. One brood of young is raised in a breeding season extending from September to December. Two or three eggs are laid in a cup-shaped nest usually placed within a metre of the ground.

HABITAT: temperate dense low heath and mature mallee stands with porcupine grass and broombush
LENGTH: 20–22 cm
DISTRIBUTION: 30,000–100,000 km²
ABUNDANCE: rare
STATUS: endangered

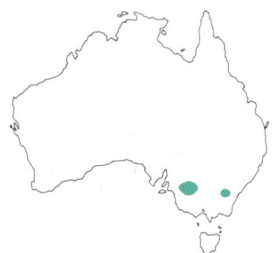

(G Chapman)

Grey Whistler

Pachycephala simplex (sim'-plex: "plain thick-head")

Despite its plain, dull plumage, the Grey Whistler resembles other whistlers in behaviour and habits, but it differs notably in its preference for the upper levels of forest canopy, where it forages slowly and deliberately for insects, often pausing motionless for minutes on end. It is sedentary, solitary, and rather difficult to observe. Its song is clear and musical but lacks the whip-like flourish at the end of certain whistled phrases that is characteristic of some other whistlers. It may sing persistently all year.

It occurs in the Moluccas and New Guinea, and is common in coastal and near-coastal regions of the Northern Territory (including Melville Island and Groote Eylandt) and in north-eastern Queensland from Cape York south to about Townsville.

Breeding may occur at any time of year, but especially from September to March. The nest is usually high but otherwise resembles that of other whistlers. Two eggs constitute the clutch.

HABITAT: tropical rainforest and adjacent eucalypt forest, mangroves, monsoon forest, melaleuca swamps
LENGTH: 14–15 cm
DISTRIBUTION: 100,000–300,000 km²
ABUNDANCE: sparse
STATUS: secure

(D & M Trounson)

Family MONARCHIDAE

(mon-ark'-id-ee: "*Monarcha*-family", after a genus of monarchs)

With a distribution centred in the Indo-Pacific region, the monarchs comprise a group of about 100 species in three large genera (*Monarcha, Myiagra, Rhipidura*) and several small peripheral groups. Their internal classification, and their relationship with other groups—notably the whistlers and the Australasian robins—are controversial: *Rhipidura*, for example, is sometimes placed in a family of its own.

Most are arboreal rainforest birds showing a number of adaptations for catching insects in midair and in the foliage of trees (feeding methods are varied): rictal bristles are well developed, the bill is broad and flattened, the tarsi are moderately short, and the wings are pointed. Most are rather solitary: communal breeding is unrecorded. Females generally resemble males, and both sexes cooperate at all stages of the nesting cycle.

Genus Arses

(ar'-sayz: "raised-voice")

The two species of this genus are closely related to *Monarcha* but have black and white plumage, conspicuous frills at the nape, and blue wattles about the eye; they also differ in foraging behaviour, spending much of their time searching the bark of trees.

Pied Monarch

Arses kaupi (kow'-pee: "Kaup's raised-voice": after J. J. Kaup, German ornithologist)

The Pied Monarch inhabits the coastal and highland rainforests of north-eastern Queensland from Mount Amos to Mount Spec. It is sedentary and fairly common.

(J Purnell)

It closely resembles the Frilled Monarch in appearance and habits but differs in having a band of black across the breast and in spending a much greater proportion of its foraging time searching the bark of trunks of trees. It is generally solitary but sometimes joins mixed feeding parties of other birds. It feeds on insects and other small invertebrates. The most characteristic call is a soft but harsh, slurred "quarrr".

Breeding occurs from October to January; only one brood is raised in a season. The nest is a delicate cup of slender twigs and vine tendrils bound with cobweb, lined with fibre and decorated with lichen, suspended from the rim in vines several metres from the ground. Two eggs form the clutch.

HABITAT: tropical rainforest and adjacent eucalypt woodland
LENGTH: 14–15 cm
DISTRIBUTION: 10,000–30,000 km²
ABUNDANCE: sparse
STATUS: possibly endangered

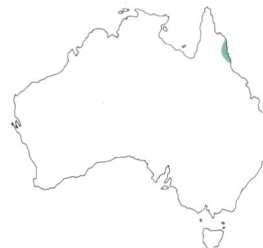

Frilled Monarch

Arses telescophthalmus (tel'-e-skof-thal'-mus: "telescope-eyed raised-voice", apparently intended to mean "spectacled-eyed", in reference to ring of bare skin around eyes)

(D & M Trounson)

Although common and widespread in New Guinea and the Aru Islands, the Frilled Monarch has an Australian distribution limited to the rainforests of northern Cape York Peninsula, where it is sedentary and fairly common.

It is closely related to the Pied Monarch of the Atherton Tableland (some researchers claiming the two to be conspecific). Like the Pied Monarch, it has a conspicuous blue eye wattle and is boldly patterned in black and white, with a broad white nuchal frill that is often erected; but the black breast band that is characteristic of the Pied Monarch is absent.

It is strictly arboreal, favouring the middle and lower levels of forest, where it searches for insects in the foliage and in crevices in the bark of limbs and trunks, often spiralling upwards like a treecreeper. Like other monarchs, it is usually solitary or in pairs.

Little is known of breeding except that (in Australia) this occurs from September to January. The nest is a frail cup of twigs and vine tendrils suspended by the rim from a horizontal tree fork at moderate heights, and two eggs form the clutch.

HABITAT: mainly tropical rainforest; occasionally adjacent eucalypt forest
LENGTH: 14–16 cm
DISTRIBUTION: 10,000–30,000 km²
ABUNDANCE: sparse to common
STATUS: secure

Genus Machaerirhynchus

(mah-kee'-ree-rink'-us: "sword-beak")

This genus consists of only two species of uncertain placement. The plumage pattern consists of areas of yellow, black and white, and the bill is remarkably wide and flattened. One species occurs in Australia and New Guinea, the other replaces it at higher altitudes in New Guinea.

Boat-billed Flycatcher

Machaerirhynchus flaviventer (flah'-vee-vent'-er: "yellow-bellied sword-beak")

A small, active but rather unobtrusive bird of the upper canopy, the Boat-billed Flycatcher (or Yellow-breasted Boatbill) is unmistakable by virtue of its yellow, black and white plumage pattern and its remarkable wide, flat (and very flexible) bill. The sexes are similar but the female is somewhat duller than the male. The posture of the Boat-billed Flycatcher is usually horizontal and it frequently cocks its tail as it gleans small, soft-bodied insects from the outer foliage. Sometimes it joins mixed feeding parties of other birds. It has a soft, buzzy song and it also utters a variety of sweet whistled notes.

It occurs in north-eastern Queensland on the Atherton Tableland and eastern and northern Cape York Peninsula, and in the lowlands and western islands of New Guinea from sea-level to an altitude of about 800 metres. It is sedentary, and generally rather uncommon. It usually lives in pairs.

Breeding occurs from September to February. The male takes about a week to construct a shallow saucer-shaped nest of vine tendrils and stems, lined with plant fibre and suspended in a slender outer fork high in a tree. Two eggs form the clutch, incubated by both parents.

HABITAT: tropical rainforest and dense riverine vegetation; occasionally wanders into adjacent eucalypt woodland
LENGTH: 11–12 cm
DISTRIBUTION: 30,000–100,000 km²
ABUNDANCE: sparse to common
STATUS: secure

(L Le Guay)

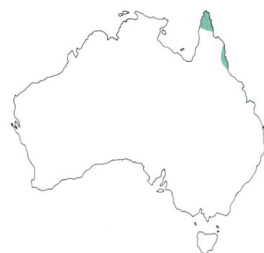

Genus Monarcha

(mon-ark'-ah: "monarch")

This is a homogeneous group of about 30 species widespread in New Guinea and island groups to the north of Australia; four species occur in Australia. Almost all live in the canopy of rainforest, where they glean insects from foliage. Most are rather solitary, quiet and unobtrusive in behaviour, but have lively whistled songs.

Black-winged Monarch

Monarcha frater (frah'-ter: "brother monarch")

Sometimes called the Pearly Flycatcher, the Black-winged Monarch has an Australian distribution restricted to rainforests in the northern part of Cape York Peninsula. Some birds may remain all year but most migrate after breeding to New Guinea, returning in October.

It closely resembles the Black-faced Monarch but has black, not grey, wings and tail. The two may prove to be conspecific, but so little is known of the Black-winged Monarch that the question remains undecided. Like the Black-faced Monarch, it is insectivorous, arboreal, rather solitary, and unobtrusive in general behaviour. Breeding occurs from October or November to January.

(RD Mackay)

HABITAT: tropical rainforest and adjacent open forest
LENGTH: 18–19 cm
DISTRIBUTION: 10,000–30,000 km^2
ABUNDANCE: sparse to common
STATUS: secure

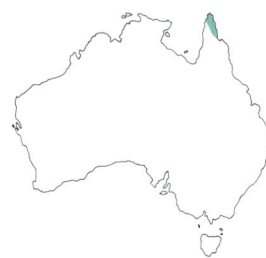

White-eared Monarch

Monarcha leucotis (luke-oh'-tis: "white-eared monarch")

The White-eared Monarch occurs in coastal and near-coastal forests of eastern Australia from Cape York to north-eastern New South Wales. It is nowhere common and seems prone to population movements that are not clearly understood: in some areas it seems sedentary, in others locally nomadic; in highland areas there may be some seasonal altitudinal migration.

It lives alone or in pairs, or sometimes small parties, spending virtually all of its time high in the upper and outer canopy, where it is often difficult to observe. It is very active when foraging, flitting fantail-like among outer foliage or hovering and prancing above the canopy, snatching up insects.

Breeding occurs from September to January; only one brood is reared in a season. Both parents build, incubate and care for the young. The nest is usually placed in a high tree fork. It is a cup of bark strips, grass and other plant material, lined with rootlets and fibre and bound with cobweb; the outside is adorned with moss and cocoons. Two eggs form the clutch. Incubation takes about 14 days, and the young fledge at about 14 days.

HABITAT: mainly warm temperate to tropical rainforest; occasionally mangroves, melaleuca swamps or streamside vegetation in eucalypt woodland
LENGTH: 13–14 cm
DISTRIBUTION: 100,000–300,000 km^2
ABUNDANCE: sparse
STATUS: vulnerable

(J Daley)

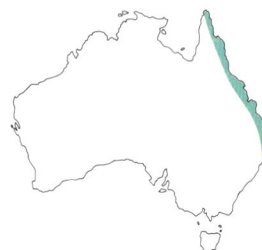

Black-faced Monarch

Monarcha melanopsis (mel'-ahn-op'-sis: "black-faced monarch")

The breeding distribution of the Black-faced Monarch extends from the Atherton Tableland southward almost to the Victorian border but it is strongly migratory, seldom occurring south of Rockhampton, Queensland, in winter. Many individuals travel to southern New Guinea. It is common and widespread.

Strictly arboreal, it tends to favour the lower canopy. It usually lives alone or in pairs, but occasionally joins mixed feeding parties of other birds. It feeds entirely on insects and other arthropods, captured mainly by gleaning but occasionally snapped up in midair. In general behaviour it is deliberate and unobtrusive, but it has a distinctive loud, cheerful whistled song.

Breeding occurs from October to January, and only one brood is raised in a season. The nest is usually a deep cup of casuarina needles or thin stems, but its form varies considerably depending on its situation; it is adorned externally with green moss and placed in a tree fork several metres above ground. Two or three eggs constitute the clutch. Both parents incubate and care for the young.

HABITAT: mainly temperate to subtropical rainforest and dense wet eucalypt forest
LENGTH: 16–19 cm
DISTRIBUTION: 300,000–1 million km²
ABUNDANCE: common
STATUS: probably secure

(G Weber)

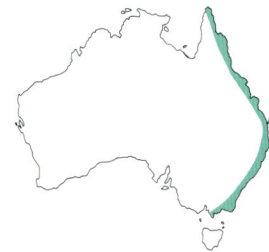

Spectacled Monarch

Monarcha trivirgatus (trie'-ver-gah'-tus: "three-striped monarch")

The Spectacled Monarch is very common in north-eastern Australia, extending sparingly southward to the vicinity of Ourimbah, New South Wales. It is resident in the north, but a breeding summer visitor (October to April) in the south. It also occurs in New Guinea, Timor and the Moluccas.

It strongly resembles the related Black-faced Monarch in being insectivorous, arboreal and rather solitary, but it is a decidedly smaller, more spritely and active bird. It often flutters its wings or tail while gleaning in outer foliage. The most common call is a scratchy, buzzy series of notes.

Breeding occurs from October to February. Placed in a low tree-fork or vine tangle, the nest is a neat cup of bark fragments, leaf skeletons and plant fibre externally adorned with green moss, lichen or spider cocoons. Two eggs form the clutch.

HABITAT: warm temperate to tropical rainforest, dense wet eucalypt forest, mangroves
LENGTH: 14–16 cm
DISTRIBUTION: 300,000–1 million km²
ABUNDANCE: common
STATUS: probably secure

(C Seller)

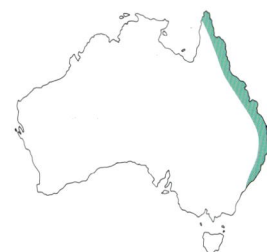

Genus Myiagra

(mie'-ee-ah'-grah: "flycatcher")

About 14 species of *Myiagra* flycatchers occur in Australia, Indonesia, New Guinea and islands of the south-western Pacific. They somewhat resemble *Monarcha* but they habitually snatch insects in flight rather than gleaning them from foliage and they are generally more active and conspicuous; the normal posture is upright rather than horizontal, and the tail is persistently trembled or shivered from side to side in a highly characteristic manner. Males often differ markedly from females in appearance.

Shining Flycatcher

Myiagra alecto (ah-lek'-toh: "Alecto flycatcher", in reference to Alekto, one of the mythological Greek Furies)

The Shining Flycatcher is resident, common and widespread in Indonesia and New Guinea, eastward to the Bismarck Archipelago, and in northern and north-eastern Australia south to Rockhampton, rarely (and erratically) south to Brisbane.

The male is entirely deep glossy black in plumage, whereas the female has only the head black; she is otherwise white below and rich chestnut above. Like other *Myiagra*

flycatchers, the Shining Flycatcher utters a variety of clear whistles, grating, rasping notes, abrupt croaks and chattering notes. It is active, noisy and conspicuous.

Unlike most other *Myiagra* flycatchers it often forages on the ground, especially the mud at the foot of mangroves, eating various small molluscs, isopods and small crabs, as well as insects.

Breeding occurs mainly from November to January, but may extend from August to March. The nest resembles that of other *Myiagra* flycatchers in construction but is usually placed low in a mangrove. Two or three eggs constitute the clutch.

HABITAT: mainly tropical mangroves; also rainforest, monsoon forest, swamps, and riverside vegetation
LENGTH: 17–18 cm
DISTRIBUTION: 300,000–1 million km²
ABUNDANCE: sparse
STATUS: probably secure

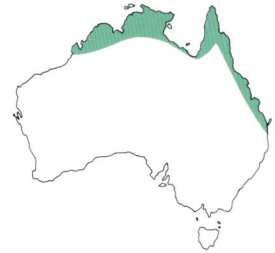

(N Chaffer)

Satin Flycatcher

Myiagra cyanoleuca (sie'-ah-noh-lue'-kah: "blue-white flycatcher")

The Satin Flycatcher occurs in New Guinea and in eastern Australia, though it is most common in the eucalypt forests of the south-east, including Tasmania. It is strongly migratory, and vagrants have wandered to New Zealand.

The breast and back of the male are deep, rich, glossy blue-back, but this species otherwise closely resembles the Leaden Flycatcher in appearance and behaviour. It has a large repertoire of calls, mostly loud, clear and strident. It spends most of its time in the high foliage of trees, snatching insects on the wing.

Breeding occurs from September to February. Often several pairs nest in close association. The nest resembles that of the Leaden Fly-catcher and is placed in similar situations. Two or three eggs form the clutch. Both parents build, incubate and rear the young: incubation takes 17 to 18 days and the young fledge at 17 to 18 days.

(K Stepnell)

HABITAT: breeds mainly in cool to warm temperate, tall, dense eucalypt forest; otherwise most wooded environments, including parks and gardens
LENGTH: 15–17 cm
DISTRIBUTION: more than 1 million km²
ABUNDANCE: sparse to common
STATUS: probably secure

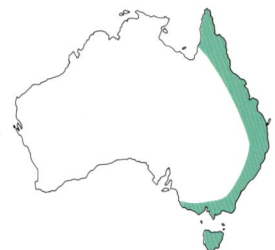

Restless Flycatcher

Myiagra inquieta (in'-kwee-ay'-tah: "restless flycatcher")

(R Slater)

The familiar Restless Flycatcher is widespread in northern, eastern and southern Australia, as well as in southern New Guinea. It is most common in coastal and near-coastal regions, and south-eastern populations are largely migratory. Some individuals disperse into the arid interior after the breeding season. Birds of the population in the Kimberley and the Top End are markedly smaller and have wider bills than eastern and southern birds, and may represent a distinct species (the Paperbark Flycatcher *Myiagra nana*).

It is frequently confused with the similarly sized (and equally active and conspicuous) Willie-wagtail, but the two are easily distinguished by the fact that the Restless Flycatcher has a white throat, whereas the Willie-wagtail's is black. Females differ little from males in appearance.

The Restless Flycatcher is strongly territorial, and is usually encountered alone or in pairs. Extremely active, noisy and conspicuous, it spends most of its time in trees, but it commonly forages in a very distinctive manner, hovering about one metre above tall grass or weed growth, on slowly fanning wings with the body arched and tail and head depressed, periodically dipping to snap up insects lurking in the rank growth below. During this behaviour it often utters a characteristic reeling, rhythmic grating note that strikingly resembles the sound produced by sharpening a steel blade against an old-fashioned sandstone grinding wheel (hence its common country name of Scissors-grinder).

Breeding occurs from July to January in the south, and from August to March in the north; often several broods are raised in succession. The nest is a cup of bark strips and grass bound with cobweb, lined with fur and fine grass and decorated with lichen and bark flakes, and placed in a fork towards the end of a horizontal branch, usually within 10 metres of the ground. Three or four eggs constitute the clutch. Incubation takes about 14 days, and the young fledge at about 14 days. Both parents cooperate in construction, incubation and rearing the young.

HABITAT: most types of temperate to tropical open eucalypt woodlands; clearings in forests; melaleuca swamps

LENGTH: 19–21 cm

DISTRIBUTION: more than 1 million km²

ABUNDANCE: sparse to common

STATUS: probably secure

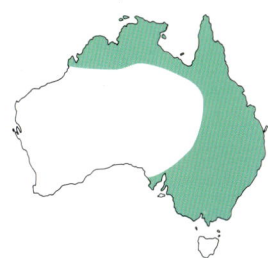

Leaden Flycatcher

Myiagra rubecula (rue-bek'-ue-lah: "reddish flycatcher")

The Leaden Flycatcher is common in New Guinea and in northern and eastern Australia from about Wyndham in Western Australia to central Victoria; it is rare in Tasmania but occasionally breeds there. It is strongly migratory in the south-east.

The upperparts and breast of the male are a distinctive shade of glossy lead-grey, but the female is dull, with a pale rufous wash on the throat. Like other *Myiagra* flycatchers it is strictly arboreal, favouring the middle and upper levels of foliage. Strongly territorial, it is usually encountered alone or in pairs. It is active, and calls frequently. It feeds on flying insects, snapped up in midair.

Nesting extends from September to February. The nest is a cup of bark strips and grass liberally bound with cobweb and decorated with lichen, placed on a high horizontal limb, well away from the trunk and usually screened from above by a larger limb. Two or three eggs form the clutch. Both parents cooperate in nest building, incubation, and raising the young. Incubation takes 14 to 15 days, and the young fledge at 12 to 15 days.

HABITAT: temperate to tropical eucalypt forest and woodland; rainforest margins, coastal scrub, mangroves
LENGTH: *c* 15 cm
DISTRIBUTION: more than 1 million km²
ABUNDANCE: sparse to common
STATUS: probably secure

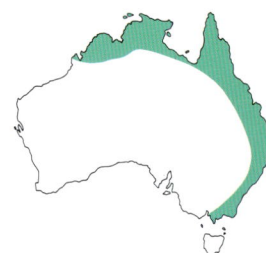

(*H & J Beste*)

Broad-billed Flycatcher

Myiagra ruficollis (rue'-fee-kol'-is: "rufous-necked flycatcher")

The Broad-billed Flycatcher is resident along the coast of northern Australia from about Derby in Western Australia, to Cooktown in Queensland, as well as on the islands of Torres Strait, the Aru Islands, Timor, and New Guinea.

In habits and appearance it is a typical *Myiagra* flycatcher, and very difficult to identify except that adults have white outer tail feathers, and the throat is a somewhat richer shade of rufous than that of related flycatchers. It is usually solitary.

Breeding occurs from October to February. The nest, constructed by both sexes, is a cup of bark strips and vine tendrils, liberally bound with cobweb and decorated with lichen and moss and placed in a vertical mangrove fork a metre or two above high-tide level. Two eggs form the clutch.

HABITAT: tropical mangroves, monsoon forest, melaleuca swamps, riverside scrub
LENGTH: *c* 15 cm
DISTRIBUTION: 10,000–30,000 km²
ABUNDANCE: sparse to common
STATUS: probably secure

(*G Chapman*)

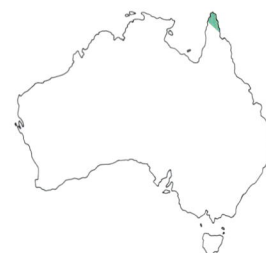

Genus *Rhipidura*

(rip'-id-ue'-rah: "fan-tail")

Fantails comprise a distinctive group of about 40 species widespread in the Indo-Pacific region. Most live in pairs in the canopy of trees; they are common, active, tame and confiding. The tail is very long and habitually fanned, and the birds indulge in active, intricate aerobatics in their pursuit of insects in the outer foliage or in midair in adjacent clearings.

Grey Fantail

Rhipidura fuliginosa (fool'-i-jin-oh'-ah: "sooty fan-tail")

The Grey Fantail occurs almost throughout Australia, including Tasmania; it also occurs in New Guinea, the Solomon Islands, Vanuatu, New Caledonia and New Zealand, and on Norfolk and Lord Howe Islands. It is partly migratory in southern Australia, nomadic in some areas and sedentary in the north.

It closely resembles the Rufous Fantail in appearance, habits, and behaviour but differs most conspicuously in being grey above, not rufous. It has a less "flamboyant" posture (the tail, although almost constantly fanned out, is held so in a much less exaggerated fashion), and it has a much wider tolerance of habitat. Although it often congregates in small parties when migrating or in its winter quarters, it is usually seen alone, almost constantly active, flitting about the outer foliage of trees or spiralling out over adjacent clearings in complex aerobatics in pursuit of insects. It is tame, inquisitive and confiding. It song is high-pitched, thin and squeaky.

Breeding occurs from July to January, when two or even three broods may be raised in succession. The nest resembles that of the Rufous Fantail but has a longer tail. Two to four eggs constitute the clutch. Incubation takes 14 days, fledging 10 to 12 days.

Both parents cooperate in all stages of the nesting cycle.

HABITAT: cool temperate to tropical in almost any wooded habitat, semiarid scrublands, mangroves, riverine vegetation, urban parks and gardens
LENGTH: 14–17 cm
DISTRIBUTION: more than 1 million km²
ABUNDANCE: common
STATUS: secure

(R Garstone)

Willie-wagtail

Rhipidura leucophrys (lue'-koh-fris: "white-browed fan-tail")

The Willie-wagtail occurs almost throughout mainland Australia, as well as in the Moluccas, New Guinea, the Bismarck Archipelago and the Solomon Islands. It is mainly sedentary or locally nomadic.

Common, ubiquitous and confiding, it usually occurs alone or in pairs and is largely terrestrial, chasing insects in quick erratic runs over open ground, or hawking for them on the wing. Its calls include a sweet whistled "sweet pretty creature" and a long scolding rattle.

Breeding may occur at any time of year but especially between July and February. Up to four broods may be raised in succession. The two to four eggs forming the clutch are deposited in a cup-shaped nest of grass and fine bark strips lined with fur or other soft material and placed on a horizontal fork, or sometimes in some artificial structure, usually within about 5 metres of the ground. Incubation takes 14 to 15 days and the young fledge at about 14 days, and both parents share in all stages of the nesting cycle.

HABITAT: almost anywhere but dense rainforest and tall wet eucalypt forest; generally prefers relatively open situations
LENGTH: 19–21 cm
DISTRIBUTION: more than 1 million km²
ABUNDANCE: common
STATUS: secure

(J Gray)

Mangrove Fantail

Rhipidura phasiana (fah'-zee-ah'-nah: "pheasant-like fan-tail")

The Mangrove Fantail occurs in southern New Guinea and in coastal mangroves of northern Australia from Shark Bay in Western Australia to the Norman River in Queensland. It resembles the Grey Fantail except in being confined to mangroves; in being somewhat smaller and paler in appearance; and in having a slightly larger bill and a shorter tail. Most breeding records fall in the period October to February, and two eggs constitute the clutch.

HABITAT: tropical mangroves
LENGTH: 15–16 cm
DISTRIBUTION: 100,000–300,000 km²
ABUNDANCE: common
STATUS: secure

(G Chapman)

Rufous Fantail

Rhipidura rufifrons (rue'-fee-fronz: "rufous-forehead fan-tail")

The Rufous Fantail occurs in Indonesia, New Guinea, the Solomon Islands and Guam, and in coastal and near-coastal districts of northern and eastern Australia. It has not been recorded in Tasmania. Strongly migratory, it is absent from Victoria and New South Wales in winter. It is rare in the Kimberley but otherwise very common in the north, numbers tending to decline southwards.

It usually occurs alone or in pairs. An extremely active bird, it constantly fidgets its wings, tail and body; the tail is almost always flamboyantly fanned and the wings are slightly drooped. It favours the middle and lower levels of forest foliage, where it feeds on small insects. The calls are very high-pitched, thin and squeaky.

Breeding occurs from October to February, and two broods may be raised in succession. Both parents cooperate in nest construction, incubation, and raising the young.

Two or three eggs, which hatch in about 14 days, are deposited in a compact cup-shaped nest of fine grass bound with cobweb, with a pendent tail about 7 to 10 centimetres long, placed in a slender fork within a few metres of the ground.

HABITAT: temperate to tropical, mainly rainforest and dense eucalypt forest; also monsoon forest, swamp and riparian woodland, mangroves; on migration may occur in open or urban situations
LENGTH: 15–17 cm
DISTRIBUTION: 300,000–1 million km²
ABUNDANCE: common
STATUS: secure

(W Lawler)

Northern Fantail

Rhipidura rufiventris (rue'-fee-vent'-ris: "rufous-bellied fan-tail")

The Northern Fantail is widespread from Indonesia through New Guinea to the Solomon Islands and, in northern Australia, from about Broome in Western Australia to the vicinity of Proserpine in Queensland. It is sedentary.

Usually alone, and strictly arboreal, it favours the middle and upper levels of foliage, where it catches insects on the wing, often returning to the same perch after several successive captures. It somewhat resembles the Grey Fantail in plumage but not in behaviour: it is larger and much more sedate and, unlike other fantails, it typically adopts an upright stance with the tail unfanned and pointed vertically downwards.

Breeding occurs from August to January. Sometimes two broods are raised in succession. One or two eggs constituting the clutch are deposited in a nest that resembles that of other fantails, placed in a slender fork usually within about 6 metres of the ground.

HABITAT: tropical eucalypt woodland, rainforest margins, monsoon forest, riverine vegetation, wooded swamps
LENGTH: 16–18 cm
DISTRIBUTION: 300,000–1 million km²
ABUNDANCE: sparse to common
STATUS: secure

(WS Peckover)

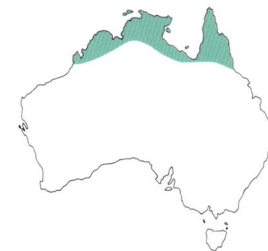

Family SYLVIIDAE

(sil'-vee-id-ee: "*Sylvia*-family", after a genus of warblers)

Some 350 species constitute the family Sylviidae. Divided into about 63 genera (nearly half of which are monotypic), the family is varied and resistant to rigorous diagnosis: most are small or very small songbirds, obscurely coloured in shades of brown, dull green or grey, and have relatively short, slender bills. Conspicuous elements of pattern are rare, and the sexes are generally similar. Most are arboreal, but many live in grassland; almost all are insectivorous. Warblers have 10 primaries, and their young have unspotted breasts.

The group is widespread in Africa, Europe and Asia, but not well represented in Australia, where only eight species occur.

Genus Acrocephalus

(ak'-roh-sef'-ah-lus: "high-head")

Nearly 30 species, widespread in Africa and Asia, constitute this genus, notorious for the difficulty of distinguishing them in the field. Most are relatively large warblers, plain brown above and somewhat paler below. Males resemble females, and there is no seasonal variation. Most inhabit dense reed beds in wetlands of various kinds, and most have rich, varied, attractive songs, delivered from dense cover. Only one species occurs commonly in Australia, though there are persistent reports of another (the Great Reed-warbler, *A. arundinaceus*) in the tropical north.

Clamorous Reed-warbler

Acrocephalus stentoreus (sten-tor'-ay-us: "loud high-head")

This common inhabitant of reed beds occurs across Australia but is rare in Tasmania and absent from most of the arid western interior; elsewhere it occurs across much of Africa and Eurasia. It is at least partly migratory but its movements remain obscure because of its skulking behaviour and its silence in winter. During the breeding season, when males defend tiny territories of less than 0.1 hectare, its loud, rich, persistent song renders it conspicuous. It feeds on insects, taken in reeds or on adjacent mudflats.

Breeding occurs from September to February. The nest is a deep cup of woven reed strips, lined with grass and feathers, and bound to reeds; three or four eggs form the clutch. Incubation takes 14 to 15 days, and the young fledge at 14 to 16 days. Only the female incubates but both parents feed the young.

HABITAT: mainly temperate, reed beds in freshwater marshes, especially *Typha* and bulrush
LENGTH: *c* 17 cm
DISTRIBUTION: more than 1 million km²
ABUNDANCE: common
STATUS: secure

(*P. Green*)

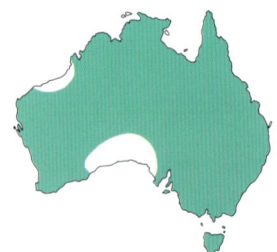

Genus *Cinclorhamphus*

(sin'-kloh-ram'-fus: "bird-bill")

This exclusively Australian genus consists of two species. Males are substantially larger than females, maintain territories by means of conspicuous song flights, and take little part in nesting. Both species are terrestrial, inhabit grasslands, and are migratory.

Brown Songlark

Cinclorhamphus cruralis (krue-rah'-lis: "[strong-]thighed bird-bill")

The Brown Songlark tends to replace the similar and closely related Rufous Songlark in more open, arid environments. It is notable among all songbirds for the pronounced difference in size between sexes, the male being about one-third longer than the female. The species is widespread across southern Australia, migrating to northern Australia in winter, the cyclic pattern of movements being complicated by much nomadism and variations in abundance from one season to the next. Breeding seldom occurs north of the Tropic of Capricorn. When not breeding, the Brown Songlark is inconspicuous, spending most of its time foraging for insects and seeds in grass tussocks. It is usually encountered alone or in small parties.

Breeding occurs mainly from September to January. As in the Rufous Songlark, the male announces his territory by means of a conspicuous display flight, which differs from that of the Rufous Songlark in being directed obliquely upwards from a conspicuous perch; he flutters upward with legs dangling, uttering a reeling, rambling, metallic series of notes, then parachutes down on open, motionless wings.

(G Chapman)

Also as in the Rufous Songlark, the female unobtrusively builds the open cup-shaped nest, incubates the clutch of three or four eggs, and rears the young unaided.

HABITAT: temperate treeless plains; low shrub steppe; croplands
LENGTH: (male): *c* 25 cm
DISTRIBUTION: more than 1 million km²
ABUNDANCE: common
STATUS: probably secure

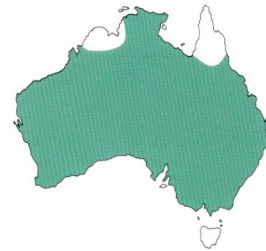

Rufous Songlark

Cinclorhamphus mathewsi (math'-ue-zee: "Mathews's bird-bill", after G. Mathews, eminent Australian ornithologist)

The rich squeaky reeling song of this common and widespread warbler is a characteristic sound of rough lightly timbered pastures and country roadsides over much of southern Australia in spring. The Rufous Songlark is migratory, and spends the winter months in northern Australia.

The diet consists of insects, captured mainly on the ground, and seeds.

The breeding male has a conspicuous display flight, moving around his territory from one vantage point to another in fluttering, horizontal flight, singing persistently as he goes. The female is quiet and skulking, and she builds the cup-shaped nest, incubates the clutch of three or four eggs, and rears the young unaided. The breeding season extends from August to February.

HABITAT: temperate to tropical, mainly open well-grassed woodland
LENGTH: 18–19 cm
DISTRIBUTION: more than 1 million km²
ABUNDANCE: common
STATUS: probably secure

(MJ Seyfort)

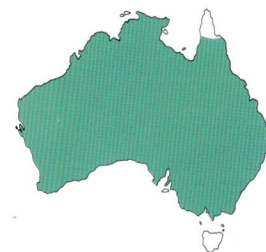

315

Genus Cisticola

(sis'-tee-koh'-lah: "*Cistus*-dweller", referring to a European flowering shrub)

This genus of about 40 species is best represented in Africa but a number of species occur across southern Asia and two extend to Australia. All are extremely similar and, in Africa, many are separable only on ecology and vocalisations. They are very small, with streaked plumage patterns and simple, persistent songs, often delivered in flight. Almost all inhabit rank grassland or similar environments. Males take little part in nesting.

Golden-headed Cisticola

Cisticola exilis (ex'-il-is: "slender *Cistus*-dweller")

The Golden-headed Cisticola is common over most of northern and eastern Australia; it is also widespread from India to China and New Guinea. Since about 1950 it has expanded vigorously south-westward into the lower Murray–Darling basin. It is mainly sedentary, but aggregations may form in prime habitats in winter. It is skulking and unobtrusive in behaviour except when breeding, and feeds on insects.

Males, females and immatures are essentially similar in appearance, but breeding males have a nuptial plumage in which the head is mainly rich golden rufous.

Breeding males announce and maintain territories by means of conspicuous fluttering song flights about 10 to 20 metres from the ground, circling over an area of grassland persistently uttering a buzzy "tchhh, s'lk s'lk"; they also sing from fence posts, weed stalks and other vantage points. The nest is an intricate affair of grass and leaves, domed, with a side entrance, and stitched into the living foliage of the supporting weed clump. The male helps build the nest but takes little part in incubation or care of the young. Three or four eggs form the clutch. The breeding season extends from September to March.

HABITAT: temperate to tropical, tall dense grassland; rank pasture; weeds and roadside vegetation
LENGTH: *c* 11 cm
DISTRIBUTION: more than 1 million km²
ABUNDANCE: common
STATUS: secure

(*N Chaffer*)

Zitting Cisticola

Cisticola juncidis (jun'-sid-is: probably misspelling of *juncinus*, hence "rush [-dwelling] *Cistus*-dweller")

The Zitting Cisticola occurs widely across Africa and Eurasia: in Australia it is confined to isolated pockets of coastal grassland from the vicinity of Rockhampton in Queensland to the Ord River in Western Australia. It is mainly sedentary, and feeds on insects.

Breeding occurs from December to April. Males are polygynous, and up to five nests have been found in the territory of a single male. Males announce and maintain territories by means of conspicuous song flights and persistent dry "tick-tick-tick" songs. The nest is a compact, domed bundle of grass, bound with cobweb, with a side entrance, hidden low in a grass tussock. Usually five eggs form the clutch.

HABITAT: warm temperate to tropical salt marshes; short coastal grassland
LENGTH: *c* 10 cm
DISTRIBUTION: 100,000–300,000 km²
ABUNDANCE: common
STATUS: secure

(*GA Cumming*)

Genus *Eremiornis*

(e'-rem-ee-or'-nis: "desert-bird")

The characteristics of the genus are those of the single species.

Spinifexbird

Eremiornis carteri (kar'-ter-ee: "Carter's desert-bird", after T. Carter, Western Australian ornithologist)

(G Chapman)

The Spinifexbird is common but locally distributed in spinifex grasslands across northern Australia. It is skulking and difficult to observe, and little is known of its habits. It is sedentary, and lives in pairs that appear to maintain permanent territories. Breeding males utter a loud, whistled song from bushes or the tops of spinifex clumps. The diet consists of insects.

Breeding occurs in response to rainfall, but is most common from August to November. Well hidden in a tussock, the nest is a deep cup of fine grass; two eggs form the clutch.

HABITAT: warm-temperate spinifex grassland
LENGTH: *c* 15 cm
DISTRIBUTION: more than 1 million km²
ABUNDANCE: sparse to common
STATUS: probably secure

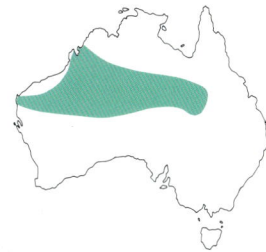

Genus *Megalurus*

(meg'-ah-lue'-rus: "large-tail")

This genus, of about five species (two in Australia), is widespread from Indonesia and New Guinea northward to the Philippines, eastward to New Caledonia and south through Australia. Like *Acrocephalus*, its members are mainly swamp-dwellers; the plumage of most is heavily streaked, at least on the back. The species are skulking in behaviour and have relatively simple songs.

Little Grassbird

Megalurus gramineus (grah-min'-ay-us: "grassy or grassland large-tail")

The Little Grassbird has been recorded over much of eastern and southern Australia but most breeding is recorded in the south-east (including Tasmania) and the extreme south-west. It seems to be largely sedentary, living in pairs that maintain territories of about 0.75 hectare. It is well known by its thin plaintive whistled note, but is skulking and very difficult to observe. It feeds on insects and other small invertebrates.

The breeding season extends from September to January. The nest is a deep cup of grass and leaves, well hidden low in swamp vegetation. Three to five eggs form the clutch.

HABITAT: cool to warm temperate dense vegetation in freshwater swamps and their margins; salt marshes; wet tussock grassland
LENGTH: *c* 14 cm
DISTRIBUTION: more than 1 million km²
ABUNDANCE: common
STATUS: probably secure

(K & B Richards)

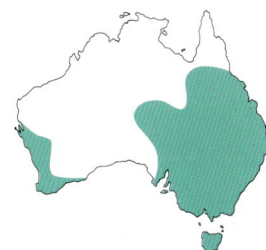

Tawny Grassbird

Megalurus timoriensis (tee'-mor-ee-en'-sis: "Timor large-tail")

The Tawny Grassbird, which has an extensive distribution from Indonesia and New Guinea northward to the Philippines, occurs in northern and eastern Australia from the Kimberley to Cape York Peninsula and south to the vicinity of Wollongong in New South Wales. In Australia it is most numerous in coastal lowlands but penetrates some distance inland in some areas. Available data on its movements are sparse and conflicting.

It is extremely skulking in behaviour, its presence usually being indicated by a loud, rich, abrupt "tchk": it also has a reeling metallic song uttered by the male in display flight. It feeds on insects.

It breeds from August to April. Well hidden low in swampy herbage, the nest is a deep cup of dry grass. Three eggs form the usual clutch.

HABITAT: warm temperate to tropical dense rank pasture, swampland, heath
LENGTH: *c* 19 cm
DISTRIBUTION: 300,000–1 million km²
ABUNDANCE: sparse to common
STATUS: secure

(J Purnell)

Family MALURIDAE

(mal-ue'-rid-ee: "*Malurus*-family", after a genus of fairywrens)

This family consists of 26 species found only in Australia and New Guinea; there are five genera, two being restricted to New Guinea; two (*Stipiturus* and *Amytornis*) to Australia, and one (*Malurus*) shared. All species are small to medium-sized birds with long tails, habitually held cocked over the back. All frequent shrubbery and undergrowth, where they are active and quick-moving, gleaning insects either from low foliage or on the ground. Some Central Australian species, however, eat seeds amounting to about half of their overall diet. Fairywrens seldom fly further than from one bush to the next, and the long tail is used more as a signal than as a rudder. Males are often brilliantly coloured, and most are much brighter in plumage than their mates. Most live in groups, and some (especially species of *Malurus*) are highly social.

Genus Amytornis

(ah'-meet-or'-nis: "Amytis-bird", after Amytis, daughter of Astyages in Greek mythology)

Members of this exclusively Australian genus are characterised by unusually shaggy plumage, streaked throughout with shafts of white. All are birds of the arid interior and some, restricted to the most remote and inhospitable areas of Australia, are very poorly known in consequence. They live in small groups, frequenting spinifex and similar vegetation and feeding on insects and seeds. They seldom fly, but are extremely nimble on the ground, usually hopping when foraging and running when avoiding disturbance, dashing rat-like into dense cover.

Grey Grasswren

Amytornis barbatus (bar-bah'-tus: "bearded Amytis-bird")

(L Robinson)

ground than other grasswrens, spending much of its time searching for seeds and insects in shrubbery. It also eats water snails. Breeding occurs in July and August, and the clutch consists of two eggs.

HABITAT: lignum and cane grass on flood plains
LENGTH: 18–20 cm
DISTRIBUTION: 30,000–100,000 km²
ABUNDANCE: rare to very sparse
STATUS: possibly endangered

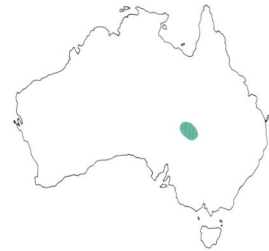

The Grey Grasswren is restricted to thickets of lignum and cane grass on the flood plains of the lower Diamantina, Cooper and Booloo River systems on the eastern fringes of the Lake Eyre basin. It was not discovered and formally described until 1967, and very little is yet known of it. Somewhat more gregarious than other grasswrens, it lives in groups that may exceed 20 or 30 individuals when concentrated by drought but which split up into territorial pairs to breed. It also forages less on the

Carpentarian Grasswren

Amytornis dorotheae (do'-ro-thay'-ee: "Dorothy's Amytis-bird", after Dorothy White, Australian naturalist)

Restricted to a small region at the head of the Gulf of Carpentaria, the Carpentarian Grasswren is the eastern representative of a trio of large, richly coloured and patterned grasswrens that occupy similar habitat in three widely separated regions of sandstone outcrops in tropical Australia. Like the other two species, it lives in small groups that forage for seeds and insects over rocky outcrops, but scuttle for cover at the least sign of disturbance. It breeds from October to March; the nest resembles that of other grasswrens, and the female incubates the two or three eggs unaided.

HABITAT: sandstone outcrops, gorges and hillsides with a cover of spinifex
LENGTH: 16–18 cm
DISTRIBUTION: 30,000–100,000 km²
ABUNDANCE: very sparse
STATUS: vulnerable

(J Whitaker)

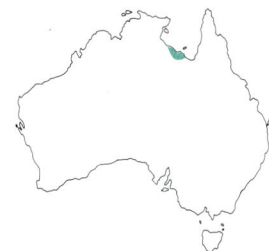

Eyrean Grasswren

Amytornis goyderi (goy'-der-ee: "Goyder's Amytis-bird", after G. D. Goyder, Australian explorer)

This grasswren is restricted to parts of the Simpson and Strzelecki Deserts in the Lake Eyre basin of South Australia. One of the least known of the Australian birds, it was discovered before 1875 but not recorded again until 1976. It lives in groups, sheltering in cane grass clumps and foraging for seeds and insects on the ground nearby. It breeds in August and September, two eggs constituting the clutch.

HABITAT: sand dunes with spinifex (*Triodia*) and cane grass (*Zygochloa*)
LENGTH: 14–16 cm
DISTRIBUTION: 100,000–300,000 km²
ABUNDANCE: very sparse
STATUS: probably secure

(D & M Trounson)

Black Grasswren

Amytornis housei (how'-see: "House's Amytis-bird", after F. M. House, Western Australian naturalist)

The Black Grasswren has a very restricted distribution in remote sandstone gorges of the western Kimberley, where it is uncommon, sedentary and communal, living permanently in groups of up to six or eight individuals. The diet consists of seeds and insects, the birds looking like rats as they scurry, bound and hop in their foraging over the jumbled rocks of their habitat. Calls include various loud ticking notes, sharp chirps and whistles. Females differ from males in their rusty rufous, not black, underparts. Almost nothing is known of the nesting cycle, except that it appears to occur from December to March.

HABITAT: tumbled sandstone outcrops, gorges and hillsides with a cover of spinifex
LENGTH: 18–21 cm
DISTRIBUTION: 30,000–100,000 km²
ABUNDANCE: rare
STATUS: probably secure

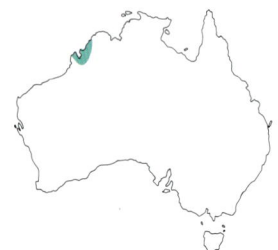

(G Chapman)

Dusky Grasswren

Amytornis purnelli (per-nel'-ee: "Purnell's Amytis-bird", after H. A. Purnell, Victorian ornithologist)

The Dusky Grasswren closely resembles the Thick-billed Grasswren in appearance and behaviour but it inhabits rocky hillsides in central Australia. It lives for most of the year in small groups, feeding on seeds and insects taken on the ground. It is active, elusive and difficult to observe.

At the start of breeding in August, groups split up into territorial pairs: the male advertises the territory by means of a song consisting of high staccato trills, delivered from prominent perches on rocks, while the female builds the nest and incubates the clutch of two or three eggs unaided. Both parents help feed the young. Sometimes two broods are raised in succession.

HABITAT: rocky hillsides with cover of spinifex
LENGTH: 15–18 cm
DISTRIBUTION: 300,000–1 million km²
ABUNDANCE: sparse
STATUS: probably secure

(J Purnell)

Striated Grasswren

Amytornis striatus (stree-ah'-tus: "striped Amytis-bird")

The Striated Grasswren is widespread across the arid interior of Australia from central New South Wales westward to the coast of Western Australia. Habitat destruction resulting from clearing and the effects of grazing by stock have caused a marked decline in range and number, especially in the east. It is sedentary and, like other grasswrens, it tends to congregate in small parties for much of the year, splitting off into territorial pairs to nest. It is active, highly strung and extremely difficult to observe, remaining almost always in dense cover.

Breeding occurs mainly from August to December. The nest is a globular structure of spinifex blades lined with fur and plant down, hidden in a spinifex clump. The clutch usually consists of two eggs, sometimes three, which hatch in 13 to 14 days. The female builds and incubates unaided but both parents rear the young, which fledge at about 11 to 12 days.

HABITAT: spinifex grassland on sand plains and under mallee; occasionally stony hillsides
LENGTH: 14–16 cm
DISTRIBUTION: more than 1 million km²
ABUNDANCE: sparse
STATUS: probably secure

(W J Labbett)

Thick-billed Grasswren

Amytornis textilis (tex-til'-us: "woven Amytis-bird")

The Thick-billed Grasswren differs most strikingly from other grasswrens in its choice of habitat: it is the only one to inhabit the shrub steppe of saltbush and bluebush that extends across the winter-rainfall arid zone of southern Australia, from central New South Wales to Shark Bay, Western Australia. It is uncommon and locally distributed: habitat destruction has caused its local extinction in many areas, possibly as a result of grazing by sheep. It lives in pairs which maintain permanent territories of about 20 to 40 hectares. The diet consists of seeds and insects gathered on the ground.

Breeding begins in August and September and may extend to February. The nest is somewhat flimsier and less definitely domed than that of other grasswrens. The clutch consists of two or three eggs. The female builds and incubates unaided while the male defends the territory with sporadic songs: both parents cooperate in rearing the young.

HABITAT: saltbush and bluebush plains
LENGTH: 18–20 cm
DISTRIBUTION: 300,000–1 million km^2
ABUNDANCE: sparse
STATUS: vulnerable

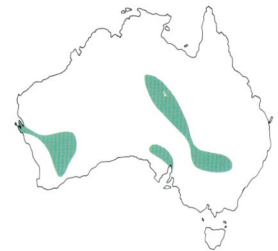

(G Chapman)

White-throated Grasswren

Amytornis woodwardi (wood'-war-dee: "Woodward's Amytis-bird", after B. H. Woodward, sometime director of the WA Museum)

Closely related to the Black Grasswren, the White-throated Grasswren is restricted to western Arnhem Land, where it is common but elusive, the birds dashing for cover and hiding in jumbled rocks at the least sign of disturbance.

The White-throated Grasswren lives in parties of six or eight birds that split up into territorial pairs to breed. It is relatively silent for much of the year but breeding males announce their territories with loud sweet songs from prominent perches on the tops of rocks. The diet consists of seeds and insects in about equal proportions. Little is known of the nesting cycle but breeding occurs between December and March and the normal clutch seems to be two eggs. The nest is a bulky domed structure of woven grass lined with finer material and placed in the top of a spinifex tussock.

HABITAT: sandstone outcrops, gorges and hillsides with a cover of spinifex
LENGTH: 20–22 cm
DISTRIBUTION: 10,000–30,000 km^2
ABUNDANCE: sparse
STATUS: vulnerable

(KH Gill)

Genus *Malurus*

(mal-ue'-rus: "soft-tail")

This group of about 13 species occurs in New Guinea and Australia. All live in undergrowth but they occupy a wide range of environments from rainforest to scrubs of the arid interior. They are sedentary and live permanently in groups. Males are brilliantly coloured (blue is especially common) whereas females are generally dull brown; there are two moults per year, and in many species all but the oldest or dominant males resume "female" plumage after breeding.

Lovely Fairywren

Malurus amabilis (ah-mah'-bil-is: "lovely soft-tail")

(H & J Beste)

Inhabiting rainforests and their margins in the north-east of Queensland, this species strongly resembles the Variegated Fairywren in appearance and behaviour but differs strikingly in the appearance of the adult female, in which the head and upperparts are greyish blue rather than brown, and the lores and face are white rather than dull red. The Lovely Fairywren is also markedly more arboreal than other fairywrens, spending most of its time in saplings and low trees and only occasionally coming to the ground.

Breeding occurs throughout the year but the peak of activity falls between July and December. The female builds the nest and incubates the (usually) three eggs unaided, and apparently receives little help in rearing the young. Incubation takes about 13 days and the young fledge at about 12 days.

HABITAT: scrubby rainforests and their margins
LENGTH: 12–13 cm
DISTRIBUTION: 100,000–300,000 km²
ABUNDANCE: common
STATUS: probably secure

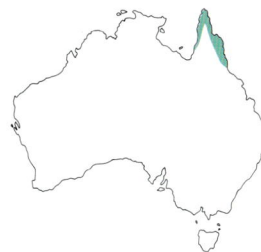

Purple-crowned Fairywren

Malurus coronatus (ko'-ron-ah'-tus: "crowned soft-tail")

The Purple-crowned Fairywren is found along streams and rivers in the Kimberley and in north-western Queensland, eastward to the head of the Gulf of Carpentaria. It has

(H & J Beste)

declined since European settlement, perhaps due to destruction of habitat caused by the trampling of cattle and buffalo coming to drink. It is sedentary, living in small parties that maintain permanent territories.

As in other fairywrens, parties are close-knit. While foraging during the day, the birds periodically gather to perch huddled together on a twig or low branch, resting and preening each other for a time before scattering again to feed. The diet consists of insects, taken mainly on the ground or in undergrowth.

The breeding season is uncertain but most activity seems to occur from September to January. The nest is a rough globe of grass and paperbark fragments placed in cane grass or in the crown of a pandanus. The usual clutch consists of three eggs.

HABITAT: pandanus and cane grass thickets at the margins of streams and rivers in melaleuca swamp and woodland
LENGTH: 13–15 cm
DISTRIBUTION: 100,000–300,000 km²
ABUNDANCE: very sparse
STATUS: endangered

Superb Fairywren

Malurus cyaneus (sie'-ah-nay'-us: "blue soft-tail")

A familiar garden bird of south-eastern Australia, the Superb Fairywren occurs in Tasmania and across most of Victoria and eastern New South Wales, northward to the vicinity of Blackall in Queensland and westward to Adelaide and the tip of Eyre Peninsula in South Australia. It is sedentary. In some respects it has profited from European settlement, expanding its distribution in many areas, but adverse factors such as feral cats and pesticides have also influenced its numbers and distribution.

Only the adult male wears the well-known brilliant blue plumage; females and immatures are mainly dull brown.

It lives permanently in parties that usually consist of a dominant blue male, his mate, and several young of the year; several such parties may coalesce in winter, forming flocks of 12 or 14 birds. The dominant female usually drives away the young of her own sex, but the dominant male tolerates other subadult males in the group, which—once reaching sexual maturity—moult into the blue plumage for the duration of the breeding season, resuming immature plumage for the winter. Active, bold and inquisitive, Superb Fairywrens feed on insects, usually gathered on open ground close to dense shrubbery, to which the birds retire for shelter when disturbed.

Breeding occurs from August to March, and usually several broods are raised in succession. The dominant female builds the nest (a loose globular structure of grass, with a side entrance, hidden in a grass tussock or low bush within a metre of the ground) alone, and may leave it unattended for a week before laying a clutch of three or four eggs. She also incubates alone for 13 to 15 days, but all members of the group help feed the young, which fledge in about 12 to 13 days.

HABITAT: dense shrubbery and undergrowth in forest and woodland
LENGTH: 13–15 cm
DISTRIBUTION: 300,000–1 million km²
ABUNDANCE: common
STATUS: secure

(JD Waterhouse)

Red-winged Fairywren

Malurus elegans (el'-e-ganz: "handsome soft-tail")

(AY Pepper)

This species is confined to far south-western Western Australia. Like other fairywrens it is communal and sedentary, frequenting dense shrubbery and feeding on insects captured mainly on the ground. Breeding occurs from September to December and two or three eggs form the clutch.

HABITAT: dense shrubbery, mainly in gullies and along creeks in eucalypt forest and woodland
LENGTH: 13–15 cm
DISTRIBUTION: 30,000–100,000 km²
ABUNDANCE: common
STATUS: vulnerable

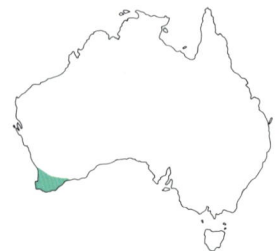

Variegated Fairywren

Malurus lamberti (lam'-bert-ee: "Lambert's soft-tail", after A. B. Lambert, British ornithologist)

The Variegated Fairywren occurs throughout mainland Australia except for the extreme south-east and south-west, the Top End and Cape York Peninsula; it is the most widespread of the fairywrens and occupies the widest range of habitats. It occurs in a number of populations, differing subtly in appearance: some of these, especially those in the Kimberley and the tropical north, are regarded by some authorities as distinct species. Like other fairywrens it lives in small parties which maintain permanent territories. It eats insects, obtained in low bushes and undergrowth, sometimes on the ground.

Breeding occurs mainly from August to December. The nest is a rough globe of coarse grass lined with feathers placed in a dense bush close to the ground. The usual clutch consists of three or four eggs. Incubation, by the female, takes 14 to 16 days, and the young fledge at 10 to 12 days.

HABITAT: dense shrubbery and undergrowth in most types of woodland from rainforest to arid acacia scrubs
LENGTH: 11–14 cm
DISTRIBUTION: more than 1 million km²
ABUNDANCE: common
STATUS: secure

(JD Waterhouse)

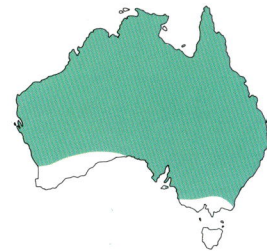

White-winged Fairywren

Malurus leucopterus (lue-kop'-te-rus: "white-winged soft-tail")

The White-winged Fairywren occurs across the arid interior of southern Australia, its distribution extending to the coast, and including some offshore islands, in the far west. It is sedentary, living in permanent groups consisting usually of an adult male (which sometimes moults into immature plumage during winter), an adult female, and several young of the year. It feeds on insects, gathered mainly in the outer foliage of small bushes but sometimes on the ground.

Breeding is strongly influenced by local rainfall but usually occurs from August to February. The nest is a small neat globe of grass lined with down and hidden in a bush within a metre of the ground. The clutch consists of two to four eggs, which hatch in about 13 days. The young fledge at about 11 days, and are fed by the female on insects brought to the nest by others of the group.

HABITAT: shrub steppe (especially saltbush, bluebush and samphire) and desert
LENGTH: 11–13 cm
DISTRIBUTION: more than 1 million km²
ABUNDANCE: common
STATUS: probably secure

(IR McCann)

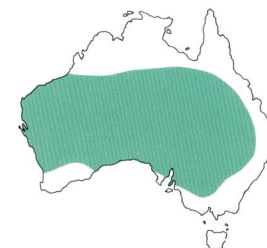

Red-backed Fairywren

Malurus melanocephalus (mel'-ah-noh-sef'-ah-lus: "black-headed soft-tail")

This abundant fairywren is characteristic of savanna grasslands throughout northern and eastern Australia, south to Port Stephens in New South Wales. It is the smallest of the fairywrens, has proportionately the shortest tail, and also differs from other Australian species in lacking blue in the plumage. Somewhat less communal than its relatives, it sometimes breeds in simple pairs as well as in groups, birds tending to roam in larger parties after breeding. Only the oldest males retain their black and red nuptial plumage all year, most moulting back into a dull brown eclipse plumage after breeding. Like other fairywrens, the Red-backed Fairywren feeds on insects taken mainly on the ground.

Though strongly influenced by local rainfall, breeding occurs more or less throughout the year in the tropical north; eastern populations breed mainly from August to March. The female builds the nest and incubates the three or four eggs unaided but is assisted in raising the young by the male, and sometimes by other group members.

(E McNamara)

HABITAT: tropical and warm temperate Blady Grass and other savannah grasslands in regions of summer rainfall
LENGTH: 10–13 cm
DISTRIBUTION: more than 1 million km²
ABUNDANCE: common
STATUS: probably secure

Blue-breasted Fairywren

Malurus pulcherrimus (pool-ke'-rim-us: "very-pretty soft-tail")

(B & B Wells)

The Blue-breasted Fairywren occurs in south-western Australia, eastward to Eyre Peninsula in South Australia. Much of its habitat has been cleared for wheat farming and its distribution has been severely diminished in consequence. It is more secretive than other fairywrens but otherwise resembles them in general behaviour, living in parties (of seldom more than five) which maintain permanent territories. The diet consists of insects, taken mostly on the ground.

Breeding occurs mainly in August and September. As in other fairywrens the female builds and incubates unaided but the group combines to rear the young. The nest resembles that of the Variegated Fairywren, and three eggs constitute the clutch.

HABITAT: sandplain heath and mallee
LENGTH: 13–15 cm
DISTRIBUTION: 100,000–300,000 km²
ABUNDANCE: sparse
STATUS: vulnerable

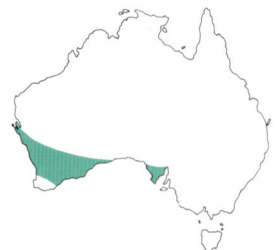

Splendid Fairywren

Malurus splendens (splen'-denz: "shining soft-tail")

The Splendid Fairywren is in many respects the arid-country representative of the Superb Fairywren, differing little from that species in general behaviour but being much less tolerant of human settlement. It occurs across southern Australia from central New South Wales to far south-western Western Australia. It is sedentary, living in groups which maintain permanent territories.

Breeding occurs from September to January. Constructed by the female and hidden close to the ground, the nest is a loose globular structure of grass, lined with feathers and with an entrance in the side. The usual clutch consists of three eggs. Incubation takes 13 to 16 days, and the young fledge at about 12 to 14 days.

HABITAT: arid acacia scrub; mallee; dense shrubbery and undergrowth in forest and woodland
LENGTH: 12–14 cm
DISTRIBUTION: more than 1 million km²
ABUNDANCE: common
STATUS: probably secure

(*JD Waterhouse*)

Genus *Stipiturus*

(stip'-it-ue'-rus: "stem-tail")

Only three species constitute the genus *Stipiturus*, or emuwrens. They much resemble fairywrens of the genus *Malurus*, living secretively in dense cover near the ground, but they differ strikingly in being mainly dun-coloured in plumage and in having only six rectrices, of unique filamentous structure. In the entire world, only two other species are restricted to six tail feathers.

Southern Emuwren

Stipiturus malachurus (mal'-ah-kue'-rus: "soft-tailed stem-tail")

The Southern Emuwren occurs in coastal and near-coastal heathlands of southern Australia from about Fraser Island to Shark Bay. It is sedentary, living in small parties that separate into territorial pairs to breed.

Although not especially shy or elusive, it seldom leaves the shelter of dense cover, only occasionally dropping to the ground to feed, or emerging to perch briefly in the crown of a bush or shrub to look about. It flies awkwardly and only with reluctance. It feeds on insects.

Breeding occurs from August to January, during which time two broods usually are raised in succession. Hidden low in dense vegetation, the nest is an oval structure of grass and sedge lined with fur and feathers, with a side entrance. Two, three or four eggs constitute the clutch. Incubation takes about 10 to 12 days, and the young fledge at 10 to 11 days. The female builds and incubates unaided, but both parents feed the young, occasionally assisted by other birds.

HABITAT: heath, wallum, sedge and other dense swampy vegetation
LENGTH: 15–19 cm
DISTRIBUTION: 100,000–300,000 km²
ABUNDANCE: sparse
STATUS: probably secure

(*R Garstone*)

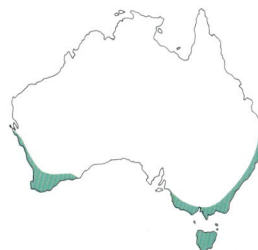

Mallee Emuwren

Stipiturus mallee (mal'-ee: "mallee stem-tail")

The Mallee Emuwren is restricted to areas of spinifex in mallee country in a small area where the boundaries of New South Wales, Victoria and South Australia come together. It closely resembles the Rufous-crowned Emuwren in habits and behaviour and shows some characteristics of the Southern Emuwren: it is probably conspecific with one or other of these birds but existing data are inadequate to determine which. Breeding occurs from September to December, and the female incubates the clutch of two or three eggs unaided.

HABITAT: mallee with an understorey of spinifex
LENGTH: 13–15 cm
DISTRIBUTION: 30,000–100,000 km²
ABUNDANCE: sparse
STATUS: probably secure

(WJ Labbett)

Rufous-crowned Emuwren

Stipiturus ruficeps (rue'-fee-seps: "rufous-headed stem-tail")

This tiny species (which, with a weight of about 5 grams, is probably the smallest Australian bird) passes its entire life in the shelter of spinifex clumps and associated bushes and shrubs, only occasionally descending to the ground to forage. It occurs across the arid interior of Australia from central Queensland to the west coast. It is sedentary and lives in pairs which often congregate in wandering flocks after breeding. It eats insects. The breeding season extends from August to October, and two or three eggs form the clutch.

HABITAT: arid spinifex sand plains and acacia scrub
LENGTH: 12–13 cm
DISTRIBUTION: 300,000–1 million km²
ABUNDANCE: sparse
STATUS: probably secure

(M Morcombe)

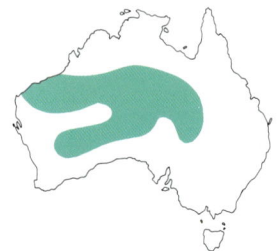

Family ACANTHIZIDAE

(ak'-an-thie'-zid-ee: "*Acanthiza*-family", after a genus of thornbills)

This family incorporates about 65 species (14 genera) of small insectivorous birds, mostly restricted to Australia. However, members of one genus (*Gerygone*) occur from the Malay Peninsula to New Caledonia; several species occur in New Guinea; and some researchers incude the endemic New Zealand genera *Mohoua* and *Finschia* within the family. Traditionally linked with the Sylviidae but now usually placed close to Maluridae, the family Acanthizidae has not yet been rigorously defined, and its closest relationships remain debatable.

Members of the Acanthizidae occupy a wide range of habitats from rainforest to scrublands of the arid interior of Australia. Most are arboreal but a number are semiterrestrial. Almost all are sedentary but several species are nomadic and one or two are migratory. Few are brightly coloured. Females generally resemble males, immatures resemble adults, and there is no seasonal variation. Most build domed nests, typically constructed by the female, which also incubates unaided; both parents rear the young. Communal breeding has been reported in many species, and some (notably in the genus *Acanthiza*) exhibit a complex social organisation.

Genus Acanthiza

(ak'-an-thie'-zah: "thornbush-dweller")

Except for one Papuan member, this genus of about 12 species is exclusively Australian. Most species are predominantly arboreal but a number live mainly in undergrowth and the lower storey of woodland: at least one species is semiterrestrial, foraging almost exclusively on the ground. All are insectivorous. Thornbills are common, gregarious and very active. In many species the colour of the rump contrasts with that of the rest of the upperparts and is a valuable field identification feature.

Yellow-rumped Thornbill

Acanthiza chrysorrhoa (krie'-soh-roh'-ah: "golden-tailed thornbush-dweller")

(G Weber)

This familiar bird is the most terrestrial of the thornbills, foraging almost entirely on open ground and avoiding tree foliage and undergrowth except for shelter.

Widespread across southern Australia, it is mainly sedentary and has a social organisation similar to that of the Buff-rumped Thornbill: it lives in clans of up to 30 individuals which occupy a permanent extended territory defended by the entire group. Breeding birds split off from the clan to form units consisting of a female and one or several males; each of these establishes and defends a nesting territory within the overall territory of the clan.

Often common on golf courses and in other urban grassed environments, the Yellow-rumped Thornbill is easily identified by its behaviour and its clear, primrose yellow rump. The contact call is an abrupt chipping note, and it has a distinctive light, twittering, warbled song.

Breeding occurs from July to December, during which period up to four broods may be raised in succession. Suspended in the outer foliage of a bush or sapling within a few metres of the ground, the nest resembles that of other thornbills except that it has a unique "false" nest, complete with egg chamber, built into the top; the entrance to the true egg chamber is concealed underneath. The clutch consists of three or four eggs, which hatch in about 19 days. The female incubates unaided, but all members of the group participate in rearing the young, which fledge in about 18 days.

HABITAT: clearings in forest and woodland; margins of shrubbery; urban parks and gardens
LENGTH: *c* 11 cm
DISTRIBUTION: more than 1 million km^2
ABUNDANCE: common
STATUS: secure

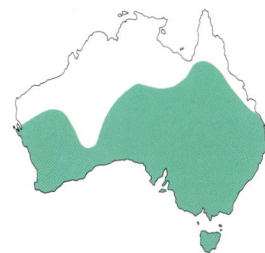

Tasmanian Thornbill

Acanthiza ewingii (ue'-ing-ee-ee: "Ewing's thornbush-dweller", after T. J. Ewing, Tasmanian naturalist)

The Tasmanian Thornbill is restricted to Tasmania and the islands of Bass Strait. It is sedentary, but spends much of the year in small flocks (often mixed with Scrubtits, Scrubwrens and Brown Thornbills), which split off into territorial pairs to breed. It forages energetically for small insects in the foliage of trees at almost any height but seldom descends to the ground. It utters a variety of thin twittering notes.

The breeding season extends from September to January. The nest is a compact domed structure with a side entrance near the top, built of grass, bark fragments and moss, lined with fur and feathers and suspended in a bush or shrub within a few metres of the ground. Three or four eggs form the clutch.

HABITAT: wet sclerophyll forest and temperate rainforest
LENGTH: *c* 10 cm
DISTRIBUTION: 30,000–100,000 km²
ABUNDANCE: common
STATUS: secure

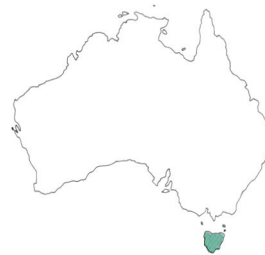

(TA Waite)

Western Thornbill

Acanthiza inornata (in'-or-nah'-tah: "plain thornbush-dweller")

Restricted to the far southwestern corner of the continent, the Western Thornbill is most closely related to the Buff-rumped Thornbill of eastern Australia, resembling it closely in habits and general behaviour. It lives in groups of a dozen or more birds that forage mainly in undergrowth in bustling, twittering bands. Breeding occurs from August to December. The nest is a globular structure of grass and bark fragments hidden close to the ground in a tree cavity or a fence post, behind loose bark on a tree trunk, or in a grass tussock. Three or four eggs constitute the clutch, and incubation is said to take 20 to 21 days.

HABITAT: mainly open eucalypt forest and woodland; also heathland
LENGTH: 9–10 cm
DISTRIBUTION: 100,000–300,000 km²
ABUNDANCE: common
STATUS: probably secure

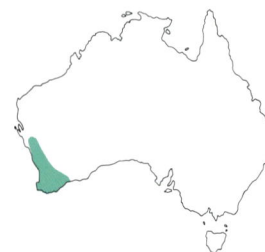

(R Green)

Slender-billed Thornbill

Acanthiza iredalei (ire'-day-lee: "Iredale's thornbush-dweller", after T. Iredale, Australian ornithologist)

Mainly sedentary or locally nomadic, the Slender-billed Thornbill lives in pairs or small parties which, like other thornbills, forage briskly for small insects in low scrub, maintaining a constant low twittering sound as they go. It differs most conspicuously from other Thornbills in its choice of habitat: samphire flats and low saltbush plains of southern Australia from the Big Desert in Victoria to Shark Bay in Western Australia. It is sedentary or locally nomadic.

Breeding occurs from July to November. A typical thornbill nest is placed in the foliage of a small bush within a metre of the ground. Three eggs constitute the clutch.

HABITAT: saltbush and bluebush steppe; low sandplain heath; samphire flats
LENGTH: 9–10 cm
DISTRIBUTION: 300,000–1 million km²
ABUNDANCE: sparse
STATUS: vulnerable

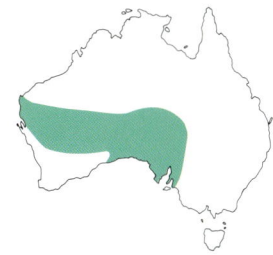

(G Chapman)

Mountain Thornbill

Acanthiza katherina (kath'-er-ee'-nah: "Katherina's thornbush-dweller", the identity of the lady being unknown)

Restricted to highlands between 400 and 1200 metres altitude in north-eastern Queensland from Cooktown to Mount Spec, the Mountain Thornbill strongly resembles the Brown Thornbill in appearance and behaviour, but it has a cream or yellowish rather than red iris, and it occurs only in the canopy of rainforest. It is active and **gregarious**, and feeds on small insects. Little is known of breeding behaviour, but the nesting season apparently extends from September to December, and two eggs constitute the clutch.

HABITAT: montane tropical rainforest
LENGTH: *c* 10 cm
DISTRIBUTION: 30,000–100,000 km²
ABUNDANCE: common
STATUS: probably secure

(R Whitford)

Striated Thornbill

Acanthiza lineata (lin-ay-ah'-tah: "lined thornbush-dweller")

(AJ Salter)

Thornbill tends to concentrate its foraging on the foliage rather than on the bark of twigs and branches; it also forages higher in the canopy.

Breeding occurs from July to December. A typical thornbill nest is suspended from the outer foliage of a eucalypt, usually about 5 to 10 metres from the ground. Three eggs form the clutch, and incubation takes about 15 to 17 days. The female builds the nest largely unaided, and incubates alone, but all group members share in raising the young.

HABITAT: most kinds of eucalypt forest and woodland
LENGTH: *c* 10 cm
DISTRIBUTION: 100,000–300,000 km²
ABUNDANCE: common
STATUS: secure

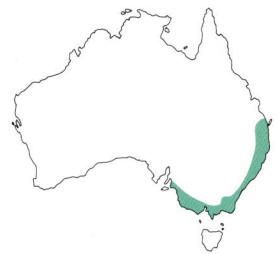

The Striated Thornbill is abundant in coastal and near-coastal regions of south-eastern Australia. It lives in clans of up to 25 birds which occupy permanent territories, several hectares in extent, and break up into units consisting of a female and one or several males for nesting. Strictly arboreal, it closely resembles the yellow Thornbill in habits and general behaviour—apart from its marked preference for eucalypts. Unlike the Brown Thornbill, with which it often occurs, the Striated

Yellow Thornbill

Acanthiza nana (nah'-nah: "dwarf thornbush-dweller")

The Yellow Thornbill is common and widespread throughout much of eastern and south-eastern Australia. It is usually encountered in small

(N & E Taylor)

groups occupying permanent territories of several hectares. It is entirely arboreal, and differs from other thornbills in generally avoiding eucalypts, instead showing a marked preference for such trees and shrubs as acacias, casuarinas, cypress pines and melaleucas. Otherwise its behaviour is typical of the group: it forages busily while maintaining a constant harsh subdued twittering.

Breeding occurs from August to December. The nest is a neat, compact dome of grass and bark fragments matted with cobweb, lined with fine grass and feathers and placed in the outer foliage of a shrub or tree several metres from the ground. Usually three eggs make up the clutch. Only the female incubates but all group members participate in rearing the young.

HABITAT: forest and woodland, especially groves of acacias, casuarinas, cypress pines and melaleucas
LENGTH: 9–10 cm
DISTRIBUTION: more than 1 million km²
ABUNDANCE: common
STATUS: secure

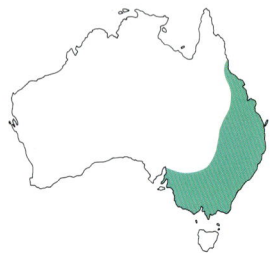

Brown Thornbill

Acanthiza pusilla (pue-sil'-ah: "dwarf thornbush-dweller")

(G Weber)

One of the most widespread of thornbills, this species occurs across southern mainland Australia (north approximately to the Tropic of Capricorn) as well as in Tasmania. Eastern populations have plain dark brown upperparts whereas those of the arid interior are grey above, with reddish rumps; the two forms were formerly regarded as distinct species, but zones of interbreeding have recently been found where they come together along the western slopes of the Great Dividing Range. The Brown Thornbill is sedentary and not especially gregarious, living usually in pairs which maintain permanent territories, although small parties sometimes congregate in winter: it occasionally joins mixed feeding flocks of other birds. Like other thornbills, it feeds mainly on small insects but, when foraging, it tends to focus most of its attention on twigs and the bark of limbs and branches, rather than on leaves, and favours undergrowth and lower storeys rather than the upper canopy of forest. It also forages at flowers, presumably for nectar. It is strongly vocal, uttering a wide variety of twittering and trilling whistled notes, as well as soft chips and churring notes.

Breeding occurs from June to December. The nest is an untidy domed structure with a side entrance near the top, built of coarse grass and bark fragments and placed in a bush or shrub within a few metres of the ground. The clutch usually consists of three (sometimes two) eggs, which hatch in 17 to 21 days. The young fledge at about 15 days. The female builds and incubates unaided but both parents feed the young.

HABITAT: shrubbery and undergrowth in almost any wooded environment from rainforest to arid acacia scrubs
LENGTH: 10–11 cm
DISTRIBUTION: 300,000–1 million km²
ABUNDANCE: common
STATUS: secure

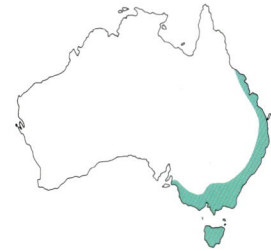

Buff-rumped Thornbill

Acanthiza reguloides (reg'-ue-loy'-dayz: "*Regulus*-like thornbush-dweller", after a genus of Old World flycatchers)

(IR McCann)

The Buff-rumped Thornbill is a common bird in the drier coastal and near-coastal woodlands of eastern Australia from the Atherton Tableland in the north-east of Queensland to the Mount Lofty Ranges of South Australia. It is a sedentary and gregarious species which lives in groups or clans of up to 20 individuals occupying a permanent communal territory of some 15 hectares or more. During the nesting season the birds dissociate into breeding units consisting of a female and sometimes more than one male, each unit occupying a smaller territory within the permanent clan territory.

In its foraging, the Buff-rumped Thornbill favours the lower levels of woodland, especially the shrub layer and undergrowth. Over much of its range it also forages to a considerable extent on the ground, energetically searching fallen logs, foliage and leaf litter for insects and other small invertebrates. However, in areas where the understorey is heavily grassed, it feeds mainly in the lower branches of trees.

Breeding occurs from August to December. The nest is a loose domed structure with a side entrance, built of grass and bark fragments, lined with fur and feathers and placed in a cavity in a tree stump, fence post or fallen log; behind loose bark on a tree trunk; or in a hollow in the ground. Four eggs constitute the usual clutch.

HABITAT: mainly open eucalypt forest and woodland
LENGTH: 10–11 cm
DISTRIBUTION: more than 1 million km²
ABUNDANCE: common
STATUS: secure

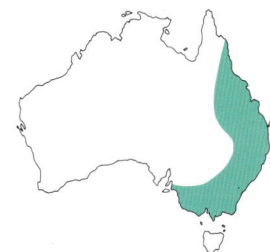

Slaty-backed Thornbill

Acanthiza robustirostris (roh'-bus-tee-ros'-tris: "strong-billed thornbush-dweller")

(J Dart)

The Slaty-backed Thornbill is widespread in the arid interior of southern Australia from the northern Simpson Desert westward to the west coast. Little is known of its behaviour but it seems sedentary and is usually encountered in pairs or, more rarely, in small parties. It forages mainly in the outer foliage of shrubs within 5 metres of the ground. Some of its calls resemble those of the Brown Thornbill but foraging birds maintain contact with a plaintive high whistled "seee".

Breeding depends on local rainfall, but occurs mainly from July to November. The nest is a domed structure with a side entrance, lined with fur and feathers and placed in a small bush. Two or three eggs form the clutch.

HABITAT: mainly Mulga scrub and similar arid scrublands
LENGTH: *c* 10 cm
DISTRIBUTION: 300,000–1 million km²
ABUNDANCE: sparse
STATUS: probably secure

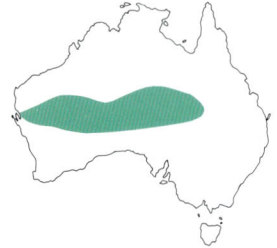

Chestnut-rumped Thornbill

Acanthiza uropygialis (ue'-roh-pi'-jee-ah'-lis: "[notable-] tail-rumped thornbush-dweller")

The Chestnut-rumped Thornbill is abundant and widespread across southern Australia west of the Great Dividing Range. It does not undertake any large-scale seasonal movements, but occupies a wide range of habitats and is locally nomadic, spending much of the year in loose wandering parties, often mixed with other thornbills. It is extremely versatile in its foraging, searching for insects and some seeds on the ground, in crevices in the bark of trees and fallen logs, and in foliage at all levels up to the crowns of trees. It forages energetically and utters a constant subdued twittering.

Breeding occurs from July to December, but the timing is influenced to some extent by local climatic conditions. The nest is a loose domed structure of grass and bark fragments, lined with fur, feathers and plant down and placed in a cavity in a tree, fallen log, fence post or similar site. Three eggs form the clutch. The young are fed by both parents, often assisted by other birds.

HABITAT: most kinds of arid scrubland, including mallee, Belah, and Mulga scrub
LENGTH: *c* 10 cm
DISTRIBUTION: more than 1 million km²
ABUNDANCE: common
STATUS: secure

(B & B Wells)

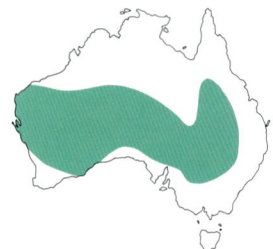

Genus *Aphelocephala*

(ah'-fel-oh-sef'-ah-lah: "sleek-head")

This genus of three species, restricted to southern Australia, strongly resembles the thornbills, but its members differ most strikingly in their stout, almost finch-like, bills and in their diet, which consists of seeds and insects in roughly equal proportions. Whitefaces are gregarious, and feed mainly on bare open ground.

Southern Whiteface

Aphelocephala leucopsis (lue-kop'-sis: "white-faced sleek-head")

The Southern Whiteface occurs across southern Australia, but in many eastern coastal areas it is somewhat less numerous than in the interior, and it has declined markedly in the vicinity of several major towns and cities. It is sedentary and gregarious, usually occurring in parties of 10 to 15 individuals: these sometimes coalesce into larger flocks of 50 or more in winter. It feeds almost entirely on bare open ground, gathering seeds and insects, but it sometimes searches the bark of dead trees and fallen logs. It is restless and very vocal, maintaining a subdued but constant twittering note when feeding.

Breeding is strongly influenced by local climatic conditions in the interior but occurs mainly from June to November. Nesting may be communal. The nest is an untidy domed structure of grass and bark fragments, lined with fur and feathers and placed in a cavity in a tree or fence post. Two to five eggs constitute the clutch.

HABITAT: open woodland and scrub
LENGTH: 11–12 cm
DISTRIBUTION: more than 1 million km²
ABUNDANCE: common
STATUS: probably secure

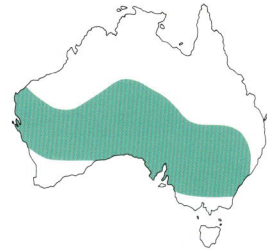

(AJ Olney)

Banded Whiteface

Aphelocephala nigricincta (nig'-ree-sink'-tah: "black-banded sleek-head")

The Banded Whiteface has an extensive distribution in the arid interior of Australia but it is little known and is nowhere common. Locally nomadic and gregarious, it feeds on seeds and insects gathered on bare open ground. Small parties often mingle with other whitefaces and, sometimes, with Crimson Chats. Breeding is heavily influenced by local rainfall but generally occurs between February and August. The nest is a globular structure built largely of small twigs, lined with feathers and with a spout-like entrance in the side, and placed in a bush within a metre or two of the ground. Two to four eggs form the clutch.

HABITAT: sparse arid scrublands, saltbush steppe, low spinifex ridges
LENGTH: 11–12 cm
DISTRIBUTION: 300,000–1 million km²
ABUNDANCE: sparse
STATUS: probably secure

(D & M Trounson)

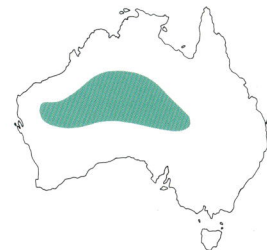

Chestnut-breasted Whiteface

Aphelocephala pectoralis (pek'-tor-ah'-lis: "[notable-] breasted sleek-head")

(L. Pedler)

Restricted to remote and inhospitable deserts on the western edge of the Lake Eyre basin in South Australia, the Chestnut-breasted Whiteface is one of the least known and least recorded of Australian birds. It is generally encountered in loose parties of up to 10 birds, often mingled with Southern Whitefaces. It feeds on insects and seeds, and is thought to be nomadic. Breeding usually commences in August; the nest is a spherical structure of twigs, lined with feathers and placed in a low bush, and four eggs constitute the clutch.

HABITAT: sparse scrublands on gibber flats
LENGTH: 11–12 cm
DISTRIBUTION: 100,000–300,000 km²
ABUNDANCE: rare
STATUS: possibly endangered

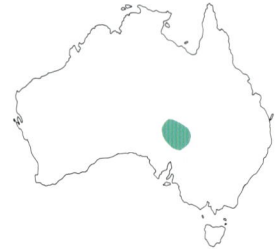

Genus *Crateroscelis*

(krah'-te-roh-skel'-is: "strong-thigh")

The genus *Crateroscelis* used to consist of three species of small terrestrial birds that inhabit New Guinea rainforests, each occupying a discrete level at lower, middle and high altitudes respectively. Formerly placed in the monotypic genus *Oreoscopus*, the Fernwren of north-eastern Queensland seems to be linked with these New Guinea birds and it is now commonly included in *Crateroscelis*.

Australian Fernwren

Crateroscelis gutturalis (gut'-er-ah'-lis: "throated strong-thigh")

The Australian Fernwren is restricted to highland forests above about 600 metres in north-eastern Queensland from about Cooktown to Mount Spec. It lives in pairs which forage quietly and unobtrusively over and through dense leaf litter for insects and other small invertebrates. It sometimes follows Chowchillas, Scrubfowls and other larger birds that rake over leaf litter. Males announce their territories by singing a simple whistled song from low vines and shrubs.

Breeding occurs from September to December. The nest is a domed structure of fern rootlets and tendrils, with a hooded side entrance, hidden under a log or in an earth bank. Two eggs constitute the clutch.

HABITAT: highland rainforest
LENGTH: 12–13 cm
DISTRIBUTION: 30,000–100,000 km²
ABUNDANCE: sparse to common
STATUS: vulnerable

(H & J Beste)

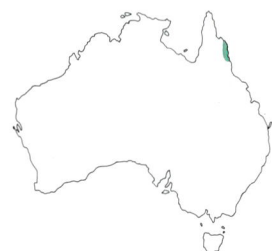

Genus *Dasyornis*

(daz'-ee-or'-nis: "hairy-bird")

The three species that constitute this genus are all secretive, sedentary birds with restricted and possibly relict distributions in south-western, southern and south-eastern Australia. All live in dense undergrowth, typically in heathland, and are difficult to observe. The sexes are similar, and there is no seasonal variation in plumage.

Eastern Bristlebird

Dasyornis brachypterus (brak'-ip-te'-rus: "short-winged hairy-bird")

(J Gray)

Breeding occurs from August to January. Well hidden in a grass tussock or similar vegetation, the nest is a domed structure of grass, with a side entrance. Two eggs form the clutch.

HABITAT: dense vegetation in coastal or highland heath
LENGTH: 20–22 cm
DISTRIBUTION: 30,000–100,000 km²
ABUNDANCE: rare to sparse
STATUS: vulnerable

The Eastern Bristlebird has a broken distribution extending, in coastal and near-coastal districts, from extreme south-eastern Queensland to far eastern Victoria. It has declined markedly since European settlement, probably as a result of clearing for settlement, burning and overgrazing.

It is sedentary, secretive and quiet except when breeding, when males announce territories (of about 1 hectare) by singing loudly and persistently. The diet consists of insects and fruit, taken mainly on the ground under dense cover, or in low undergrowth.

Rufous Bristlebird

Dasyornis broadbenti (brawd'-bent-ee; "Broadbent's-hairy-bird", after K. Broadbent, Australian naturalist)

The Rufous Bristlebird is still common in coastal south-eastern Australia from the Otway Ranges of Victoria westward to Lake Alexandrina, but a population in south-western Australia is almost certainly extinct, not having been recorded since 1906. Sedentary, secretive and difficult to observe, it resembles other bristlebirds in behaviour and feeds on insects, seeds and berries. Breeding occurs from September to December; the nest resembles that of other bristlebirds, and two eggs form the clutch.

HABITAT: coastal dune scrub
LENGTH: 23–26 cm
DISTRIBUTION: 30,000–100,000 km²
ABUNDANCE: very sparse
STATUS: endangered

(L Pedler)

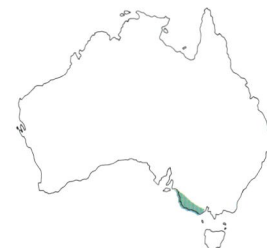

337

Western Bristlebird

Dasyornis longirostris (lon'-jee-ros'-tris: "long-billed hairy-bird")

The Western Bristlebird is now very rare, with an extremely restricted distribution in the vicinity of Albany, Western Australia: formerly it occurred northward around the coast to Perth. It is sedentary, living in permanent pairs which maintain territories of about 6 hectares. Males sing persistently all year but the birds are otherwise quiet, secretive, and very difficult to observe, seldom leaving the shelter of dense cover. The diet consists of insects, seeds and berries, gathered mainly on the ground.

Breeding occurs in August and September. The nest is a rough dome of grass, with a side entrance. Two eggs form the clutch.

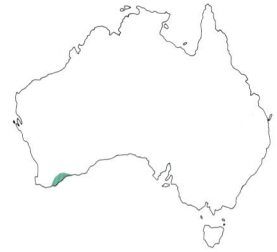

HABITAT: coastal heath
LENGTH: 18–20 cm
DISTRIBUTION: 10,000–30,000 km²
ABUNDANCE: rare
STATUS: endangered

Genus *Gerygone*

(je'-ri-goh'-nay: "sound-born", presumably a suggestion that the sound of these birds is their birthright or their reason for existence)

Notable in particular for their light, elfin and extremely attractive songs, the 19 or so members of this genus are distributed from Malaya to the islands of the south-west Pacific, including Australia and New Zealand. Most are relatively subdued in colour and pattern (often white or yellow below and grey or dull green above). All are arboreal and insectivorous, living in pairs and foraging actively but unobtrusively in the upper foliage of trees.

Green-backed Warbler

Gerygone chloronota (klor'-oh-noh'-tah: "green-backed sound-born")

The Green-backed Warbler is common in its restricted habitat in the Kimberley and the Top End: it is also widespread in New Guinea. It is sedentary and rather solitary and resembles other members of the genus in general behaviour, spending much of its time gleaning insects in the upper foliage of trees.

It breeds sporadically throughout the year but mainly between October and April. The nest is a compact sphere with a pendent tail and a hooded side entrance. Two or three eggs form the clutch; incubation takes about 12 days, and the young fledge at about 10 days.

HABITAT: monsoon rainforest; mangroves
LENGTH: 9–10 cm
DISTRIBUTION: 100,000–300,000 km²
ABUNDANCE: common
STATUS: secure

(I Morris)

Western Warbler

Gerygone fusca (fus'-kah: "dusky sound-born")

The Western Warbler is the only member of its genus to inhabit arid woodlands: it is widespread across mainland Australia but its distribution is not continuous. There are three isolated populations: in the east, the centre and the south-west.

It seems to be largely sedentary except in the south-west, where it may be partly migratory. It lives in territorial and apparently permanent pairs which glean insects in the upper and outer foliage of trees. It is generally quiet and unobtrusive but breeding males have a sweet, wavering song.

Breeding occurs mainly from August to January. The nest is a compact dome with a tail and a hooded side entrance suspended in leafy branches several metres from the ground: it is constructed of grass, bark fragments and fibres, matted with cobweb and lined with feathers. Two or three eggs constitute the clutch, incubation takes about 12 days, and the young fledge at about 10 days. Nest construction and incubation are largely by the female, but both parents feed the young.

HABITAT: mainly temperate and tropical eucalypt woodland
LENGTH: 10–11 cm
DISTRIBUTION: more than 1 million km²
ABUNDANCE: sparse to common
STATUS: secure

(*WR Taylor*)

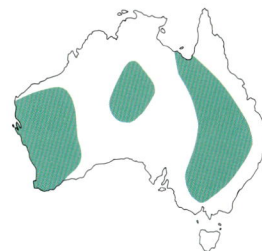

Mangrove Warbler

Gerygone laevigaster (lee'-vee-gah'-ster: "light-bellied sound-born")

The Mangrove Warbler is common in southern New Guinea and along the coast of northern and eastern Australia from the vicinity of Broome in Western Australia to Newcastle in New South Wales. Although mainly sedentary, it is spreading steadily southwards and has been recorded in the Sydney region. Living in pairs, it forages for insects in foliage at almost any height, maintaining contact with soft chattering notes. It also has a sweet wavering song.

Breeding occurs from September to April. The nest resembles that of other members of the genus but is more compact and neatly made than some: it has a pendent tail up to 10 centimetres long, and a hooded side entrance. The clutch consists of two or three eggs, which hatch in about 12 days; the young fledge at about 10 days.

HABITAT: mangroves; coastal scrub and woodland; urban parks and gardens
LENGTH: 10–11 cm
DISTRIBUTION: 100,000–300,000 km²
ABUNDANCE: sparse to common
STATUS: secure

(*J Purnell*)

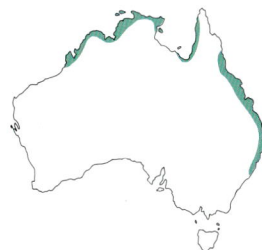

339

Large-billed Warbler

Gerygone magnirostris (mag'-nee-ros'-tris: "great-beaked sound-born")

The Large-billed Warbler occurs in New Guinea and along the coasts of northern and north-eastern Australia. It is sedentary, living in pairs which actively but unobtrusively glean insects from the inner foliage of trees (resembling others of its genus in general behaviour). It sometimes joins mixed feeding parties of other birds. Breeding occurs from September to April. Assisted only casually by the male, the female builds a long untidy nest suspended over water, in which she lays and incubates a clutch of two or three eggs for about 12 to 14 days. The young, which are fed by both parents, fledge at about 10 to 11 days.

HABITAT: mangroves; rainforest bordering rivers and streams; melaleuca swamps
LENGTH: 10–11 cm
DISTRIBUTION: 100,000–300,000 km^2
ABUNDANCE: common
STATUS: secure

(J Estbergs)

Brown Warbler

Gerygone mouki (mue'-ki: "mouki sound-born", meaning unknown, said to be Aboriginal Australian)

The Brown Warbler is one of the commonest of birds in the rainforests of north-eastern Queensland on the Atherton and Eungella Tablelands, and in south-eastern Australia from the vicinity of Gympie to eastern Victoria. It is sedentary and usually encountered in small parties which maintain an incessant light twittering as they forage actively through the foliage in the middle to upper levels of the canopy of trees.

Breeding occurs from September to February. The nest is a slender oval structure of plant fibres bound with cobweb and lined with plant down, with a hooded side entrance and a pendent tail about 10 centimetres long; it is suspended by the roof from twigs in outer foliage up to 15 metres from the ground. Two or three eggs constitute the clutch. The female builds and apparently incubates largely unaided, but both parents feed the young. Incubation takes 16 to 19 days and the young fledge at about 15 to 16 days.

HABITAT: rainforest
LENGTH: 10–11 cm
DISTRIBUTION: 100,000–300,000 km^2
ABUNDANCE: common
STATUS: secure

(N Chaffer)

White-throated Warbler

Gerygone olivacea (ol'-iv-ah'-say-ah: "olive sound-born")

(P Green)

The White-throated Warbler is common in south-eastern New Guinea and in coastal and near-coastal regions of northern and eastern Australia from about Broome in Western Australia to extreme south-eastern South Australia. Northern populations are sedentary but those in the east are migratory, arriving back on their breeding grounds in the south-east early in September. It lives in pairs or sometimes family parties, gleaning insects high in the canopy of trees. It is generally rather quiet and unobtrusive but breeding males persistently utter an attractive rippling song.

Breeding occurs mainly from September to January. Constructed by the female with only casual assistance from the male, the nest is a compact, elongated oval structure with a pendent tail and a hooded side entrance, suspended in the outer foliage of a small tree or sapling several metres from the ground. Construction may take up to two weeks. The clutch consists of two or three eggs which, incubated by the female, take 12 days to hatch. Fed by both parents, the young fledge at about 10 days.

HABITAT: tall open eucalypt woodland
LENGTH: 10–11 cm
DISTRIBUTION: more than 1 million km²
ABUNDANCE: sparse to common
STATUS: secure

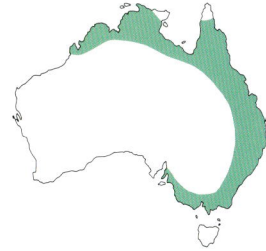

Fairy Warbler

Gerygone palpebrosa (pahl'-pe-broh'-sah: "eye-lidded sound-born")

The Fairy Warbler is common in New Guinea and in north-eastern Australia from Cape York to Fraser Island. It is sedentary and, although often encountered alone or in pairs, it is more gregarious than other members of its genus, frequently occurring in small parties. In other aspects of its behaviour it much resembles those species but it is unusual in that the male differs in plumage from the female, at least in the population on Cape York Peninsula: the male has a blackish face and throat and a white chin whereas the female has a plain whitish throat.

It breeds sporadically throughout the year but mainly from September to March. The nest is a compact sphere with a pendent tail and a hooded side entrance. The clutch consists of two or three eggs, which hatch in about 12 days. The young fledge at about 10 days.

HABITAT: mainly lowland rainforests and their margins; sometimes mangroves
LENGTH: 10–11 cm
DISTRIBUTION: 100,000–300,000 km²
ABUNDANCE: common
STATUS: secure

(D & M Trounson)

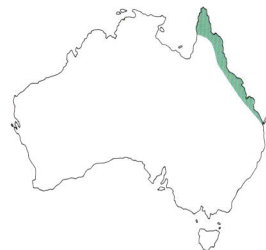

341

Dusky Warbler

Gergone tenebrosa   (ten'-eb-roh'-sah: "dark sound-born")

Least-known of Australian warblers, the Dusky Warbler is restricted to mangroves along the north-west coast from Shark Bay to King Sound. It is usually encountered in small groups, gleaning insects in the upper foliage of trees, but it occasionally forages on the ground. Breeding occurs from September to March.

HABITAT: mangroves
LENGTH: *c* 12 cm
DISTRIBUTION: 100,000–300,000 km²
ABUNDANCE: sparse to common
STATUS: probably secure

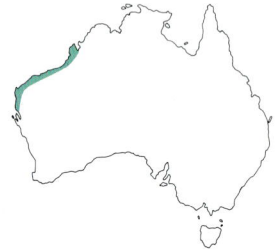

(G Chapman)

Genus Origma

(o-rig'-mah: "cave [-bird]")

The characteristics of the genus are essentially those of the single species.

Rock Warbler

Origma solitaria   (sol'-it-ar'-ee-ah: "solitary cave-bird")

The Rock Warbler (or Origma) occurs in Hawkesbury Sandstone and associated limestone areas of east-central New South Wales from about Scone to Bermagui. It is generally sedentary but some birds may wander locally. It lives in pairs, searching for insects and seeds on rock faces, cliffs and the rocky banks of streams.

Breeding occurs from August to December. The nest is a globular structure of matted rootlets, grass, and bark fragments, lined with fur, feathers and plant down, and suspended by cobweb from the roof of a dark cave. Both the nest and its site are unique among Australian birds. Three eggs constitute the usual clutch.

HABITAT: mainly the rocky margins of streams running through sandstone or limestone formations
LENGTH: 13–14 cm
DISTRIBUTION: 30,000–100,000 km²
ABUNDANCE: sparse to common
STATUS: vulnerable

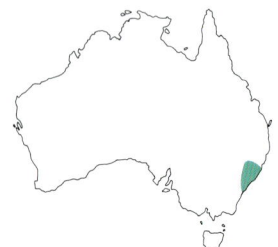

(G Weber)

Genus *Pycnoptilus*

(pik'-nop-til'-us: "thick-feather")

The characteristics of the genus are essentially those of the single species.

Pilotbird

Pycnoptilus floccosus (flok-oh'-sus: "downy thick-feather")

The Pilotbird is a common and sedentary inhabitant of dense forests in south-eastern Australia from the Blue Mountains west to Canberra and south to the Dandenong Ranges of Victoria, living in pairs which maintain permanent territories. It is generally quiet and unobtrusive but not especially difficult to observe. Foraging birds methodically search the ground and leaf litter for insects, worms and other ground-living invertebrates, and they sometimes follow lyrebirds, taking advantage of the newly raked-over litter left by the larger birds. The flight is low, feeble, and seldom sustained for more than a few metres. The common call is a piercing whistle often transliterated as "guinea-a-week".

Breeding occurs from August to January. The nest is a rough dome of grass, lined with feathers and with a side entrance. One or two eggs form the clutch. The female builds and incubates (for about 20 to 22 days) unaided, but both parents feed the young, which fledge in about 15 to 18 days.

HABITAT: mainly dense eucalypt forest
LENGTH: *c* 17 cm
DISTRIBUTION: 100,000–300,000 km^2
ABUNDANCE: common
STATUS: probably secure

(JD Waterhouse)

Genus *Sericornis*

(se'-rik-or'-nis: "silk-bird")

In terms of species, this genus is almost equally divided between Australia and New Guinea. Its internal relationships are debatable and some aberrant Australian taxa (such as those often placed in the genera *Hylacola*, *Chthonicola* and *Calamanthus*) are now often incorporated within the genus. Its members are generally dark brown or dull olive in colour, plain in pattern and unremarkable in proportions. Most are sedentary and typically live in small groups. Most are energetic, inquisitive but furtive birds of forest undergrowth, and they utter a variety of harsh but subdued chattering and scolding notes upon being disturbed. All are insectivorous.

Tropical Scrubwren

Sericornis beccarii (bek-ah'-ree-ee: "Beccari's silk-bird', after O. Beccari, Italian explorer–naturalist)

The Tropical Scrubwren is a little-known species inhabiting New Guinea and the rainforests of northern Cape York Peninsula. It is closely related to the Large-billed Scrubwren and occupies a similar niche, gleaning insects in the foliage of lower and middle storeys of forest; the two forms may hybridise where their ranges meet in the vicinity of Cooktown. The breeding season extends from October to December.

HABITAT: rainforest
LENGTH: 10–11 cm
DISTRIBUTION: 30,000–100,000 km²
ABUNDANCE: common
STATUS: secure

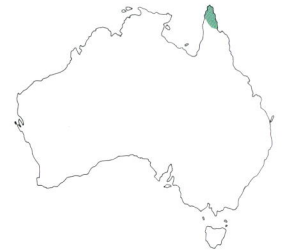

(D & M Trounson)

Redthroat

Sericornis brunneus (broon-ay'-us: "brownish silk-bird")

The Redthroat inhabits the arid interior of Australia, where it is common and widespread across the southern part of the continent. Its distribution has, however, contracted since European settlement, presumably as a result of clearing and overgrazing. It is sedentary and lives mainly in territorial pairs, but small parties or even small flocks may congregate under some circumstances. It forages mainly on the ground, gathering seeds and small insects. It is an accomplished songster with a repertoire which includes mimicry of other birds. The sexes are similar in appearance, except that the female lacks the reddish throat patch of the male.

Breeding occurs from August to December. The nest is a domed structure of grass and other vegetation, lined with feathers and with a side entrance, hidden in a bush or tree within a metre of the ground. Three or four eggs form the clutch.

HABITAT: acacia scrubs and saltbush plains
LENGTH: 11–12 cm
DISTRIBUTION: more than 1 million km²
ABUNDANCE: sparse
STATUS: vulnerable

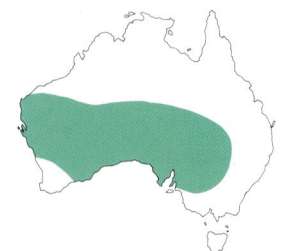

(WJ Labbett)

Rufous Fieldwren

Sericornis campestris (kam-pes'-tris: "field silk-bird")

The Rufous Fieldwren inhabits various arid treeless environments in a patchy and local distribution extending across southern Australia from central New South Wales to the west coast. It is sedentary, living mainly in pairs which maintain permanent territories. It is generally quiet, unobtrusive and rather timid. Foraging birds seldom fly and scurry rapidly for cover if disturbed. Breeding males announce their territories by means of a light warbling song delivered from a conspicuous perch in the top of a bush. Breeding occurs mainly from July to November. The nest resembles that of other *Sericornis* species, and three or occasionally four eggs constitute the clutch.

HABITAT: arid heathland, saltbush and bluebush steppe, samphire flats
LENGTH: 12–13 cm
DISTRIBUTION: more than 1 million km²
ABUNDANCE: sparse to common
STATUS: probably secure

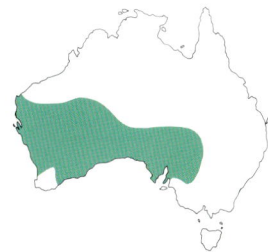

(E McNamara)

Shy Hylacola

Sericornis cauta (kaw'-tah: "shy silk-bird")

The Shy Hylacola replaces the closely related Chestnut-rumped Hylacola in mallee heathlands of southern Australia. The two species approach each other very closely in the vicinity of Bendigo in Victoria, but hybridisation is unknown and the two are kept apart by rigid habitat preferences. They are difficult to distinguish in the field but the Shy Hylacola has a small white patch in the wing, absent from the Chestnut-rumped Hylacola. Breeding occurs from August to November, and two or three eggs constitute the clutch.

HABITAT: dense whipstick mallee; coastal dune scrub; sandplain heath
LENGTH: 13–14 cm
DISTRIBUTION: 300,000–1 million km²
ABUNDANCE: sparse
STATUS: vulnerable

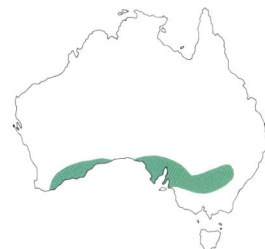

(WJ Labbett)

Yellow-throated Scrubwren

Sericornis citreogularis (sit'-ray-oh-gue-lar'-is: "yellow-throated silk-bird")

The Yellow-throated Scrubwren occurs in coastal and highland forests in two widely separated regions of eastern Australia: from Cooktown to Mount Spec in north-eastern Queensland, and from about Gympie south to southern New South Wales. It is sedentary, living in permanent territorial pairs. It forages almost exclusively on the ground, feeding on insects and other small invertebrates, picking them mainly from the surface of the litter rather than raking it aside. The common contact call is a sharp "tik", but both sexes also utter a rich and complex song that often includes mimicry of other birds.

Breeding occurs from August to March, during which period several broods are often raised in succession. The nest is very variable, but is usually an almost shapeless mass of stems, rootlets and other local vegetation, with a side entrance and, often, several interior egg chambers: usually each is lined and renovated in turn for each successive brood. In a typical site the nest is suspended from twigs and branches hanging low over a stream, in which situation the whole assembly much resembles a mass of flood debris. Both parents build, incubate and care for the young. Usually three eggs form the clutch; incubation takes about 21 days, and the young fledge at 21 days.

HABITAT: rainforest
LENGTH: 12–14 cm
DISTRIBUTION: 100,000–300,000 km
ABUNDANCE: common
STATUS: probably secure

(G Weber)

White-browed Scrubwren

Sericornis frontalis (fron-tah'-lis: "[notable-] forehead silk-bird")

The White-browed Scrubwren is abundant and widespread in eastern and southern Australia from the vicinity of Cairns in the east to Shark Bay in the west. It is sedentary, living in pairs or small parties which occupy permanent territories up to about 5 hectares in extent. It is active and inquisitive but reluctant to leave the shelter of dense cover. It utters a variety of harsh churring, twittering and scolding notes and has a simple, clear whistled song. Most foraging is on the ground or in undergrowth, the diet including a variety of small insects.

Breeding occurs from July to December or January. The nest is a loose, rough domed structure with a side entrance. Built of grass, bark fragments and other vegetation and lined with feathers, it is well hidden on or near the ground in dense vegetation. Two or three eggs form the clutch. The female apparently builds and incubates unaided, but all members of the group participate in rearing the young.

HABITAT: shrubbery and undergrowth in almost any wooded environment
LENGTH: 12–14 cm
DISTRIBUTION: more than 1 million km²
ABUNDANCE: common
STATUS: secure

(JD Waterhouse)

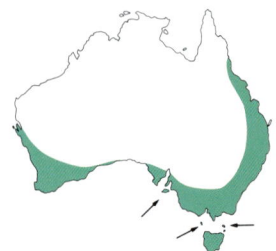

Striated Fieldwren

Sericornis fuliginosus (fool'-i-jin-oh'-sus: "sooty silk-bird")

The Striated Fieldwren (or Calamanthus) replaces the Rufous Fieldwren in south-eastern Australia from south coastal New South Wales to extreme south-eastern South Australia; it is especially common in Tasmania. The two species are very closely related, differing most strikingly in their habitat preferences; the Striated Fieldwren is also somewhat larger and more conspicuously streaked.

Breeding occurs from August to December, and three or four eggs constitute the clutch.

HABITAT: tussock grasslands and coastal swampy heath
LENGTH: 13–14 cm
DISTRIBUTION: more than 1 million km²
ABUNDANCE: sparse to common
STATUS: probably secure

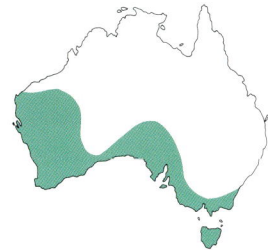

(JR Napier)

Atherton Scrubwren

Sericornis keri (ker'-ee: "[Bellenden] Ker silk-bird")

The Atherton Scrubwren is so similar to the Large-billed Scrubwren in appearance and habits that it was not distinguished as a species until 1964. The two species coexist but the Atherton Scrubwren is found only in highland rainforests above an altitude of 650 metres in north-eastern Queensland, from the Windsor Tableland to Mount Spec. The two differ most strikingly in behaviour: the Atherton Scrubwren is solitary and spends virtually all of its time on the ground, where it feeds on insects and other small invertebrates. Breeding occurs from August to December, and two eggs constitute the clutch.

HABITAT: highland rainforest
LENGTH: 12–13 cm
DISTRIBUTION: 30,000–100,000 km²
ABUNDANCE: sparse
STATUS: probably secure

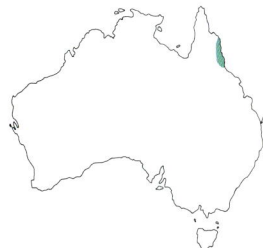

(T & P Gardner)

Large-billed Scrubwren

Sericornis magnirostris (mag'-nee-ros'-tris: "great-beaked silk-bird")

(DH Jeans)

The Large-billed Scrubwren occurs in coastal and near-coastal districts of eastern Australia from the vicinity of Cooktown in Queensland to eastern Victoria. On the Atherton Tableland in the north-east of Queensland it coexists with the related and closely similar Atherton Scrubwren, which is solitary and terrestrial. The Large-billed Scrub-wren lives in small parties of up to about eight birds, which break up into territorial pairs to breed. Unlike other scrubwrens it spends little time in the undergrowth and seldom descends to the ground; instead it forages mainly in the foliage of the middle and lower storeys of forest, feeding on insects and other small invertebrates. It often joins mixed foraging parties of other birds.

The breeding season extends from July to January. In some areas the Large-billed Scrubwren builds its own nest—a dome of plant stems and grass lined with fur and feathers—but in other areas it regularly uses the abandoned nests of the Yellow-throated Scrubwren. Three or four eggs constitute the clutch.

HABITAT: rainforest
LENGTH: 10–12 cm
DISTRIBUTION: 100,000–300,000 km²
ABUNDANCE: common
STATUS: probably secure

Scrubtit

Sericornis magnus (mag'-nus: "great silk-bird")

The Scrubtit is a sedentary species restricted to dense forests of Tasmania, including King Island in Bass Strait. It lives in territorial pairs during the breeding season but may wander locally in winter. Most of its time is spent in the lower levels of forest, where it tends to concentrate its foraging on the trunks and stems of trees and saplings rather than the foliage, methodically probing cracks and crevices in the bark for insects and other small invertebrates.

Breeding occurs from September to January. The nest is a loose, bulky domed structure of grass and other vegetation, lined with feathers and plant down, with an entrance in the side, and hidden in dense vegetation on or near the ground. Three or four eggs constitute the clutch.

HABITAT: dense forest; subalpine scrub; melaleuca scrub (on King Island)
LENGTH: 11–12 cm
DISTRIBUTION: 30,000–100,000 km²
ABUNDANCE: common
STATUS: secure

(TA Waite)

Chestnut-rumped Hylacola

Sericornis pyrrhopygius (pie-roh-pi'-jee-us: "fire-rumped silk-bird")

The Chestnut-rumped Hylacola occurs in south-eastern Australia from south-eastern Queensland to Kangaroo Island and the Flinders Ranges of South Australia. It is sedentary and lives mainly in pairs, or occasionally in small parties. Like other scrubwrens, it forages mainly on the ground for insects and small seeds. It is active but very secretive, and difficult to observe. Males announce their territories with one of the most attractive of Australian bird songs—a varied, sustained performance of silvery warbles and trills interwoven with mimicry of other birds.

The breeding season extends from June to November. The nest is a compact domed structure of grass and other vegetation, lined with fur and feathers and with a spout-shaped side entrance, and hidden in dense vegetation on or near the ground. Incubated by the female, the usual clutch consists of three eggs.

HABITAT: dense heathlands and shrubberies
LENGTH: 13–14 cm
DISTRIBUTION: 300,000–1 million km²
ABUNDANCE: sparse to common
STATUS: probably secure

(IR McCann)

Speckled Warbler

Sericornis sagittatus (sa'-jit-ah'-tus: "arrowed silk-bird")

The Speckled Warbler occurs in south-eastern Australia from the vicinity of Rockhampton to central Victoria, extending to the western slopes of the Great Dividing Range in many areas. It is sedentary, territorial and insectivorous, living in pairs or trios, or small family parties. Generally quiet and unobtrusive, it forages mainly on the ground, seldom wandering far from the shelter of bushes and shrubs. It sometimes joins mixed feeding parties of other birds. Its calls include a soft churr in alarm, and a quiet warbling song; it sometimes mimics other birds.

Breeding occurs from August to January. The nest resembles that of other scrubwrens, and three or four eggs constitute the clutch. Both parents cooperate in rearing the young.

HABITAT: shrubs and undergrowth in open eucalypt woodland
LENGTH: 12–13 cm
DISTRIBUTION: 300,000–1 million km²
ABUNDANCE: common
STATUS: probably secure

(WR Moore)

Genus Smicrornis

(smie-kror'-nis: "small-bird")

The characteristics of the genus are those of the single species.

Weebill

Smicrornis brevirostris (brev'-ee-ros'-tris: "short-billed small-bird")

This tiny, sedentary bird is common almost throughout mainland Australia. Usually encountered in loose flocks of up to about 10 individuals, it forages energetically but methodically through the outer foliage of trees, sometimes joining mixed feeding flocks of other birds. It has a frequent and characteristic trick of hovering briefly before outer sprays of foliage, snapping up insects flushed from the leaves (a habit occasionally shared by some thornbills). Feeding birds maintain contact by means of sharp, buzzy notes, frequently uttered; the Weebill also has a simple but loud and vivacious song.

The breeding season varies with locality but generally begins in August in the south-east; October in the north. Usually two broods are raised in succession. The nest is a compact sphere of grass and other vegetation, lined with feathers and closely bound with cobweb, with a spout-like side entrance near the top. It is usually suspended from twigs in the outer foliage of a shrub or sapling up to 10 metres from the ground. Two eggs constitute the usual clutch. The female builds and incubates (for about 12 days) almost unaided, but both parents feed the young, which fledge in about 10 days.

HABITAT: most kinds of temperate and tropical eucalypt woodland
LENGTH: 8–9 cm
DISTRIBUTION: more than 1 million km²
ABUNDANCE: common
STATUS: secure

(B & B Wells)

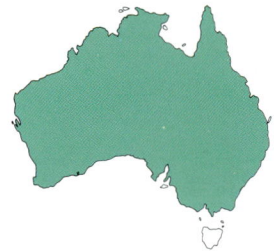

Family NEOSITTIDAE

(nee'-oh-sit'-id-ee: "*Neositta*-family", after a genus of nuthatches)

This family consists of two or three species of gregarious, arboreal birds grouped in a single genus, *Daphoenositta*. The very distinct Pink-faced Sittella inhabits highland cloud and moss forests at 2000 to 3500 metres altitude in New Guinea. The remaining two forms may or may not be conspecific: the Papuan Sittella occurs in highland forest in New Guinea, and the Varied Sittella ranges widely across mainland Australia, favouring eucalypt and acacia woodland.

Extremely acrobatic, foraging sittellas explore upper branches of trees for bark-haunting insects in a distinctive, rapid, spiralling pattern of progress. Breeding is communal. The sexes differ slightly in appearance, and immatures resemble adults.

Genus *Daphoenositta*

(dah-fee'-noh-sit'-ah: "gory-nuthatch")

The characteristics of the genus are those of the family.

Varied Sittella

Daphoenositta chrysoptera (krie-sop'-te-rah: "golden-winged gory-nuthatch")

The passage of a flock of sittellas imparts an unmistakable air of bustle and urgency to the woodland scene. Common and widespread, sittellas live in groups (generally of about 10 or 12 birds) that together roam over an extensive home range. Rarely descending to the ground, they scuttle nimbly about on the bark of upper twigs and branches of trees, seldom staying long in one before flying to the next. They feed on insects and small spiders hidden in the bark, and have been seen using twigs as tools to facilitate the extraction of insect larvae from crevices.

The Varied Sittella occurs across mainland Australia in several populations so distinct that they were formerly regarded as species. However, they interbreed so widely where their respective ranges abut or overlap that this notion cannot be maintained. Concise description is difficult but, in general, populations in the arid interior and far north have large white flashes in the wings whereas birds in the more humid east and south-east have dull orange wing flashes; the head may be variously black, grey, white or streaked, according to locality. Several widely used common names reflect this variability: Orange-winged Sittella, White-headed Sittella, Black-capped Sittella, Striated Sittella, White-winged Sittella.

Breeding occurs mostly from August to January. Only the female incubates but the male, assisted by others of the group, contributes to nest construction and the care of the young. The nest is small, neat and cup-shaped, constructed of bark fragments and spider-silk and cocoons; it is usually placed in an upright fork at a considerable height. Three eggs constitute the usual clutch; these hatch in about 14 days, and the young fledge at about 18 days.

HABITAT: mainly eucalypt forest and woodland; arid scrubland; rarely rainforest
LENGTH: *c* 13 cm
DISTRIBUTION: more than 1 million km²
ABUNDANCE: common
STATUS: secure

(JD Waterhouse)

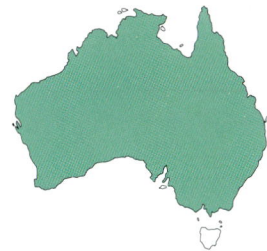

Family CLIMACTERIDAE

(klee'-mak-te'-rid-ee: *Climacteris*-family", after a genus of Australian treecreepers)

In forests and woodlands around the world there are birds that spend most of their time on tree trunks, hitching their way steadily upwards from the base in short spasmodic hops, searching for insects hidden in the bark. Aptly known as treecreepers, most are quite small, mainly brown in colour, and streaked below. Six Australian species conform to this description but they (together with a close relative in New Guinea) are unrelated to the superficially similar birds elsewhere in the world: together they constitute the family Climacteridae. Their true affinities are obscure: the family has been associated with the thornbills and honeyeaters but current understanding links it more closely with the bowerbirds, lyrebirds and scrub-birds.

Sittellas (Neosittidae) also forage on the bark of trees, but they are active and acrobatic, and generally favour the topmost twigs and branches. Australian treecreepers tend to focus their foraging activities on the trunk and major limbs within about 30 metres of the ground; most species visit the ground at least occasionally to feed, and some do so persistently.

Australian treecreepers feed mainly on ants, but other insects, small spiders, nectar and mellitose are also eaten. All are largely sedentary and most live in groups that cooperate in nesting activities. The White-throated Treecreeper is a conspicuous exception in this respect, and is often isolated in its own genus, *Cormobates* (all other species being included in the genus *Climacteris*).

Genus *Climacteris*

(klee'-mak-te'-ris: "staircase [-bird]")

The characteristics of this genus are essentially those of the family, the distinction between *Climacteris* and *Cormobates* resting largely on differences in behaviour (*Cormobates* is solitary, *Climacteris* markedly communal).

White-browed Treecreeper

Climacteris affinis (ah-fin'-is: "related staircase-bird")

An uncommon, solitary and quiet inhabitant of the southern arid zone, the White-browed Treecreeper resembles the White-throated Treecreeper of humid eastern forests in much of its behaviour: it lives in pairs that nevertheless see little of each other; it occupies permanent territories; and it is largely arboreal, spending less than a quarter of its time on the ground. Its distribution widely overlaps those of the Rufous and Brown Treecreeper, but it occupies very different habitat, favouring denser thickets of belah and mulga. Like other treecreepers, it feeds largely on ants.

Often two broods are raised in a breeding season lasting from August to November. The female alone incubates, but little is known of other aspects of nesting behaviour. The nest is in a tree cavity, usually quite low; the clutch is usually two or three eggs, and the young fledge at about 23 days.

HABITAT: temperate arid woodlands, especially of casuarina, acacia (Mulga), callitris and eucalypts
LENGTH: *c* 15 cm
DISTRIBUTION: 300,000–1 million km^2
ABUNDANCE: sparse
STATUS: probably secure

(*RW Fidler*)

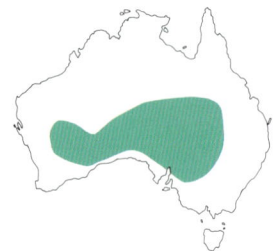

Red-browed Treecreeper

Climacteris erythrops (e-rith'-rops: "red-faced staircase-bird")

A rather uncommon species, the Red-browed Treecreeper inhabits eastern Australia from the vicinity of Melbourne in Victoria to near Gympie in Queensland, mainly along the coast and associated highlands. It frequently occurs alongside the White-throated Treecreeper but, being somewhat less flexible in its choice of habitat, it tends more to be restricted to cool dense eucalypt woodland. Even where the two species occupy overlapping territories, direct competition is avoided by the Red-browed Treecreeper's preference for smooth-barked eucalypts (the White-throated Treecreeper favours rough-barked trees), where it spends much time rummaging in tangles of partly shed bark. It also often descends to the ground to forage on fallen logs (the White-throated Treecreeper is more rigidly arboreal). Both species feed largely on ants.

The Red-browed Treecreeper lives in groups that occupy permanent territories; immature males remain with their parents for several years, although young females usually leave during their first year. The female incubates and initiates nest construction but she is otherwise assisted by her mate and others of the group. Built of fur and bark fragments, the nest is situated in a tree cavity. The usual clutch consists of two eggs, which hatch in 18 days; the young fledge at about 26 days.

HABITAT: mainly moist cool temperate eucalypt forest
LENGTH: *c* 16 cm
DISTRIBUTION: 300,000–1 million km²
ABUNDANCE: sparse to common
STATUS: probably secure

(J Purnell)

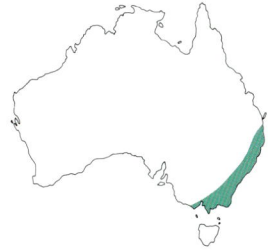

Black-tailed Treecreeper

Climacteris melanura (mel'-ahn-ue'-rah: "black-tailed staircase-bird")

This tropical treecreeper inhabits northern Australia from the Pilbara to the vicinity of Mount Isa, Queensland, in a distribution interrupted by the Great Sandy Desert. It is a common, conspicuous and easily identified species, but its habits have been little studied. Like the Brown Treecreeper it spends much time on the ground, and it lives in groups that occupy permanent territories. It feeds mainly on ants. The female resembles the male except in having a white throat.

Breeding has been recorded from September to January. The nest is in a tree cavity, and two or three eggs make up the usual clutch.

HABITAT: tropical eucalypt woodland
LENGTH: *c* 20 cm
DISTRIBUTION: 300,000–1 million km²
ABUNDANCE: sparse to common
STATUS: probably secure

(J Purnell)

353

Brown Treecreeper

Climacteris picumnus (pik-um'-nus: "woodpecker staircase-bird")

The distribution of the Brown Treecreeper extends across most of eastern mainland Australia from Spencer Gulf in the south to Cape York and the Gulf of Carpentaria. Northern birds are very much darker than southern birds and were formerly distinguished as a separate species, the Black Treecreeper.

By far the least arboreal of treecreepers, it spends at least half of its time on the ground, foraging in leaf litter and on fallen limbs and trunks. It is most numerous in eucalypt woodland but extremely flexible in its choice of habitat. It often occurs together with other species of treecreeper but prefers more open environments than are favoured by most of these and has therefore benefited from land clearing since European settlement. In winter it sometimes forms the nucleus of mixed-species foraging parties. The diet consists mainly of ants and other insects.

It is the largest of treecreepers, and easy to identify. Immature birds resemble adults, and the sexes are only subtly different in appearance: the female is distinguished by a small area of rufous streaking on the centre of the breast. The characteristic note is a loud, abrupt "pink", uttered alone or in lengthy series.

The Brown Treecreeper is sedentary, and lives in groups of about five or six birds that jointly defend and occupy permanent territories of about 5 to 10 hectares. Such groups involve a mated pair and several immature birds, almost always male.

Breeding generally occurs from June to January, and two broods may be raised in succession. Breeding is communal: the female alone incubates but she is assisted by the group in nest construction and care of the young. Territoriality is relaxed while young are fed, and it is not uncommon for helpers at one nest to also contribute substantially to the care of young at nests in adjacent territories. The nest, situated in a tree cavity at almost any height, is constructed of bark fragments, grasses, and feathers. The two or three eggs hatch in about 17 days and the young fledge in about 26 days.

HABITAT: mainly temperate to subtropical eucalypt woodland
LENGTH: *c* 19 cm
DISTRIBUTION: more than 1 million km²
ABUNDANCE: common
STATUS: secure

(G Chapman)

Rufous Treecreeper

Climacteris rufa (rue'-fah: "rufous staircase-bird")

The Rufous Treecreeper is the only treecreeper to occur in south-

(G Chapman)

western Australia. Its distribution extends eastwards to the Eyre Peninsula in South Australia but it is rare on the intervening Nullarbor Plain. It is locally common but seems sensitive to changes in land management and may be declining.

In many respects it is the south-western representative of the eastern Brown Treecreeper, resembling it in many of its habits. It is sedentary and communal (trios seem most common), and spends much time foraging on the ground, as well as on tree trunks.

Breeding occurs from June to January. The nest is similar to that of other treecreepers, and two or

three eggs constitute the usual clutch.

HABITAT: temperate forest, woodland and scrubland, including tall mallee
LENGTH: *c* 18 cm
DISTRIBUTION: 300,000–1 million km²
ABUNDANCE: common
STATUS: secure

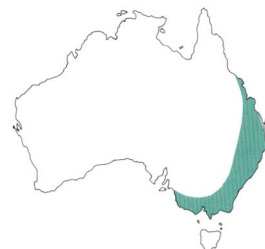

Genus *Cormobates*

(kor'-moh-bah'-tayz: "stump-climber")

The characteristics of the genus are those of the single species.

White-throated Treecreeper

Cormobates leucophaea (lue'-koh-fee'-ah: "dusky-white stump-climber")

This treecreeper is readily identified by its conspicuous white throat and the lack of a distinct eyebrow stripe. The female resembles the male but has a small rusty mark on the face; immature birds differ only in being duller, but the juvenile female has a bright gingery rufous rump.

A common woodland species, the

(T & P Gardner)

White-throated Treecreeper inhabits south-eastern mainland Australia from extreme south-eastern South Australia to the vicinity of Rockhampton; isolated populations occur in the Mount Lofty Ranges in South Australia and the Clarke Range and the Atherton Tableland in Queensland; a closely related (possibly conspecific) form occurs in New Guinea.

Strongly sedentary, it lives in pairs that together defend a permanent territory but it is an intensely solitary species and even mated pairs spend little time actually together, usually foraging and roosting independently. Like other treecreepers, it forages in a distinctive manner, methodically hitching its way up a tree trunk from near the ground to some 10 or 20 metres above it, then swooping to the base of some nearby tree to repeat the process. It feeds on bark-haunting insects, especially ants. Its high-pitched staccato piping notes, often uttered in lengthy series, are among the commonest bird sounds in south-eastern eucalypt forests.

Breeding occurs from August to

January; sometimes two broods are raised in a season. The female constructs the nest and incubates alone but the male helps care for the young. For nesting, a tree cavity is selected (at almost any height) and lined with bark fragments and fur. Two or three eggs form the clutch. Incubation takes about 21 days, and the young fledge at 25 days.

HABITAT: temperate forest and woodland, including rainforest
LENGTH: *c* 18 cm
DISTRIBUTION: 300,000–1 million km²
ABUNDANCE: common
STATUS: secure

355

Family EPHTHIANURIDAE

(ef'-thee-an-ue'-rid-ee: "*Ephthianura*-family", after a genus of chats)

Five species in two genera constitute the family Ephthianuridae. Australian chats are small, gregarious, nomadic, terrestrial, insectivorous songbirds that inhabit open, treeless environments; several are supremely well adapted to desert conditions. The family is uniquely Australian, and possession of a brush-tipped tongue (and other anatomical details) strongly indicates a close relationship with the honeyeaters.

Unusually among Australian passerines, males of several species have distinct nuptial and eclipse plumages, and breeding adult males are among the most brilliantly coloured of Australian birds. Adult females and immatures are very much duller, but generally resemble males sufficiently closely in pattern not to seriously impede identification in the field.

Genus Ashbyia

(ash'-bee-ah: "Ashby [-bird]", after E. Ashby, South Australian ornithologist)

The characteristics of the genus are those of the single species.

Gibberbird

Ashbyia lovensis (luv-en'-sis: "Love's Ashby-bird", after the Rev. J. R. B. Love, inland missionary)

The Gibberbird, most desert-adapted of Australian songbirds, is confined to the barren stony plains of the Lake Eyre basin in the southern Simpson Desert, wandering from this region only under the most extreme conditions of drought or flood. It is sedentary or locally nomadic. It resembles other chats in its terrestrial, insectivorous ecology, but it is much less gregarious: flocks of more than 10 are rare, and it is usually encountered alone or in pairs.

It differs from other chats in having dull plumage and a plain brownish rump; females differ little from males. A characteristic idiosyncrasy is its habit of perching alertly on small rocks or pebbles. Uniquely among chats, the male advertises his territory by means of a song flight.

The breeding season is from June to December or after rain, and breeding is sometimes loosely colonial. The nest is a flimsy cup of grass on the ground in the shelter of a tussock. Two to four eggs constitute the clutch. Both parents incubate.

HABITAT: warm temperate treeless gibber plains
LENGTH: 12–13 cm
DISTRIBUTION: 100,000–300,000 km²
ABUNDANCE: sparse
STATUS: probably secure

(*T & P Gardner*)

Genus *Ephthianura*

(ef'-thee-an-ue'-rah: "decreasing-tail")

The characteristics of the genus are essentially those of the family.

White-fronted Chat

Ephthianura albifrons (al'-bee-fronz: "white-forehead decreasing-tail")

The White-fronted Chat is much more sedentary than other Australian chats. Terrestrial, gregarious, and almost exclusively insectivorous, it is typical of its group except in one conspicuous respect: it is not desert-adapted. It is a bird of samphire flats, estuarine wastelands, salt marshes, sand dunes, bald paddocks and other impoverished habitats of southern Australia, from Shark Bay in the west to mid-coastal New South Wales. It is the only chat found in Tasmania.

Adult males are nearly unmistakable by virtue of their mainly white heads and a bold black band across the breast. Dingy greyish in colour, females and immatures are undistinguished in appearance but usually show enough of a smudgy dark band across the breast to identify them. The characteristic note is a terse, high, light "tang".

Breeding occurs mainly from July to January. As in other chats, the female builds the nest, the male defends the territory. The nest is a cup-shaped structure of grass and small twigs lined with fur and placed near the ground in a tussock or similar site. The clutch consists of four or five white eggs, freckled with brown; they hatch in about 14 days.

(*IL Morgan*)

The young fledge at about 14 to 15 days.

HABITAT: low, flat, temperate plains; samphire flats, bare pasture
LENGTH: 11–12 cm
DISTRIBUTION: more than 1 million km²
ABUNDANCE: common
STATUS: probably secure

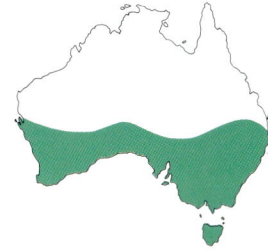

Orange Chat

Ephthianura aurifrons (or'-ee-fronz: "golden-forehead decreasing-tail")

Like the Crimson Chat, the Orange Chat is a nomadic and desert-adapted species; the two closely resemble each other in distribution, habits and behaviour. However, the Orange Chat is somewhat less prone to form flocks of more than 50 individuals and, since it seems to be even more closely adapted to arid conditions, its distribution is rather more stable in consequence (it seldom breeds north of 22°S). It also favours slightly different habitat: canegrass swamps, spinifex plains, samphire flats and herb steppe, especially around salt lakes. Nevertheless, the two species sometimes form mixed flocks.

The adult male is generally brilliant orange with a black mask, but females and young males are very much duller and lack the black mask. The species could easily be confused with the Yellow Chat but, whereas that bird has a near-white iris, the Orange chat has rich brown or reddish eyes.

Breeding occurs mainly from August to December, when wandering flocks break up into territorial pairs. Constructed of grass, small twigs and other plant material, the cup-shaped nest is placed low in a saltbush or similar site.

(*IR McCann*)

Incubation, by both parents, takes about 12 days; the young fledge at about 10 days.

HABITAT: arid temperate to subtropical plains, saltbush, canegrass, lignum, and herb steppe
LENGTH: 10–12 cm
DISTRIBUTION: more than 1 million km²
ABUNDANCE: sparse to common
STATUS: probably secure

Yellow Chat

Ephthianura crocea (kroh-say'-ah: "yellow decreasing-tail")

This chat is widespread across northern Australia but its habitat requirements are so precise that it occurs only in a few widely scattered localities. An irruptive nomad, it inhabits the sedges, reeds and

(I. Pedler)

samphire flats around swamps, especially those prone to drying out, roaming from one to another as each becomes uninhabitable. The widespread installation of bores and other artificial sources of permanent water since European settlement has perhaps stabilised its distribution, and almost certainly expanded it: it has now been recorded southward into South Australia.

Like other chats, it is gregarious, although it seldom forms groups larger than about 10 individuals. It is terrestrial and feeds mostly on insects. It has not been intensively studied, and little is known of its breeding behaviour. Only the adult male possesses the diagnostic black partial breast band.

HABITAT: reed beds, sedges, and samphire flats at the margins of tropical and warm temperate swamps
LENGTH: 11–12 cm
DISTRIBUTION: 300,000–1 million km²
ABUNDANCE: rare
STATUS: possibly endangered

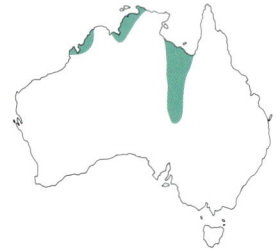

Crimson Chat

Ephthianura tricolor (trie'-kol-or: "three-coloured decreasing-tail")

The Crimson Chat is a nomadic and desert-adapted species of the arid southern and western interior of Australia. Periodically and erratically, it occurs far beyond the boundaries of its core distribution but there are few, if any, breeding records north of about 19°S. The brilliant adult male in breeding plumage is unmistakable, but non-breeding males and adult females

are plain, dull brown, and with only a suggestion of crimson mottling on the breast; immatures resemble adult females.

It is strongly gregarious, and usually encountered in small parties or in flocks of up to 100 birds. It is almost entirely terrestrial, feeding on insects and fallen seeds; it sometimes harvests nectar from low herbs. Like other chats, it has a

distinctive flight: swift, erratic, jerky and undulating.

Breeding takes place in loose colonies, each male defending an area of about 20 metres radius around the nest. The season is very variable, but generally from August to December, at least in the south. The nest is a cup of grass, small twigs and other plant material placed a few centimetres from the ground in a low bush or grass tussock; three or four eggs form the clutch. Both sexes incubate. The eggs hatch in about 12 days and the young fledge at about 14 days.

HABITAT: arid temperate to subtropical shrubland and scrub; saltbush plains and herb steppe
LENGTH: 10–12 cm
DISTRIBUTION: more than 1 million km²
ABUNDANCE: sparse
STATUS: secure

(B & B Wells)

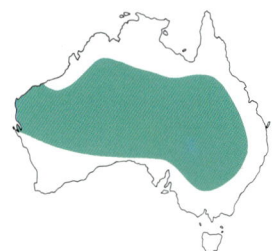

Family PARDALOTIDAE

(par'-dah-loh'-tid-ee: "*Pardalotus*-family", after the genus of pardalotes)

Five (perhaps only four) species of small, arboreal songbirds, placed in a single genus, constitute the exclusively Australian family Pardalotidae.

Pardalotes have short, stubby bills and tails, relatively long pointed wings, and complex, colourful plumage patterns. There is only slight sexual difference in plumage and no seasonal variation; immatures resemble adults. They are strongly gregarious, and notable for their intimate association with eucalypts, feeding almost entirely on various species of lerps, an insect family more or less confined to eucalypts.

Although they spend virtually all of their time high in the canopy of trees, all species nest, at least occasionally, in tunnels in the ground; at least three species do so almost invariably. Otherwise, they nest in tree cavities. Both sexes contribute to nest construction, incubation, and care of the young.

Genus *Pardalotus*

(par'-dah-loh'-tus: "spotted [-bird]")

The characteristics of the genus are those of the family.

Spotted Pardalote

Pardalotus punctatus (punk-tah'-tus: "spotted spotted-bird")

The Spotted Pardalote is common wherever there are eucalypt trees in eastern Australia from Cairns to Adelaide, in Tasmania, and in south-western Australia. It is sedentary and rather solitary in habits but wandering flocks sometimes congregate in winter. It can be distinguished from other pardalotes by its bright red rump, its lemon yellow throat, and the neat round white spots (in females and young dull yellow) on its black crown.

Although it nests in tunnels in such sites as roadside cuttings and sand heaps, it otherwise spends its time in treetop foliage, where it is inconspicuous except for its persistent, monotonous call, "slee-p ba-bee". It feeds on insects, especially lerps.

Breeding extends from June to January, and usually two broods are raised in succession. The nest is a domed structure of bark fragments placed at the end of a tunnel. The clutch consists of three to six unmarked, lustrous white eggs, which hatch in about 14 days.

HABITAT: cool temperate to tropical eucalypt forest and woodland; suburban parks, gardens and boulevards; occasionally heathland
LENGTH: *c* 9.5 cm
DISTRIBUTION: more than 1 million km^2
ABUNDANCE: common
STATUS: secure

(*G Chapman*)

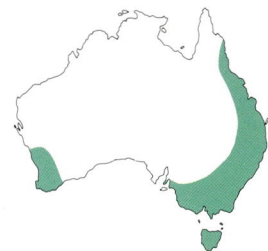

Forty-spotted Pardalote

Pardalotus quadragintus (kwod'-rah-jin'-tus: "forty-spotted-bird")

(D Watts)

It breeds from August to December, and two broods are normally raised in succession. The nest is a domed structure of dry grass and bark fragments in a tree cavity, usually in the stump of a eucalypt. Three or four eggs constitute the clutch. The incubation period is 16 days, and the young fledge at about 25 days.

HABITAT: cool temperate dry eucalypt forests, especially of Manna Gum
LENGTH: *c* 10 cm
DISTRIBUTION: 10,000–30,000 km²
ABUNDANCE: sparse
STATUS: endangered

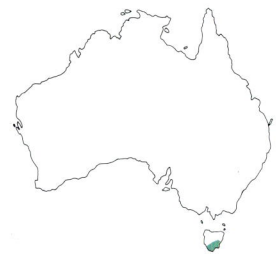

The Forty-spotted Pardalote is restricted to Tasmania and Flinders Island, Bass Strait; it formerly also occurred on King Island. A rare, sedentary and declining species, it lives in loose colonies that are now effectively confined to peninsulas and offshore islands; its total population is only a few thousand individuals and its continued survival is a matter of grave concern. Its decline has been associated with forest clearing and competition from the two other species of pardalotes that occur in Tasmania.

Like other pardalotes, it forages mainly in the foliage of the upper canopy, feeding largely on lerps.

Red-browed Pardalote

Pardalotus rubricatus (rue'-brik-ah'-tus: "reddened spotted-bird")

This dry-country representative of the pardalotes is common across northern and central Australia, south to about Shark Bay in the west and north-western New South Wales in the east. It is easily identified by its pale iris and scarlet and buff eyebrow.

It is sedentary and, like other pardalotes, strongly arboreal. It feeds mainly on insects.

Breeding occurs from June to February. The nest is a domed structure of grass and bark fragments at the end of a tunnel excavated by the breeding pair in the earthen bank of a dry stream or watercourse. Two or three eggs form the clutch.

HABITAT: warm temperate and tropical semiarid eucalypt woodland, riparian woodland, and acacia (Mulga) scrubland
LENGTH: *c* 12 cm
DISTRIBUTION: more than 1 million km²
ABUNDANCE: common
STATUS: secure

(WJ Labbett)

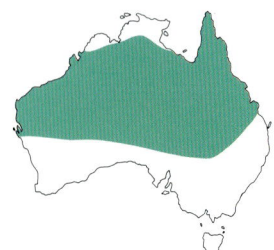

Striated Pardalote

Pardalotus striatus (stree-ah'-tus: "striped spotted-bird")

The Striated Pardalote is common throughout most of Australia and Tasmania. In general, northern populations are characterised by a black cap, while the southern populations have a black cap narrowly streaked with white. Northern populations are mainly sedentary while the southern populations show a tendency to congregate in large roving flocks outside the breeding season; the Tasmanian population is strongly migratory.

In this latter characteristic it differs strikingly from the Spotted Pardalote, with which its distribution broadly overlaps. Spotted Pardalotes are sedentary, living more or less permanently in pairs that maintain well-spaced territories. Striated Pardalotes, on the other hand, do not have established territories even when breeding (defending only the area immediately around the nest), and are strongly nomadic and sociable for the rest of the year. Like all pardalotes, it is strongly arboreal and feeds largely on lerps. The most characteristic call is a persistent, abrupt, high-pitched "chip-chip".

Breeding occurs from June to January, and two broods are normally raised in succession. The nest is usually placed in a cavity in a tree, an earthen bank, or a similar site. The two to five plain white eggs hatch in about 14 days.

HABITAT: cool temperate to tropical eucalypt forest and woodland
LENGTH: *c* 11.5 cm
DISTRIBUTION: more than 1 million km²
ABUNDANCE: common
STATUS: secure

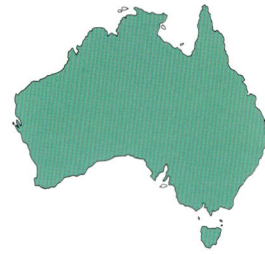

(JD Waterhouse)

Yellow-rumped Pardalote

Pardalotus xanthopygus (xan'-thoh-pie'-gus: "yellow-rumped spotted-bird")

The Yellow-rumped Pardalote occurs across southern Australia from the south-west corner of Western Australia eastwards to central Victoria and south-western New South Wales. It is generally common, but is apparently declining in some areas; it is mainly sedentary or locally nomadic.

Except for its very strong association with mallee woodland it much resembles the Spotted Pardalote in habits, behaviour and appearance (except that it has a yellow, not red, rump). It hybridises with the Spotted Pardalote in some areas, and some authorities regard the two forms as conspecific.

HABITAT: mainly arid to semiarid temperate mallee woodland and whipstick; occasionally casuarina woodland

LENGTH: *c* 10 cm
DISTRIBUTION: 100,000–300,000 km²
ABUNDANCE: sparse
STATUS: vulnerable

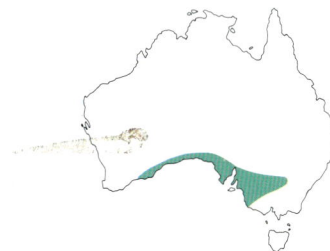

(T & P Gardner)

Family MELIPHAGIDAE

(mel'-ee-fah'-jid-ee: "*Meliphaga*-family", after a genus of honeyeaters)

This family of more than 170 species is almost confined to the Australasian region, only two species marginally extending beyond the generally accepted boundaries of the region. Some species are widespread in Polynesia, Micronesia and New Zealand, and about 63 species inhabit New Guinea, but the group reaches its maximum diversity in Australia, where about 68 species occur (about 16 being shared with New Guinea). Active, conspicuous and intensely aggressive, the honeyeaters constitute one of the prominent elements of the Australian avifauna.

Honeyeaters vary greatly in size but are generally unremarkable in proportions. Most are coloured in neutral shades of green, brown or grey, arranged in a wide variety of patterns, but some are very brightly coloured. Some have naked heads or naked wattles on the face. An especially characteristic feature is a plume or contrasting patch of feathers (usually white or yellow) on the cheek. The male is often slightly larger than the female, but in most genera the sexes are essentially similar in appearance.

The structure of the tongue is unique: strongly modified to suit a diet of nectar, it is long and greatly extensible. The sides are curled up at the base to form two deep grooves, and the tip is deeply cleft to form four delicately fringed, brush-like lobes. In feeding, capillary action assists the transfer of nectar to the tip of the tongue, which is then retracted and squeezed against protuberances on the upper palate, forcing the nectar down the grooved basal part of the tongue into the mouth to be swallowed.

Some Australian honeyeaters are generalists, but many show pronounced specialisations facilitating the exploitation of three major sources of food: nectar, insects and fruit. Many honeyeaters that inhabit tropical rainforests eat mainly fruit, whereas many from heaths or arid interior scrubs are nectarivorous, and many species that live in eucalypt forest and woodland, especially in the south and east, are primarily insectivorous.

Many honeyeaters are sedentary, some are strongly nomadic, and a few are migratory. Honeyeaters are generally arboreal and most are gregarious. Many species breed communally and some exhibit very advanced social structures. Usually, the female builds and incubates unaided but the male assists her in rearing the young. As a general trend, communal breeding seems most prominent in insectivorous species, least marked in nectarivorous honeyeaters.

Genus *Acanthagenys*

(ah-kan'-thah-jen'-is: "spiny-cheek")

The characteristics of the genus are essentially those of the single species.

Spiny-cheeked Honeyeater

Acanthagenys rufogularis (rue'-foh-gue-lar'-is: "rufous-throated spiny-cheek")

(G Little)

This medium-sized honeyeater is generally common and widespread across mainland Australia, especially in scrublands of the arid interior, but sparsely distributed or absent in sandy deserts. It reaches the southern coast from near Esperance in Western Australia to about Melbourne in Victoria (it also occurs on Kangaroo Island), but it is absent along the east coast and in the northern tropics from the Kimberley to Cape York. It is mainly sedentary, but locally nomadic or partly migratory.

Active, conspicuous and aggressive, and easily identified by its pale rump, buff throat and bicoloured bill, it is strongly gregarious except when breeding; it usually occurs in small parties and sometimes in very large flocks. It spends most of its time foraging in the outer foliage of trees, but it occasionally feeds on the ground or flies up to catch insects in midair. Its varied diet includes nectar, fruit, insects, small reptiles and nestling birds. It frequently calls from exposed perches or in display flight, and often in chorus with other birds. Its extensive repertoire includes a range of whistling, gurgling and bubbling notes, as well as an abrupt "tock".

Breeding occurs at any time of year but there is a peak of activity from June to January. Usually placed in a bush or tree within a few metres of the ground, the nest is a deep suspended cup of fibre, plant stems and grasses bound with cobwebs and sometimes lined with feathers or other soft material. The clutch consists of two or three eggs, which hatch in about 14 days. The young fledge at about 15 days. The female incubates alone but both parents participate in raising the young.

HABITAT: mainly arid woodland, mallee and acacia scrubland, especially with an understorey of spinifex
LENGTH: *c* 26 cm
DISTRIBUTION: more than 1 million km²
ABUNDANCE: common
STATUS: secure

Genus *Acanthorhynchus*

(ah-kan'-thoh-rink'-us: "spine-bill")

The two very similar spinebills (one in eastern Australia, the other in the west) are both small, intensely active, sedentary and rather solitary honeyeaters with extremely long, slender, and slightly downcurved bills. They forage mainly low in shrubbery and rely heavily on nectar, a diet supplemented with insects taken from flowers or snatched in midair. Relatively brightly coloured, they have intricate plumage patterns but the sexes are essentially similar in appearance.

Western Spinebill

Acanthorhynchus superciliosus (sue'-per-sil'-ee-oh'-sus: "eyebrowed spine-bill")

(*G Rogerson*)

The Western Spinebill is confined to the south-western corner of Australia, north to about Geraldton and east to Israelite Bay, where it is common and mainly rather sedentary.

In general behaviour it very closely resembles the Eastern Spinebill, being rather solitary and almost ceaselessly active, mainly foraging low in flowering shrubs, and depending largely on nectar for food, although it frequently hawks for insects in midair. Its calls include weak disyllabic notes and a metallic twittering song.

Usually two successive broods are raised during a breeding season extending from August to January. The nest is a small cup-shaped structure of bark and twigs, bound with cobweb, lined with plant down, and suspended by its rim in a slender horizontal fork of a tree or bush within a few metres of the ground. One or two eggs constitute the clutch.

HABITAT: heath, woodland, forest, suburban parks and gardens
LENGTH: *c* 15 cm
DISTRIBUTION: 100,000–300,000 km²
ABUNDANCE: common
STATUS: probably secure

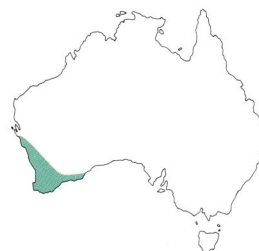

Eastern Spinebill

Acanthorhynchus tenuirostris (ten'-ue-ee-ros'-tris: "narrow-beaked spine-bill")

(*JD Waterhouse*)

One of the commonest of eastern honeyeaters, the Eastern Spinebill occurs from near Cooktown in Queensland south to Tasmania and west to Kangaroo Island and the vicinity of Adelaide in South Australia, mainly along the coast and associated highlands. Most populations are rather sedentary, but in mountainous regions in the south-east there is a pronounced dispersal to adjacent lowlands after breeding.

The sexes are similar in appearance but the female is noticeably duller and has a bluish grey, not black, crown.

Usually rather solitary, the Eastern Spinebill depends almost entirely on nectar, foraging mainly low in flowering shrubs in almost ceaseless activity, and occasionally snatching insects in midair. Its flight is swift and darting, with characteristically noisy wing beats. Its calls include a loud piping whistle uttered in lengthy series, and harsh twittering notes.

Two successive broods are usually raised during a breeding season extending from August to March. The nest is a small cup-shaped structure of bark and grass lined with feathers and slung from a vertical fork at almost any height. Two or three eggs constitute the clutch. Occasionally assisted by the male, the female builds and incubates but both parents participate in rearing the young. Incubation takes about 14 days.

HABITAT: forest, woodland, rainforest, heath, suburban parks and gardens
LENGTH: *c* 17 cm
DISTRIBUTION: 300,000–1 million km²
ABUNDANCE: sparse to common
STATUS: probably secure

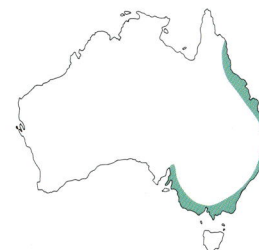

Genus *Anthochaera*

(an'-thoh-kee'-rah: "flower-rejoicer")

The wattlebirds, so called because of the possession of a naked wattle just below the eye (round and red in one species; long, pendent and yellow in another), make up a genus of four species confined to southern Australia, mainly in coastal districts. They are large to very large honeyeaters, long-tailed and slender in build, with dull brown plumage characterised by numerous fine white streaks. The sexes are similar in appearance but males are substantially larger than females. They are nomadic, gregarious, noisy and intensely pugnacious. They feed largely on nectar.

Red Wattlebird

Anthochaera carunculata (kah-run'-kue-lah'-tah: "fleshy-lobed flower-rejoicer", in reference to the fleshy wattle)

The Red Wattlebird is common and widespread across southern mainland Australia from about Geraldton in Western Australia to south-eastern Queensland. It is migratory or locally nomadic, and it has occurred as a casual visitor in New Zealand.

A red wattle on the side of the face is not always conspicuous, and it is best distinguished from the closely similar Little and Brush Wattlebirds (with which it sometimes mingles) by its larger size and a strong tinge of clear yellow on the abdomen. The sexes are similar.

Extremely active, noisy, quarrelsome and conspicuous, it usually occurs in small parties, rarely in large flocks, and frequently associates with other honeyeaters. Occasionally it feeds in the undergrowth or even on the ground, but it spends most of its time in the middle and upper foliage of trees. It feeds mainly on nectar but also snatches insects in midair or gleans them from foliage or from the bark of limbs and branches: it also eats fruit. Its calls are loud, harsh and varied, and include a distinctive loud, abrupt "kwok".

Usually only one brood is raised during a breeding season extending from July to December. The nest is a bulky shallow cup of twigs, grass and bark fragments, sometimes lined with wool, placed in a tree several metres from the ground. The clutch of two or three hatches in about 15 days, and the young fledge at about 15 days.

HABITAT: mainly eucalypt forest, but often common in open woodland, heath, orchards and suburban parks and gardens
LENGTH: *c* 36 cm
DISTRIBUTION: 300,000–1 million km²
ABUNDANCE: common
STATUS: secure

(J Gray)

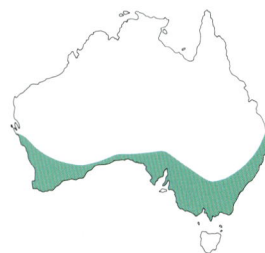

Brush Wattlebird

Anthochaera chrysoptera (kries-op'-te-rah: "golden-winged flower-rejoicer")

The Brush Wattlebird is common in Tasmania and throughout the coastal lowlands and associated highlands of south-eastern Australia from Kangaroo Island and the vicinity of Adelaide in South Australia, eastwards and northwards to about Cooloola in Queensland. It is strongly nomadic, or locally migratory in some areas.

It closely resembles the Red Wattlebird in appearance and behaviour but it is much smaller, lacks yellow on the belly, and reveals a dull rufous panel on the outer wing in flight. It lacks facial wattles. There is also a marked difference in habitat preferences between the two species: the Brush Wattlebird is most common on heathland, whereas the Red Wattlebird favours tall forest.

The Brush Wattlebird is occasionally solitary but it is usually encountered in loose parties or flocks, and freely associates with other honeyeaters. On flowering heathlands it is sometimes the most numerous and conspicuous bird. It feeds largely on nectar and forages mainly in flowering shrubs, but it also eats seeds and fruits, and frequently hawks for insects in midair. It is active, noisy, conspicuous, and extremely pugnacious; parties frequently squabble amongst themselves, and persistently chase smaller honeyeaters. Its calls include a wide range of loud, harsh, discordant notes, sometimes uttered in duet.

Breeding is sometimes loosely colonial, and may occur at any time of the year, but nesting activity tends to reach a peak between July and December. Up to three broods are sometimes raised in succession. The nest is a loose bulky cup of twigs and grass, lined with wool or feathers, and placed in thick cover on or near the ground, or high in the foliage of a tree. The clutch consists of two or three eggs, which hatch in about 12 days. The female incubates unaided but both parents participate in rearing the young, which fledge at about 16 days.

HABITAT: mainly heathland, especially where banksias are numerous; also woodland, forest, urban parks and gardens
LENGTH: *c* 31 cm
DISTRIBUTION: 100,000–300,000 km²
ABUNDANCE: common
STATUS: secure

(J Gray)

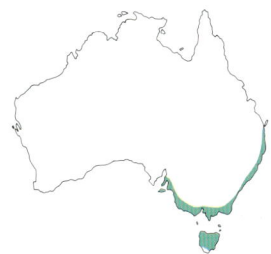

Little Wattlebird

Anthochaera lunulata (lue'-nue-lah'-tah: "crescent[-marked] flower-rejoicer")

The Little Wattlebird is confined to far south-western Australia from near Moora to the vicinity of Esperance, where it is common and locally nomadic. It so closely resembles the eastern Brush Wattlebird in appearance and behaviour that the two forms have only recently been distinguished as species: the Little Wattlebird has a more slender bill, a reddish brown (not bluish grey) iris, and its plumage is somewhat plainer and less conspicuously streaked with white. There are numerous other subtle differences: for example, its bubbling song phrases are more complex and sustained, and it lays only one egg. Breeding occurs mainly between July and November. Like the Brush Wattlebird, it is noisy and conspicuous, gregarious and pugnacious, feeds mainly on nectar, and is most numerous in the vicinity of flowering shrubs.

HABITAT: mainly heathland and forest with much shrubbery; also suburban parks and gardens
LENGTH: 28–30 cm
DISTRIBUTION: 30,000–100,000 km²
ABUNDANCE: common
STATUS: probably secure

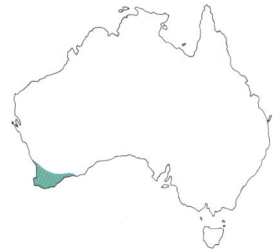

(G Chapman)

Yellow Wattlebird

Anthochaera paradoxa (pah'-rah-dox'-ah: "paradoxical flower-rejoicer")

Among the largest of all honeyeaters, the Yellow Wattlebird is confined to Tasmania and King Island, where it is common but possibly declining in some areas. It is generally nomadic, but there is also a pronounced annual dispersal from higher altitudes to coastal lowlands in winter.

Noisy, active, acrobatic, conspicuous and aggressive, it is usually encountered in small parties, sometimes pairs or large flocks, foraging mainly in foliage high in the canopy of eucalypt forest, but it often searches for insects on bark and sometimes on the ground. It frequently visits garden bird feeders and associates freely with other honeyeaters. Its calls are mostly brief, raucous, and guttural, and include extraordinary gargling phrases.

Breeding occurs from August to December, and sometimes two broods are raised in succession, especially in coastal lowlands. The nest is a loose bulky cup of twigs, bark and leaves, lined with plant down, feathers and fine grass, and usually placed in a vertical tree fork more than 5 metres above the ground. Two or three eggs constitute the clutch. The pair jointly selects the nest site but only the female builds (although the male may escort her as she works, and sometimes brings material) and incubates; both parents care for the young. Incubation takes 14 to 16 days, and the young fledge at about 18 to 21 days.

HABITAT: mainly mature eucalypt forest; also woodland, heath, urban parks and gardens
LENGTH: c 47 cm
DISTRIBUTION: 30,000–100,000 km²
ABUNDANCE: common
STATUS: probably secure

(TA Waite)

Genus Certhionyx

(ser'-thee-on'-ix: "clawed-bird")

The relationships of the three species in this exclusively Australian genus are uncertain, but their inclusion in a single genus is supported by their pronounced similarities in ecology and behaviour. They are almost exclusively nectarivorous and strongly nomadic, wandering in flocks in constant search of flushes of flowering of shrubs and trees in open woodland throughout the northern and central interior of Australia. They show some links with *Myzomela*, especially in their pronounced sexual dimorphism: adult males are boldly black and white, but females are plain, brownish and nondescript.

Black Honeyeater

Certhionyx niger (nie'-jer: "black clawed-bird")

The Black Honeyeater is widespread across interior mainland Australia but is rather uncommon and erratically distributed, its movements and way of life being intimately linked with the flowering cycles of various species of *Eremophila*. Like the Banded Honeyeater of more tropical woodlands, it wanders in flocks, following the blossoming of these and similar plants in response to local rainfall, dependent largely on nectar and foraging mainly in foliage. It is active and conspicuous, frequently hawks for insects in midair, and associates freely with other honeyeaters at flowering trees. Its most characteristic calls include a high-pitched disyllabic note and a faint, plaintive "chee". The sexes are very different in appearance: the male is black and white, the female nondescript pale brown, with a faint pale superciliary.

Breeding is heavily influenced by local rainfall, but occurs usually from September to December. Males announce their territories by means of a simple song uttered during high display flights.

The nest is a rough shallow cup of twigs and plant stems bound with cobweb and suspended in a slender fork of a low bush. Two or three eggs constitute the clutch. Incubation takes about 16 days and, reared by both parents, the young fledge at about 18 days.

HABITAT: mainly acacia and other arid scrublands, mallee
LENGTH: *c* 12 cm
DISTRIBUTION: more than 1 million km^2
ABUNDANCE: sparse
STATUS: probably secure

(G Chapman)

Banded Honeyeater

Certhionyx pectoralis (pek'-tor-ah'-lis: "chest [-marked] clawed-bird")

The Banded Honeyeater occurs across northern Australia from the Kimberley to Cape York Peninsula, extending southward to about Charters Towers. It is strongly nomadic and erratic in occurrence, subject to dramatic irruptions associated with widespread flowering of eucalypts, melaleucas and other trees and shrubs. Associating freely with other honeyeaters, it forages mainly in the upper foliage of trees and depends primarily on nectar, but it also hawks after flying insects. It is strongly gregarious, usually gathering in loose flocks of about 20 to 50 or more individuals, which travel from one source of nectar to another in high, undulating, jerky flight somewhat reminiscent of that of a chat. It is active and conspicuous, and flocks maintain contact with a variety of sharp, chirping notes.

Breeding occurs mainly from October to April. Banded Honey-eaters are thought to mate for life, and pairs drop out of their flocks to establish isolated nesting territories, announced by the male in a clear, descending, chattering song delivered during high display flights. The nest is a fragile suspended cup of grass and bark, bound with cobweb and decorated with spider egg sacs, and placed in the outer foliage of a tree or bush several metres from the ground. Two eggs constitute the clutch.

HABITAT: mainly tropical open eucalypt woodland
LENGTH: *c* 14 cm
DISTRIBUTION: 300,000–1 million km^2
ABUNDANCE: sparse to common
STATUS: probably secure

(B & B Wells)

Pied Honeyeater

Certhionyx variegatus (var'-ee-gah'-tus: "variegated clawed-bird")

The Pied Honeyeater is widespread across arid interior mainland Australia but, like the Black Honeyeater, it is strongly nomadic and erratic in occurrence, following the flowering cycles of eremophilas and similar plants. It is generally more common in the west than in the east. It usually travels in small groups, forages mainly in the crowns of bushes and trees, and feeds largely on nectar supplemented by flying insects. The adult male is black and white, and easily identified by its white rump and patches in the wings, but the female is very drab, mouse brown and nondescript.

Breeding is strongly influenced by local rainfall but occurs mainly from September to February. Males announce their territories by means of spectacular display flights, in which they climb vertically 20 to 30 metres above the surrounding vegetation, then dive headfirst with tail fanned and wings closed, singing as they drop. The nest is a small loose cup of twigs and grass bound with cobweb and suspended in a slender horizontal fork within a few metres of the ground. Two to four eggs constitute the clutch. Both parents build, incubate and rear the young. Incubation takes 13 days, and the young fledge at about 10 days.

HABITAT: mainly acacia and other arid scrublands; also mallee and dry heathland
LENGTH: *c* 20 cm
DISTRIBUTION: more than 1 million km^2
ABUNDANCE: sparse
STATUS: probably secure

(G Chapman)

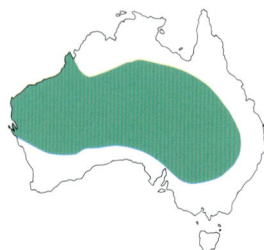

Genus Conopophila

(koh'-noh-pof'-il-ah: "gnat-lover", i.e., feeder upon gnats)

The three members of this genus seem to be more closely related to one another than to any other honeyeater (except possibly *Grantiella*), but their relationships are very uncertain and a rigid definition of the group is lacking. They are small, active, and relatively short-billed honeyeaters with a mainly tropical distribution. One species occurs also in New Guinea but the group is otherwise Australian.

Rufous-banded Honeyeater

Conopophila albogularis (ahl'-boh-gue-lar'-is: "white-throated gnat-lover")

Mainly coastal in distribution, the Rufous-banded Honeyeater occurs in New Guinea and the Aru Islands; in northern Australia it is found from near Port Keats to about Groote Eylandt, and on Cape York Peninsula, south on the west coast to about the Mitchell River and on the east coast to Princess Charlotte Bay. It is very common in the Northern Territory but much less numerous in the east. It is locally nomadic.

Extremely active, it occurs alone, in pairs or in small parties, foraging at all levels in riverside vegetation and mangroves, visiting flowers for nectar, gleaning insects from leaves and snapping them up in midair, while maintaining an almost constant stream of twittering notes. It often associates with other honeyeaters.

Breeding occurs mainly from September to November. Usually suspended in foliage over water, the nest is a deep cup of bark and fibre, bound with cobweb and lined with plant down. The clutch consists of two or three eggs, which hatch in about 12 days. Fed and cared for by both parents, the young fledge at about 10 to 11 days.

HABITAT: mainly open woodland near water: urban parks and gardens
LENGTH: *c* 13 cm
DISTRIBUTION: 10,000–30,000 km²
ABUNDANCE: sparse to common
STATUS: secure

(J Estbergs)

Rufous-throated Honeyeater

Conopophila rufogularis (rue'-foh-gue-lar'-is: "rufous-throated gnat-lover")

The Rufous-throated Honeyeater is widespread across northern Australia from the Kimberley to Cape York Peninsula and from about Cairns south to the Burnett River. It is generally common but is very much less numerous on the east coast than in the north.

It spends much of the year in small flocks, wandering in search of flowering shrubs. Nectar constitutes the major part of the diet but the Rufous-throated Honeyeater also gleans insects from foliage and snatches them in midair: occasionally it feeds on the ground. Its calls include soft twittering and harsh chattering notes.

Breeding occurs mainly from September to May, when wandering flocks fragment into isolated pairs which establish small nesting territories. Constructed by the female over about 14 days, the nest is a deep cup of grass and wild cotton suspended by its rim from a slender horizontal fork several metres from the ground. The clutch consists of two or three eggs which, incubated by the female, hatch in about 14 days. Reared by both parents, the young fledge at about 12 days.

HABITAT: mainly open forest and woodland; mangroves, suburban parks and gardens
LENGTH: *c* 14 cm
DISTRIBUTION: 300,000–1 million km^2
ABUNDANCE: common
STATUS: probably secure

(G Chapman)

Grey Honeyeater

Conopophila whitei (whie'-tee: "White's gnat-lover", after A. H. E. White, son of H. L. White, Australian ornithologist)

One of the least known of Australian birds, the Grey Honeyeater is a small nondescript grey species that is widespread across the arid western interior of Australia but is rare, erratic in occurrence, highly nomadic, and seldom observed.

It usually occurs alone or in small parties, but it frequently associates with thornbills, gerygones, white-faces or other small birds. It forages mainly for insects gleaned in low foliage but also feeds on nectar and small fruits. It is generally quiet, unobtrusive and easily overlooked, but its calls include several loud, sharp notes and a light jingling song.

Breeding is strongly influenced by local rainfall but usually occurs from August to November. The nest is a small frail cup of fibre and hair, lined with plant down and suspended from its rim in the outer foliage of a bush within a metre or two of the ground. One or two eggs constitute the clutch. Both parents cooperate in incubation and in rearing the young.

HABITAT: acacia scrubland and woodland, especially Mulga
LENGTH: *c* 12 cm
DISTRIBUTION: more than 1 million km^2
ABUNDANCE: very sparse to sparse
STATUS: possibly endangered

(B & B Wells)

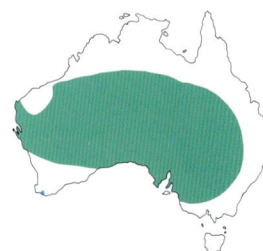

Genus Entomyzon

(en'-toh-mie'-zon: "insect-sucker")

The characteristics of the genus are essentially those of the single species.

Blue-faced Honeyeater

Entomyzon cyanotis (sie'-ah-noh'-tis: "blue-eared insect-sucker")

The Blue-faced Honeyeater occurs in southern New Guinea and is widespread in northern and eastern mainland Australia from the Kimberley to Cape York and south to the vicinity of Adelaide, but it is absent from central southern New South Wales and eastern Victoria. It occurs in most kinds of open woodlands and is often especially numerous in locally disturbed or damaged environments. Abundant in the tropics, its numbers decline sharply southwards, and it is very much more local and less numerous in the south. It is sedentary or locally nomadic in the north, possibly migratory in the south.

It is active, noisy and conspicuous, usually occurring in small parties.

Strongly arboreal, it forages mainly on the bark and limbs of trees, but often on the ground or in undergrowth; it also hawks for insects in midair. The varied diet includes fruits, insects and nectar, and it frequently raids the nests of small birds for eggs and chicks. It also sometimes damages cultivated fruits such as pawpaws and bananas. The extensive repertoire of calls include a loud metallic "tink", persistently repeated, and many harsh mewing and chattering notes.

It often breeds in loose colonies, two or three successive broods sometimes being raised during a breeding season that extends from June to January. Usually placed in a tree fork several metres from the ground, the nest is a large untidy cup of bark strips, fibre and rootlets, lined with plant down, but the Blue-faced Honeyeater often neglects to build a nest, instead appropriating the abandoned nest of another species, especially the Grey-crowned Babbler. Two or three eggs constitute the clutch. The female incubates unaided but both parents participate in rearing the young, frequently assisted by other birds.

HABITAT: mainly eucalypt, melaleuca and pandanus woodland; also riverine woodland, gardens, parks, orchards
LENGTH: *c* 32 cm
DISTRIBUTION: more than 1 million km²
ABUNDANCE: common
STATUS: secure

(G Weber)

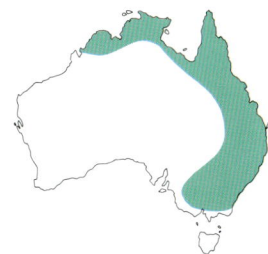

Genus *Glycichaera*

(glis'-ee-kee'-rah: "sweet-rejoicer")

The characteristics of the genus are essentially those of the single species.

Green-backed Honeyeater

Glycichaera fallax (fal'-ax: "false sweet-rejoicer")

This little known honeyeater is widespread in New Guinea and in the Aru Islands: in Australia, it is confined to rainforests in the northern part of Cape York Peninsula, southward to the McIlwraith Range. It is sedentary and rather uncommon.

Small, greenish, nondescript and rather unobtrusive in behaviour, it is easily overlooked or mistaken for a gerygone, except for its white eyes and short, slender, pointed bill. It is usually encountered in pairs or small parties, foraging mainly in foliage in the upper canopy. It feeds on insects. Its calls are soft and simple, sometimes run together as a twittering song.

The nest and eggs have never been described, and almost nothing is known of its breeding behaviour.

(C Cameron)

HABITAT: mainly monsoon rainforest
LENGTH: *c* 12 cm
DISTRIBUTION: 10,000–30,000 km^2
ABUNDANCE: common
STATUS: secure

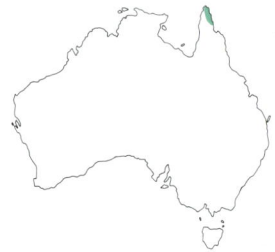

Genus *Glyciphila*

(glis'-ee-fil'-ah: "sweet-lover")

The characteristics of the genus are essentially those of the single species.

Tawny-crowned Honeyeater

Glyciphila melanops (mel'-ahn-ops: "black-faced sweet-lover")

Widespread but not especially common on coastal heaths and in low open woodland across southern Australia, the Tawny-crowned Honeyeater occurs from about Geraldton to Israelite Bay in the south-west, and from Kangaroo Island and Eyre Peninsula to the Richmond River, New South Wales, in the east, including Tasmania and the islands of Bass Strait. It is mainly sedentary or locally nomadic.

In much of its behaviour it resembles honeyeaters of the genus *Phylidonyris* but it is much less gregarious, being usually encountered alone or in pairs. Like these honeyeaters it is largely dependent on the nectar of low flowering shrubs and it frequently hawks for insects in midair: it also often forages on the ground, taking nectar from low flowers or catching crawling insects. Breeding males announce their territories by an intricate, prolonged song flight high over the heath.

Sometimes loosely colonial, the Tawny-crowned Honeyeater may breed at any time of year, but tends to reach peaks of activity from July to March and again from September to November. Well hidden in dense cover close to the ground, the nest is a cup of twigs and grass bound with cobweb. The clutch consists of two to four eggs.

(J Handel)

HABITAT: mainly low heath and open woodland with an understorey of flowering shrubs
LENGTH: *c* 17.5 cm
DISTRIBUTION: 300,000–1 million km^2
ABUNDANCE: common
STATUS: probably secure

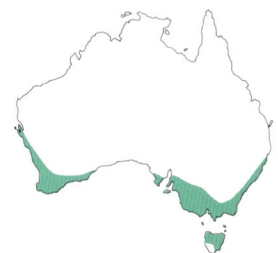

Genus *Grantiella*

(grahn'-tee-el'-ah: "little-Grant[-bird]", after W. R. O. Grant, British ornithologist)

The characteristics of the genus are essentially those of the single species.

Painted Honeyeater

Grantiella picta (pik'-tah: "painted little-Grant-bird")

The Painted Honeyeater is widespread but generally uncommon throughout most of eastern mainland Australia. It is locally nomadic and strongly migratory, breeding mainly in the interior southeast and dispersing northward to northern Queensland and the Northern Territory to spend the winter. It is unusual among honeyeaters in its almost complete dependence upon the berries or drupes of mistletoes of the genus *Amyema*, and its wanderings follow the flowering and fruiting cycles of these plants. It forages mainly in the upper canopy of trees, occasionally eating insects and nectar.

Males generally arrive on their breeding grounds a week or more ahead of the females, and establish territories by means of loud vivacious songs delivered from high perches or during towering song flights, or by quiet chases between rivals through the trees. The most distinctive call is a loud, penetrating "georgie", but it also utters a variety of other calls.

Sometimes two successive broods are raised during a breeding season extending from October to March. The nest is a frail shallow dish constructed of grass and rootlets bound with cobweb and suspended in the outer foliage of a bush or tree from 3 to 20 metres from the ground. The clutch consists of two eggs, which hatch in about 15 days. The young fledge at about 14 days. Both parents participate in nest building, incubation and rearing the young.

HABITAT: open forest, woodland and scrubland with large mistletoe populations
LENGTH: *c* 16 cm
DISTRIBUTION: more than 1 million km^2
ABUNDANCE: sparse
STATUS: probably secure

(G Weber)

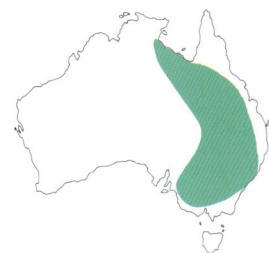

Genus *Lichmera*

(lik-may'-rah: "tongue-player", in reference to the mobility of the long tongue)

Members of this genus are mainly small, very plain brownish honeyeaters with a small patch of stiff, white or pale yellow feathers behind the eye. They are active, aggressive and nomadic, frequenting shrubbery and the margins of forest. Although they may congregate at flowering trees, they are usually rather solitary. Most have loud, rich, relatively complex songs. Essentially tropical in distribution, the genus includes 11 or 12 species, only one of which occurs in Australia.

Brown Honeyeater

Lichmera indistincta (in'-dis-tink'-tah: "indistinct tongue-player")

The Brown Honeyeater occurs in New Guinea and eastern Indonesia and is common and widespread across northern Australia and southward along the west coast to about Esperance, and along the east coast to Sydney. Locally nomadic, it occupies a wide range of habitats from mangroves to arid interior scrubs, but wherever it occurs, the woodland is both low and very dense.

It is usually encountered in pairs or small flocks, frequently congregating at flowering bushes or trees. Strongly arboreal, it feeds mainly on nectar but readily switches to insects when nectar is temporarily scarce, gleaning them from foliage or snatching them in midair. Active, noisy and conspicuous, it utters a variety of strong, clear warbling song phrases.

Breeding occurs from April to November. The nest is a suspended cup of bark and vine stems bound with cobweb, decorated with spider egg cocoons, lined with soft fibres, and placed in the outer foliage of a bush, usually within a metre or two of the ground. Two eggs constitute the clutch. Incubation takes about 14 days, and the young fledge at about 14 days. The male defends the territory but apparently takes little or no active part in the nesting cycle.

HABITAT: forest, woodland, heath, scrubland, mangroves, riverine timbers, suburban parks and gardens
LENGTH: *c* 16 cm
DISTRIBUTION: 30,000–100,000 km²
ABUNDANCE: common
STATUS: probably secure

(*MJ Seyfort*)

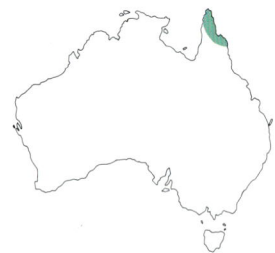

Genus *Manorina*

(mah'-noh-ree'-nah: "thin-nose")

Confined to Australia, the miners constitute a group of four species notable for the extreme sophistication of their social structures. Intensely gregarious and aggressive but unremarkable in general appearance, they are plain brownish or greenish birds with rather short yellow bills that live in complex communities of up to several hundred individuals which jointly defend a territory from which all avian trespassers are excluded. Mature females initiate breeding, but each is assisted by several adult males in rearing her brood of young. Miners are essentially generalists in their foraging, but they rely heavily on insects gleaned from foliage, on the bark of trees, or from the ground.

Yellow-throated Miner

Manorina flavigula (flah'-vee-gue'-lah: "yellow-throated thin-nose")

The Yellow-throated Miner occurs almost throughout mainland Australia west of the Great Dividing Range, wherever there are trees, but it is absent from eastern Arnhem Land, western Gulf of Carpentaria and Cape York Peninsula. It is mainly sedentary. In the east its distribution abuts on and to some extent overlaps that of the Noisy Miner, when the two species are most easily distinguished by the Yellow-throated Miner's white rump; birds in south-western Australia, however, have dark grey rumps.

So far as is known, the Yellow-throated Miner resembles the Noisy Miner in behaviour although its social structure seems to be somewhat less complex and communities tend to be much smaller. Like the Noisy Miner, it forages almost indiscriminately on the ground or in trees, feeding on insects, nectar and fruits. Its calls are similar to those of the Noisy Miner, and include a musical pre-dawn song uttered by the community in chorus.

Breeding is colonial and communal, and extends from July to December. The nest is a loose cup-shaped structure of twigs and grasses, lined with wool, fur, feathers and plant down, and usually placed in a tree fork several metres from the ground. Three or four eggs constitute the clutch.

HABITAT: forest, woodland, mallee, heath, partly cleared cultivated land and urban gardens
LENGTH: *c* 28 cm
DISTRIBUTION: more than 1 milllion km²
ABUNDANCE: common
STATUS: secure

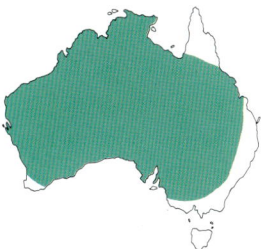

(*G Chapman*)

Noisy Miner

Manorina melanocephala (mel'-ah-noh-sef'-ah-lah: "black-headed thin-nose")

(MJ Seyfort)

The Noisy Miner is common and widespread in eastern Australia from near Adelaide in South Australia to the vicinity of Laura in northern Queensland, but it is much less numerous in the far north than in the south. It is sedentary. It generally avoids coastal areas of high rainfall in the south-east but has successfully colonised several major urban centres, such as Sydney, since the 1970s. It has been introduced to the Solomon Islands.

Living in large communities of up to several hundred members, the Noisy Miner has an intricate social structure in which the relationship of one individual to another depends to some extent on the current activity. Essentially, each mature female serves as a focus for several males; together they form a group that functions as a discrete unit within the larger clan when breeding, but which may dissolve to form temporary associations with other clan members in some other activities such as group foraging or defence of the clan territory. The Noisy Miner is extremely aggressive, and all avian trespassers onto the clan territory are persistently and vigorously challenged and evicted. It is also unusually flexible in its foraging: searching for food almost indiscriminately on the ground, on trunks and limbs of trees, or in foliage. It is mainly insectivorous but its varied diet also includes nectar and fruits. Its extensive repertoire of calls includes various piping, chattering, and peevish scolding notes, as well as a melodious song uttered as a pre-dawn communal chorus.

Breeding is colonial and communal, and several successive broods are raised during a nesting season extending from July to December. The nest is a loose cup of bark strips, grass and fibres placed 1 to 18 metres from the ground in the fork of a tree or bush. The clutch varies from one to four eggs. Incubation takes 15 to 16 days, and the young fledge at about 16 days. The female builds and incubates unaided, but all members of her group assist in raising the young.

HABITAT: mainly dry open eucalypt forest and woodland; also suburban parks and gardens
LENGTH: *c* 28 cm
DISTRIBUTION: more than 1 million km²
ABUNDANCE: common
STATUS: secure

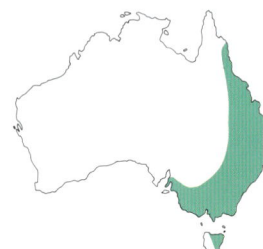

Bell Miner

Manorina melanophrys (mel'-ah-noh-fris: "black-browed thin-nose")

This familiar bird is widespread in south-eastern mainland Australia from near Melbourne to the vicinity of Gympie in Queensland, mainly in coastal and near-coastal districts. It is sedentary and locally common.

The Bell Miner lives in large communities and is only rarely encountered alone or in small parties. It is strongly arboreal, foraging mainly in the upper foliage of trees, but it sometimes descends to the undergrowth and occasionally forages on the ground. It feeds mainly on insects. It is strongly aggressive, and persistently challenges and evicts all avian trespassers into the communal territory. It utters a variety of harsh churring and chattering notes but its most distinctive call is an unmistakable loud, ringing, metallic "tink-tink" note, bell-like and far-carrying.

Breeding is colonial and communal, and occurs from July to February and from April to June. Often two broods are raised in succession. The nest is a loose cup-shaped structure suspended by its rim from the foliage of a low bush, constructed of grass and fine twigs and lined with softer material. Two or three eggs constitute the clutch. Incubation takes about 15 days, and the young fledge at about 15 days. The female builds and incubates unaided, but all group members participate in rearing the young.

(JD Waterhouse)

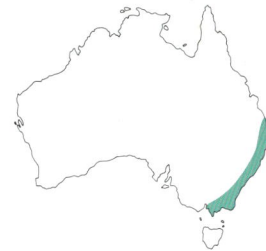

HABITAT: mainly mature tall wet eucalypt forest
LENGTH: *c* 19 cm
DISTRIBUTION: 300,000–1 million km^2
ABUNDANCE: common
STATUS: probably secure

Black-eared Miner

Manorina melanotis (mel'-ahn-oh'-tis: "black-eared thin-nose")

Restricted to mallee regions in the vicinity of the border between South Australia, Victoria and New South Wales, the Black-eared Miner is extremely rare, if not already extinct. The decline has been attributed to clearing of mallee and to inter-breeding with the closely related species, the Yellow-throated Miner. It is sedentary.

It is very similar to other miners in general appearance and behaviour but its community structure is simple and it usually occurs in small parties, foraging on the ground, on the trunks and limbs of trees, or in foliage. It is mainly insectivorous but it also eats fruits and nectar. Its vocalisations, like those of other miners, include various piping, chattering and scolding notes and a pre-dawn song uttered in chorus.

Breeding occurs in September and October. The nest is a bulky cup of twigs and grasses, lined with softer material and placed in a fork of a mallee eucalypt or similar-sized tree. Two or three eggs constitute the clutch. Incubation takes about 14 days, and the young fledge at about 14 days.

(J Purnell)

HABITAT: mallee woodland
LENGTH: *c* 26 cm
DISTRIBUTION: 30,000–100,000 km^2
ABUNDANCE: sparse
STATUS: possibly endangered

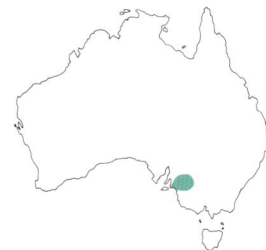

Genus *Meliphaga*

(mel'-ee-fah'-gah: "honey-eater")

This genus of about 12 species is best represented in New Guinea but four species occur in Australia. Inhabiting rainforest almost exclusively, they are plain greenish, medium-sized honeyeaters characterised mainly by a contrasting patch of pale yellow on the cheek. They are notable for their extreme similarity in appearance: although very distinct in calls, habitat and behaviour, the various species are almost impossible to tell apart in the field. They are generally rather solitary, forage in canopy foliage, and rely mainly on fruits for food.

White-lined Honeyeater

Meliphaga albilineata (al'-bee-lin-ay-ah'-tah: "white-lined honey-eater")

This little-known honeyeater is fairly common but is restricted to tropical northern and north-western Aus-

(AL Hertog)

tralia, from western Arnhem Land south to Katherine Gorge and west to the Kimberley. It is sedentary.

It is usually encountered alone or in pairs, foraging in the foliage of trees and feeding on fruits, nectar and insects. Its calls include a variety of loud rapid chipping notes.

Breeding apparently occurs from September to January, but few nests have been found and little is known of the breeding cycle. Two eggs constitute the clutch, deposited in a small cup-shaped nest of fine twigs and leaves, bound with cobweb lined with fine grass and suspended several metres from the ground.

HABITAT: mainly scrubby rainforest fringing sandstone escarpments
LENGTH: *c* 20 cm
DISTRIBUTION: 10,000–30,000 km²
ABUNDANCE: sparse
STATUS: probably secure

Yellow-faced Honeyeater

Meliphaga chrysops (kries'-ops: "golden-faced lichen-mouth")

Almost ubiquitous in eucalypt woodlands throughout south-eastern Australia, the Yellow-faced Honeyeater is common from extreme south-eastern South Australia to the Atherton Tableland in Queensland: it has been recorded as a vagrant in Tasmania, on Lord Howe Island and in New Zealand. Some populations are sedentary but most in the southeast are strongly migratory.

It is usually encountered alone or in pairs, but it congregates in small

groups or large flocks on migration: at some points on their migration routes birds can be observed streaming northward in May at the rate of thousands per hour. It is ordinarily strongly arboreal, foraging mainly in the upper canopy of trees and feeding largely on insects, although it also eats fruits and nectar. Its calls include a terse, repeated "chip", and a loud "chick-up" note, frequently running into a song.

Breeding occurs from July to

January. Slung from its rim in the fork of a tree, the nest is a small cup of grass and bark fibre, bound with cobwebs, decorated occasionally with green moss, and lined with plant down. The clutch consists of two or three eggs, which hatch in about 14 days. The young fledge at about 17 days.

HABITAT: forest, woodland, heath, mangroves, orchards, parks and gardens
LENGTH: *c* 18 cm
DISTRIBUTION: 300,000–1 million km²
ABUNDANCE: common
STATUS: secure

(JD Waterhouse)

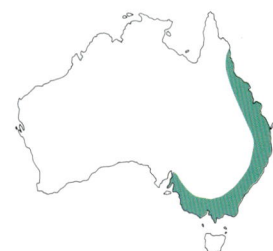

Purple-gaped Honeyeater

Meliphaga cratitia (krah-tit'-ee-ah: "wattled lichen-mouth")

(MJ Seyfort)

The Purple-gaped Honeyeater occurs across southern Australia in two populations that are separated by the Nullarbor Plain: one in southern Western Australia, except for the exreme south-west; the other in the south-east from Eyre Peninsula and Kangaroo Island to north-western Victoria. Strongly associated with dense whipstick mallee heath, it is common and mainly sedentary.

It lives in pairs which maintain more or less permanent territories. It forages mainly in low foliage, feeding largely on insects, manna and honeydew, but also often visits flowers for nectar. Its calls include a series of loud, abrupt notes, and it announces its territory with harsh repeated chattering notes uttered from an exposed perch in the top of a bush.

Breeding occurs mainly from August to December. The nest is a small cup of bark and fine grass, bound with cobweb and decorated with spider egg cocoons, and suspended by its rim in a slender horizontal fork seldom more than a metre or two from the ground. Two eggs constitute the clutch.

HABITAT: mainly whipstick mallee; also heath, forest, woodland, suburban parks and gardens
LENGTH: *c* 19 cm
DISTRIBUTION: 300,000–1 million km²
ABUNDANCE: sparse
STATUS: probably secure

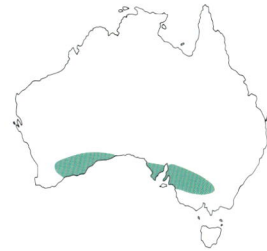

Mangrove Honeyeater

Meliphaga fasciogularis (fah'-see-oh-gue-lar'-is: "throat-banded lichen-mouth")

(K Ireland)

The Mangrove Honeyeater is restricted to coastal lowlands of eastern Australia from near Townsville in Queensland to extreme north-eastern New South Wales. It is sedentary but seems to be gradually extending its range southward.

It may be conspecific with the Varied Honeyeater, which it closely resembles and which replaces it in similar habitat to the north. It lives in small groups. Although foraging mainly in the lower foliage of trees, it sometimes descends to the ground, and frequently hawks for insects in midair. It feeds on nectar, fruits, insects, and small marine invertebrates such as crabs. Active, noisy and conspicuous, it utters a variety of loud, rollicking calls.

Most breeding occurs from August to December, and two broods are usually raised in succession. Placed low in the foliage of mangroves, the nest is an open cup of grass bound with cobweb, lined with rootlets, hair or other fine material and decorated with spider egg cocoons. Two eggs constitute the clutch.

HABITAT: mainly mangroves and associated vegetation; also open woodland, parks and gardens, offshore islands
LENGTH: *c* 21 cm
DISTRIBUTION: 100,000–300,000 km²
ABUNDANCE: common
STATUS: probably secure

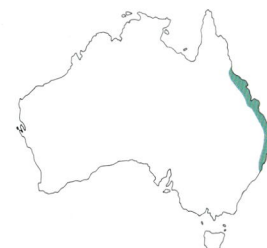

Yellow-tinted Honeyeater

Meliphaga flavescens (flah-ves'-enz: "yellowish lichen-mouth")

This tropical honeyeater occurs in southern New Guinea and in northern Australia from the Kimberley eastward to the Atherton Tableland in Queensland, but it does not occur on Cape York Peninsula. Sedentary and locally common, its distribution abuts on that of the closely similar Fuscous Honeyeater. The two forms are so closely related that they probably constitute a single species.

Like the Fuscous Honeyeater, it lives in small communities that maintain permanent joint territories. It forages mainly in the upper foliage of trees and feeds on insects, manna, and nectar. It is noisy, conspicuous and aggressive.

Breeding may occur at any time of year, but activity reaches a peak between July and December. The nest resembles that of the Fuscous Honeyeater, and two or three eggs constitute the clutch. Incubation takes about 12 days, and the young fledge at about 14 days.

HABITAT: mainly tropical eucalypt woodland
LENGTH: *c* 16 cm
DISTRIBUTION: 300,000–1 million km²
ABUNDANCE: common
STATUS: probably secure

(*J Estbergs*)

Yellow-throated Honeyeater

Meliphaga flavicollis (flah'-vee-kol'-is: "yellow-necked lichen-mouth")

The Yellow-throated Honeyeater is confined to Tasmania and the islands of Bass Strait, where it is common and widespread in a variety of habitats, but especially eucalypt forests. It is mainly sedentary, but coastal populations are augmented in winter by migrants dispersing from higher altitudes, which sometimes form wandering flocks.

It is essentially an island representative of the mainland White-eared Honeyeater, which it much resembles in general appearance and behaviour. Like that species, it forages in lower foliage and on the bark of limbs and branches, relying more on manna, honeydew and insects than on nectar or fruit, although it occasionally visits flowering trees. It lives in pairs which occupy apparently permanent territories, but females and immatures disperse in winter. It is active and conspicuous, and utters a variety of loud, rich notes.

Breeding occurs from August to December. The nest is a bulky cup of bark and grass, lined with wool and hair, and placed in a shrub or grass tussock seldom more than a metre or two from the ground. The clutch consists of two or three eggs, which hatch in about 15 days. The young fledge at about 14 days. Males reportedly evict both mate and newly fledged young to promptly take another mate and father another brood in the same territory.

HABITAT: mainly eucalypt forest and woodland; heath and suburban parks and gardens
LENGTH: *c* 20 cm
DISTRIBUTION: 30,000–100,000 km²
ABUNDANCE: common
STATUS: secure

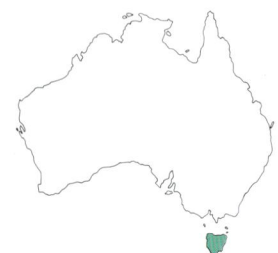

(*J Purnell*)

Yellow Honeyeater

Meliphaga flava (flah'-vus: "yellow lichen-mouth")

This honeyeater is common and widespread on Cape York Peninsula, south on the western side to the Flinders River and along the east coast to Broad Sound. It is sedentary. Except for its strongly yellow coloration, it is very similar to the White-gaped Honeyeater in general appearance and behaviour, and occupies similar habitats. Although the two species overlap broadly in distribution, they tend to replace each other: where one is rare the other is common.

The Yellow Honeyeater lives in pairs or small parties but large numbers may congregate with other honeyeaters at flowering trees. Noisy, aggressive, and conspicuous, it forages mainly in the foliage of trees and shrubs but sometimes descends to the ground, and frequently hawks for insects in midair. It feeds on fruits, nectar and insects. It utters a variety of loud metallic whistles and scolding chattering notes, and has a loud, rich, rollicking song.

Breeding occurs mainly from October to March. The nest is a small suspended cup of bark or coconut fibre, bound with cobwebs and lined with fine grass, and placed in outer foliage several metres from the ground in a tree or bush. The clutch consists of two eggs.

HABITAT: woodlands, especially riverside vegetation; also parks, gardens, orchards and melaleuca swamps and mangroves
LENGTH: *c* 19 cm
DISTRIBUTION: 300,000–1 million km²
ABUNDANCE: common
STATUS: secure

(T & P Gardner)

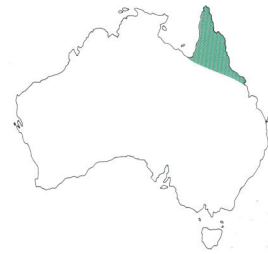

Bridled Honeyeater

Meliphaga frenata (fren-ah'-tus: "bridled lichen-mouth")

The Bridled Honeyeater is common but restricted to the highlands of north-eastern Queensland from Mount Amos south to Mount Spec near Townsville, mainly above about 450 metres altitude, but some individuals descend to adjacent lowlands in winter. It is otherwise mainly sedentary.

It is usually encountered alone or in pairs but flocks often congregate at flowering trees. It forages mainly in the limbs and foliage of trees but also visits garden feeders, and scrounges scraps at forest picnic grounds. It eats fruits, nectar and insects.

Breeding occurs from September to January. Suspended by its rim from a tree fork several metres from the ground, the nest is a small cup of vine tendrils lined with softer materials. Two eggs constitute the clutch.

HABITAT: mainly rainforest
LENGTH: *c* 22 cm
DISTRIBUTION: 30,000–100,000 km²
ABUNDANCE: common
STATUS: probably secure

(N Chaffer)

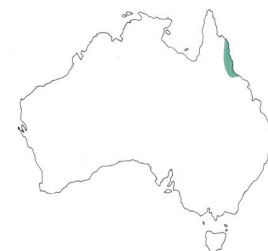

Fuscous Honeyeater

Meliphaga fusca (fus'-kus: "dusky-brown lichen-mouth")

The Fuscous Honeyeater is common and widespread in eastern Australia from south-eastern South Australia eastwards and northwards along drier sections of the coast as well as in the highlands and on the western slopes of the Great Dividing Range to about Townsville in Queensland. In the north, its distribution overlaps with that of the Yellow-tinged Honeyeater, with which it is probably conspecific.

It is gregarious and mainly sedentary, living in communities that jointly defend a permanent territory, although territories at higher altitudes in the south-east are usually temporarily abandoned during winter, their previous occupants forming wandering flocks that disperse to adjacent lowlands and return in spring. The Fuscous Honeyeater forages mainly in the upper foliage of trees, its varied diet including insects, manna, honeydew, nectar and fruits. It is an active, aggressive and conspicuous species which utters a wide range of dry, rattling, harsh or whistled calls, as well as a distinctive, rapid, bubbling song, delivered from a high perch in foliage, or in flight.

Breeding is communal and loosely colonial, and occurs mainly from July to December. The nest is a cup of bark strips and grass, bound with cobweb and decorated with spider egg cocoons, and suspended in the outer foliage of a tree at almost any height from the ground. The clutch consists usually of three eggs, which take about 14 days to hatch. The female incubates unaided, but both parents participate in rearing the young, often assisted by other birds.

HABITAT: mainly eucalypt forest and woodland; also heathland, acacia scrub, suburban parks and gardens
LENGTH: *c* 18 cm
DISTRIBUTION: 300,000–1 million km²
ABUNDANCE: common
STATUS: probably secure

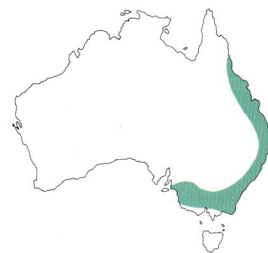

(IR McCann)

Graceful Honeyeater

Meliphaga gracilis (grah'-sil-is: "slender honey-eater")

This small olive honeyeater often occurs alongside the Lesser Lewin's Honeyeater, and the two species are extremely difficult to distinguish. However, the Graceful Honeyeater has a smaller, more flared yellowish ear patch and a slightly longer, more slender bill. It is common in lowland rainforests in New Guinea, the Aru Islands, and north-eastern Queensland from near Townsville northward to Cape York. It is sedentary.

It is usually solitary, but sometimes congregates at flowering trees. Strongly arboreal, it is active and acrobatic but rather quiet and unobtrusive. It feeds mainly on fruit, frequently supplemented with nectar and insects. Its most distinctive call is a loud single "click" or "tick", but it also utters a reedy whistling note and a series of high ascending whistles.

Breeding occurs from September to February. The nest is a bulky cup of bark strips and green moss, lined with soft material and suspended from a fork usually 2 to 6 metres from the ground. The clutch consists of two eggs, which hatch in about 14 days. The young fledge at about 14 days.

HABITAT: rainforest, open forest, mangroves; suburban parks and gardens
LENGTH: *c* 17 cm
DISTRIBUTION: 100,000–300,000 km²
ABUNDANCE: sparse
STATUS: probably secure

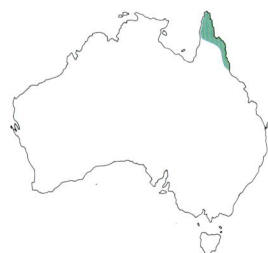

(WS Peckover)

Eungella Honeyeater

Meliphaga hindwoodi (hiend'-wood-ee: "Hindwood's lichen-mouth", after K. A. Hindwood, Australian zoologist)

(T Lindsey)

Not discovered until the 1970s, this honeyeater is restricted to highland rainforests in the Clarke and Sarina Ranges near Mackay in central eastern Queensland. It is common but little is known of its movements.

It is often solitary and inconspicuous, but noisy, active, and aggressive congregations often gather at flowering trees. It forages mainly in the foliage of the upper canopy, feeding on nectar and insects. Its vocabulary includes a short, sharp "chip" or "churr" and a short series of metallic notes, sometimes run together to form a song.

Very little is known of its breeding behaviour.

HABITAT: mainly rainforest
LENGTH: *c* 21 cm
DISTRIBUTION: less than 10,000 km²
ABUNDANCE: sparse
STATUS: vulnerable

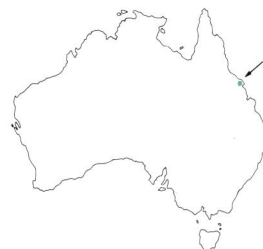

Grey-headed Honeyeater

Meliphaga keartlandii (keert'-land-ee-ee: "Keartland's lichen-mouth", after G. D. Keartland, Australian naturalist)

(B & B Wells)

The Grey-fronted Honeyeater is common, but somewhat patchily distributed, across northern Australia, mostly north of the Tropic of Capricorn but excluding the Kimberley and monsoon areas of the Northern Territory. It occupies a range of habitats but is especially characteristic of low shrublands in rocky breakaway country or on desert sand dunes.

It usually occurs in small groups, which often congregate into wandering flocks (sometimes exceeding 100 individuals) after breeding. In general behaviour it much resembles its close relative, the Purple-gaped Honeyeater of southern Australia: like the latter, it forages mainly in the canopy foliage of trees, feeds largely on insects, manna and honeydew, and announces its territory with loud chattering notes uttered from an exposed perch in the crown of a tree.

Breeding is heavily influenced by local rainfall but commonly begins in July or August, extending to November. Usually placed within a metre or two of the ground in the foliage of a bush or sapling, the nest is a small suspended cup of plant fibre and grass, bound with cobwebs and spider egg cocoons and lined with rootlets and plant down. Two eggs constitute the clutch.

HABITAT: Mulga scrubland, mallee woodland, savannah with flowering bloodwoods and hakeas
LENGTH: *c* 16 cm
DISTRIBUTION: more than 1 million km²
ABUNDANCE: sparse to common
STATUS: probably secure

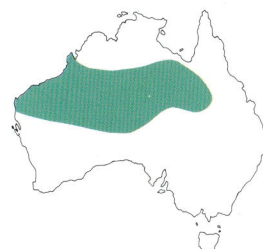

White-eared Honeyeater

Meliphaga leucotis (lue-koh'-tis: "white-eared lichen-mouth")

This is one of the commonest of honeyeaters across much of southern and eastern mainland Australia, north in the east to near Hughenden in Queensland and, in the west, to near Geraldton in Western Australia. It is mainly sedentary, but marked altitudinal migrations occur in south-eastern montane regions.

Notably less gregarious than other honeyeaters, it lives in pairs which maintain apparently permanent territories. It seldom joins mobs of honeyeaters at flowering trees, being more dependent on manna, honeydew, and insects (gleaned mainly from foliage and from the bark of limbs and branches) than on fruit or nectar. It prefers shrubbery and the lower foliage levels of trees. Its calls are rich, loud and varied.

Breeding occurs mainly from August to December. The nest is a small cup of grass and fine twigs, lined with fur and plant down, and usually placed in a grass tussock or low in a leafy shrub. The clutch consists of two or three eggs which hatch in about 14 days. The young fledge at about 14 to 15 days.

HABITAT: mainly eucalypt forest and woodland; also mallee and heath
LENGTH: *c* 21 cm
DISTRIBUTION: more than 1 million km²
ABUNDANCE: common
STATUS: secure

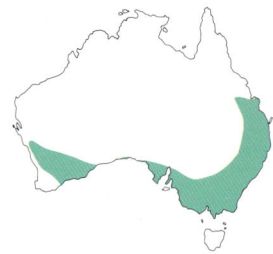

(JD Waterhouse)

Lewin's Honeyeater

Meliphaga lewinii (lue'-win-ee-ee: "Lewin's honey-eater", after J. W. Lewin, Australian naturalist)

Lewin's Honeyeater is widespread in eastern Australia from the McIlwraith Range in Queensland to the vicinity of Melbourne in Victoria. It is sedentary and very common, occurring at all altitudes in the south but mainly restricted to altitudes above about 600 metres in the tropics.

Usually rather solitary, it forages in the foliage and on the trunks of trees, especially in the upper and middle levels. It feeds mostly on fruits, but it also eats insects and nectar. Its most distinctive and persistent call is a loud staccato rattle of notes.

Breeding occurs mainly from September to January. Suspended by its rim in dense foliage, the nest is a large untidy cup of fibre and moss (sometimes paper), bound with cobweb and lined with plant down. Two or three eggs constitute the clutch. Incubation takes 14 to 15 days, and the young fledge at 14 to 15 days.

HABITAT: mainly rainforest and wet sclerophyll forest; also woodland, heath, suburban parks and gardens
LENGTH: *c* 22 cm
DISTRIBUTION: 300,000–1 million km²
ABUNDANCE: common
STATUS: secure

(IP Rowles)

Yellow-tufted Honeyeater

Meliphaga melanops (mel'-ahn-ops: "black-faced lichen-mouth")

(RF Kenyon)

The Yellow-tufted Honeyeater is widespread in eastern mainland Australia. A very distinct sub-species, once regarded as a species (the Helmeted Honey eater, *L. m. cassidix*), was formerly widespread in eastern Victoria but is now reduced to perhaps 100 to 150 individuals, confined to a minute fragment of eucalypt woodland along Woori Yallock Creek near Yellingbo. Elsewhere the species is common.

The Yellow-tufted Honeyeater has a varied diet including nectar and fruits, but it is essentially a lerp specialist, feeding primarily on leaf-eating insects and the sugary secretions that eucalypt foliage produces in response to insect damage. The yellow-tufted Honey-eater forages in foliage at almost any height, frequently descending to the undergrowth and sometimes feeding on the ground. It is noisy, active and conspicuous, and utters a variety of loud, rich call-notes.

Breeding is colonial and communal, and occurs mainly from July to January; sometimes two or three broods are raised in succession. Suspended in the outer foliage of a tree at almost any height, the nest is a cup-shaped structure of bark fragments and grass, bound with cobweb, decorated with spider egg cocoons and lined with fine grass and fur or feathers. The clutch consists of two or three eggs, which hatch in about 14 days. Fed by both parents with the assistance of other birds, the young fledge at about 13 days.

HABITAT: mainly eucalypt forest and woodland, especially along rivers and creeks; also heath, mangroves, gardens, mallee, and acacia scrubland
LENGTH: *c* 23 cm
DISTRIBUTION: 300,000–1 million km²
ABUNDANCE: sparse
STATUS: probably secure

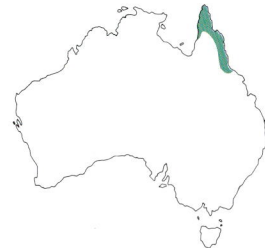

Lesser Lewin's Honeyeater

Meliphaga notata (noh-tah'-tah: "marked honey-eater")

This species (sometimes called the Yellow-spotted Honeyeater) is common in coastal lowlands of north-eastern Queensland, including islands in Torres Strait, south on the west coast of Cape York Peninsula to about the Nassau River and, on the east coast, south to near Townsville. It also occurs in New Guinea. It is sedentary.

Usually solitary and unobtrusive, it forages mainly in the foliage of trees, feeding largely on fruits. Its most distinctive call is a liquid "chip", but it also utters a series of urgent querulous notes and a clear piercing whistle.

Breeding occurs from August to February. Suspended by the rim in dense foliage several metres from the ground, the nest is a deep cup of palm fibre, bark strips and dry leaves bound with cobweb, decorated with green lichen, and lined with plant down. The clutch consists of two eggs, which hatch in about 15 days; the young fledge at about 15 days.

(T & P Gardner)

HABITAT: mainly rainforest and its margins; also riverine woodland and forest, coastal scrubs, mangroves; suburban parks and gardens
LENGTH: *c* 20 cm
DISTRIBUTION: 100,000–300,000 km²
ABUNDANCE: common
STATUS: probably secure

Yellow-plumed Honeyeater

Meliphaga ornata (or-nah'-tus: "decorated lichen-mouth")

(*G Chapman*)

Strongly associated with wetter mallee districts of southern Australia, the Yellow-plumed Honeyeater occurs from the vicinity of Temora in south-central New South Wales westwards to Perth. It is sedentary and usually fairly common.

It seldom congregates in flocks, and is usually encountered alone or in pairs, foraging mainly in foliage in the crowns of trees and feeding largely on insects, manna and honeydew. It utters a variety of loud and somewhat harsh notes, and announces its territory with a rapid trilling song, uttered from a conspicuous high perch or in the course of a near-vertical display flight.

Breeding occurs mainly from July to January. Usually placed close to the ground in the foliage of a bush or shrub, the nest is a suspended cup of bark or grass, bound with cobweb, decorated with spider egg cocoons, and lined with plant down. Two or three eggs constitute the clutch.

HABITAT: mainly mallee and acacia scrub
LENGTH: *c* 18 cm
DISTRIBUTION: 300,000–1 million km^2
ABUNDANCE: sparse to common
STATUS: probably secure

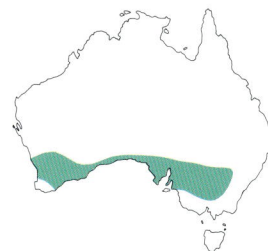

White-plumed Honeyeater

Meliphaga penicillata (pen'-is-il-ah'-tus: "brushed lichen-mouth")

(*G Chapman*)

The White-plumed Honeyeater occurs in tall eucalypt woodland throughout most of mainland Australia except for the tropical north and regions of high rainfall in the extreme south-west and south-east. It seems to be steadily extending its range and it is now common in some east-coast urban areas where it was formerly uncommon.

Mainly sedentary, it lives in small communities which maintain permanent territories. It forages mainly in the upper and outer foliage of trees. Like most *Meliphaga* honeyeaters, it is essentially a gleaner, a plucking insects from leaves and relying heavily on manna and honeydew, but it also feeds on nectar and fruits. Active and conspicuous, it utters a wide variety of abrupt calls, a piping alarm note, and a rapid song, delivered from foliage or during a vertical song flight.

Breeding is communal and loosely colonial: it may occur at any time of year, but reaches a peak of activity between August and December. The nest is a small suspended cup of grass bound with cobweb, lined with feathers, hair and wool and placed in the outer foliage of a bush or tree. Two or three eggs constitute the clutch. The female builds the nest and incubates unaided for about 14 days, but her mate and all group members assist in raising the young, which fledge at about 14 days.

HABITAT: mainly eucalypt forest and woodland; also heath, suburban parks and gardens
LENGTH: *c* 18 cm
DISTRIBUTION: more than 1 million km^2
ABUNDANCE: common
STATUS: secure

Grey-fronted Honeyeater

Meliphaga plumula (plue'-mue-lus: "little-plumed lichen-mouth")

The Grey-fronted Honeyeater occurs in arid scrub almost throughout mainland Australia. In the south its distribution abuts on, or partly overlaps, that of the related and very

similar Yellow-plumed Honeyeater, but the two species seldom occur together: in such areas, the Yellow-plumed Honeyeater favours dense wet mallee, leaving arid, scrubby, or stunted mallee to the Grey-fronted Honeyeater.

The Grey-fronted Honeyeater lives in small groups and is locally nomadic, following flushes of flowering bushes and shrubs. It forages mainly in the upper foliage, feeding on nectar, insects and fruits. The vocal repertoire includes a variety of loud, sharp calls, and it announces its territory with a series of rapid trilled whistles, uttered from a high perch within foliage or during a near-vertical display flight.

Breeding is heavily influenced by local rainfall, but usually begins in July and extends to January. Usually suspended in the outer foliage of a bush or shrub within a metre or two

(*G Chapman*)

of the ground, the nest is a small cup of bark fragments and grass bound with cobwebs and decorated with spider egg cocoons. The clutch consists of two or three eggs, which hatch in about 14 days. The young fledge at about 14 days.

HABITAT: eucalypt woodland, mallee and acacia scrub
LENGTH: *c* 17 cm
DISTRIBUTION: more than 1 million km²
ABUNDANCE: sparse to common
STATUS: probably secure

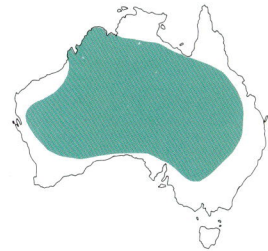

White-gaped Honeyeater

Meliphaga unicolor (ue'-nee-kol-or: "single-coloured lichen-mouth")

The White-gaped Honeyeater is common and widespread in tropical Australia from the vicinity of Broome in Western Australia eastward to Cape York Peninsula (except for the northern part) and south to near Townsville. It is mainly sedentary and is most numerous in coastal lowlands.

(*G Chapman*)

It lives in pairs or small groups which forage mainly in the foliage and on the trunks and limbs of trees, feeding on insects, fruits and nectar. A noisy and conspicuous species, it has the unusual habit—for a honeyeater—of frequently cocking its tail. It utters a variety of rich, loud, fluted whistled phrases often run

together as a rollicking song uttered in duet, as well as terse, harsh, repetitive chirps.

Breeding occurs mainly from September to January. Placed in outer foliage at almost any height from the ground, the nest is a deep cup constructed of grasses and melaleuca bark bound with cobweb and lined with plant down. Two eggs constitute the clutch.

HABITAT: mangrove, riverine forest and woodland, coastal thickets and melaleuca swamps
LENGTH: *c* 22 cm
DISTRIBUTION: 300,000–1 million km²
ABUNDANCE: sparse to common
STATUS: probably secure

Varied Honeyeater

Meliphaga versicolor (ver'-see-kol'-or: "variously-coloured lichen-mouth")

Often considered to be merely a subspecies of the Mangrove Honeyeater, the Varied Honeyeater occurs in New Guinea and in north-eastern Australia from the Torres Strait islands south to about Townsville. It is common, sedentary, and almost exclusively coastal in distribution.

Strongly arboreal, it lives in small groups that forage mainly in foliage but occasionally on the ground. It feeds mainly on insects, nectar and fruits. An active, aggressive, conspicuous and noisy species, its calls include a scolding chatter and many loud, varied and musical phrases, often uttered in duet.

Breeding occurs from August to November. The female builds and incubates but both parents rear the young, often assisted by other birds. Slung by its rim from foliage several metres from the ground, the nest is an open cup of rootlets and fibres bound with cobweb and lined with softer material; two eggs constitute the clutch.

HABITAT: mainly mangroves and associated vegetation; open forests and woodland; parks and gardens; offshore islands
LENGTH: *c* 21 cm
DISTRIBUTION: 100,000–300,000 km^2
ABUNDANCE: sparse to common
STATUS: probably secure

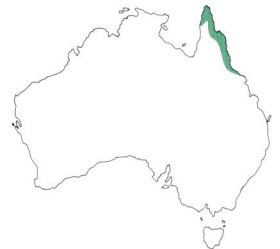

(*D & M Trounson*)

Singing Honeyeater

Meliphaga virescens (vi-res'-enz: "greenish lichen-mouth")

The most widespread of Australian honeyeaters, this species occurs in scrubland and shrubberies almost throughout the mainland west of the Great Dividing Range. It is mainly sedentary, or locally nomadic.

A common, active and conspicuous species, it lives in pairs or small parties and forages mainly in shrubbery and lower foliage but sometimes feeds on the ground and often hawks for insects in midair. Its varied diet includes nectar, fruits, insects and, occasionally, birds' eggs. It utters a variety of calls, including a rollicking song delivered from a conspicuous high perch, often in duet.

Usually two successive broods are raised during a breeding season extending mainly from July to February. The nest is a cup-shaped structure of grass bound with cobweb and lined with plant down, placed in the foliage of a tree or bush

seldom more than a few metres from the ground. The clutch consists of two or three eggs, which hatch in about 12 to 15 days. The female incubates unaided but both parents participate in rearing the young, which fledge at about 13 days.

HABITAT: mainly arid woodland and scrubs, including mallee and acacia; also coastal scrubland, mangroves and suburban parks and gardens
LENGTH: *c* 22 cm
DISTRIBUTION: more than 1 million km^2
ABUNDANCE: common to abundant
STATUS: secure

(*AJ Olney*)

Genus *Melithreptus*

(mel'-ee-threp'-tus: "honey-fed", hence "honey-eater")

Possibly allied to *Meliphaga* but very distinct, this genus consists of six or seven species characterised by a plumage in which the underparts are white, the back is dull green, and the head is mainly black; there is usually a narrow band of white across the hindneck and a patch of naked, brightly coloured skin around the eye. The group is widespread in Australia and one species occurs also in New Guinea. These honeyeaters frequent the crowns of trees and feed almost exclusively on insects. They are mostly nomadic and extremely gregarious, living in permanent groups in which mutual preening and communal roosting and nesting are common.

Black-headed Honeyeater

Melithreptus affinis (ah-fin'-is: "related honey-eater")

Closely related to the mainland White-naped Honeyeater, the Black-headed Honeyeater is confined to Tasmania and the islands of Bass Strait, where it is common and widespread. It lives in small close-knit groups that are more or less

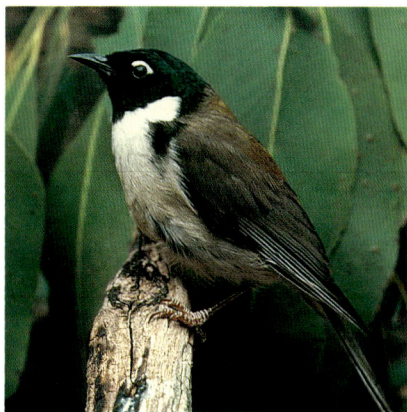

(RH Green)

sedentary through much of the year but often coalesce with other groups and wander more widely in winter, frequently joining mixed feeding flocks of other species. Almost exclusively arboreal, it forages mainly in foliage in the upper canopy of trees, feeding on insects, manna and honeydew; it rarely visits flowering trees for nectar.

Sometimes two or three successive broods are raised during a breeding season extending from October to December. The nest resembles that of other *Melithreptus* honeyeaters and is similarly placed. The clutch usually consists of three eggs, which hatch in about 15 to 16 days. The young fledge at about 15 days. Both parents build the nest and incubate, and they are assisted by

other group members in rearing the young.

HABITAT: mainly eucalypt forest and woodland; also orchards, suburban parks and gardens
LENGTH: *c* 15 cm
DISTRIBUTION: 30,000–100,000 km²
ABUNDANCE: common
STATUS: probably secure

White-throated Honeyeater

Melithreptus albogularis (ahl'-boh-gue-lar'-is: "white-throated honey-eater")

The White-throated Honeyeater occurs in southern New Guinea, and

(AL Herzog)

is common and widespread across northern Australia from the Kimberley to Cape York Peninsula, and southward, mainly in coastal lowlands, into north-eastern New South Wales. It is mainly sedentary and lives in small groups. Closely resembling other *Melithreptus* honeyeaters in behaviour, it forages mainly in foliage, preferring the upper canopy of trees, and feeding largely on insects. Its calls are brief and simple.

Breeding occurs from March to January. The nest is a small neat cup of bark fragments and grass, bound with cobweb, decorated with spider egg cocoons and lined with plant down, and suspended several metres from the ground in the outer foliage of a tree. Two eggs constitute the

clutch, and both parents rear the young, usually assisted by other birds.

HABITAT: mainly eucalypt forest and woodland, especially along rivers and streams; occasionally rainforest
LENGTH: *c* 15 cm
DISTRIBUTION: more than 1 million km²
ABUNDANCE: common
STATUS: secure

Brown-headed Honeyeater

Melithreptus brevirostris (brev'-ee-ros'-tris: "short-billed honey-eater")

The Brown-headed Honeyeater is common in arid eucalypt woodlands across southern Australia and north, in the east, to the vicinity of Rockhampton. It lives permanently in close-knit groups of 10 to 20 individuals, which together wander over a home range several square kilometres in extent, moving rapidly through the trees, foraging mainly in foliage in the crown, and feeding largely on insects, spiders and other small invertebrates. Its most common call is a simple, terse "chip", uttered in flight or while foraging.

Breeding occurs from June to January. The nest is a small neat cup of grass, fur, and spider egg cocoons, bound with cobweb, lined with plant down, and suspended in the outer foliage of a tree or bush. Two or three eggs constitute the clutch. All members of the group participate in nest building, incubation and care of the young.

(JD Waterhouse)

HABITAT: mainly dry eucalypt forest and woodland, including mallee and low heath; suburban parks and gardens
LENGTH: *c* 14 cm
DISTRIBUTION: more than 1 million km²
ABUNDANCE: common
STATUS: probably secure

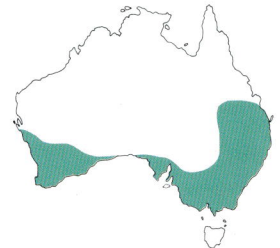

Black-chinned Honeyeater

Melithreptus gularis (gue-lar'-is: "[notable-]throated honey-eater")

The Black-chinned Honeyeater has an extensive distribution across northern Australia, extending southward in the east, mainly inland of the Great Dividing Range, to south-eastern South Australia. It is generally uncommon and thinly distributed. Birds in the tropical north are brighter in plumage and are sometimes treated as a distinct species, the Golden-backed Honeyeater, *M. laetior*.

It is mainly sedentary, and lives in small groups which jointly maintain permanent, extensive territories. Active and forever on the move, parties roam their territories, foraging high in the canopy and feeding mainly on insects, manna and honeydew, a diet sometimes supplemented with nectar and fruits; the birds also probe the bark of limbs, branches and twigs for insects. Calls include a variety of loud, rich notes.

Usually two successive broods are raised during a breeding season extending from July to December. Usually suspended high in the outer foliage of a tree, the nest is a small, neat, deep and closely felted cup of grass and bark fragments bound with cobweb and lined with plant down. The clutch consists of two or three eggs, which hatch in about 15 days. The young fledge at about 14 days. Helpers at the nest have been recorded.

(WJ Labbett)

HABITAT: mainly eucalypt forest and woodland
LENGTH: *c* 16 cm
DISTRIBUTION: more than 1 million km²
ABUNDANCE: sparse
STATUS: probably secure

White-naped Honeyeater

Melithreptus lunatus (lue-nah'-tus: "crescent [-marked] honey-eater")

(K Stepnell)

The White-naped Honeyeater is common in eucalypt woodland in south-western Australia and along the east coast from the Atherton Tableland southward to Yorke Peninsula and Kangaroo Island in South Australia. It does not occur in Tasmania. Over much of its range it is mainly sedentary but, in the south-east, it is strongly migratory, congregating in April and May in flocks of about 50 individuals to move northward to wintering grounds, possibly in south-eastern Queensland, and returning by different routes in the spring. It migrates by day, and in its northward migrations it frequently associates with the Yellow-faced Honeyeater, which is also migratory.

Except on migration, it usually occurs in small parties which, like those of other *Melithreptus* honey-eaters, move quickly through the trees, foraging mainly in the upper canopy and gleaning manna and honeydew from leaves. Its calls are brief and simple.

Two successive broods are often raised during a breeding season extending from July to February. Usually suspended 2 to 6 metres from the ground in the outer foliage of a eucalypt, the nest resembles that of other *Melithreptus* honeyeaters. The clutch consists of two or three eggs, which hatch in about 14 days. The young fledge at about 15 days. All members of the group assist in nest defence and in rearing the young.

HABITAT: mainly eucalypt forest, woodland and heath
LENGTH: *c* 14 cm
DISTRIBUTION: 300,000–1 million km²
ABUNDANCE: common
STATUS: secure

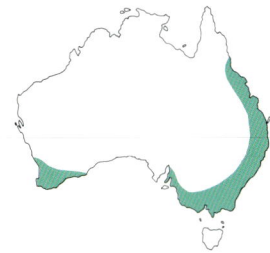

Strong-billed Honeyeater

Melithreptus validirostris (val'-id-ee-ros'-tris: "strong-billed honey-eater")

Closely related to the Black-chinned Honeyeater of the Australian mainland, the Strong-billed Honey-eater is restricted to Tasmania and the islands of Bass Strait, where it is common and widespread in eucalypt forests. It is nomadic and lives in small groups. It forages mainly on the limbs and branches of trees, probing and tearing at the bark for insects, but it also gleans

insects, manna and honeydew from foliage; it rarely visits flowers for nectar. Active, noisy and conspicuous, groups maintain contact with a variety of loud, staccato calls.

Breeding occurs from August to December, often in loose groups. The nest is a bulky suspended cup of bark fragments, grass and spider egg cocoons, lined with fine grass, fur and feathers, and placed in the outer

foliage of a tree some distance from the ground. The clutch consists of two or three eggs, which hatch in about 15 days. The young fledge at about 15 days. Both parents participate in nest building, incubation and rearing the young.

HABITAT: mainly eucalypt forest and woodland
LENGTH: *c* 17 cm
DISTRIBUTION: 30,000–100,000 km²
ABUNDANCE: common
STATUS: secure

(TA Waite)

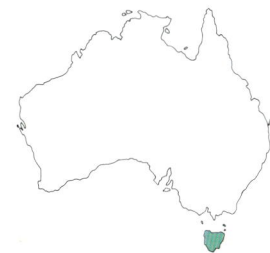

Genus Myzomela

(mie'-zo-may'-lah: "honey-sucker")

With about 24 member species, this is the largest genus of honeyeaters. The group is widespread in Indonesia, New Guinea and islands of the south-west Pacific, but only three species occur in Australia. All are very small, vivacious, and active honeyeaters with a close superficial resemblance to sunbirds, darting and flitting about flowering shrubs or in the canopy of trees, taking nectar from blossoms and snatching insects in midair. Members of this genus are unusual among honeyeaters in that males usually differ strikingly from females in appearance. Females are usually dull brown, plain and nondescript whereas adult males of most species are boldly patterned in various combinations of black, white and vivid red.

Red-headed Honeyeater

Myzomela erythrocephala   (e-rith'-roh-sef'-ah-lah: "red-headed honey-sucker")

(*G Chapman*)

The Red-headed Honeyeater occurs in eastern Indonesia, New Guinea and the islands of Torres Strait, and in coastal lowlands of northern Australia from about Broome in Western Australia to about Princess Charlotte Bay in Queensland. Virtually confined to mangroves, it is generally common in the west, uncommon in the east. It is mainly sedentary.

The adult male is a tiny blackish honeyeater with a brilliant red head and rump, but females and immature birds are pale brown, nondescript, and very difficult to distinguish from female Scarlet Honeyeaters except by habitat and distribution.

Active, vivacious and inquisitive, it usually occurs alone or in pairs, occasionally in flocks. It feeds on insects and nectar, and forages mainly at flowers and in the upper foliage of trees. Its calls include shrill whistled notes, squeaks, and a lively, jingling, twittering song.

Breeding occurs at any time of year, but especially from October to January. The nest is a frail cup of bark and rootlets, lined with plant down and suspended from a slender fork in the outer foliage of a small tree or bush. Two or three eggs constitute the clutch.

HABITAT: mainly mangroves; occasionally adjacent forest and woodland, and melaleuca swamps
LENGTH: *c* 13 cm
DISTRIBUTION: 30,000–100,000 km^2
ABUNDANCE: sparse to common
STATUS: secure

Dusky Honeyeater

Myzomela obscura (ob-skue'-rah: "dusky honey-sucker")

A sombre greyish brown honeyeater lacking any distinguishing features, the Dusky Honeyeater occurs in New Guinea and eastern Indonesia, and in the coastal lowlands of northern Australia from Port Keates and Melville Island to Port Bradshaw in the Northern Territory, and on Cape York Peninsula and south along the east coast nearly to Southport. Mainly sedentary, it is abundant in the tropics, but numbers steadily decline southwards.

It is usually encountered alone or in pairs, but it occasionally occurs in flocks. Intensely active, tame and confiding, it forages mainly at flowers and in the foliage of bushes and trees, feeding mainly on nectar but also taking insects and fruit. It utters a variety of harsh, single calls, often run together in a series, as well as a twittering song of short, sharp notes.

Breeding occurs mainly from September to December. The nest is a frail cup of fibre and grass, bound with cobweb and lined with finer grass, and suspended in the outer foliage of a bush or tree. Two eggs constitute the clutch.

(WS Peckover)

HABITAT: mangroves, forest, rainforest, coastal scrubland and melaleuca swamps
LENGTH: *c* 15 cm
DISTRIBUTION: 300,000–1 million km²
ABUNDANCE: common
STATUS: secure

Scarlet Honeyeater

Myzomela sanguinolenta (sang'-win-oh-len'-tah: "blood-coloured honey-sucker")

(D & V Blagden)

The Scarlet Honeyeater occurs in Sulawesi, the Moluccas and Lesser Sundas, New Caledonia, and coastal lowlands and associated ranges of eastern Australia from Iron Range in Queensland south to the vicinity of Melbourne. It is much more abundant in the north than in the south, and numbers decline abruptly south of about Sydney. It is strongly nomadic.

Smallest of Australian honey-eaters, the adult male has black wings and tail, a whitish abdomen, and vivid scarlet breast, head and upperparts: the female is pale brown and nondescript. It is active and vivacious, usually occurring in loose groups. Largely dependent on nectar for food, it forages at flowers and in foliage, favouring the upper canopy in woodland, but often snatches insects in midair. Males announce their territories with a pleasant, tinkling, aimless song, delivered from a high perch.

Two successive broods are often raised during a breeding season extending from July to January. The nest is a flimsy cup of bark and rootlets, lined with fine grass and suspended in outer foliage at almost any height in a tree or bush. The usual clutch consists of two eggs, which hatch in about 12 days. The female incubates alone, but both parents participate in rearing the young.

HABITAT: forest, rainforest, woodland, melaleuca swamps, orchards, acacia shrubland, heath, parks and gardens
LENGTH: *c* 11 cm
DISTRIBUTION: 300,000–1 million km²
ABUNDANCE: sparse to common
STATUS: secure

Genus *Philemon*

(fil'-ay-mon: "loving [-bird]")

Typical members of this genus are notable for their extraordinary appearance: the naked head is covered with black, leathery skin (hence an old name, "leatherhead", for the group). However, some species have conventionally feathered heads, the naked areas being reduced to patches on the face. Friarbirds are medium-sized to large honeyeaters with plain plumage (mainly greyish brown above, off-white below), conspicuously lanceolate feathers around the lower neck, and extraordinarily loud raucous calls; most species have a prominent protuberance or casque on the upper mandible. Juveniles have yellowish throats and pale-scalloped mantles. With about 16 species, *Philemon* is one of the largest of the meliphagid genera: four (possibly five) species occur in Australia.

Silver-crowned Friarbird

Philemon argenticeps (ar-jen'-tee-seps: "silver-headed loving-bird")

The Silver-crowned Friarbird is common but local and erratic in occurrence across northern Australia from the Kimberley eastward to the vicinity of Townsville in Queensland. It is mainly a bird of coastal lowlands but, in some areas, its distribution extends inland some distance to semiarid regions, especially along major rivers. It is mainly sedentary.

It is very similar to the Helmeted Friarbird, but somewhat smaller and paler, with a silvery crown and hindneck. Like the Helmeted Friarbird it is noisy, boisterous and aggressive, and spends most of its time in the upper canopy of forest, feeding on fruits, nectar and insects. Often uttered in duet, its calls are loud and raucous, with a distinctive metallic, clanking quality; one characteristic call is often transliterated as "tobacco, uh, more tobacco, uh".

Breeding occurs from September to February. The nest is a large cup suspended by its rim in the outer foliage of a tree, constructed of bark strips, fibre and cobwebs and lined with grass and plant down. Two or three eggs constitute the clutch. The female builds and incubates, but both parents care for the young.

HABITAT: forest (especially of eucalypts and melaleucas), woodland, scrubland, suburban parks and gardens
LENGTH: *c* 32 cm
DISTRIBUTION: more than 1 million km²
ABUNDANCE: common
STATUS: probably secure

(R Whitford)

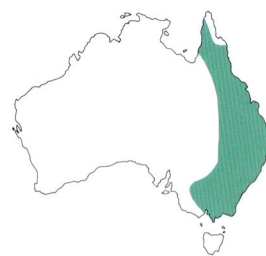

Helmeted Friarbird

Philemon buceroides (bue'-se-roy'-dayz: "*Buceros*-like loving-bird", after a genus of hornbills)

(N Chaffer)

The Helmeted Friarbird is widespread in eastern Indonesia and New Guinea, and in northern and north-eastern Australia, in the Northern Territory from Bynoe Harbor to Yirrkala including Melville and other nearby islands, and in Queensland from Cape York to near Sarina. Most common in coastal lowlands, it is mainly sedentary or locally nomadic. Birds belonging to the population on Melville Island and the adjacent mainland coast are somewhat smaller and occupy a different habitat: it has been argued that these constitute a distinct species.

The Helmeted Friarbird lives mainly in pairs, but small flocks sometimes congregate at flowering trees, where they associate freely with other honeyeaters. It forages mainly in the upper canopy of trees, where it feeds largely on fruits, a diet frequently supplemented with nectar and insects. Like other friarbirds, it is often noisy, conspicuous and boisterous in general behaviour. Its vocabulary includes a range of loud, harsh crackles and squawks, guttural wailing and metallic notes.

Breeding occurs from August to February. The nest is a bulky cup of bark strips, twigs and plant fibre suspended from foliage up to 14 metres from the ground; nests are often located near those of Spangled Drongos, Figbirds or Shining Starlings. Usually three or four eggs form the clutch. The female builds and incubates alone, but both parents cooperate in raising the young.

HABITAT: mainly rainforest edges; also eucalypt forest, woodland, mangroves, orchards, swamps and suburban parks and gardens
LENGTH: *c* 37 cm
DISTRIBUTION: 100,000–300,000 km²
ABUNDANCE: common
STATUS: secure

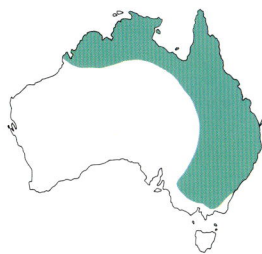

Little Friarbird

Philemon citreogularis (sit'-ray-oh-gue-lar'-is: "lemon-throated loving-bird")

This common and widespread friarbird inhabits northern and eastern Australia from south-eastern South Australia to the vicinity of Broome in Western Australia, mainly inland of the Great Dividing Range.

(WS Peckover)

It is migratory in the south, nomadic in the north, and numbers tend to decline southward. It also occurs in eastern Indonesia and in southern New Guinea.

It differs from other Australian friarbirds in that only the face (not the entire head) is naked, and the skin is bluish grey, not black; it also lacks the characteristic friarbird casque on the bill. In its general behaviour it resembles other friarbirds. Noisy, active and pugnacious, it frequents the upper canopy of woodland in pairs or small parties. Somewhat more dependent on nectar than other friarbirds, it is most numerous and conspicuous at flowering trees, but it also eats fruits and insects. Its varied diet includes a relatively musical song.

Two successive broods are usually raised during a breeding season extending from August to April. The nest is a suspended cup of bark strips and fibre, lined with grass, and placed 2 to 10 metres from the ground in the outer foliage of a tree. Two to four eggs constitute the clutch. The female incubates alone but both parents rear the young.

HABITAT: mainly open woodland; also mangroves, suburban parks and gardens, orchards
LENGTH: *c* 29 cm
DISTRIBUTION: more than 1 million km²
ABUNDANCE: common
STATUS: secure

Noisy Friarbird

Philemon corniculatus (korn-ik'-ue-lah'-tus: "little-horned loving-bird")

The Noisy Friarbird inhabits south-eastern New Guinea and eastern Australia from Cape York Peninsula to the vicinity of Melbourne, rarely extending westward along major rivers into South Australia. It is migratory in the south, nomadic in the north, and numbers tend to decline northwards.

It usually occurs in pairs or small parties, but it often congregates in large numbers at groves of flowering trees, when it is even more noisy, boisterous and aggressive than other friarbirds. It is strongly arboreal, preferring the upper canopy and outer foliage, but it occasionally forages on the ground and frequently hawks for insects in midair. Its diet includes fruits and berries (both wild and cultivated), nectar, insects, and eggs and nestling birds. Its extraordinary calls are extremely varied, and often uttered in duet or chorus. They include many loud, raucous, and harsh shouts and cackling notes; most distinctive is a rollicking "four-o'clock" or "chok-chok".

Sometimes two broods are raised in succession during a breeding season extending from July to February. Suspended by its rim up to 10 metres from the ground in the outer foliage of a tree, the nest is a bulky cup-shaped structure of grass, twigs, bark fragments, string, spider egg cocoons and similar miscellaneous material, lined with wool and plant down. Two to four eggs constitute the clutch. The female incubates alone but both parents rear the young.

HABITAT: mainly open eucalypt woodland; heath; suburban parks and gardens
LENGTH: *c* 36 cm
DISTRIBUTION: 300,000–1 million km^2
ABUNDANCE: common
STATUS: secure

(J-P Ferrero)

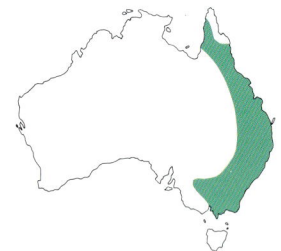

Genus *Phylidonyris*

(fil'-id-on'-i-ris: pleasure-lover)

The so-called "yellow-winged" honeyeaters constitute a cluster of four species in which the plumage is mainly white, very heavily streaked with black, and there is a large conspicuous panel of bright yellow in the wing. All are similar in size and proportions, rely heavily on nectar for food, and are essentially nomadic inhabitants of shrubby habitats of southern Australia, where several species often occur together. The four species tend to be isolated ecologically, respectively inhabiting tall dense humid forest, open sclerophyll woodland, heathland, and arid interior scrubs. They congregate at flowering trees but, although very high densities are sometimes reached, there is little evidence of communalism. All are intensely active and aggressive, and have sharp, metallic calls.

White-fronted Honeyeater

Phylidonyris albifrons (ahl'-bee-fronz: "white-forehead pleasure-lover")

The White-fronted Honeyeater is widespread across most of Australia south of the tropics and west of the Great Dividing Range. It is strongly nomadic and erratic in its occurrence but is generally somewhat more numerous in the southern interior than in the north. The arid-zone representative of the *Phylidonyris* honeyeaters, it strongly resembles the other species in general appearance and behaviour, being active, gregarious, aggressive and conspicuous, foraging mainly in flowering bushes and shrubs and largely dependent on nectar.

Breeding is usually loosely colonial, and its timing is strongly influenced by local rainfall and the flowering of trees and shrubs, but tends to reach a peak between August and December. Usually well hidden within a metre or two of the ground, the nest is an open cup of fine grass and bark bound with cobweb and lined with plant down. The clutch consists of two or three eggs, which hatch in about 14 days. The young fledge at about 14 days.

HABITAT: arid woodland and scrubs with an understorey of flowering shrubs
LENGTH: *c* 18 cm
DISTRIBUTION: more than 1 million km²
ABUNDANCE: sparse
STATUS: probably secure

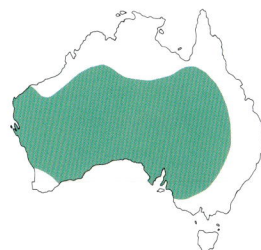

(WJ Labbett)

White-cheeked Honeyeater

Phylidonyris nigra (ni'-grah: "black pleasure-lover")

Essentially a bird of coastal heaths, the White-cheeked Honeyeater inhabits eastern Australia from near Cairns in Queensland to about Jervis Bay in New South Wales, and in south-western Australia from near Geraldton to about Eyre. It is very common in the east but somewhat scarce and patchy in distribution in the west. It is strongly nomadic, following the flowering cycles of various shrubs.

Like the New Holland Honeyeater with which it often occurs, it is intensely active, noisy, aggressive and conspicuous. It feeds mainly on nectar but often hawks for insects in midair and sometimes forages in foliage, gleaning insects and manna. Breeding males announce their small territories (usually encompassing only the immediate vicinity of the nest and a few adjacent song-perches) by means of a high-pitched, rapid, musical whistled song, delivered either from a conspicuous high perch or during an erratic, spiralling, near-vertical song flight.

Breeding, which is often loosely colonial, occurs at any time of year, but tends to reach two peaks of activity: from August to November and again from March to May. Well hidden in dense foliage seldom more than a metre from the ground, the nest is a bulky cup of fibre, bark and grass, bound with cobweb and lined with plant down. The clutch consists of one to three eggs, which hatch in about 15 days. Reared by both parents, the young fledge at about 15 days.

HABITAT: mainly heathland and eucalypt forest and woodland with a dense understorey of flowering shrubs; suburban parks and gardens
LENGTH: *c* 20 cm
DISTRIBUTION: 300,000–1 million km²
ABUNDANCE: common
STATUS: secure

(JD Waterhouse)

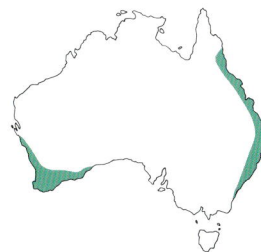

397

New Holland Honeyeater

Phylidonyris novaehollandiae (noh'-vee-hol-and'-ee-ee: "New-Holland pleasure-lover")

(J Handel)

The New Holland Honeyeater is abundant and widespread in south-eastern Australia, from the vicinity of Brisbane south to Tasmania and west into south-eastern South Australia, also in extreme south-western Western Australia. It is more numerous in the south than in the north, and it is essentially a bird of the coastal lowlands and associated highlands. It is strongly nomadic, following the flowering cycles of various shrubs, especially banksias.

It is often confused with the White-cheeked Honeyeater, which closely resembles it in appearance and general behaviour, but the New Holland Honeyeater has a white iris, conspicuous against a mainly black face and cheeks. It also tends to occupy a wider range of habitats, but the two species often occur together.

Usually encountered in pairs or small flocks, it often reaches very high densities on heathland at peak flowering times. In forest, it forages mainly in the middle and lower levels, especially in the shrub layer. Like other *Phylidonyris* honeyeaters, it depends primarily on nectar, but it also frequently hawks for insects. Intensely active, noisy, aggressive and conspicuous, it utters a variety of terse, harsh notes and rapid shrill whistled phrases.

Sometimes loosely colonial, breeding may occur at any time of year, but tends to reach a peak from July to December and again from March to May, and several broods may be raised in succession. The nest is a cup-shaped structure of bark and grass, bound with cobwebs and lined with softer material. It is usually placed in an emergent bush or small tree somewhat above the level of surrounding vegetation. Two or three eggs form the clutch. Incubation takes about 18 days. The female incubates alone but both parents participate in rearing the young, which fledge at about 16 days.

HABITAT: mainly dense tall heathland; also forest and woodland with an understorey of flowering shrubs; suburban parks and gardens
LENGTH: *c* 18 cm
DISTRIBUTION: more than 1 million km²
ABUNDANCE: common
STATUS: secure

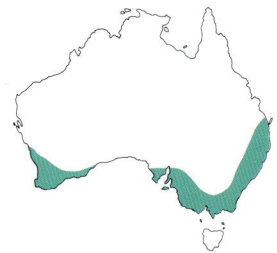

Crescent Honeyeater

Phylidonyris pyrrhoptera (pi'-rop-te'-rah: "fire-colour-winged pleasure-lover")

(J Gray)

The Crescent Honeyeater closely resembles other *Phylidonyris* honeyeaters in general appearance and behaviour, but it inhabits much denser and wetter sclerophyll forests than the others, and is especially associated with higher altitudes. It is widespread in south-eastern Australia from Kangaroo Island and the Mount Lofty Ranges of South Australia through Victoria and Tasmania and north to the Gloucester Tops region of New South Wales. It is abundant in Tasmania but somewhat less numerous on the mainland. It is mainly sedentary but, at higher altitudes, there is a marked coastwards dispersal to lower levels during winter.

It lives mainly in pairs but congregates at flowering trees and shrubs. Strongly arboreal, it forages at all levels but tends to favour the middle storeys of forest. Like other *Phylidonyris* honeyeaters it depends primarily upon nectar, but it also feeds on insects (gleaned from foliage or snapped up in midair), manna and honeydew. It is active, noisy, aggressive and conspicuous. Its most distinctive call is a loud, sharp "E-gypt!".

Breeding may occur at any time but tends to peak between July and January and again in March and April. Usually placed in a bush or grass tussock within a metre or two of the ground, the nest is an open cup of bark fragments and twigs, lined with plant down. The clutch consists of two or three eggs, which hatch in about 14 days. The female builds and incubates unaided but both parents participate in raising the young, which fledge at about 14 days.

HABITAT: mainly wet eucalypt forest with a dense understorey, especially in highlands; also heaths, suburban parks and gardens
LENGTH: *c* 16 cm
DISTRIBUTION: 100,000–300,000 km²
ABUNDANCE: common
STATUS: secure

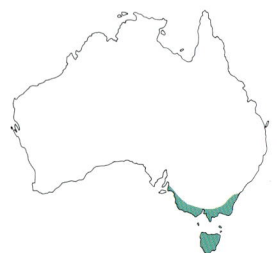

Genus *Plectorhyncha*

(plek'-toh-rink'-ah: "twisted-bill" probably intended to be *plectrorhyncha*, "pointed-bill")

The characteristics of the genus are essentially those of the single species.

Striped Honeyeater

Plectorhyncha lanceolata (lahn'-say-oh-lah'-tah: "lance-shaped pointed-bill")

This distinctively marked honeyeater is common and widespread in southern and eastern Australia north to the vicinity of Charters Towers in Queensland, but it avoids high-rainfall coastal districts in much of

(GK Taylor)

southern Victoria and in eastern New South Wales north to about Toukley. It is mainly sedentary or locally nomadic.

It lives in small groups, seldom congregating in large flocks, and foraging mainly in thick outer foliage of trees. Though vigorous in defence of its territory, it is markedly less aggressive to other species than many honeyeaters. Its diet is varied but it depends more upon fruits and insects than on nectar. Often uttered in duet or chorus, the melodious song is a rollicking, bubbling series of notes.

Breeding occurs mainly from August to January. Suspended in outer foliage seldom more than a few metres from the ground, the nest is a cup of grass and fibre, lined with fine grass and sometimes decorated with Emu feathers. The clutch varies from two to five eggs, and both parents participate in raising the young, often assisted by other birds.

HABITAT: mainly open woodland and acacia scrubland; melaleuca swamps, mangroves, suburban parks and gardens
LENGTH: *c* 23 cm
DISTRIBUTION: more than 1 million km²
ABUNDANCE: common
STATUS: secure

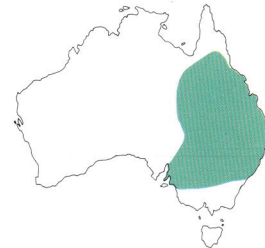

Genus *Ramsayornis*

(ram'-zee-orn'-is: "Ramsay-bird", after E. P. Ramsay, Australian ornithologist)

Members of this genus are notable for building domed nests very different in structure from the suspended cup-shaped nests of other honeyeaters. Otherwise the characteristics of the genus are essentially those of the two very similar included species.

Bar-breasted Honeyeater

Ramsayornis fasciatus (fah'-see-ah'-tus: "banded Ramsay-bird")

The Bar-breasted Honeyeater is widespread in northern Australia from the vicinity of Broome, Western Australia, to near Rockhampton in Queensland. It is common and locally nomadic: largely dependent on nectar, it follows the flowering cycles of various trees and shrubs.

(W Lawler)

It also eats insects. Its distribution overlaps that of the closely related Brown-backed Honeyeater in some areas, and both species are strongly associated with melaleuca swamps and streamside vegetation: where they meet, the Bar-breasted Honeyeater tends to favour drier, scrubbier habitats.

Breeding may occur at any time of year but reaches a peak of activity between October and March. The nest is a domed structure with a hooded side entrance, constructed mainly of bark fragments and lined with plant down and fibre, usually suspended several metres over water. Three or four eggs constitute the clutch.

HABITAT: mainly open woodland, melaleuca swamps, and riverside scrub
LENGTH: *c* 15 cm
DISTRIBUTION: 300,000–1 million km²
ABUNDANCE: sparse to common
STATUS: probably secure

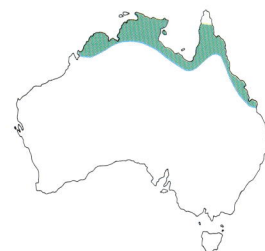

Brown-backed Honeyeater

Ramsayornis modestus (mod-est'-us: "modest Ramsay-bird")

Mainly plain brown above and white below, this small nondescript honeyeater is widespread in New Guinea and occurs in north-eastern Queensland from Cape York Peninsula south to about Weipa in the west and Bowen in the east, where it is very common but strongly associated with melaleuca swamps and similar riverside vegetation. It is migratory, wintering in New Guinea and largely absent from Australia from about May to August. It is thought to migrate in flocks at night.

Usually encountered in pairs, it is very active but rather unobtrusive, foraging in woodland at all levels from undergrowth to the upper canopy and feeding largely on insects gleaned from leaves or snapped up in midair. It also visits flowering shrubs for nectar. Its calls are brief and simple, the most distinctive being a sharp, repeated "chik" or "mick".

Sometimes loosely colonial, breeding occurs mainly from November to April. The nest is a domed structure with a hooded side entrance suspended in the outer foliage of a bush or tree seldom more than a few metres from the ground, and usually overhanging water. It is built mainly of bark fragments and lined with bark fibre. Two or three eggs constitute the clutch. Incubation and fledging periods are both about 15 days. Both parents cooperate in building the nest and rearing the young, but only the female incubates.

HABITAT: woodland, rainforest edges, forest, mangrove and melaleuca swamps
LENGTH: *c* 13 cm
DISTRIBUTION: 100,000–300,000 km²
ABUNDANCE: common
STATUS: secure

(N Chaffer)

Genus *Trichodere*

(trik'-oh-day'-ray: "hairy-neck")

The characteristics of the genus are essentially those of the single species.

White-streaked Honeyeater

Trichodere cockerelli (cok'-er-il-ee: "Cockerell's hairy-neck", after J. T. Cockerell, sometime collector for John Gould)

The White-streaked Honeyeater is confined to northern Cape York Peninsula south to near Weipa in the west and to Shiptons Flat near Cooktown in the east. It is common and apparently locally nomadic.

Strongly territorial, it is usually encountered alone or in pairs, but it often congregates in flocks at flowering trees. It feeds on nectar and insects, gathered mainly in the upper foliage of woodland and heath. It is active, noisy and conspicuous.

Breeding occurs from January to April. The nest is a frail suspended cup of rootlets bound with cobwebs, placed in the fork of a tree or bush within a metre or two of the ground. Two eggs constitute the clutch.

HABITAT: mainly heathland and melaleuca swamps; also forest, rainforest, woodland, mangroves, and riverside scrub
LENGTH: *c* 20 cm
DISTRIBUTION: 30,000–100,000 km²
ABUNDANCE: sparse to common
STATUS: probably secure

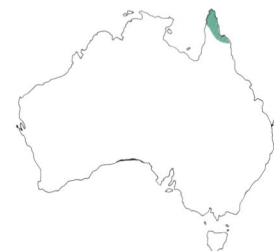

(D & M Trounson)

Genus *Xanthomyza*

(ksan'-thoh-mie'-zah: "yellow-sucker")

The characteristics of the genus are essentially those of the single species.

Regent Honeyeater

Xanthomyza phrygia (fri'-jee-ah: "embroidered yellow-sucker")

The Regent Honeyeater is widespread in south-eastern Australia, approximately from Adelaide to Rockhampton. Once common, it has suffered a marked decline since the early 1900s and its range is apparently contracting. Now virtually confined to Victoria and New South Wales, it is most numerous along the western slopes and outliers of the Great Dividing Range. The cause of the decline is unknown, but it is perhaps associated with the widespread clearing, for agriculture, of eucalypt forest in valleys. It is nomadic.

Gregarious when not breeding, it is usually encountered in small parties (although, in the past, it sometimes occurred in large flocks). Noisy, aggressive and conspicuous, it forages mainly in flowers and foliage in the upper canopy, but it descends to the ground to bathe in ponds or roadside puddles. Its diet includes nectar, fruits and insects. It has an extensive repertoire of calls, including quiet metallic tinkling notes and loud rollicking yelps.

Breeding occurs from August to January. The nest is a compact cup of bark strips bound with cobweb, lined with plant down, and built into a fork of a tree 1 to 20 metres above ground. Two or three eggs form the clutch. The female incubates unaided but both parents participate in rearing the young.

HABITAT: mainly eucalypt forest and woodland
LENGTH: *c* 22
DISTRIBUTION: 300,000–1 million km²
ABUNDANCE: very sparse
STATUS: vulnerable

(*G Chapman*)

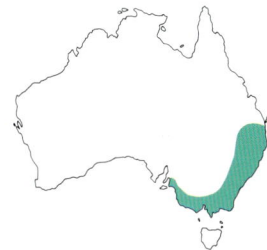

Genus *Xanthotis*

(ksan-thoh'-tis: "yellow-ear")

Confined to Australia and New Guinea, and closely related to the genus *Meliphaga*, this genus is distinguished by the rather long slender bills, unusually mottled and streaked plumage, and naked face patches of its members. There are three or four species (two in Australia), all of which inhabit rainforest (especially in lowlands) and eat insects, fruits and nectar.

Tawny-breasted Honeyeater

Xanthotis flaviventer (flah'-vee-ven'-ter: "yellow-bellied yellow-ear")

The Tawny-breasted Honeyeater inhabits New Guinea and the northern part of Cape York Peninsula (south to the McIlwraith Range on the east coast, and Malaman Creek on the west), mainly in coastal lowlands. It is sedentary and locally common.

Arboreal, rather solitary and relatively undemonstrative, it strongly resembles Macleay's Honeyeater in general behaviour. It feeds mainly on insects, but also eats fruits and nectar. It is very vocal, and its calls include a variety of loud whistles and chattering notes. It breeds mainly from November to April. The nest is a cup of bark strips, lined with fine rootlets and bark, and suspended by its rim in a tree fork 3 to 15 metres from the ground. One or two eggs form the clutch. Incubation takes about 15 days and the young fledge at about 14 days. As in other honeyeaters, the female builds and incubates unaided, but both parents rear the young.

HABITAT: mainly rainforest, especially margins, clearings and damaged areas
LENGTH: *c* 22 cm
DISTRIBUTION: 30,000–100,000 km²
ABUNDANCE: sparse
STATUS: secure

(D & M Trounson)

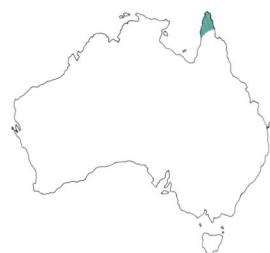

Macleay's Honeyeater

Xanthotis macleayana (mak-lay'-ah'-nah: "Macleay's yellow-ear", after W. Macleay, Australian biologist)

Macleay's Honeyeater is a widespread, common and familiar honeyeater within its relatively restricted distribution in north-eastern Queensland, approximately from Cooktown south to Paluma, including Hinchinbrook Island. It is sedentary, and somewhat more numerous in highland forests than in the coastal lowlands.

It is usually encountered alone or in pairs but flocks sometimes congregate at rich food sources. It is strongly arboreal, foraging mainly in foliage, in epiphytes and on the bark of trees, favouring the middle to lower canopy levels. Especially characteristic is its habit of rummaging noisily in pendent tangles of dead leaves. Compared to most honeyeaters, it is relatively unobtrusive, methodical, and undemonstrative in behaviour but it is often noisy. It feeds mainly on insects and spiders but also eats nectar and fruits, both native and cultivated.

Breeding occurs from October to January. Usually hidden in a vine tangle several metres from the ground, the nest is a small cup of dead leaves and bark bound with cobwebs, decorated with spider egg cocoons and lined with finer plant fibres. Two eggs form the clutch.

HABITAT: mainly rainforest; also occasionally open forest; mangroves; suburban parks and gardens
LENGTH: *c* 21 cm
DISTRIBUTION: 30,000–100,000 km²
ABUNDANCE: common
STATUS: probably secure

(A Evans)

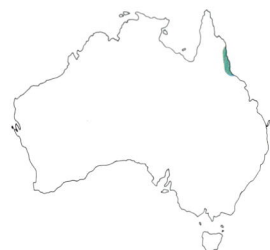

Family DICAEIDAE

(die-see'-id-ee: "*Dicaeum*-family", after a genus of flower-peckers)

Containing about 55 species, the family Dicaeidae constitutes a group of small, arboreal, mainly nectar- and fruit-eating birds, widespread across southern Asia; one species occurs in Australia. Similarities in the structure of the tongue and other features indicate a relationship with the sunbirds, but the family has no other close relatives in the Australian region. The pardalotes (Pardalotidae) were formerly included in the Dicaeidae, but their similarities are now considered to be superficial, and not to indicate a relationship.

Several aberrant genera inhabit specialised environments in New Guinea, the Philippines and the Indonesian region, but the majority of species are included in two closely related genera, *Prionochilus* and *Dicaeum*; the latter has one Australian member.

Genus Dicaeum

(die-see'-um: "flower-pecker")

The characteristics of the genus are discussed under the family heading.

Mistletoebird

Dicaeum hirundinaceum (hi'-roon-din'-ah-say'-um: "swallow-like flower-pecker")

The life histories of the Mistletoebird and the many species of Australian mistletoes are intimately entwined in an unusual commensal relationship: the bird feeds almost exclusively on the berries of these plants, voiding the seeds undamaged; the plants in turn profit by their widespread dissemination. Also present in eastern Indonesia, the Mistletoebird occurs across mainland Australia wherever the plants occur, but is rare or absent where these are absent. It is strongly nomadic.

Tiny, dumpy and short-tailed, the Mistletoebird is unmistakable. The adult male is glossy blue-black above and white below, with a bright red breast and undertail coverts; the female is dull brown above and pallid below, with dusky pink undertail coverts. Young birds resemble adult females.

The Mistletoebird is usually encountered alone or in pairs. Strictly arboreal, it generally forages in the upper canopy. It has a distinctive high piercing call-note, and its flight is swift and erratic, usually high over the treetops.

Most breeding occurs from September to March. The nest is a neat purse-shaped structure of plant down and cobwebs felted together, and suspended in the outer foliage of a tree or shrub. often within a metre or two of the ground. The clutch consists of three eggs. The female alone incubates, but both adults care for the young, which fledge in about 15 days.

HABITAT: all woodland habitats, wherever there is mistletoe
LENGTH: *c* 11
DISTRIBUTION: more than 1 million km²
ABUNDANCE: common
STATUS: secure

(JD Waterhouse)

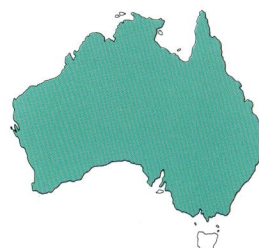

Family NECTARINIIDAE

(nek'-tar-in-ee'-id-ee: "*Nectarinia*-family", after a genus of sunbirds)

Mainly tropical in distribution, sunbirds are colourful, vivacious and intensely active small birds that dart and hover around flowers, feeding mainly on nectar and small insects. About 117 species constitute the family, which is best represented in Africa but has numerous Asian members, one extending to Australia. There are several aberrant species and species groups (such as the spiderhunters of South-East Asia), but most belong in the genus *Nectarinia*, one of the largest of avian genera.

In most respects the single Australian species is typical of the group as a whole, although somewhat duller in plumage than many. The family is closely related to the flowerpeckers (Dicaeidae) but seems very distinct from the honeyeaters (Meliphagidae) and white-eyes (Zosteropidae).

Genus Nectarinia

(nek'-tar-in'-ee-ah: "nectar[-bird]")

The characteristics of the genus are essentially those of the single Australian species.

Yellow-bellied Sunbird

Nectarinia jugularis (jue'-gue-lar'-is: "necked nectar[-bird]")

A common and familiar bird in tropical parks and gardens, the Yellow-bellied Sunbird inhabits north-eastern Queensland from Cape York to about Gladstone; there is some evidence of a steady southward expansion. Elsewhere it is widespread from South-East Asia to the Solomon Islands. It is mainly a bird of coastal lowlands but it is also present, locally and in small numbers, in adjacent highlands such as the Atherton Tableland. It is sedentary.

It is usually encountered in pairs, but small parties may form after the breeding season. Feeding on nectar and small insects and spiders, it forages at flowers and foliage, favouring undergrowth and lower levels of forest. Its flight is swift, agile and darting, and it frequently hovers at flowers.

The sexes are similar except that the female's underparts are entirely yellow, whereas the adult male has a brilliant metallic blue throat; immatures resemble adult females.

Almost impossible to confuse with any other bird, it is small, vivacious and intensely active, and so confiding that it freely enters houses to investigate cut flowers. Verandas and outhouses are frequently selected as sites for the long, pendent nest.

Breeding may occur at any time of year, but especially from August to March; often several broods are raised in succession. The elaborate nest is constructed of fibre, plant down and bark fragments, and has a hooded side entrance and a long dangling tail. Two or three eggs are laid, and incubation and fledging both take about 15 days. The female builds the nest (sometimes casually assisted by the male) and incubates, but both parents feed the young.

HABITAT: tropical woodland, rainforest, and mangroves; parks and gardens
LENGTH: *c* 13 cm
DISTRIBUTION: 100,000–300,000 km²
ABUNDANCE: common
STATUS: secure

(*T & P Gardner*)

Family ZOSTEROPIDAE

(zos'-ter-oh'-pid-ee: "*Zosterops*-family", after a genus of white-eyes)

About 80 species of small songbirds comprise the unusually homogeneous family Zosteropidae, widespread in Africa, southern Asia and the Australasian region. About 20 species are aberrant, belong to small genera, and inhabit various islands and island groups in the Oriental region and Micronesia, but all others belong to the genus *Zosterops*, one of the largest of all avian genera, and representative of the family as a whole.

Members of the Zosteropidae are widely known as white-eyes or silvereyes because almost all species have a conspicuous narrow band of white feathers rimming the eye. The various species differ little in plumage, and species limits within the group are still uncertain. The sexes are similar, there is no seasonal variation, and immatures resemble adults. Both parents cooperate in nest building, incubation, and the care of the young. White-eyes are arboreal, gregarious, common and widespread.

Genus Zosterops

(zos'-ter-ops: "banded-face")

The characteristics of the genus are discussed in the account of the family.

Pale White-eye

Zosterops citrinella (sit'-rin-el'-ah: "little-yellow banded-face")

The Pale White-eye occurs on a number of small islands in Torres Strait and south along the east coast to near Cooktown in Queensland; elsewhere it occurs on Tanimbar and the Lesser Sunda Islands in Indonesia. Sedentary and locally common, it closely resembles the Silvereye in appearance but its head and back are light olive yellow and its throat is lemon yellow.

Like other white-eyes it is gregarious and arboreal, but the islands that it inhabits are seldom visited, and little is known of its habits. It eats insects and berries.

The breeding season is uncertain. The nest resembles that of the Silvereye, and two to four eggs form the usual clutch.

HABITAT: low trees and scrubby thickets, always on small tropical islands

LENGTH: *c* 12 cm

DISTRIBUTION: 10,000–30,000 km^2

ABUNDANCE: common

STATUS: probably secure

(VA Drake)

Silvereye

Zosterops lateralis   (lat'-er-ah'-lis: "[notable-]sided banded-face")

A small greenish bird of low foliage and garden shrubbery, the Silvereye is among the most familiar of small native songbirds to most Australians. It is easily identified by its conspicuous white eye ring; the amount of grey on the back and of chestnut tinge on the flanks varies markedly among the various populations.

It is common to abundant in eastern and southern Australia from north of Cooktown in Queensland to the vicinity of Shark Bay in Western Australia. It also occurs in New Zealand and on many islands of the south-western Pacific, including Norfolk and Lord Howe Islands. In most areas it is largely sedentary, but Tasmanian birds migrate to the south-eastern mainland every autumn, returning in the spring.

It usually occurs in loose wandering flocks, from which pairs, mated for life, periodically drop out to breed, rejoining the flock later. It is mainly arboreal but it often feeds on the ground. Its most characteristic calls include a thin high peevish "cheee" and a long, rambling, twittering song. It eats insects, small spiders, and fruit; in some areas it sometimes causes damage in orchards and vineyards.

Breeding occurs from September to February, and usually two broods are raised in succession. The nest is a small cup of grass and other plant material bound with cobweb, suspended by the rim from a slender tree fork. Two to four eggs form the clutch; the incubation period is 10 to 13 days, and the young fledge at about 9 to 12 days.

(MF Soper)

HABITAT: cool temperate to tropical forest and woodland, especially shrubbery, heathland, suburban parks and gardens
LENGTH: *c* 12.5 cm
DISTRIBUTION: more than 1 million km²
ABUNDANCE: common
STATUS: secure

Yellow White-eye

Zosterops lutea   (lue-tay'-ah: "yellow banded-face")

The Yellow White-eye inhabits the coastal lowlands of tropical Australia from the Peron Peninsula in Western Australia to the vicinity of Weipa on the west coast of Cape York Peninsula; a small population also occurs near Ayr in north-eastern Queensland.

It closely resembles the Silvereye in general appearance and habits but differs in having plumage that is strongly tinged with yellow. It is mainly sedentary, but tends to disperse locally in winter.

It feeds mainly on insects and small spiders, also taking small fruits and berries.

Breeding occurs from October to January. The nest is a small cup of grasses and other plant material, suspended by its rim in mangroves. The usual clutch consists of two to four eggs.

(B & B Wells)

HABITAT: mainly tropical mangroves; sometimes adjacent scrubland and sand dunes
LENGTH: *c* 12 cm
DISTRIBUTION: 100,000–300,000 km²
ABUNDANCE: common
STATUS: secure

Family ESTRILDIDAE

(es-tril'-did-ee: "*Estrilda*-family", after a genus of waxbills)

This is a family of about 137 species in 27 genera widespread in the Old World, especially in the tropics. Their classification is controversial, but the family can be roughly divided into three groups: waxbills, grassfinches, and mannikins. Waxbills occur in Africa and India, but the grassfinches and mannikins are well represented in Australia, where 17 species occur naturally, and one species (the Nutmeg Mannikin, *Lonchura punctulata*) has been introduced and is thriving.

Members of the family are generally very small, colourful, gregarious and active birds that are, as the name "grassfinch" suggests, strongly associated with grasslands of various kinds. They feed mainly on seeds, sometimes supplemented with insects. Plumage patterns are usually intricate, the sexes are similar or nearly so, there is no seasonal variation, and immatures resemble adults. The pair bond is usually permanent, and both sexes cooperate fully in the nesting cycle. Body contact and mutual preening between mated pairs are common, and many species build nests specifically for the purpose of roosting communally at night.

Genus Aidemosyne

(ie'-dem-oh'-sin-ay: "modest [-bird]")

The characteristics of the genus are those of the single species.

Plum-headed Finch

Aidemosyne modesta (mod-est'-ah: "modest modest-bird")

The Plum-headed Finch is a strongly nomadic inhabitant of rank grasslands on the western slopes and tablelands of the Great Dividing Range from the Atherton Tableland south to the vicinity of Canberra. Its movements and status are difficult to assess, but it seems rarer in the north than in the south and its numbers may be declining. In behaviour and ecology it strongly resembles the Star Finch, living and breeding in groups of about 20 to 50 birds, which often coalesce into larger flocks in winter. The pair bond is permanent. Roost-nests are not built. The Plum-headed Finch eats seeds, gathered on the ground or taken directly from the plant, and insects.

Breeding occurs mainly from September to January. The nest is spherical, with a side entrance but no entrance tunnel. It is built of green grass and usually placed close to the ground in a dense bush. The clutch usually consists of five or six eggs. Incubation takes 12 to 14 days, and the young fledge at about 21 days. Nest construction is mainly by the female (the male fetches material) but both parents incubate and rear the young.

HABITAT: tall grass in dry temperate eucalypt woodlands
LENGTH: 10–11 cm
DISTRIBUTION: 300,000–1 million km²
ABUNDANCE: sparse to common
STATUS: probably secure

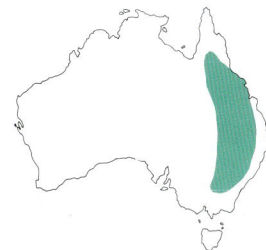

(*P Sellers*)

Genus Emblema

(em'-blay'-mah: "tiled [-bird]", in reference to tile-like patterning of plumage)

Members of this genus are characterised by their bright red rumps and mainly southern distribution. Most inhabit dense heaths, shrubbery, and forest undergrowth, but the Painted Firetail is an exception, being adapted to an arid environment. The conventional inclusion of the Red-browed Firetail is followed here, but it probably should be isolated in a separate genus, *Aegintha*.

Beautiful Firetail

Emblema bellum (bel'-um: "beautiful tiled-bird")

(RH Green)

This shy, solitary and elusive inhabitant of south-eastern heathlands reaches its maximum abundance in Tasmania, but it also has a fragmented mainland distribution extending from the vicinity of Sydney in New South Wales to near Adelaide in South Australia (including Kangaroo Island). It may be declining, being threatened by feral cats, wildfires and land clearing. It lives in permanent pairs, frequenting dense cover and feeding on insects and seeds. At night it sleeps in roost-nests specially constructed for the purpose. In winter it sometimes congregates in small flocks.

Breeding occurs from October to January. Built of small sticks and grass, the nest is a globular structure with a tunnel leading to an entrance in the side. The clutch varies from two to six eggs. Both parents incubate and rear the young.

HABITAT: mainly cool temperate heaths; melaleuca thickets; swamps
LENGTH: 12–13 cm
DISTRIBUTION: 100,000–300,000 km²
ABUNDANCE: very sparse
STATUS: vulnerable

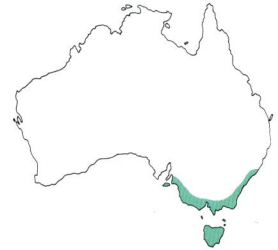

Diamond Firetail

Emblema guttata (gut-ah'-tah: "spotted tiled-bird")

The Diamond Firetail is widespread in south-eastern Australia, especially west of the Great Dividing Range. It is uncommon and patchily distributed in coastal districts and has declined markedly in some areas, probably as a result of land clearing and illegal trapping for aviculture. It is generally sedentary. Not especially sociable, it lives in pairs or small groups that may coalesce into somewhat larger flocks in winter. It feeds mainly on seeds gathered on the ground but also eats insects.

Breeding occurs mostly from August to January, and several pairs may nest in loose association. The nest is a bulky, flask-shaped structure with a side entrance approached by a woven tunnel. It is built of grass and lined with feathers and placed in dense foliage in a bush or mistletoe clump, usually several metres from the ground. Four to seven eggs form the clutch, and they take 12 to 14 days to hatch. The young fledge at about 10 to 12 days. Both sexes build the nest and incubate and rear the young.

(CL Gill)

HABITAT: temperate eucalypt woodland; mallee; agricultural land
LENGTH: 12–13 cm
DISTRIBUTION: more than 1 million km²
ABUNDANCE: sparse
STATUS: vulnerable

Red-eared Firetail

Emblema oculata (ok'-ue-lah'-tah: "eye-marked tiled-bird", in reference to eye-like spots in plumage)

(*G Rogerson*)

Shy, solitary and elusive, the Red-eared Firetail is closely related to the Beautiful Firetail and occupies the same niche in south-western heathlands as does the Beautiful Firetail in the east. It closely resembles that species in most aspects of its behaviour.

Breeding occurs from September to January. Built of small sticks and grass, the nest is a globular structure with a tunnel leading to an entrance in the side. The clutch, which is incubated by both parents, varies from two to six eggs.

HABITAT: mainly temperate heaths; melaleuca thickets; swamps
LENGTH: 12–13 cm
DISTRIBUTION: 30,000–100,000 km²
ABUNDANCE: sparse
STATUS: vulnerable

Painted Firetail

Emblema picta (pik'-tah: "painted tiled-bird")

The Painted Firetail lives in rocky gorges and along escarpments in the arid northern interior of Australia, never far from water. It is sedentary and lives permanently in pairs or small groups. It forages almost entirely on the ground or on rocks, its diet consisting largely of the seeds of spinifex.

The pair bond is not especially strong: body contact is avoided, and mated birds do not roost together. The Painted Firetail does not build roost-nests. It also differs from other Australian grassfinches by drinking in the "scoop-tilt-and-swallow" technique used by most birds—other grassfinches drink by sucking water into the throat.

Breeding is influenced by local rainfall, and may occur at any time of year. The nest is a globular structure with a side entrance, constructed of twigs, stalks of spinifex and other plant material. Three or four eggs form the clutch, which is incubated by both parents.

HABITAT: mainly warm temperate rock outcrops, gorges and hillsides in spinifex grassland and acacia scrub
LENGTH: 11–12 cm
DISTRIBUTION: more than 1 million km²
ABUNDANCE: sparse to common
STATUS: probably secure

(*D & M Trounson*)

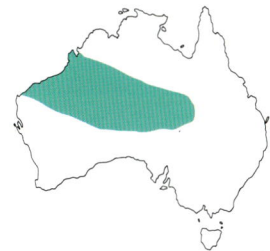

Red-browed Firetail

Emblema temporalis (tem'-por-ah'-lis: "templed tiled-bird")

This familiar finch is common along the coast and associated highlands of eastern Australia from Cape York to the vicinity of Adelaide, including Kangaroo Island but not Tasmania. It has been introduced to Tahiti and a feral population persists precariously in the Darling Ranges in south-western Australia.

After breeding it congregates in flocks that may wander locally, but it is generally sedentary.

Red-browed Firetails feed mainly on the ground, the flock scattering to gather small seeds and, sometimes, insects and small fruits, and flying up to shelter in low bushes if disturbed. Flocks build special roosting nests in which to spend the night. The pair bond is permanent and very strong; even in winter flocks the members of a pair stay close, roosting together and frequently preening each other.

Breeding occurs from August to December in the south, December to April in the north. Constructed mainly of grass and lined with feathers, the nest is a flask-shaped structure with a hooded side entrance; it is often placed in a thorny bush within a few metres of the ground. Four to six eggs form the clutch. Incubation (by both sexes) takes 13 to 14 days, and the young fledge at about 15 to 17 days.

HABITAT: mainly clearings in cool temperate to tropical rainforest and wet sclerophyll forest; suburban parks and gardens
LENGTH: 11–12 cm
DISTRIBUTION: 300,000–1 million km²
ABUNDANCE: sparse to common
STATUS: probably secure

(JD Waterhouse)

Genus *Erythrura*

(e'-rith-rue'-rah: "red-tail")

Often known as parrotfinches, members of this genus have a distinctively shaped bill and are usually brightly coloured. Many species inhabit Indonesia and New Guinea, but only two occur in Australia, the generic allocation of one (the Gouldian Finch) being controversial. Most inhabit the shrub layer in rainforest.

Gouldian Finch

Erythrura gouldiae (gule'-dee-ee: "Gould's red-tail", after J. Gould, eminent British ornithologist)

The Gouldian Finch is widespread across tropical Australia from the Kimberley to Cape York but it has declined drastically during the 20th century and is now rare east of the Gulf of Carpentaria. Fire and trapping for the cagebird trade have been implicated in its decline. It has been suggested that, like the Pictorella Mannikin, the Gouldian Finch is migratory, dispersing southward to breed during the summer monsoon and retiring to coastal lowlands during the dry season but available data are inconclusive. It is strongly gregarious in all activities but birds do not preen one another and do not build roost-nests. While breeding, it feeds largely on insects, especially swarming termites; at other times it feeds on seeds harvested from plants, not gathered on the ground.

Alone among Australian grass-finches, the Gouldian Finch is polymorphic: about three-quarters of the population have black heads, about one-quarter have red heads, and a few individuals have dull yellow heads. The sexes are virtually identical in appearance.

Breeding occurs from December to April. Uniquely among Australian grassfinches, the Gouldian Finch nests in tree cavities or termite mounds, occasionally lining these with a rudimentary woven structure of grass. The clutch varies from four to eight eggs. Incubation takes 12 to 13 days, and the young fledge at about 21 days. Both parents incubate and care for the young.

HABITAT: tropical savannah woodland
LENGTH: 13–14 cm
DISTRIBUTION: 300,000–1 million km²
ABUNDANCE: sparse
STATUS: endangered

(D & M Trounson)

Blue-faced Finch

Erythrura trichroa (trie-kroh'-ah: "three-coloured red-tail")

The Blue-faced Finch is widespread in Indonesia, New Guinea, the Solomon Islands and Vanuatu, but it has only a small population in Australia, restricted to rainforests of the Atherton Tableland in north-eastern Queensland. It is a shy and elusive species that is not often seen. It lives in small groups, feeding on insects and seeds gathered in shrubs and the lower foliage of trees; it is especially fond of seeding bamboos and tends to be locally nomadic in response to the flowering cycles of these plants.

Little is known of its nesting habits: breeding occurs from November to April, the clutch consists of three or four eggs, and these hatch in about 12 to 14 days.

(D & M Trounson)

HABITAT: tropical rainforest
LENGTH: 12–13 cm
DISTRIBUTION: 30,000–100,000 km^2
ABUNDANCE: very sparse
STATUS: secure

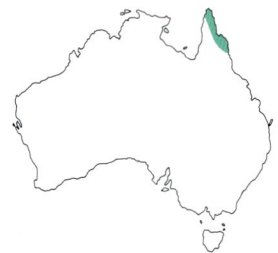

Genus *Lonchura*

(lonk-ue'-rah: "spear-tail")

Mannikins are well represented in Indonesia and New Guinea but only three species occur naturally in Australia (a fourth is introduced). All are strongly gregarious birds of reed beds and similar environments. They eat seeds, usually harvested directly from plants rather than gathered on the ground. Mutual preening is rare, and they do not build roost-nests.

Chestnut-breasted Mannikin

Lonchura castaneothorax (kah'-stah-nay'-oh-thor'-ax: "chestnut-breasted spear-tail")

The Chestnut-breasted Mannikin is common and widespread in New Guinea and in northern and eastern Australia from the Kimberley east to Cape York and south to the vicinity of Nowra in New South Wales. It seems locally nomadic in the far north-west but is possibly sedentary in the east. Its ancestral habitat consists of extensive reed beds fringing swamps and lagoons but it has successfully invaded various cultivated crops in many areas and is sometimes an agricultural pest. It is strongly gregarious in all activities but individuals neither preen one another nor use roost nests.

Breeding may occur at any time, but especially in late summer, and is colonial. Nests are usually only a few metres apart and within a metre of the ground; they are constructed of grass and are flask-shaped, with a downward-pointing entrance tunnel or hood. Four to six eggs make up the clutch; incubation takes 12 to 14 days, and the young fledge at about 21 days. Both parents incubate and rear the young.

HABITAT: reed beds and rank grass at the margins of temperate to tropical lagoons and swamps; croplands
LENGTH: 11–12 cm
DISTRIBUTION: more than 1 million km^2
ABUNDANCE: sparse to common
STATUS: secure

(D & M Trounson)

Yellow-rumped Mannikin

Lonchura flaviprymna (flah'-vee-prim'-nah: "yellow-rumped spear-tail")

The Yellow-rumped Mannikin is a bird of the swampy coastal flood plains of the far north-west, notably those of the Ord, Daly, Victoria and Roper Rivers. It is sedentary or locally nomadic. It is very closely related to the Chestnut-breasted Mannikin, with which it often occurs in mixed flocks. Hybridisation is common and the offspring are probably fertile. In all respects but appearance it closely resembles the Chestnut-breasted Mannikin.

Breeding occurs from January to March. Four or five eggs form the clutch. Incubation, by both parents, takes 12 to 14 days and the young fledge at about 22 days.

(D & M Trounson)

HABITAT: rank grassland on tropical coastal flood plains of major rivers; reed beds; croplands
LENGTH: 11–12 cm
DISTRIBUTION: 100,000–300,000 km²
ABUNDANCE: sparse to common
STATUS: probably secure

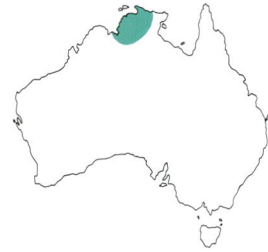

Pictorella Mannikin

Lonchura pectoralis (pek'-tor-ah'-lis: "breasted spear-tail")

The Pictorella Mannikin occurs across northern Australia from the Kimberley to Cape York Peninsula but it has declined drastically during the 20th century and is now very uncommon east of Arnhem Land. Frequent fires have been implicated in its decline. Nomadic and irruptive, it lives in small flocks that tend to disperse inland to breed during the summer monsoon, retiring to the coast during the dry season, at which time they associate freely with other finches. It tends to avoid proximity to human habitation.

Breeding occurs from January to May. The nest is a clumsy flask-shaped structure with a side entrance, built of grass and twigs and placed in a low bush or grass tussock close to the ground. Four to six eggs constitute the clutch, incubated by both parents.

HABITAT: arid tropical acacia savannah; spinifex grassland; margins of coastal swamps
LENGTH: 11–12 cm
DISTRIBUTION: 300,000–1 million km²
ABUNDANCE: sparse to common
STATUS: probably secure

(D & M Trounson)

413

Genus *Neochmia*

(nay-ok'-mee-ah: "change", possibly implying difference from a more familiar species)

Like *Emblema*, members of this genus have red rumps, but their distribution is tropical. They inhabit dense grass in tropical savannah woodland.

Crimson Finch

Neochmia phaeton (fee'-ton: "shining *Neochmia*")

The Crimson Finch formerly occupied dense grassland near water in tropical woodland, but it is now also a familiar inhabitant of canefields and pineapple plantations across northern Australia. It is most characteristic of the coastal plain but occurs in places at some distance inland. It also occurs in New Guinea. Mainly sedentary, it is territorial and aggressive, and seldom forms flocks. Pairs and small parties move restlessly in dense grass, eating seeds mainly taken from the plant, seldom on the ground. The Crimson Finch also eats insects. Roost-nests are not built.

Breeding occurs mostly from January to April in the north, September to May in the north-east. The nest is a bulky domed structure with a side entrance but no entrance tunnel. It is built of coarse grass and lined with feathers. The male alone fetches material but both sexes build; both also incubate and care for the young. Five to eight eggs form the clutch. Incubation takes 12 to 14 days, and the young fledge at 20 to 21 days.

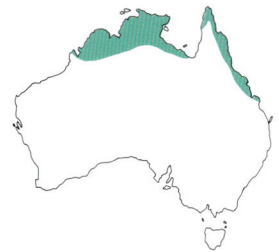

(D & M Trounson)

HABITAT: mainly tall grass in tropical woodland; crops and agricultural land; canefields; rank roadside vegetation
LENGTH: 13–14 cm
DISTRIBUTION: 300,000–1 million km²
ABUNDANCE: common
STATUS: probably secure

Star Finch

Neochmia ruficauda (rue'-fee-kaw'-dah: "rufous-tailed *Neochmia*")

The distribution of the Star Finch once extended across northern Australia from Shark Bay in the west to northern New South Wales in the east, but it has declined markedly during the 20th century, and may be endangered; it is now reported with any frequency only west of the Gulf of Carpentaria. Little is known of its movements but it sometimes forms wandering flocks when not breeding.

It lives in small groups, feeding mainly on grass seeds gathered from the fruiting head rather than from the ground. Members of mated pairs remain in close contact, frequently preening each other and roosting together (although roost-nests are not built).

Most breeding occurs in March and April. The nest is a globular structure of grass, lined with feathers and placed in a tussock or low bush up to several metres from the ground. Three to six eggs constitute the clutch. Both parents incubate (for 12 to 14 days) and care for the young (which fledge at about 15 to 17 days).

(K Ireland)

HABITAT: mainly tall rank streamside grass and rushes in warm temperate to tropical swamps and woodland
LENGTH: 11–12 cm
DISTRIBUTION: 300,000–1 million km²
ABUNDANCE: rare to common
STATUS: probably secure

Genus Poephila

(poh'-ay-fil'-ah: "grass-lover")

Almost exclusively Australian, this genus consists of several highly social species that live in undergrowth in savannah woodland and acacia scrubs. Plumage patterns are often intricate but bright colours are rare.

Long-tailed Finch

Poephila acuticauda (ah-kue'-tee-kaw'-dah: "sharp-tailed grass-lover")

The Long-tailed Finch is common in tropical Australia from the Kimberley to the head of the Gulf of Carpentaria. Living in sedentary groups of 10 to 30 birds, it feeds mainly on the ground, drinks in the morning and evening, and roosts communally at night in nests built for the purpose high in eucalypt trees. The pair bond is permanent and extremely close: members of a mated pair are seldom more than a metre or two apart, even when not breeding. The diet consists of seeds and insects.

Breeding occurs from January to May. The nest is a globular structure with a side entrance approached by a woven tunnel, built of grass and lined with feathers, and usually placed in the upper foliage of a eucalypt or pandanus. Both parents build, incubate and care for the young. The usual clutch consists of four or five eggs, which take about 13 days to hatch. The young fledge at about 22 days but remain with their parents for several weeks after fledging.

HABITAT: tropical savannah woodland
LENGTH: 14–15 cm
DISTRIBUTION: 300,000–1 million km²
ABUNDANCE: common
STATUS: probably secure

(D & M Trounson)

Double-barred Finch

Poephila bichenovii (bee'-shen-oh'-vee-ee: "Bicheno's grass-lover", after J. E. Bicheno, botanist and sometime Colonial Secretary of Tasmania)

The Double-barred Finch is widespread in northern and eastern Australia and has expanded southward across a broad front from the coast of southern New South Wales to the Murray–Darling river system. It is a common, sedentary, and highly social species that is never found far from water. In its general behaviour and ecology it resembles the Zebra Finch, differing most conspicuously in its preference for denser woodland in regions of relatively high rainfall. Its diet, like that of the Zebra Finch, consists of seeds and insects, gathered mainly on the ground. The call is a light, nasal "tak".

Breeding occurs from January to March in the north-west, mainly July to November elsewhere. The nest is a globular structure with a side entrance approached by a woven tunnel, built of grass and lined with feathers, and usually placed in a bush a metre or two from the ground. Both parents build, incubate and care for the young. Four or five eggs form the clutch. Incubation takes about 12 to 14 days and the young fledge at about 21 days.

HABITAT: temperate to tropical eucalypt woodland and acacia scrub with a grassy understorey; agricultural land; suburban parks and gardens
LENGTH: 10–11 cm
DISTRIBUTION: more than 1 million km²
ABUNDANCE: common
STATUS: probably secure

(K Ireland)

Black-throated Finch

Poephila cincta (sink'-tah: "banded grass-lover")

This grassfinch is in many respects the eastern representative of the

(D & M Trounson)

Long-tailed Finch: the two species are closely similar in habits and ecology and both are especially notable for the extraordinarily close pair bond. In appearance, the Black-throated Finch differs most conspicuously in its black bill. It has declined drastically during the 20th century, especially in the south, and is now very seldom recorded in New South Wales, where it was once fairly common.

Breeding may occur in any month. The details of the nesting cycle closely resemble those of the Long-tailed Finch.

HABITAT: tropical and warm temperate savannah woodland
LENGTH: 13–14 cm
DISTRIBUTION: 300,000–1 million km²
ABUNDANCE: rare to sparse
STATUS: vulnerable

Zebra Finch

Poephila guttata (gut-ah'-tah: "spotted grass-lover")

The abundant Zebra Finch occurs almost throughout mainland Australia, occupying a vast range of habitats from barren deserts to coastal croplands, wherever there is some grass and a few trees or bushes for nesting and roosting. Nearby fresh water is essential. It is sedentary, living permanently in flocks, breeding in loose colonies, and roosting communally in nests built for the purpose. The pair bond is permanent. The diet consists of seeds and insects, gathered mainly on the ground.

Breeding is opportunistic and almost continuous, halted only by drought or cold. The nest is a flask-shaped structure of grass with a side entrance approached by a woven tunnel, placed in a grass tussock or low bush. Both parents build the nest (the male fetching, the female constructing), incubate and care for the young, and both sleep in the nest at night. The clutch usually consists of four or five eggs. Incubation takes about 12 to 14 days and the young fledge at about 21 days. Young birds attain adult plumage in about 90 days and are then capable of breeding.

HABITAT: most kinds of dry, lightly timbered grasslands; agricultural land
LENGTH: *c* 10 cm
DISTRIBUTION: more than 1 million km²
ABUNDANCE: common
STATUS: secure

(D & M Trounson)

Masked Finch

Poephila personata (per'-son-ah'-tah: "masked grass-lover")

The Masked Finch is widespread across tropical Australia from the Kimberley to Cape York Peninsula. Like the Double-barred Finch, it is a common, sedentary, and very social species that is never found far from water. It lives in groups of about 20 to 50 birds which may coalesce into flocks of thousands in winter. The pair bond is permanent and very close. Insects form a substantial part of the diet, especially during the breeding season; otherwise it eats seeds, gathered mainly on the ground.

Breeding is timed to coincide with the end of the summer monsoon (generally March to June), and is loosely colonial. The nest is a globular structure with a side entrance approached by a woven tunnel, built of grass and lined with feathers, and usually placed in a bush a metre or two from the ground. Both parents build, incubate and care for the young. Incubation of the four to six eggs takes about 13 to 14 days and the young fledge at about 22 days.

HABITAT: mainly rank grass in tropical eucalypt woodland
LENGTH: 13–14 cm
DISTRIBUTION: 300,000–1 million km²
ABUNDANCE: common
STATUS: secure

(*D & M Trounson*)

Family ORIOLIDAE

(o'-ree-oh'-lid-ee: "*Oriolus*-family", after the genus of orioles)

Containing about 25 species, the family Oriolidae is exclusively Old World in distribution (the North American orioles—e.g., the familiar Baltimore Oriole—are not related) and best represented in the tropics. All but one species are included in a single genus, *Oriolus*.

Representatives of the genus *Oriolus* are strictly arboreal, frequenting the upper canopy levels in forest and woodland. They have penetrating, persistent, melodious but relatively simple songs and calls. Yellow, black and dull green are common plumage colours. Most feed on fruit and insects. They are relatively solitary, living in dispersed pairs.

The single species in the genus *Sphecotheres* differs in being markedly gregarious, strongly sexually dimorphic, and exclusively fruit-eating. It has a much shorter, blunter bill and a large area of naked skin around the eye.

Both genera are represented in Australia.

Genus *Oriolus*

(o'-ree-oh'-lus: "oriole")

The characteristics of the genus are discussed in the account of the family.

Yellow Oriole

Oriolus flavocinctus (flah'-voh-sink'-tus: "yellow-banded oriole")

The Yellow Oriole is common in forest and woodland along rivers and streams in the coastal lowlands of tropical Australia from the Kimberley to north-eastern Queensland; it also occurs in New Guinea and the Aru Islands. It is sedentary, living in pairs that maintain permanent territories.

Strictly arboreal, it is unobtrusive except for its distinctive bubbling calls. It feeds on fruit and insects. Breeding occurs from October to March. Nesting behaviour is similar to that of the Olive-backed Oriole.

(J Estbergs)

HABITAT: tropical gallery rainforest and monsoon vine-thicket
LENGTH: 26–28 cm
DISTRIBUTION: 300,000–1 million km^2
ABUNDANCE: sparse
STATUS: secure

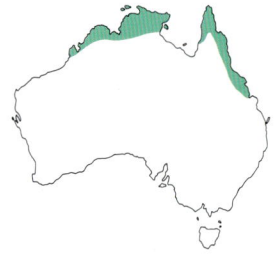

Olive-backed Oriole

Oriolus sagittatus (sa'-jit-ah'-tus: "arrow-marked oriole")

This common woodland speeies occurs in eastern and northern Australia from Victoria to the vicinity of Broome in Western Australia; vagrants reach the Mount Lofty Ranges in South Australia. it is also found in southern New Guinea. Northern birds are rather sedentary or locally nomadic but in the south most birds are migratory, and the species is rare south of about 27°S in winter.

The Olive-backed Oriole lives in pairs that spend almost all of their time in the upper foliage of trees. Fruit is the main item of diet, but it also eats insects gleaned from foliage. The distinctive call has been aptly described as "rolling, mellow, bubbled."

Breeding occurs from September to January. The nest is a pouch-shaped structure of wool, hair and miscellaneous plant materials, suspended by the rim from a high slender fork. Two to four eggs constitute the clutch. Incubation takes 17 to 18 days, and the young fledge in 15 to 16 days. The female builds and incubates, but both parents feed the young; the young leave the territory very soon after fledging. Sometimes the female yields total care of the young to the male while she nests again.

(J Gray)

HABITAT: temperate and tropical eucalypt woodland and forest
LENGTH: 26–28 cm
DISTRIBUTION: more than 1 million km^2
ABUNDANCE: common
STATUS: secure

Genus *Sphecotheres*

(sfek'-oh-thay'-rayz: "wasp-hunter")

The characteristics of the genus are those of the single species.

Figbird

Sphecotheres viridis (vi'-rid-is: "green wasp-hunter")

The Figbird has an extensive distribution in eastern Indonesia, New Guinea, and northern and eastern Australia from the Kimberley and the Top End to the Illawarra district of New South Wales. It is locally nomadic.

Southern birds differ in appearance from northern birds so markedly (southern males are grey below, where northern birds are yellow) that they were until recently considered to be separate species, but an extensive zone of hybridisation exists between Townsville and Mackay in Queensland. Females are dull brown above and strongly streaked below, and immatures resemble adult females.

Figbirds live in groups that remain linked through the breeding season and often coalesce into larger flocks during winter. They feed almost exclusively on fruits, especially figs, but also those of many other native and exotic trees including lantana, mulberry, and cultivated crops such as bananas and pawpaws. Strictly arboreal, they are noisy, active and conspicuous, and are often common in towns and cities.

The breeding season extends from October to February. The nest is a flimsy saucer of vine tendrils and twigs, suspended by the rim from a high slender horizontal tree fork. The usual clutch consists of three eggs. Both parents build, incubate and care for the young, sometimes assisted by other birds.

HABITAT: warm temperate to tropical rainforest and its margins; eucalypt forest; urban parks and gardens
LENGTH: 28–29 cm
DISTRIBUTION: 300,000–1 million km²
ABUNDANCE: common
STATUS: secure

(H & J Beste)

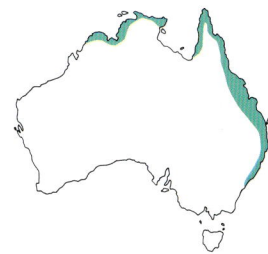

Family STURNIDAE

(stern'-id-ee: "*Sturna*-family", after a genus of starlings)

About 113 species distributed in some 25 genera constitute the family Sturnidae, which (except for artificial introductions by humans) is exclusively Old World in distribution. The family is especially well represented in Africa, where a number of species are brilliantly coloured. Otherwise, typical starlings are glossy black in colour, although some are brown, or have patches of white in the plumage. Insects constitute the bulk of the diet. Some tropical species are strongly arboreal but most are terrestrial. Starlings *walk* on the ground: they do not hop.

The Common Starling, *Sturnus vulgaris*, is abundant across northern Eurasia and has been widely introduced elsewhere in the world, including Australia. Following artificial introduction, Australia is also now the home of another species, the Common or Indian Mynah, *Acridotheres tristis*, a member of a genus largely restricted to the Indian subcontinent. With about 24 species, *Aplonis* is the largest genus; strictly arboreal and mainly fruit-eating, it has many representatives in New Guinea and islands in the south-western Pacific eastward to Tahiti; one of these species migrates from New Guinea to Australia to breed, and is the only starling naturally occurring in Australia. Another species, now extinct, once occurred on Norfolk Island.

Genus Aplonis

(ah-ploh'-nis: possibly misspelling of *Apollonis*, referring to Apollo)

The characteristics of the genus are discussed in the account of the family.

Shining Starling

Aplonis metallica (met-al'-ik-ah: "metallic *Aplonis*")

The Shining (or Metallic) Starling occurs from eastern Indonesia through New Guinea to the Solomon Islands as well as in north-eastern Queensland from Cape York southward nearly to Mackay. It is mainly a bird of coastal lowlands but also occurs, for example, on the Atherton Tableland. Although abundant in summer, it is usually rare in winter (when presumably most migrate to New Guinea) but knowledge of its movements is uncertain; traffic across Torres Strait has not been detected.

The sexes are similar, but immatures differ markedly, being white below, streaked with black.

Strongly gregarious, the Shining Starling roosts and nests communally. A typical site for a nesting colony is a huge emergent rainforest tree on the edge of a clearing; sometimes containing nearly 200 nests, and used year after year, such colonies are extremely conspicuous. Overnight roosts are also conspicuous, and are a familiar feature of gardens, parks and urban boulevards in the tropics. The Shining Starling feeds largely on fruit.

Nesting extends from August to December or January. Packed closely together, nests are bulky masses of rootlets and strips of palm leaves, with a side entrance that is often extended into a spout. Three or four eggs form the usual clutch.

HABITAT: tropical rainforests and their margins; mangroves; suburban parks and gardens
LENGTH: 21–24 cm
DISTRIBUTION: 100,000–300,000 km²
ABUNDANCE: common
STATUS: secure

(J Purnell)

Family DICRURIDAE

(die-krue'-rid-ee: "*Dicrurus*-family", after a genus of drongos)

The term "drongo" is a familiar element of Australian colloquial language, but it originated as the indigenous Malagasy name for a member of this family, a group of 20 species widespread in the Old World tropics. All belong to the genus *Dicrurus*, with the exception of an aberrant small rainforest species in the highlands of New Guinea (*Chaetorhynchus*).

Drongos are medium-sized songbirds with a conspicuously forked or fish-shaped tail; in a few species the outermost feathers are extended in flamboyant streamers or racket-shaped structures. The bill is crow-like, and rictal bristles are prominent, but the legs are short and relatively unsuited to locomotion. All have glossy black plumage; the sexes are similar and immatures resemble adults. Alert, aggressive and active in flight, they hawk for large flying insects from high exposed perches.

Genus Dicrurus

(die-krue'-rus: "double-tail")

The characteristics of the genus are essentially those of the family.

Spangled Drongo

Dicrurus hottentottus (hot'-en-tot'-us: "Hottentot double-tail", a confusing name based on a misapprehension that the first specimen came from South Africa)

In the coastal lowlands of northern and eastern Australia the Spangled Drongo ranges from the Kimberley to the Victorian border, but it is rare and erratic in occurrence south of Sydney in New South Wales (where it is mainly a winter visitor), and seldom breeds south of about 31°S. Elsewhere it is widespread from India to New Guinea and the Solomon Islands, though the relationships of extralimital populations are uncertain, and the Australian population (*D. h. bracteatus*) may prove to be a distinct species. It is partly migratory in eastern Australia, and mainly resident in the north.

It tolerates a variety of woodland habitats, including gardens, parks and urban boulevards, but it is especially characteristic of rainforest edges, where it selects prominent perches from which to sally out to capture flying insects in brisk, agile aerobatics. Occasionally it takes beetles and similar insects on the ground, or plucks them from foliage; it also eats fruit and, sometimes, small birds. It has a variety of metallic chattering calls. It is usually alone or in pairs, but not infrequently forms small flocks.

Breeding occurs mainly from September to March. A typical nest is a flimsy saucer of vine tendrils and other plant material slung from a slender horizontal fork up to 25 metres from the ground. Three to five eggs form the clutch. Both parents incubate and care for the young.

HABITAT: warm temperate to tropical woodland, including rainforest, vine thickets, mangroves; urban parks and gardens
LENGTH: 30–32 cm
DISTRIBUTION: more than 1 million km²
ABUNDANCE: common
STATUS: secure

(WS Peckover)

Family ARTAMIDAE

(ar-tah'-mid-ee: "*Artamus*-family", after the genus of woodswallows)

Woodswallows are stocky, long-winged and short-tailed birds of open woodland. They spend most of their time perched high on exposed dead twigs and branches or catching aerial insects in sailing, fluttering flight (most also occasionally consume nectar). Their legs are short and they are clumsy on the ground. They are almost unique among passerines in several respects: for example, they commonly soar, and they possess powder-down feathers. Their affinities are uncertain but they evidently originated in Australia, where six of the 10 to 15 species occur; all are placed in the genus *Artamus*. All are gregarious, some very strongly so, and they generally breed in loose colonies. The sexes are similar or nearly so, there is no seasonal variation, and immatures resemble adults.

Genus *Artamus*

(ar-tah'-mus: "butcher")

The characteristics of the genus are those of the family.

Black-faced Woodswallow

Artamus cinereus (sin-er-ay'-us: "ashy butcher")

The Black-faced Woodswallow is common across most of mainland Australia except for south-eastern coastal regions. It seems to be mainly sedentary and, of all woodswallows, is the one most closely associated with arid, sparsely timbered country; it is often encountered perched on

(D & V Blagden)

roadside fences or telephone wires in grassland or desert. It usually occurs in small parties which frequently huddle closely together; mutual preening is common and the birds roost in a packed cluster in a tree cavity. Like other woodswallows it feeds largely on flying insects but sometimes imbibes nectar.

It might easily be confused with the Dusky Woodswallow except for its preference for arid environments and its plain dark wings. The sexes are similar.

The breeding season is variable but is mainly from August to January. Nests, which are flimsy saucers of grass and twigs, are placed in stubs or on fence posts, or in bushes; three or four eggs form the clutch. Incubation takes about 14 to 15 days and fledging about 18 days. Both parents rear the young, sometimes assisted by other birds.

HABITAT: temperate to tropical eucalypt and acacia woodland; grassland
LENGTH: 18–19 cm
DISTRIBUTION: more than 1 million km²
ABUNDANCE: common
STATUS: secure

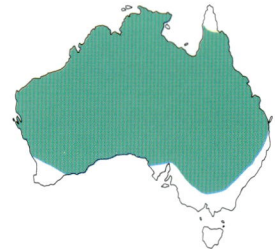

Dusky Woodswallow

Artamus cyanopterus (sie'-an-op'-te-rus: "blue-winged butcher")

Largely confined to southern and eastern coastal regions and the adjacent subinterior, the Dusky Woodswallow represents the genus

(MJ Seyfort)

Artamus in areas of relatively high rainfall: it is the only woodswallow common in Tasmania. In some areas it is migratory, in others resident. It lives in flocks sometimes exceeding 100 individuals but usually less than 20 or 30. Group members frequently preen each other, huddle together, and roost in clusters in tree cavities. Perched birds frequently rotate their tails.

The Dusky Woodswallow is easily identified by its diagnostic white bar extending along the outer primaries, frequently obvious even in the folded wing.

Breeding occurs from August to January. Nests, placed on fence posts or in bushes or trees up to 20 metres from the ground, tend to be clumped, and all flock members cooperate in challenging disturbance or predators at any one nest. Three or four eggs form the clutch; incubation takes

about 16 days, and the young fledge at 15 to 20 days. Both parents build, incubate and care for the young.

HABITAT: cool temperate to tropical moist eucalypt woodland; forest clearings
LENGTH: 17–18 cm
DISTRIBUTION: more than 1 million km²
ABUNDANCE: common
STATUS: secure

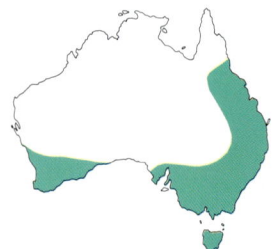

White-breasted Woodswallow

Artamus leucorhynchus (lue'-koh-rink'-us: "white-billed butcher")

The White-breasted Woodswallow occurs in northern Western Australia and the Northern Territory and virtually throughout eastern Australia but is absent from Tasmania and the far south and local in south-eastern coastal regions; there are some indications that it is expanding its distribution in central Australia, and it is largely migratory south of about latitude 26°S. Elsewhere it occurs from South-East Asia eastwards through New Guinea to islands of the south-western Pacific. It is much more closely associated with water than other woodswallows, and its typical habitat is dead timber at the margins of swamps and slow rivers but, in the tropics, it is also a common sight on roadside telephone wires in sugarcane and other farming country.

It is readily identified by its dark grey head and white underparts, and the absence of any white in the tail. It is strongly gregarious, and resting flocks often huddle closely together; mutual preening is common, and roosting is communal. The diet consists of flying insects.

Breeding generally occurs from August to January although, in the tropics, although it tends to follow the monsoon. The nest is a compact, sturdy bowl of sticks and twigs in a tree fork or crotch 10 metres or more from the ground. Three or four eggs form the clutch, which is incubated by both parents.

(MJ Seyfort)

HABITAT: warm temperate to tropical woodland, especially near water; farmland; mangroves
LENGTH: 17–18 cm
DISTRIBUTION: more than 1 million km²
ABUNDANCE: common
STATUS: secure

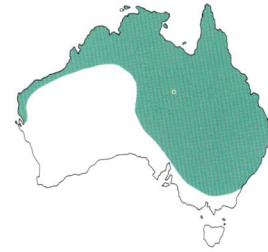

Little Woodswallow

Artamus minor (mie'-nor: "smaller butcher")

Especially characteristic of gorges in rocky country in the tropics and the arid interior, the Little Woodswallow is identified by its plain "dark chocolate" plumage. It is generally sedentary but is sometimes locally nomadic in marginal habitats. It usually occurs in flocks of about 20, and feeds mainly on flying insects. Group members frequently huddle, mutually preen, and cluster-roost in tree cavities or rock crevices; like miniature Dusky Woodswallows, they persistently rotate their tails.

Breeding may occur at any time after rain, but is usually from August to January. Constructed of grass and twigs, the nest is usually placed in a rock crevice. Three eggs form the clutch.

(G Little)

HABITAT: warm temperate to tropical rocky gorges; acacia woodland; spinifex; tussock grassland
LENGTH: 12–13 cm
DISTRIBUTION: more than 1 million km²
ABUNDANCE: sparse
STATUS: probably secure

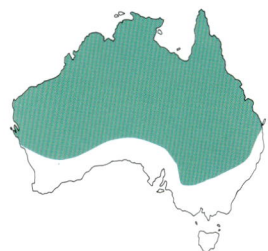

Masked Woodswallow

Artamus personatus (per'-son-ah'-tus: "masked butcher")

The range of this woodswallow extends across mainland Australia but it breeds more commonly in the south than in the north: characteristically it occurs in mixed flocks with the White-browed Woodswallow, the White-browed Woodswallow predominating in the east, the Masked Woodswallow in the west. It is also characteristic of the species that it occurs un-predictably in huge numbers at any given locality, breeding rapidly then moving on; sometimes the birds move on even before breeding has been completed, leaving large numbers of young in the nests to starve. Mainly aerial in habits, it feeds largely on flying insects, and occasionally nectar.

Breeding occurs from August to December and is colonial but not communal. Nests, which are cup-shaped structures of grass and twigs and are often crammed close together, are built on fence posts, in dead trees or in bushes. The two or three eggs constituting the clutch hatch in about 12 days. Both parents incubate.

HABITAT: mainly cool temperate woodland
LENGTH: 19–20 cm
DISTRIBUTION: more than 1 million km²
ABUNDANCE: common
STATUS: secure

(R Drummond)

White-browed Woodswallow

Artamus superciliosus (sue'-per-sil'-ee-oh'-sus: "eyebrowed butcher")

The only Australian woodswallow with a distinct white eyebrow, this species is strongly nomadic and usually occurs in vast swarms. It closely resembles the Masked Woodswallow in all aspects of its life history, but is almost exclusively eastern in distribution (the other species predominates in the west) and occurs occasionally in Tasmania. Mixed colonies are common but hybrids are rare. Both parents build, incubate and rear the young; eggs hatch in about 12 days and the young fledge at about 13 to 14 days.

HABITAT: mainly cool temperate woodland
LENGTH: 19–20 cm
DISTRIBUTION: more than 1 million km²
ABUNDANCE: common
STATUS: secure

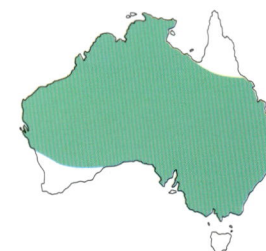

(J Gray)

Family CORCORACIDAE

(kor'-kor-ah'-sid-ee: "*Corcorax*-family", after a genus of choughs)

This endemic Australian family contains only two species, each in its own genus (*Corcorax* and *Struthidea*). Formerly it was thought also to include the Magpielark and the Torrentlark (of the highlands of New Guinea), an idea strengthened by the remarkable similarity between their distinctive mud-built nests, but it has recently been show that these two species are not related and they have been isolated in their own family, Grallinidae.

The family is notable for the extreme development of communal behaviour in its members.

Genus Corcorax

(kor-kor'-ax: possibly "raven-like", from Greek *korax*, raven or crow)

The characteristics of the genus are those of the single species.

White-winged Chough

Corcorax melanorhamphos (mel'-ah-noh-ram'-fos: "black-billed *corcorax*")

The White-winged Chough is widespread in eucalypt woodlands throughout eastern Australia, from about Cairns to the Eyre Peninsula, but it does not occur in Tasmania or on Kangaroo Island. It lives in permanent groups of four to 10 birds that may coalesce briefly into larger flocks in winter. It is sedentary: breeding groups defend a territory of about 20 hectares in summer and roam a home range of up to 1000 hectares in winter.

The diet consists mainly of insects and other small invertebrates. Groups forage mostly on the ground, advancing in formation, chattering and chuckling softly and methodically searching the leaf litter, shuffling dead leaves aside with the bill or probing soft earth.

When disturbed the birds fly up into lower branches of trees, piping and churring loudly and revealing bold white flashes in the outer wing (mainly hidden when at rest). They roost together in trees.

Breeding extends from August to December. All members of the group are involved, together building a large bowl-shaped nest of mud and fibre (which is usually constructed on a horizontal limb 5 to 10 metres from the ground), incubating the eggs and caring for the young; sometimes several females lay in the same nest. The normal clutch is three to five eggs, which hatch in 19 days. The young fledge at about 25 days, but remain dependent on the group for much of their food throughout the following winter. Sexual maturity is reached in the fourth year.

Male and female are virtually identical, but immatures differ in having brown, not red, eyes.

HABITAT: temperate eucalypt woodland; mallee; callitris scrub; sometimes enters pine plantations
LENGTH: 44–47 cm
DISTRIBUTION: more than 1 million km^2
ABUNDANCE: common
STATUS: probably secure

(IR McCann)

425

Genus Struthidea

(strue-thid'-ay-ah: "bird-form", significance unknown)

The characteristics of the genus are those of the single species.

Apostlebird

Struthidea cinerea (sin'-e-ray'-ah: "ash-coloured bird-form")

The Apostlebird is widespread in eastern Australia but absent from much of Victoria and coastal regions of New South Wales. It may have expanded in range since 1900, especially in northern Queensland; an isolated population in the Top End appears to have established itself since about 1922. Sedentary and strictly communal, it lives in permanent, close-knit groups of about eight to 18 birds that together occupy a territory of 10 to 15 hectares. It needs constant access to water, and all territories contain a spring, a pond, or some other watering point.

Active, playful and garrulous, Apostlebirds spend most of their time on the ground. They squabble frequently, and indulge in follow-my-leader chases through the trees.

Mutual preening is common during dust-bathing, or when birds are lined up close together on a low branch. When alarmed they move up into trees, preferring to bound upwards from branch to branch rather than fly. They roost in trees.

In summer the diet consists mainly of insects; in winter, seeds (gleaned from the ground) predominate.

Breeding occurs from August to February. The nest is a solid bowl-shaped structure of mud and fibre, lined with grass and plastered to a horizontal limb several metres from the ground. Two to five eggs constitute the clutch. The incubation period is about 18 days, and the young fledge in 13 to 18 days. All members of the group participate in nest building, incubation and care of the young.

HABITAT: temperate eucalypt, acacia and callitris woodland, especially Mulga, mallee, and brigalow
LENGTH: 29–33 cm
DISTRIBUTION: more than 1 million km²
ABUNDANCE: common
STATUS: probably secure

(JR McCann)

Family GRALLINIDAE

(grah-lin'-id-ee: "*Grallina*-family", after the genus of the Magpielark)

This family consists of two species: the familiar Magpielark of Australia and the Torrentlark of New Guinea. The latter bird, which inhabits mountain streams, has generally been placed in its own genus *Pomareopsis*, but further research may well demonstrate that it is insufficiently distinct to deserve this separation, a conclusion that would place both species in the single genus *Grallina*.

These birds were formerly included in the family Corcoracidae, largely on the basis of striking similarities in behaviour and the distinctive structure of their nests, but recent research strongly indicates that they are more nearly related to the monarchs (Monarchidae).

Genus Grallina

(grah-lee'-nah: "little-stilt-bird")

The characteristics of the genus are essentially those of the single species.

Magpielark

Grallina cyanoleuca (sie'-ah-noh-lue'-kah: "blue-white little-stilt-bird")

This conspicuous species is among the most widespread and familiar of Australian songbirds. It requires trees in which to nest, mud for nest construction, and open areas in which to feed. Where these requirements are met, it may be found almost anywhere across mainland Australia, even in the heart of towns and cities. It is merely a casual vagrant in Tasmania. Elsewhere it occurs in southern New Guinea, apparently as a recent (*c* 1970?) immigrant.

Established pairs maintain permanent territories some 8 to 10 hectares in extent, but young birds roam widely in flocks, which are often quite large. There is evidence of migratory movements in some areas.

Adult males, females and immatures differ distinctly (though not obviously) in appearance. Males have a white eyebrow and black throat; females have a white forehead and throat but lack the male's white eyebrow; and immatures have a white throat *and* a white eyebrow.

Almost all feeding takes place on the ground: the Magpielark is especially fond of foraging along the muddy margins of shallow ponds and lagoons. In city parks and gardens it walks about on open lawns (the head bobbing back and forth in a characteristic manner), and may approach picnickers for scraps. The diet includes insects and their larvae, earthworms, and small molluscs. The calls are loud and ringing, and often uttered in antiphonal duets.

Breeding may occur at any time, but there is a distinct peak in activity from August to December. Often, several broods are raised in succession. The nest is a bowl of plant fibre bound liberally with mud, usually plastered (with little attempt at concealment) on a horizontal bough 3 to 10 metres from the ground. Three to five eggs constitute the usual clutch. Incubation, by both sexes, takes 17 to 18 days. The young fledge at 18 to 23 days.

HABITAT: almost the entire mainland but especially open woodland; urban and suburban areas; avoids rainforest
LENGTH: 26–30 cm
DISTRIBUTION: more than 1 million km^2
ABUNDANCE: common to abundant
STATUS: secure

(M Cohen)

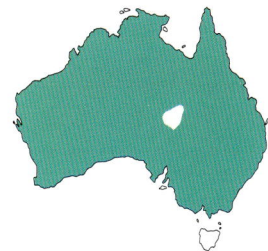

Family CRACTICIDAE

(krak-tis'-id-ee: "*Cracticus*-family", after the genus of butcherbirds)

Confined to Australia and New Guinea, the common family Cracticidae consists of nine or 10 woodland species grouped in three very distinct genera. All have varied diets in which large insects figure prominently; most forage mainly on the ground.

Butcherbirds (*Cracticus*) are relatively small; the common name is derived from their habit of impaling on thorns prey which is too large to swallow: this is then torn apart with the bill. Two species are confined to New Guinea but the remaining four are Australian. Most are patterned in grey, black and white; the sexes are similar and there is no seasonal variation.

The Australian Magpie (*Gymnorhina*) is medium-sized, black and white in plumage, and communal in habits. It is quite unrelated to the European Magpie, which is a member of the Corvidae.

Largest of the family, the three species of currawongs (*Strepera*) are exclusively Australian, occurring mainly in the south and east. They breed in scattered pairs in summer but flock in winter. The sexes are similar, there is no seasonal variation, and immatures resemble adults; a conspicuous white flash in the wing is a common element of plumage pattern.

Genus Cracticus

(krak'-tik-us: "screamer")

The characteristics of the genus are discussed in the account of the family.

Black-backed Butcherbird

Cracticus mentalis (men-tah'-lis: "[notable-] chinned screamer")

In Australia the Black-backed Butcherbird is confined to Cape York Peninsula but it also occurs in New Guinea. Sedentary, and living in pairs on permanent territories, it is in many respects the tropical counterpart of the Grey Butcherbird, except that it often feeds in trees as well as pouncing on prey on the ground. It breeds from June to November, and three or four eggs form the clutch. It has been studied little.

HABITAT: mainly tropical open eucalypt woodland
LENGTH: 25–27 cm
DISTRIBUTION: 100,000–300,000 km²
ABUNDANCE: sparse
STATUS: secure

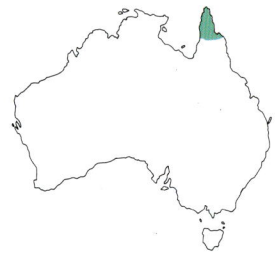

(D & M Trounson)

Pied Butcherbird

Cracticus nigrogularis (nig'-roh-gue-lar'-is: "black-throated screamer")

This species is in many respects the dry-country counterpart of the equally familiar and somewhat smaller Grey Butcherbird. Its pure, fluted calls are among the loveliest of Australian bird sounds. It is widespread across mainland Australia, except for south-eastern coastal regions, and is slowly expanding in the south-west. It is mainly sedentary, and very similar to the Grey Butcherbird in diet, behaviour and general habits except that it sometimes lives in small parties.

Breeding occurs from May to November in the tropics, August to December further south. The nest is a cup of sticks and twigs in a vertical tree fork, and the clutch consists of three to five eggs, incubated by the female. Sometimes aided by other birds, both parents feed the young.

HABITAT: temperate to tropical, mainly open woodland
LENGTH: 32–36 cm
DISTRIBUTION: more than 1 million km²
ABUNDANCE: common
STATUS: secure

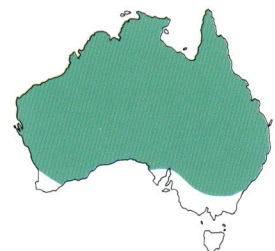

(IR McCann)

Black Butcherbird

Cracticus quoyi (kwoy'-ee: "Quoy's screamer", after J. R. Quoy, French zoologist)

The Black Butcherbird occurs in the Top End and in north-eastern Queensland south to the vicinity of Mackay, as well as in New Guinea. Largest of butcherbirds, it is entirely black in plumage but the population living between the Herbert and Endeavour Rivers in Queensland has a dull rufous plumage which perhaps represents immatures. It is sedentary, established pairs maintaining permanent territories some 10 hectares in extent. A jungle species, it is quiet, furtive and unobtrusive in general behaviour and not easy to observe. It eats almost any small animal that it can subdue and has a special fondness for nestling birds; it also feeds on fruit. It forages on the ground or at any level in the foliage.

It breeds from September to February. The nest, built of sticks and twigs, is placed in a low tree fork. Two to four eggs form the clutch. Both parents build, incubate and feed the young.

(H & J Beste)

HABITAT: tropical rainforest; vine scrub; mangroves
LENGTH: 40–44 cm
DISTRIBUTION: 100,000–300,000 km²
ABUNDANCE: common
STATUS: secure

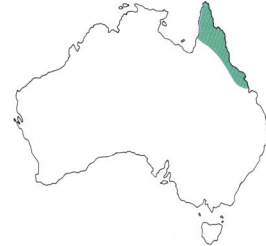

Grey Butcherbird

Cracticus torquatus (tor-kwah'-tus: "collared screamer")

A familiar bird to many southern suburban dwellers, the Grey Butcherbird has rich, complex, carolling calls (often uttered in a duet between the sexes) and often comes to garden feeders. It occurs across Australia but is much more numerous in the south than in the north: it is rare or absent over much of the arid interior, and there is an isolated population in the Kimberley and Top End. It is generally common but may be declining in some areas. Established pairs maintain permanent territories and the species is mainly sedentary.

The sexes are similar (the white throat is a useful field character), but immatures are mainly dusky brown above, pale below.

It often occurs in proximity to the Pied Butcherbird, separated only by subtle habitat distinctions: this species favours dense foliage, the Pied Butcherbird open woodland. Most prey (mainly large insects) is taken on the ground, pounced on from a vantage point a few metres high. The diet also includes nestlings and small birds, reptiles and mammals, and occasionally seeds and fruit.

Breeding occurs from July to January, and only one brood is raised per season. The nest, a compact cup of sticks and twigs lined with grass and fibre, is placed in a vertical tree fork up to 10 metres from the ground. Three to five eggs form the clutch, which hatches in 24 to 26 days. The female alone incubates, but she is fed on the nest by the male. Fed by both parents, the young fledge at about 28 days.

(RB Legge)

HABITAT: cool temperate to subtropical forest and woodland, including suburban parks and gardens
LENGTH: 26–30 cm
DISTRIBUTION: more than 1 million km²
ABUNDANCE: common
STATUS: secure

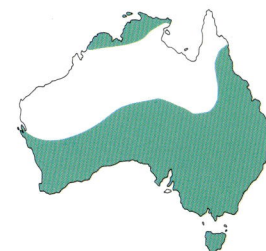

Genus Gymnorhina

(jim'-noh-rie'-nah: "naked-nose")

The characteristics of the genus are those of the species.

Australian Magpie

Gymnorhina tibicen (tib'-ee-sen: "flute-playing naked-nose")

(G Weber)

The Australian Magpie occurs across Australia and in southern New Guinea; it has been introduced to New Zealand. It requires open country on which to feed and clumps of trees in which to roost and breed, conditions that are often met in most kinds of agricultural land, as well as in urban parks and gardens. Perhaps once uncommon, it seems to have thrived since European settlement and is now one of Australia's most common birds. It is sedentary.

There are three chief populations, so strikingly different that each was formerly regarded as a distinct species: essentially, south-eastern birds have white backs, northern and eastern birds have black backs, and south-western birds mottled backs. The bold white patch on the nape is a unique plumage feature. The sexes are similar, and immatures resemble adults except in being duller.

Magpies live in stable groups of up to about 20 individuals, which jointly occupy permanent territories used for feeding, roosting and nesting. The size of the territory depends on the quality of the habitat, but is normally about 10 to 20 hectares. The species feeds mainly on ground-dwelling insects.

Breeding occurs from June to December. The nest is a rough basket of sticks in a tree. The usual clutch is three or four eggs, which hatch in 20 days; fledging takes about 24 days. Females build, incubate and rear their young largely unaided, though it is uncommon for those females not under the protection of the dominant male to succeed in raising young. After fledging, all group members contribute to the care and guardianship of the young.

HABITAT: almost entire mainland but particularly open eucalypt woodland; farmland; urban and suburban parks and gardens
LENGTH: 38–44 cm
DISTRIBUTION: more than 1 million km^2
ABUNDANCE: common
STATUS: secure

Genus Strepera

(strep'-er-ah: "noise-maker")

The characteristics of the genus are discussed in the account of the family.

Black Currawong

Strepera fuliginosa (fool'-i-jin-oh'-sah: "sooty noise-maker")

(WB Gibson)

The Black Currawong is confined to Tasmania and islands in Bass Strait. Like the mainland Pied Currawong, it breeds in scattered pairs in dense highland eucalypt forest, flocking to the lowlands to avoid excessive snowfall in winter, at which season it is often a familiar scavenger at

HABITAT: cool temperate dense highland eucalypt forest; heath (in winter) and woodland; urban and suburban parks and gardens
LENGTH: 48–49 cm
DISTRIBUTION: 30,000–100,000 km^2
ABUNDANCE: common
STATUS: secure

picnic grounds and in urban parks and gardens. It feeds mainly on the ground, and the diet is varied.

Breeding occurs from August to December. The nest is a rough basket of sticks high in a tree. Two to four eggs form the clutch.

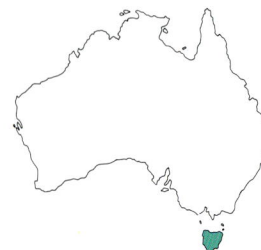

Pied Currawong

Strepera graculina (grak'-ue-lee'-nah: "jackdaw-like noise-maker")

The Pied Currawong occurs along the coast and associated highlands of eastern Australia from Cape York Peninsula to Victoria; it also occurs on Lord Howe Island. It is largely sedentary, but highland birds move in winter to flock in the lowlands. In the south-east at this season, many move westwards along the Murray–Darling River system, returning in spring. In many coastal districts it has expanded in range and number in recent decades and it is now common in several coastal cities where it was previously virtually unknown.

It breeds in scattered pairs but at other seasons forms flocks, which may disperse to forage by day, but which gather at dusk in communal roosts. In summer in its ancestral habitat of highland eucalypt forest it feeds largely on stick insects but its diet is otherwise very varied, and includes fruit and small vertebrates, especially nestling birds. It feeds on the ground or in trees, and often scavenges in parks or at picnic grounds. It utters a variety of loud, ringing, yodelling calls.

Breeding occurs from August to January. The nest is a rough basket of sticks lined with grass and bark fragments in a tree, usually at a considerable height. Usually three eggs are laid, which are incubated for 28 days by the female while the male feeds her; both parents feed the young, which fledge in about 30 days.

HABITAT: temperate to subtropical dense highland eucalypt forest; woodland; urban and suburban parks and gardens
LENGTH: 44–48 cm
DISTRIBUTION: 300,000–1 million km²
ABUNDANCE: common
STATUS: secure

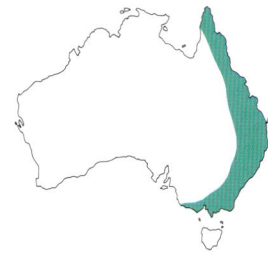

(HJ Pollock)

Grey Currawong

Strepera versicolor (ver'-see-kol'-or: "varicoloured noise-maker")

The Grey Currawong closely resembles the Pied Currawong in many respects but is much scarcer, more solitary, and less versatile in choice of habitat. The most obvious distinction in appearance is its dark grey (not white) rump. Mainly sedentary, it occurs in several discrete populations, some of which were formerly regarded as species: one is restricted to Tasmania, another is strongly associated with the mallee belt across southern Australia, and another occurs in the south-west. It has a striking ringing call, and feeds mainly on insects found under bark in trees or in leaf litter on the ground, though it also eats a range of items including seeds, fruit and nestling birds.

Breeding occurs from July to November. The nest is a rough, shallow bowl of sticks, lined with grass, and placed high in a tree. The two or three eggs are incubated by the female.

HABITAT: temperate eucalypt forest and woodland; mallee; coastal heath
LENGTH: 45–51 cm
DISTRIBUTION: more than 1 million km²
ABUNDANCE: sparse to common
STATUS: probably secure

(PD Munchenberg)

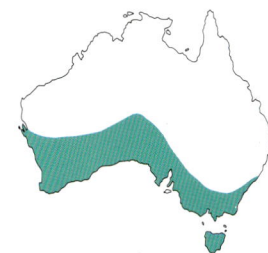

Family PTILONORHYNCHIDAE

(tie'-lon-oh-rink'-id-ee: "*Ptilonorhynchus*-family", after a genus of bowerbirds)

In most respects, bowerbirds are unremarkable songbirds of medium size, generally solitary, unobtrusive, arboreal, and fruit-eating. They are found only in Australia and New Guinea, mainly in rainforest. Most have some conspicuous element of bright colour in their plumage, and a few are spectacular; there is a general tendency for bright colours to be on the dorsal rather than the ventral surface (in the closely related Paradisaeidae this general trend is reversed).

With a few exceptions, mating is polygynous (i.e., males mate with more than one female), the male taking no part in nest construction, incubation or the rearing of young. It is in their mating behaviour that bowerbirds are remarkable: males build elaborate structures of twigs and sticks on the ground, to which itinerant females are enticed for copulation by means of songs or intricate displays. The structure of these display bowers varies widely among species, but two basic architectural styles are evident: an "avenue" bower consists essentially of twin parallel walls on a platform, whereas a "maypole" bower consists of a cone of sticks built around the base of a suitable sapling (only one Australian species builds a maypole bower). Bowers are painted with a mixture of saliva and masticated plant material, and decorated with an assortment of miscellaneous items including feathers, fruits, snail shells, vertebrae, pebbles and (near human habitation) litter such as drinking straws and bottle tops. Different species select different ranges of items, and generally show strong colour preferences. At least during the breeding season, males spend the greater part of their time in the maintenance and decoration of these structures.

About 18 species are usually recognised, allocated to seven or eight genera: about nine species are confined to New Guinea, seven to Australia, and two are shared.

Genus Ailuroedus

(ayl'-ue-ree'-dus: "cat-voice")

Members of this genus are characterised by bright green upperparts, and are unique among bowerbirds in having (apparently) abandoned the bower-building habit and reverted to monogamous breeding. The Tooth-billed Bowerbird of northern Queensland is an exception in being brown above, in clearing an arena on the forest floor (though it does not build a bower), and in being polygamous. (In recogniton of these differences it is sometimes isolated in its own genus, *Scenopoeetes*). There are three species: one is restricted to New Guinea, one to Australia, and one is shared.

Tooth-billed Catbird

Ailuroedus dentirostris (den'-tee-ros'-tris: "tooth-beaked cat-voice")

The Tooth-billed Catbird is restricted to highland rainforests between 600 and 1400 metres altitude on the Atherton Tableland in north-eastern Queensland. It is sedentary but individuals sometimes wander below 500 metres. It spends almost all of its time in the forest canopy, feeding on fruit, leaves and insects; unusually among birds, leaves figure prominently in the diet.

During the breeding season (from September to January) the male clears a circular arena on the forest floor up to 3 metres in diameter; all litter is swept away and over the resulting space up to 20 selected leaves are carefully arranged, all placed underside up, being replaced as they wither. The male sings and displays to visiting females, enticing them to the stage, where copulation takes place. The song is an extraordinary rambling medley of clear liquid notes, whistles, churrs, chatters and, possibly, mimicry.

Two eggs are laid in a loose, fragile nest saucer of sticks and twigs, usually hidden high in a vine tangle.

(MS Sharland)

HABITAT: warm temperate highland rainforest
LENGTH: 26–27 cm
DISTRIBUTION: 10,000–30,000 km²
ABUNDANCE: common
STATUS: probably secure

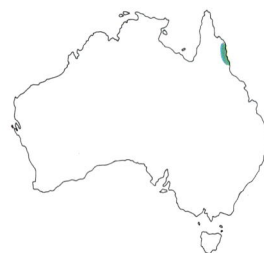

Green Catbird

Ailuroedus melanotis (mel'-an-oh'-tis: "black-eared cat-voice")

The Green Catbird occurs in New Guinea and in the rainforests of northern Cape York Peninsula, at altitudes above about 500 metes on the Atherton Tableland and in. the mid-eastern rainforest block from about Maryborough in Queensland south to the vicinity of Narooma in New South Wales. Northern birds

(MJ Seyfort)

have darker heads and are more distinctly spotted below than southern birds and, until recently, were widely regarded as constituting a distinct species, called the Spotted Catbird.

The Green Catbird is sedentary, apparently living in pairs that maintain permanent territories. It is usually encountered alone, but small parties sometimes congregate at fruiting trees. Strictly arboreal, it feeds on fruit (especially figs), seeds and insects; it also takes eggs and nestling birds. Two distinctive calls, uttered by both sexes, consist of an abrupt metallic "tck", and a drawn-out, abrasive wail that somewhat resembles the miaow of a cat.

Breeding occurs from September to January. The nest is a rough loose bowl of sticks, leaves and vine tendrils hidden in a vine tangle, the crown of a tree fern, or dense tree

foliage at almost any height from the ground. Usually two eggs constitute the clutch, hatching in 23 to 24 days; the young fledge at about 21 days. The male defends the territory and assists in feeding young, but the female alone builds and incubates.

HABITAT: mainly tropical rainforest; occasionally dense eucalypt forest
LENGTH: 24–32 cm
DISTRIBUTION: 100,000–300,000 km²
ABUNDANCE: common
STATUS: probably secure

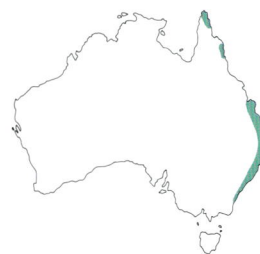

Genus *Chlamydera*

(klah'-mee-day'-rah: "cloaked-neck")

Bowerbirds are essentially birds of dense rainforest but members of this genus are unique in inhabiting arid environments ranging from desert to savannah woodland, mangroves and the margins of rainforest. The plumage is mainly brown or grey, intricately spotted, and the sexes are similar. Bowers are complex and of the avenue type. There are five species: one restricted to New Guinea, three to Australia and one shared.

Fawn-breasted Bowerbird

Chlamydera cerviniventris (ser-vin'-ee-vent'-ris: "fawn-bellied cloaked-neck")

In Australia the Fawn-breasted Bowerbird is restricted to the northern part of Cape York Peninsula, but it is widespread in New Guinea. It is sedentary, solitary, strongly arboreal and mainly a fruit-eater.

Adult males build complex bowers that are usually reconstructed

(MJ Seyfort)

annually on the same site. Two parallel walls of woven twigs are constructed on a substantial rectangular platform, also of twigs, that may be up to 40 centimetres high. The walls are about 30 centimetres high and 40 centimetres long, and built so close together as to only barely allow the bird's passage. The inner surfaces of the walls are painted with a mixture of saliva and plant pulp, and the platform is decorated with green berries.

Breeding occurs sporadically all year but mainly from September to December. One egg (very rarely two) is laid, deposited in a loose, shallow saucer of sticks and twigs, usually hidden high in dense foliage in a tree fork.

HABITAT: tropical rainforest margins; coastal scrub; vine scrub; mangroves
LENGTH: 26–28 cm
DISTRIBUTION: 10,000–30,000 km²
ABUNDANCE: sparse
STATUS: secure

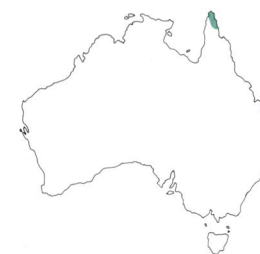

Western Bowerbird

Chlamydera guttata (gut-ah'-tah: "spotted cloaked-neck")

(*RW Fidler*)

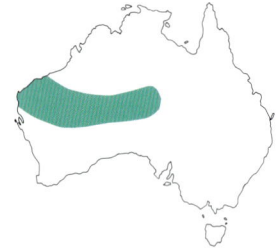

The Western Bowerbird is the western counterpart of the Spotted Bowerbird: opinions differ as to its taxonomic status. It is smaller than the eastern bird, with relatively shorter bill and tail, and is much more richly coloured. There are few differences in habits and behaviour. Their respective ranges approach each other in the northern Simpson Desert, but they are apparently not quite in contact and do not hybridise.

HABITAT: mainly arid savannah woodland
LENGTH: 25–27 cm
DISTRIBUTION: 300,000–1 million km²
ABUNDANCE: sparse to common
STATUS: probably secure

Spotted Bowerbird

Chlamydera maculata (mak'-ue-lah'-tah: "spotted cloaked-neck")

The Spotted Bowerbird is a familiar inhabitant of homestead gardens almost throughout the arid zone of eastern Australia, extending in the south down the Darling River drainage almost to the Murray River: it has declined markedly in the south during the 20th century. It is largely sedentary but, impelled by drought, individuals sometimes travel far. It seldom occurs far from permanent water, and it eats fruit, nectar, and insects and other arthropods.

The bower is of the avenue type: two parallel walls of woven twigs are constructed on a platform of twigs in a cleared area about a metre in diameter. The walls may be up to 50 centimetres high, 70 centimetres long, and 20 centimetres apart. The bower is surrounded with neat piles of small objects, usually white in colour: bones, shells, pebbles and spent cartridge cases are prominently represented.

Breeding occurs from September to February. The nest is a loose flimsy saucer of twigs and leaves in a horizontal fork several metres from the ground. Two or three eggs form the clutch.

HABITAT: mainly warm temperate savannah woodland
LENGTH: 28–30 cm
DISTRIBUTION: 300,000–1 million km²
ABUNDANCE: rare to common
STATUS: probably secure

(*MJ Seyfort*)

Great Bowerbird

Chlamydera nuchalis (nue-kah'-lis: "[notable-] naped cloaked-neck")

The Great Bowerbird, which occurs in the Kimberley, in the Top End and on Cape York Peninsula, is essentially the tropical representative of the Spotted Bowerbird, which it much resembles in habits and behaviour. It is sedentary (occasionally forming small wandering flocks when not breeding) and solitary. It forages mainly in trees rather than on the ground, and feeds largely on fruit.

Constructed along avenue lines, each bower is the exclusive property of a single male, but males and their bowers are loosely associated in clans. A cleared apron in front of and behind the bower is thickly carpeted with white objects, especially shells, vertebrae and small pebbles. Some males sometimes build new bowers each year. On the approach of an itinerant female, the male displays and cavorts with harsh churrs, cocked tail, raised crest, lowered wings, and frequent flicks of the head, revealing his bright yellow mouth.

Breeding occurs mainly from October to January. The nest is a rough loose bowl of sticks and leaves, placed several metres from the ground in a tree. There is usually only one egg.

HABITAT: tropical rainforests and their margins; savannah woodland; mangroves; gardens
LENGTH: 33–37 cm
DISTRIBUTION: 300,000–1 million km^2
ABUNDANCE: sparse to common
STATUS: probably secure

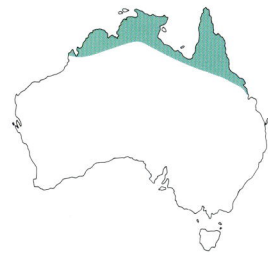

(D & V Blagden)

Genus *Prionodura*

(pree'-on-oh-due'-rah: "saw-tail")

The characteristics of the genus are those of the single species.

Golden Bowerbird

Prionodura newtoniana (nue'-tun-ee-ah'-nah: "Newton's saw-tail", after A. Newton, British zoologist)

The Golden Bowerbird is restricted to highland rainforest at altitudes between 900 and 1500 metres on the Atherton Tableland in north-eastern Queensland. It is sedentary. Adult males are brown and brilliant golden yellow, but females are dull brown. Solitary, quiet and unobtrusive in general behaviour, the Golden Bowerbird forages at all levels in the forest, feeding mainly on fruit, a diet occasionally supplemented with insects.

Males live in loose clans; upon reaching maturity, a male establishes a place in the group, building a bower that he will maintain for the rest of his life, constantly repairing and adding to it. On a sloping hillside he selects two saplings, about a metre apart and linked by a low branch, around each of which he constructs unequal pyramids of sticks and twigs between 1 and 3 metres high; the linking branch is also draped with sticks and twigs, glued in place with saliva, and decorated with lichens, moss, flowers and fruit. He mates with itinerant females, hovering about the bower, flaring his crest, spreading his tail, and hopping from side to side in order to entice a female to the bower, where copulation takes place.

Breeding occurs from September to January. Selecting a secluded site some distance from the males' clan, mated females build, incubate and rear their young alone, laying one or two eggs in a deep rough cup of twigs, vines and dead leaves placed in a tree fork up to 3 metres from the ground.

HABITAT: dense tropical highland rainforest
LENGTH: 24–25 cm
DISTRIBUTION: 30,000–100,000 km^2
ABUNDANCE: common
STATUS: secure

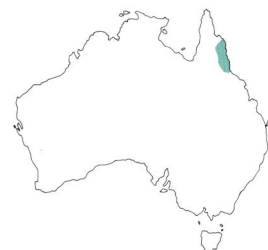

(P Green)

Genus *Ptilonorhynchus*

(tie'-lon-oh-rink'-us: "feathered-bill")

The characteristics of the genus are those of the single species.

Satin Bowerbird

Ptilonorhynchus violaceus (vee'-oh-lah'-say-us: "blue feather-bill")

The Satin Bowerbird is common in coastal regions and associated highlands of eastern Australia from the Atherton Tableland in north-eastern Queensland to the Otway Ranges in Victoria. Once their bowers are established, adult males are mainly sedentary but some highland birds move to adjacent lowlands in winter, and young males wander the countryside in bands of 30 to 40 or more until they attain adult plumage at six or seven years. The adult male is deep, rich blue in colour, but immatures and adult females are dull green above and finely barred below. The diet is mainly fruit.

The bower is of the avenue type, consisting of two parallel walls of interwoven twigs, about 30 centimetres high and 15 centimetres apart, constructed on a platform (also of woven twigs) extending in the form of an apron in front of and behind the walls. A range of items—feathers, fruit, and (in areas near human habitation) drinking straws, bottle tops and other miscellaneous litter—are arranged on one or both aprons. These items are almost always blue in colour. Most bowers are oriented north and south. Males spend much of their time arranging and rearranging these items, or perched watchfully in low foliage nearby; theft is common. As is the case with other bowerbirds, males entice itinerant females to the bower (where copulation takes place) with a variety of calls and antics. Once mated, females leave the area to rear the young alone.

The nest is a rough, shallow bowl of twigs and dry leaves at almost any height in a tree. The clutch consists of two eggs (occasionally one or three), which hatch in 21 to 22 days. Fledging at about 20 days, the young are then cared for away from the nest for a further eight or nine weeks. The breeding season extends from September to January.

HABITAT: mainly temperate dense eucalypt forest, rainforest
LENGTH: 28–33 cm
DISTRIBUTION: 100,000–300,000 km^2
ABUNDANCE: common
STATUS: probably secure

(N Chaffer)

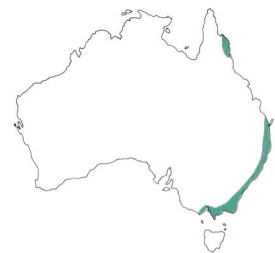

Genus *Sericulus*

(se-rik'-ue-lus: "little-silken [-bird]")

Resplendently clad in brilliant yellow, orange, and black, adult males of this genus are among the most colourful of bowerbirds; they build correspondingly less complex bowers of the avenue type. Females are mainly grey and brown, though intricately patterned. There are three species: two in New Guinea and one in Australia.

Regent Bowerbird

Sericulus chrysocephalus (krie'-soh-sef'-ah-lus: "golden-headed little-silken")

This brilliant bowerbird inhabits east coast forests from near Sydney in New South Wales to the vicinity of Rockhampton, and in the Clarke Range near Mackay, Queensland. It is mainly sedentary but wandering parties, consisting mainly of females and immatures, often form in winter.

The species is strongly arboreal and mainly eats fruit.

Adult males build relatively simple bowers on the forest floor, usually concealing them under bushes, fern thickets or vine tangles. The bowers somewhat resemble those of the Satin Bowerbird in structure but may be attended by several adult and immature males. Ornaments, arranged between the parallel walls, are usually red-brown in colour, and may include berries, leaves, snail shells and other items.

As in other bowerbirds, mated females build, incubate and rear their young alone. Often quite close to the bower, the nest is a shallow bowl of twigs, vines and dead leaves, usually concealed in a vine tangle from 3 to 20 metres or more from the ground. The two eggs (rarely three) are incubated for 18 days. The breeding season is October to January.

HABITAT: mainly warm temperate rainforest
LENGTH: 25–28 cm
DISTRIBUTION: 100,000–300,000 km²
ABUNDANCE: sparse to common
STATUS: vulnerable

(G Threlfo)

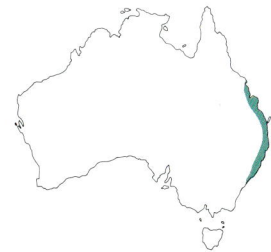

Family PARADISAEIDAE

(pa'-rah-dis-ee'-id-ee: "*Paradisea*-family", after a genus of birds of paradise)

Among the most spectacular of birds, the birds of paradise constitute a group of 43 species (in 20 genera) with a distribution centred on New Guinea: only a few occur on islands of the Indonesian archipelago and only four species in two genera (*Ptiloris* and *Manucodia*) occur in Australia. Their relationship with the bowerbirds (family Ptilonorhynchidae) is widely accepted but disputed by some authorities.

Notable for the extreme development of plumes and other display devices, male birds of paradise usually differ strikingly from females of the same species, which are usually dull brown and unremarkable in appearance. Although the group is notable for diversity of mating systems, males usually display to attract females, which mate promiscuously and raise the young unaided, no pair bond being formed. In many species, males congregate to compete in communal displays. These usually involve exhibition and manipulation of the spectacular plumage. Some species display on the ground or close to it, but most display in trees.

Genus Manucodia

(man'-ue-koh'-dee-ah: "bird-of-the-gods", a Latinisation of Malay *manuk dewata*, which has this meaning)

Members of this genus (the manucodes) are unusual among birds of paradise in that the sexes form pair bonds and cooperate in raising the young. The sexes are virtually identical in appearance, the males lacking the flamboyant display plumes otherwise characteristic of the family. Males have a highly modified trachea, lying coiled between the skin and the pectoral muscles. Mildly gregarious, the species are arboreal and feed largely on fruit. The single Australian species (four others occur in New Guinea) is found only at the tip of Cape York Peninsula.

437

Trumpet Manucode

Manucodia keraudrenii (ke-roh'-dren-ee-ee: "Keraudren's bird-of-the-gods", after P. M. Keraudren, French navigator)

In Australia the Trumpet Manucode occurs only at the tip of Cape York

(WS Peckover)

Peninsula, but it is widespread in New Guinea where it mainly inhabits hill forest between 1000 and 2000 metres above sea-level.

Its common name is an allusion to its remarkable calls, which are varied, loud and penetrating. It spends most of its time in the canopy, feeding mainly on fruit but occasionally extending its diet to include insects and spiders. Groups of five or six may congregate at fruiting trees; otherwise it lives in pairs that maintain apparently permanent territories.

It breeds mainly from October to January. The shallow, flimsy cup-shaped nest is built of twigs and vine stems and placed in a tree fork about 10 to 20 metres from the ground. One or two eggs form the clutch. Incubation, by both parents, takes about 15 to 16 days.

HABITAT: tropical rainforest
LENGTH: 28–32 cm
DISTRIBUTION: 30,000–100,000 km²
ABUNDANCE: sparse to common
STATUS: secure

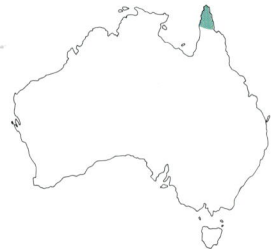

Genus Ptiloris

(til-o'-ris: "feathered-bill")

The characteristics of this genus are essentially those of the species. Although treated here as distinct species, all three riflebirds are very similar, and some researchers urge their combination in a single species. Females differ strongly from males in appearance, being mainly brown above and barred below, and are somewhat smaller. Riflebirds are polygynous, arboreal, and rather solitary in behaviour; they inhabit rainforests.

Magnificent Riflebird

Ptiloris magnificus (mag-nif'-ik-us: "magnificent feathered-bill")

(D & M Trounson)

The Magnificent Riflebird is abundant but somewhat local in distribution in the rainforests of New Guinea and within its restricted Australian range at the tip of Cape York Peninsula. In New Guinea it seems relatively more common in highlands than in lowlands.

Arboreal, solitary, and inconspicuous in behaviour, it resembles the Paradise Riflebird in appearance and life history. It breeds from September to February, perhaps sporadically throughout the year. It has been seen to bathe in forest pools.

HABITAT: tropical rainforest
LENGTH: (males) 30–33 cm
DISTRIBUTION: 10,000–30,000 km²
ABUNDANCE: sparse to common
STATUS: secure

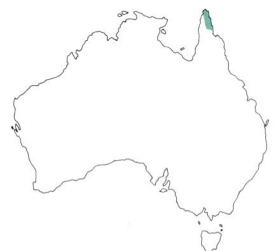

Paradise Riflebird

Ptiloris paradiseus (pa'-rah-diz'-ay-us: "paradise feathered-bill")

The Paradise Riflebird occurs in rainforests from the Barrington Tops region of New South Wales north to the vicinity of Bundaberg in Queensland. With the destruction of forests in the coastal lowlands, its range has contracted since European settlement, and it is now largely a bird of the highlands. It is sedentary.

Except in display, it is relatively quiet and inconspicuous in behaviour. Strictly arboreal, it forages in the canopy (somewhat like a large

(*T & P Gardner*)

treecreeper), methodically exploring limbs and branches, probing in cracks and crevices with the long curved bill for insects and spiders, or rummaging acrobatically in epiphytes and dangling clumps of leaves. It also eats fruits. The call is distinctive, a harsh long-drawn "yaa-a-a-sssss".

Like other birds of paradise, the Paradise Riflebird is polygynous: adult males (apparently at least three years old) defend one or several display perches—usually a high horizontal limb—at which they spend much time calling, or in solitary display. In display, the wings are fanned stiffly upwards, the head is thrown back to present the iridescent throat shield, and the bill is held wide open to reveal the bright yellow mouth; in this posture the head is swung rhythmically from side to side and the wings raised and lowered, producing a distinctive rustling sound. Females are attracted by the display, mating takes place, and the female promptly leaves; no pair bond is formed.

Breeding takes place from September to January. Working alone, the female builds a bulky cup-shaped nest of leaves, sticks and twigs in a tangle of leaves or vines up to 30 metres from the ground. The outside is often decorated with moss,

orchids or snakeskin. Usually two eggs are laid: these hatch in about 15 to 16 days; the young fledge at about 28 days.

HABITAT: subtropical to warm temperate rainforests, including *Nothofagus* forest; occasionally, dense eucalypt forest
LENGTH: 28–30 cm
DISTRIBUTION: 30,000–100,000 km²
ABUNDANCE: sparse to common
STATUS: probably secure

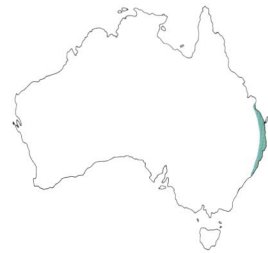

Victoria's Riflebird

Ptiloris victoriae (vik-tor'-ee-ee: "[Queen] Victoria's feathered-bill")

Victoria's Riflebird inhabits rainforests of north-eastern Queensland, approximately from Mount Spec near

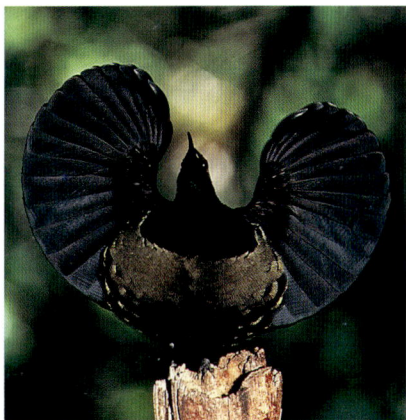

(*H & J Beste*)

Townsville to the vicinity of Cooktown; it is common in lowland and highland forests, and on some coastal islands.

Arboreal, rather solitary, and usually inconspicuous in behaviour, it resembles the Paradise Riflebird in life history and appearance (although markedly smaller). It eats much fruit, and may occasionally cause damage in orchards. Adult males are sedentary, but females and young males are non-territorial and wander locally, sometimes—especially in winter—in small flocks. September to January is the breeding season.

HABITAT: tropical rainforests
LENGTH: 23–25 cm
DISTRIBUTION: 10,000–30,000 km²
ABUNDANCE: sparse to common
STATUS: probably secure

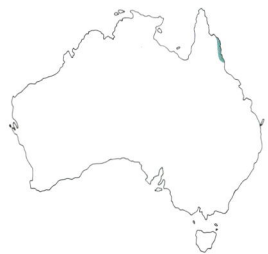

Family CORVIDAE

(kor'-vid-ee: "*Corvus*-family", after a genus of crows)

The family Corvidae is a cosmopolitan group of some 113 species, usually arranged in about 23 genera. A now abandoned division into two subfamilies (Corvinae, the crows and ravens; and Garrulinae, the jays and magpies) is superficially useful but has proven taxonomically unsound. Jays and magpies are widespread in the Old and New Worlds but none occurs in Australasia. *Corvus* is a cosmopolitan genus, whose members are variously named crows or ravens. The distinction has no biological significance but there is a worldwide tendency to reserve the term "raven" for the larger species, "crow" for the smaller species; most generalisations apply indiscriminately to both, and the general term "corvid" is favoured by ornithologists for use in contexts where it is tedious and unnecessary (and potentially misleading) to imply a distinction.

Most corvids (and all Australian species) are entirely black in plumage; there is little sexual, seasonal or age variation. In consequence most species are extremely difficult to identify in the field, and the five Australian species are most usefully distinguished by their calls. Adults of all species have white eyes.

In general, corvids are remarkably adaptable, most species having flexible requirements for food, habitat and climate; many are omnivorous. Common and successful, one or more species are usually prominent in local avifaunas throughout the world and often abundant, even in towns and cities.

Genus Corvus

(kor'-vus: "crow")

The characteristics of the genus are discussed in the account of the family.

Little Crow

Corvus bennetti (ben'-et-ee: "Bennett's crow", after K. H. Bennett, Australian ornithologist)

The Little Crow is the corvid most strongly associated with the arid interior of mainland Australia; it is common in many desert towns and around homesteads, and reaches the coast only in Western Australia. It is nomadic, strongly sociable, and flexible in choice of habitat. It is omnivorous, its diet consisting of about 26 per cent carrion, 48 per cent insects and 26 per cent plant material.

Little Crows commonly breed in loose colonies, nests being 20 to 50 metres apart. Breeding may occur at any time, but is generally from July to October. The nest, which is placed in a tree or bush or on a telegraph pole, is a bulky basket of sticks, lined with fur. Mud is used in its construction, a practice unique among Australian corvids. The clutch is three to six (usually four) eggs, incubated by the female for 17 to 19 days. The young fledge at 29 to 31 days.

HABITAT: arid temperate open country, scrub, desert
LENGTH: 45–47 cm
DISTRIBUTION: more than 1 million km²
ABUNDANCE: common
STATUS: secure

(*P. Klapste*)

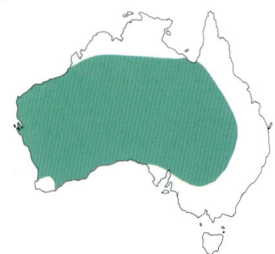

Australian Raven

Corvus coronoides (ko'-ron-oy'-dayz: "raven-like crow")

The Australian Raven is common throughout much of south-eastern Australia and in the south-west; breeding is rare north of about 20°S but vagrant parties occasionally reach the head of the Gulf of Carpentaria. Young birds join wandering flocks but adults mate for life and are sedentary, occupying territories of about 100 hectares.

Its distribution overlaps that of several other crows and ravens; all are very similar in appearance and field identification is difficult. However, the Australian Raven has a diagnostic call: a long-drawn, falling and fading wail. It also caws like other crows and ravens.

In summer it eats mainly insects, shifting to carrion at other seasons. Changing patterns of land use since European settlement have probably benefited it considerably.

Breeding occurs from July to September. The nest is an untidy bowl of sticks high in a tree. The clutch consists of four or five greenish eggs, which are incubated by the female for 20 days while she is fed at the nest by the male. The young fledge at about 43 days.

HABITAT: temperate to subtropical open areas with large trees; mainly pastoral country and eucalypt woodland; urban and suburban areas—always with available fresh water
LENGTH: 45–48 cm
DISTRIBUTION: more than 1 million km²
ABUNDANCE: common
STATUS: secure

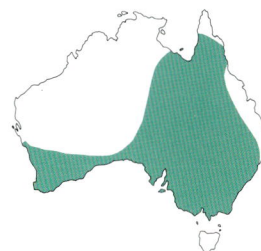

(*G Weber*)

Little Raven

Corvus mellori (mel'-or-ee: "Mellor's crow", after W. J. Mellor, Australian ornithologist)

This common corvid is restricted to south-eastern mainland Australia; it is nomadic and strongly gregarious, forming flocks of several hundred in winter and breeding semicolonially in spring; nesting territories average 2 to 3 hectares per pair. It is omnivorous.

The Little Raven and the Australian Raven are so similar in appearance that their distinctness was not recognised until the 1960s, but they are very different in ecology: the latter occupies scattered permanent territories in prime habitat, while the Little Raven arranges its annual cycle to exploit temporary abundance of food and other resources in marginal habitat. In consequence, although the two species overlap widely, the Australian Raven tends to dominate in the south-eastern coastal strip, the Little Raven in the more arid interior. The Little Raven tends to occupy more open country, and feeds more on insects, less on carrion.

The breeding season is flexible, but is generally in August and September. The nest is a bulky basket of sticks lined with fur, placed in a tree or bush or on a telegraph pole. The clutch is four or five eggs, incubated by the female for 19 to 20 days. The young fledge at 35 to 41 days.

HABITAT: open temperate eucalypt woodland and acacia scrubs; alpine heaths and moorland; pastures and grazing land
LENGTH: 48–50 cm
DISTRIBUTION: 300,000–1 million km²
ABUNDANCE: common
STATUS: secure

(*P Klapste*)

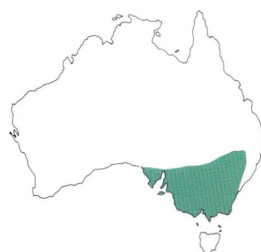

Torresian Crow

Corvus orru (o'-rue: meaning unknown)

The Torresian Crow is especially characteristic of the tropical north (it is the only corvid in the Top End and on Cape York Peninsula) but its distribution extends southward to about 30°S. Elsewhere it occurs in New Guinea and Indonesia. Immature birds are nomadic and sociable but adults mate for life and occupy permanent territories of about 100 to 200 hectares. It is omnivorous; nearly half of its diet consists of insects, and about a quarter of carrion. Its calls are distinctively terse, nasal, and high-pitched.

Peak breeding activity occurs from August to November. The nest is a rough basket of twigs and sticks in a tree, usually shaded by foliage and within 5 to 10 metres of the ground. Four eggs form the clutch, incubated by the female.

HABITAT: open warm temperate to tropical savannah woodland, forest, farmland, beaches
LENGTH: 48–53 cm
DISTRIBUTION: more than 1 million km²
ABUNDANCE: common
STATUS: secure

(K Ireland)

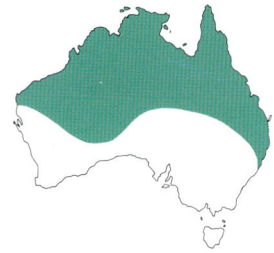

Forest Raven

Corvus tasmanicus (tas-mah'-nik-us: "Tasmanian crow")

The Forest Raven is found in Tasmania (where it is the only corvid); on Wilsons Promontory and in the Otway Ranges in Victoria; and on the New England Tableland and in the Myall Lakes district in New South Wales.

As in the Australian Raven, adults mate for life and occupy permanent territories, but Forest Ravens form roving flocks in winter; where the two species occur together, breeding pairs exclude one another from their respective territories. The Forest Raven is omnivorous. The most useful identification feature rests in its deep, abrupt cawing notes; unlike other corvids it commonly forages under the forest canopy as well as in clearings and open country.

It breeds from August to November. Four eggs form the usual clutch, incubated by the female.

HABITAT: cool temperate highland eucalypt forest and woodland; Tasmanian population more flexible, foraging often on beaches, pasture land, heaths
LENGTH: 53–54 cm
DISTRIBUTION: 100,000–300,000 km²
ABUNDANCE: common
STATUS: secure

(T Howard)

Introduced Species

Mallard

Anas platyrhynchos (plat'-ee-rink'-os: "flat-beaked duck")

Closely related to, and interbreeding with, the Pacific Black Duck, the Mallard is native to Eurasia, north-western Africa and North America. Although introduced to Australia in the 1860s and 1870s, it seems not to have become firmly established as a feral species until the 1950s. It is social, living in groups in coastal and inland waters, particularly in ponds associated with urban and suburban parks and gardens. It usually feeds in shallow water, dabbling and up-ending for tubers, aquatic invertebrates, tadpoles and frogs: it also eats seeds, buds and food scraps.

The male is readily distinguished by the iridescent green colour of the head and neck and the white collar.

Breeding males are extremely aggressive and rather indiscriminate, often mating with the Pacific Black Duck. The nest is a single bowl of vegetation on the ground. Eight to 10 eggs are laid, hatching after about four weeks.

HABITAT: temperate wetlands and pools, usually close to human settlement
LENGTH: 47–61 cm
DISTRIBUTION: 300,000–1 million km²
ABUNDANCE: sparse to abundant
STATUS: secure

(T & P Gardner)

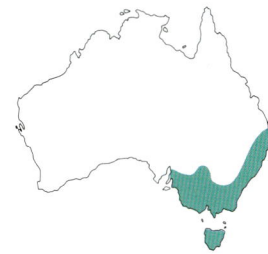

Genus *Lophortyx*

(loh-for'-tix: "crested quail")

The two species of North American quails in this genus are plump, gregarious seed-eaters. A crest or plume of feathers on the top of the head is characteristically club-shaped.

California Quail

Lophortyx californicus (kal'-if-orn'-ik-us: "Californian crested-quail")

Native to North America, the California Quail was introduced into Victoria in the 1860s. It appears to have died out on the mainland but birds introduced to King Island in the 1930s may have established a feral population. It is readily identified by a club-shaped black crest on the top of the head. Mainly terrestrial, it moves about in groups, feeding on seeds and insects.

Mating occurs from September to January. The nest, usually under cover, is a single, saucer-shaped scrape, roughly lined with vegetation. Eight to 10 eggs are laid. The male takes no part in incubating or in care of the young.

HABITAT: cool temperate grassland, cropland and woodland
LENGTH: *c* 24 cm
DISTRIBUTION: less than 10,000 km²
ABUNDANCE: sparse
STATUS: vulnerable

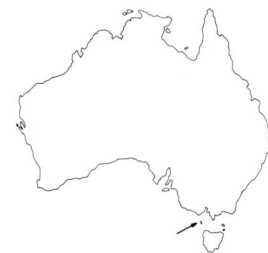

(B Chudleigh)

Genus *Phasianus*

(fah'-zee-ah-nus: "pheasant")

The characteristics of this genus are those of the single species.

Ringneck Pheasant

Phasianus colchicus (kol'-kik-us: "Colchian pheasant", in reference to Colchis, situated between the Black and Caspian Seas)

Originating from Eastern Europe to Asia Minor, the Ringneck Pheasant has been widely translocated as a game bird. It was first introduced from Great Britain to Victoria in the 1860s and many subsequent attempts have been made to establish it in various areas. It now exists in feral populations on King Island (Bass Strait) and Rottnest Island (off Fremantle, WA). It is primarily a terrestrial bird, feeding on seeds, fruits, insects and other invertebrates. It seldom flies, except when disturbed or when moving into a tree to roost at night. It is readily recognised by its size, long tail and mottled plumage: the male has a blue neck above a white collar.

The Ringneck Pheasant is polygamous. Without any assistance from a male, the female makes a single nest on the ground and incubates up to 17 eggs.

HABITAT: temperate woodland and heath
LENGTH: 55–89 cm
DISTRIBUTION: less than 10,000 km²
ABUNDANCE: sparse to very sparse
STATUS: vulnerable

(*JD Waterhouse*)

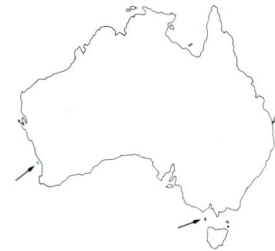

Rock Pigeon

Columba livia (liv'-ee-ah: "blue-grey dove")

Probably native to arid cliff country in the Middle East, the Rock Pigeon began to be translocated through Europe more than 4000 years ago. Trained to nest in pigeonhouses or dovecotes, it provides a significant source of meat. We do not know when it was first introduced to Australia but this was probably in the 18th century (and many times subsequently). The Rock Pigeon is gregarious and prefers to nest on ledges or in niches of cliffs: multi-storied buildings provide a very acceptable alternative. Feral populations show considerable individual variation, due to admixture of "fancy" breeds with the wild type. Seeds are the preferred diet but, around human habitation, the Rock Pigeon also eats food scraps, spilt grain, shoots, berries, worms, insects and other invertebrates.

Sexual maturity is reached as early as six months. The female lays a clutch of two eggs in a rough nest of sticks and faeces. Both parents incubate the eggs and feed the young. Under favourable conditions, breeding is continuous.

HABITAT: cool temperate to sub-tropical environments with access to fresh water: mostly around cities and other habitations
LENGTH: *c* 33 cm
DISTRIBUTION: more than 1 million km²
ABUNDANCE: common to abundant
STATUS: secure

(*T Howard*)

Genus Streptopelia

(strep'-toh-pel'-ee-ah: "necklaced dove")

The turtledoves of this genus are more slenderly built than most members of the Columbidae, with a rather small head and long neck. The bill is straight and slender. Two species are feral in Australia. A third, the Collared Dove, *S. decaocto*, has spread with remarkable rapidity from Asia into Europe over the past 60 years.

Spotted Turtledove

Streptopelia chinensis (chin-en'-sis: "Chinese necklaced-dove")

(MF Soper)

This species, native to southern Asia and Indonesia, was introduced into Australia from Great Britain in the 1860s. In general it has remained around human habitations but it is often found in country areas where wheat is spilled from trucks and trains or thrown out to poultry. Nowhere is it common. Much of its diet consists of food scraps but seeds comprise about one-third of the average intake.

Breeding takes place throughout the year but with a peak in spring and summer. The clutch of two eggs is laid in a frail nest of twigs. Eggs hatch in about two weeks.

HABITAT: cool temperate to tropical cities and towns
LENGTH: 27–32 cm
DISTRIBUTION: 300,000–1 million km²
ABUNDANCE: common
STATUS: secure

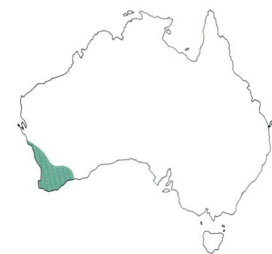

Senegal Turtledove

Streptopelia senegalensis (sen'-eg-ahl-en'-sis: "Senegal necklaced-dove")

Closely related to the Spotted Turtledove, and native to southern Asia and Africa, this species, also known as the Laughing Dove, was released around Perth at the end of the 19th century. Like the Spotted Turtledove, it appears to rely upon human activities for much of its food but it has spread a little further into agricultural land. It feeds on the ground, on food scraps, seeds and grain.

The frail nest of twigs is placed in a tree. Breeding takes place throughout the year, with a peak in spring and early summer. Two eggs are laid and incubated by both parents. Eggs hatch in about two weeks; young fledge when about three weeks old.

HABITAT: temperate woodland and grassland, mostly close to human activities
LENGTH: 25–27 cm
DISTRIBUTION: 100,000–300,000 km²
ABUNDANCE: common
STATUS: secure

(MF Soper)

Family MUSCICAPIDAE

(mus'-kee-kah'-pid-ee: "*Muscicapa*-family", after a genus of Old World flycatchers)

This large family comprises the thrushes and Old World flycatchers. The group is extremely diverse and difficult to define on appearance but their toes are separate, and they tend to have slender bills and to be rather more insectivorous than frugivorous. Many species have an elaborate song.

Genus Turdus

(ter'-dus: "thrush")

Extending over Eurasia and North America, this genus includes many familiar singing species, including the inappropriately named American Robin. Many species live very successfully in parks, gardens and farmland.

Blackbird

Turdus merula (me'-rue-lah: "blackbird thrush")

Native to Europe, the Blackbird was successfully introduced into Victoria, South Australia and Tasmania in the 1860s. It has

(MF Soper)

extended its mainland range to include most of Victoria and southern New South Wales but its distribution is largely restricted to towns, gardens and orchards (where it has become a serious pest of soft fruits). Fruits are eaten· when available but the diet otherwise includes snails, slugs, worms, arthropods and seeds, mostly taken on the ground.

Breeding takes place from September to January, during which period the male utters the warbling song for which the species was brought to Australia. The nest is an open cup of grasses, held together with mud and placed in a shrub or low tree. Three to five (usually four) eggs are laid.

HABITAT: cool temperate woodland or parkland, usually in the vicinity of human settlement
LENGTH: 25–26 cm
DISTRIBUTION: 300,000–1 million km²
ABUNDANCE: common
STATUS: secure

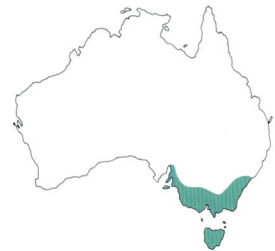

Song Thrush

Turdus philomelos (fil'-oh-mel'-os: "song-loving thrush")

Widespread throughout central Asia and Europe, and wintering in north Africa and Arabia, the Song Thrush is a familiar and popular bird. It was introduced into Australia in the 1860s and 1870s but, despite many

(MF Soper)

attempts to extend its range, it has been established as a feral population only in southern Victoria. It is largely restricted to public gardens and parks, particularly those with deciduous trees. It feeds on slugs, worms, other invertebrates and berries.

It was brought to Australia for its complex song, a territorial advertisement uttered by the male during courtship and through the period during which it attends the nest. Breeding occurs from September to January. The female lays four or five eggs in a well-constructed, cup-shaped nest of twigs and grasses, usually situated in the fork of a low branch or in dense shrubbery.

HABITAT: cool temperate parks and gardens, particularly with deciduous trees
LENGTH: *c* 23 cm
DISTRIBUTION: 30,000–100,000 km²
ABUNDANCE: very sparse
STATUS: probably secure

Family PYCNONOTIDAE

(pik'-noh-noh'-tid-ee: "*Pyconotus*-family", after a genus of bulbuls)

More than 100 species of these rather dull-coloured and short-winged birds occur from Africa to South-East Asia. Most feed on berries and fruits and seldom descend to the ground: their legs are short and their feet small. Many species are crested.

Genus *Pycnonotus*

(pik'-noh-noh'-tus: "thick-back")

The bulbuls that comprise this family tend to have rather dense or tangled-looking plumage on the back. Although the plumage is usually sombre, it may be enlivened by patches of white, yellow or red.

Red-whiskered Bulbul

Pycnonotus jocosus (joh-koh'-sus: "jesting thick-back")

Native to southern and eastern Asia, the species was brought to Australia as a cagebird. Individuals were released as early as the 1880s but it seems that feral populations did not become established until early in the 20th century. It is now common in coastal areas around Sydney and Coffs Harbour, less so around Melbourne. The diet comprises insects, berries and fruits. It is readily recognised by the pointed crest, red ear patches and white cheeks.

It breeds from August to March. The nest is an open, rather rough cup of bark, twigs, rootlets and other plant material, usually placed at a height of 2 to 3 metres in the fork of a tree or in a bush. Two to four eggs constitute the clutch.

HABITAT: temperate, well-watered suburban gardens and densely vegetated woodland, particularly in gullies where privet is established
LENGTH: 20–23 cm
DISTRIBUTION: 30,000–100,000 km²
ABUNDANCE: sparse to common
STATUS: secure

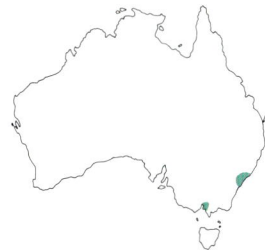

(JD Waterhouse)

Nutmeg Mannikin

Lonchura punctulata (punk'-tue-lah'-tah: "spotted spear-tail")

Native to south-eastern Asia and Indonesia, the Nutmeg Mannikin seems not to have existed in a feral state prior to the 1930s: existing populations appear to have been founded by escaped cagebirds. It is a rather sombre-coloured bird with pale underparts having a delicate brown reticulate pattern. It moves in pairs or flocks, feeding upon food scraps, grass seeds and carrion. Groups nest communally, with close body contact and mutual grooming.

The sexes are similar. Breeding occurs throughout the year in the tropics; in spring and summer in the southern part of the range. The nest is a well-built, flask-shaped structure of grass and leaves, with a side entrance. It is placed at a height of 1 to 6 metres in a tree or shrub, often in close proximity to other nests. Four to seven eggs are incubated by both parents.

(M Trenerry)

HABITAT: tropical to temperate well-watered grassland, weedy cropland and wasteland, usually close to human settlements
LENGTH: 10–12 cm
DISTRIBUTION: 100,000–300,000 km²
ABUNDANCE: common
STATUS: secure

Family PASSERIDAE

(pahs'-er-id-ee: "*Passer*-family", after the genus of sparrows)

This small family of small, social, dull-plumaged birds originally from Africa and Eurasia includes several species that have followed humans to almost every part of the inhabited world. The feed mainly upon seed (or food scraps) foraged on the ground. The bill is strong and pointed. There are only nine flight feathers on the wing (usually 10 in other birds).

Genus Passer

(pahs'-er: "sparrow")

The small genus of social, dull-plumaged, ground-feeding birds has the dual distinction of being one of the first to have been given a scientific name and to have had that name applied to the largest assemblage of living birds, the Passeriformes.

House Sparrow

Passer domesticus (dom-est'-ik-us: "domestic sparrow")

(B Chudleigh)

HABITAT: cool temperate to subtropical urban areas, parklands, suburban gardens, farms and a variety of woodland environments close to permanent water
LENGTH: 14–16 cm
DISTRIBUTION: more than 1 million km²
ABUNDANCE: abundant
STATUS: secure

Native to much of Eurasia and North Africa, the House Sparrow has followed humans into most parts of the world. It was introduced into Australia in the 1860s by British settlers for sentimental reasons and, although largely concentrated around areas of human settlement, it is now widely distributed over the eastern part of the continent, except in the driest and most tropical areas. Extreme measures are taken in an effort to prevent its spread into the western half of the continent.

It is social, moving in small flocks and feeding, mainly on the ground, upon food scraps, seeds, cereals and some insects. The male is more brightly patterned than the female and has a black "bib" on the upper chest and throat. It resembles the Tree Sparrow but differs in having a dark grey to dull brown crown.

Breeding can occur throughout the year but is mostly in spring and summer. The nest is a rough, domed structure of straw or grass, lined with down and usually with a side entrance. It is placed under eaves, in other spaces in buildings, in tree holes and in a variety of similar situations. The clutch of three to six eggs is incubated, mostly by the female, for 12 to 14 days. Both parents feed the young, which fledge at about two weeks of age.

448

Tree Sparrow

Passer montanus (mon-tah'-nus: "mountain sparrow")

Native to Eurasia, the Tree Sparrow was introduced to Australia from Great Britain in the 1860s. It is now established in eastern Victoria and the Riverina district. The sexes are similar and distinguishable from the House Sparrow by having a chestnut crown and a black ear patch. It moves in small groups, feeding mostly on the ground, on seeds and insects. It sometimes flocks with the House Sparrow.

Breeding occurs from September to January. The nest is an untidy domed structure, lined with down, usually with a side entrance and situated in a tree hole. Four to six eggs are incubated by both parents for 12 to 14 days. Fledging occurs at about two weeks. Two broods may be raised in a season.

(J Fennell)

HABITAT: generally around human habitations in cities, suburbs and farms in temperate, well-watered areas
LENGTH: *c* 14 cm
DISTRIBUTION: 100,000–300,000 km²
ABUNDANCE: sparse
STATUS: probably secure

Family FRINGILLIDAE

(frin-jil'-id-ee: "*Fringilla*-family", after a genus of finches)

The "true" finches that comprise this family are small, stocky birds which, like members of the Passeridae, have only nine flight feathers on the wing. The short, conical beak is adapted to a diet of seeds.

Genus *Carduelis*

(kar'-due-ay'-lis: "thistle[-bird]")

Members of this genus have a typical "finch-like" appearance, with a compact body and short, conical bill.

Goldfinch

Carduelis carduelis (kar'-due-ay'-lis: "thistle[-bird]")

Native to Eurasia and Northern Africa, the Goldfinch was introduced from Great Britain in the 1860s. It is now established on the south-eastern mainland and most of Tasmania. The sexes are similar and the species is readily recognisable by its red face, enclosing a white bill and black-rimmed eyes. It is social, moving in large flocks during the colder part of the year. It feeds mostly on seeds, particularly thistles, other plants of the family Asteraceae, and grasses. Some insects are eaten or taken as food for nestlings.

Breeding occurs from October to March. The nest is a compact open cup of rootlets and grasses, lined with fine plant or animal fibres. It is usually placed in a conifer or an introduced deciduous tree at a height of 2 to 12 metres. Three to seven eggs are incubated by the female for about 13 days: young fledge at about 12 days but continue to be fed until about one month old.

(MF Soper)

HABITAT: temperate, well-watered grassland near human habitations, farmland, orchards and weedy wasteland, usually with introduced deciduous trees and conifers
LENGTH: *c* 13 cm
DISTRIBUTION: 300,000–1 million km²
ABUNDANCE: sparse to common
STATUS: secure

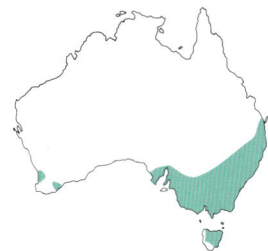

Greenfinch

Cardeulis chloris (klo'-ris: "yellow-green thistle[-bird]")

Native to Europe, western Asia and North Africa, the Greenfinch was introduced to Australia in the 1860s. Its overall colour is olive green with a yellowish edge to the wings and tail: it lacks the red face and distinct markings of the Goldfinch. The Greenfinch moves in small groups, feeding on seeds—including those of conifers—which it cracks with its strong bill. It is largely confined to the south-eastern coast of the mainland and to much of Tasmania.

Breeding occurs from October to January. The nest is a bulky open cup of twigs, lined with down or hair. Typically, it is placed in a conifer or shrub at a height of 2 to 12 metres.

(*MF Soper*)

HABITAT: cool temperate to temperate well-watered parks, suburban gardens, orchards, farmlands and pine plantations
LENGTH: *c* 15 cm
DISTRIBUTION: 100,000–300,000 km²
ABUNDANCE: sparse to common
STATUS: probably secure

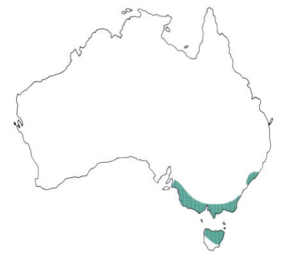

Family ALAUDIDAE

(ah-law'-did-ee: "*Alauda*-family", after a genus of larks)

Most of the 70 or so species in this family have long, pointed wings and a rather short tail, permitting rapid, turning flight. Courtship and territorial advertisement often involve a soaring flight with elaborate calls.

Genus Alauda

(ah-law'-dah: "lark")

This genus is typical of the family Alaudidae.

Skylark

Alauda arvensis (ar-ven'-sis: "farmland lark")

Native to Eurasia and North Africa, this species was introduced into Australia in the 1850s for the attractive song uttered by the male when he soars above his territory in a display flight. Most of its life is spent on the ground searching among grassland or swamp vegetation for seeds, shoots, insects and other small invertebrates. It is often found in altered, weedy environments, farmland and golf courses. On the ground, it is an inconspicuous bird with mottled plumage and a small erectile crest. In flight, the pale edges of the wings and tail may be prominent.

Breeding takes place from September to January. Three to five eggs are laid in a rough, cup-shaped nest built on the ground amid vegetation. Eggs hatch after about 11 days.

(*B Chudleigh*)

HABITAT: cool temperate to temperate well-watered open woodland, heathland, farmland, grassland and swamps
LENGTH: 17–19 cm
DISTRIBUTION: 300,000–1 million km²
ABUNDANCE: sparse to common
STATUS: secure

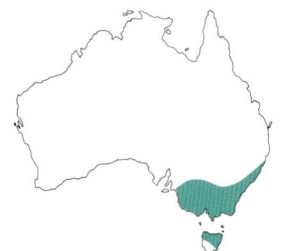

Family STURNIDAE

(ster'-nid-ee: "*Sturna*-family", after a genus of starlings)

The starlings, mynas and oxpeckers that comprise this family are all sturdy, strong-legged birds, usually with a very glossy plumage. The bill is usually slender and straight. Many species are strongly gregarious.

Genus Sturnus

(ster'-nus: "starling")

Perhaps the most characteristic feature of starlings is their glossy (metallic) plumage and their gregarious behaviour.

Common Starling

Sturnus vulgaris (vul-gar'-is: "common starling")

Introduced from England in the mid-nineteenth century, this very adaptable bird has spread over most of south-eastern Australia, including Tasmania. Considerable vigilance is employed to prevent its establishment in Western Australia and shooters are called out whenever there is evidence of a stray bird. This caution is well-based: although the Starling feeds largely on ground-dwelling insects and other invertebrates, it eats fruit when it is available and can therefore be a serious pest of orchards. Moving about in large flocks, it can cause great damage in a short time.

The Starling nests in any available holes or hollows. In cities, it commonly uses the eaves of buildings and very large numbers may congregate at night in the crowns of ornamental palms, making a great noise as they settle in for the night. It also nests in tree-holes, often to the detriment of small native species that require these sites for breeding.

A glossy black, with an iridescent sheen, the Starling is readily recognised, except immediately after moulting, when the buff-coloured tips of the feathers give it a mottled appearance.

Breeding occurs from August to December. The nest is a roughly constructed mass of twigs and grasses, often supplemented by paper. Four to eight eggs are brooded by both parents for about 12 days. The young are fledged at the age of about three weeks.

(JD Waterhouse)

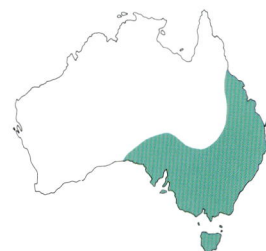

HABITAT: Cool to warm temperate woodland, grassland, pastures and parks

LENGTH: 21–23 cm

DISTRIBUTION: more than 1 million km²

ABUNDANCE: common

STATUS: secure

451

Genus Acridotheres

(ak-rid-oh-thay'-rayz: "locust-eater")

The several species in this small genus are confident, gregarious, ground-feeding frugivores (or scavengers) which run or hop on well-developed legs.

Common Myna

Acridotheres tristis (tris'-tis: "dull-coloured locust-eater")

Also known as the Indian Myna, this species is native to eastern and southern Asia, where it is a common cagebird. It was introduced in the 1860s and is now established around several cities of eastern and southern Australia. It is recognisable by its yellow legs, bill and eye patch, as well as its confident, jaunty behaviour. It moves about in pairs or small groups during the day, feeding mostly on the ground upon food scraps, berries and small fruits. At night, it roosts communally in a tree or under cover of a human construction.

Breeding occurs from October to March. The nest is a rough construction of twigs, grass, leaves, feathers and man-made materials. The clutch comprises three to six (usually four or five) eggs.

HABITAT: mainly urban areas but also suburbs and vicinity of some country towns: in canefields of northern Queensland
LENGTH: 23–26 cm
DISTRIBUTION: 100,000–300,000 km²
ABUNDANCE: sparse to abundant
STATUS: secure

(B Chudleigh)

CLASS AVES

ORDER STRUTHIONIFORMES
    Family Dromaiidae
      Genus *Dromaius* Emu
    Family Casuariidae
      Genus *Casuarius* cassowaries

ORDER PODICIPEDIFORMES
    Family Podicipedidae
      Genus *Podiceps* grebes
      Genus *Poliocephalus* grebes
      Genus *Tachybaptus* grebes

ORDER SPHENISCIFORMES
    Family Spheniscidae
      Genus *Eudyptes* rockhopper penguins
      Genus *Eudyptula* Little Penguin

ORDER PROCELLARIIFORMES
    Family Diomedeidae
      Genus *Diomedea* albatrosses
      Genus *Phoebetria* albatrosses
    Family Procellariidae
      Genus *Calonectris* shearwaters
      Genus *Daption* Cape Petrel
      Genus *Fulmarus* fulmars
      Genus *Halobaena* Blue Petrel
      Genus *Lugensa*, Kerguelen Petrel
      Genus *Macronectes* giant-petrels
      Genus *Pachyptila* prions
      Genus *Procellaria* petrels
      Genus *Pseudobulweria* petrels
      Genus *Pterodroma* petrels
      Genus *Puffinus* shearwaters
    Family Hydrobatidae
      Genus *Fregetta* stormpetrels
      Genus *Oceanites* stormpetrels
      Genus *Pelagodroma* White-faced stormpetrel
    Family Pelecanoididae
      Genus *Pelecanoides* diving-petrels

ORDER PELECANIFORMES
    Family Pelecanidae
      Genus *Pelecanus* pelicans
    Family Anhingidae
      Genus *Anhinga* darters

    Family Sulidae
      Genus *Morus* gannets
      Genus *Sula* boobies
    Family Phalacrocoracidae
      Genus *Phalacrocorax* shags
    Family Fregatidae
      Genus *Fregata* frigatebirds
    Family Phaethontidae
      Genus *Phaethon* tropicbirds

ORDER CICONIIFORMES
    Family Ardeidae
      Genus *Ardea* herons
      Genus *Botaurus* bitterns
      Genus *Bubulcus* Cattle Egret
      Genus *Butorides* Mangrove Heron
      Genus *Egretta* egrets
      Genus *Ixobrychus* bitterns
      Genus *Nycticorax* night-herons
    Family Ciconiidae
      Genus *Xenorhynchus* Black-billed Stork
    Family Threskiornithidae
      Genus *Platalea* spoonbills
      Genus *Plegadis* Glossy Ibis
      Genus *Threskiornis* ibises

ORDER ANSERIFORMES
    Family Anatidae
      Genus *Anas* ducks
      Genus *Anseranas* Magpie-goose
      Genus *Aythya* Hardhead
      Genus *Biziura* Musk Duck
      Genus *Cereopsis* Cape Barren Goose
      Genus *Chenonetta* Maned Duck
      Genus *Cygnus* swans
      Genus *Dendrocygna* whistling-ducks
      Genus *Malacorhynchus* Pink-eared Duck
      Genus *Nettapus* pygmy-geese
      Genus *Oxyura* Blue-billed Duck
      Genus *Stictonetta* Freckled Duck
      Genus *Tadorna* shelducks

ORDER ACCIPITRIFORMES
    Family Pandionidae
      Genus *Pandion* Osprey

Family Accipitridae
 Genus *Accipiter* hawks etc.
 Genus *Aquila* eagles
 Genus *Aviceda* bazas
 Genus *Circus* harriers
 Genus *Elanus* kites
 Genus *Erythrotriorchis* Red Goshawk
 Genus *Haliaeetus* sea eagles
 Genus *Haliastur* kites
 Genus *Hamirostra* Black-breasted Buzzard
 Genus *Hieraaetus* eagles
 Genus *Lophoictinia* Square-tailed Kite
 Genus *Milvus* kites
Family Falconidae
 Genus *Falco* falcons

ORDER GALLIFORMES
Family Megapodiidae
 Genus *Alectura* Brush Turkey
 Genus *Leipoa* Malleefowl
 Genus *Megapodius* scrubfowls
Family Phasianidae
 Genus *Coturnix* quails

ORDER GRUIFORMES
Family Turnicidae
 Genus *Turnix* buttonquails
Family Pedionomidae
 Genus *Pedionomus* Plains Wanderer
Family Rallidae
 Genus *Eulabeornis* Chestnut Rail
 Genus *Fulica* coots
 Genus *Gallinula* native-hens
 Genus *Poliolimnas* White-browed Crake
 Genus *Porphyrio* swamphens
 Genus *Porzana* crakes
 Genus *Rallina* rails
 Genus *Rallus* rails
 Genus *Tricholimnas*
Family Gruidae
 Genus *Grus* cranes
Family Otididae
 Genus *Ardeotis* bustards

ORDER CHARADRIIFORMES
Family Jacanidae
 Genus *Irediparra* Lotusbird
Family Burhinidae
 Genus *Burhinus* stone-curlews

 Genus *Esacus* stone-curlews
Family Rostratulidae
 Genus *Rostratula* Painted Snipe
Family Haematopodidae
 Genus *Haematopus* oystercatchers
Family Glareolidae
 Genus *Glareola* pratincoles
 Genus *Stiltia* Australian Pratincole
Family Charadriidae
 Genus *Charadrius* plovers
 Genus *Erythrogonys* Red-kneed Dotterel
 Genus *Peltohyas* Inland Dotterel
 Genus *Pluvialis* plovers
 Genus *Vanellus* lapwings
Family Recurvirostridae
 Genus *Cladorhynchus* Banded Stilt
 Genus *Himantopus* stilts
 Genus *Recurvirostra* avocets
Family Scolopacidae
 Genus *Arenaria* turnstones
 Genus *Calidris* stints, sandpipers, knots
 Genus *Gallinago* snipes
 Genus *Limicola* Broad-billed Sandpiper
 Genus *Limnodromus* Asian Dowitcher
 Genus *Limosa* goodwits
 Genus *Numenius* curlews
 Genus *Phalaropus* phalaropes
 Genus *Philomachus* Ruff
 Genus *Tringa* tattlers, sandpipers
 Genus *Tryngites* Buff-breasted Sandpiper
Family Stercorariidae
 Genus *Catharacta* skuas
 Genus *Stercorarius* jaegers
Family Laridae
 Genus *Anous* noddies
 Genus *Chlidonias* terns
 Genus *Gelochelidon* terns
 Genus *Gygis* White Tern
 Genus *Hydroprogne* Caspian Tern
 Genus *Larus* gulls
 Genus *Procelsterna* Grey Noddy
 Genus *Sterna* terns

ORDER COLUMBIFORMES
Family Columbidae
 Genus *Chalcophaps* doves
 Genus *Columba* pigeons
 Genus *Ducula* fruitpigeons
 Genus *Geopelia* doves

Genus *Geophaps* pigeons
Genus *Leucosarcia* Wonga Pigeon
Genus *Lopholaimus* Topknot Pigeon
Genus *Macropygia* pigeons
Genus *Ocyphaps* Crested Pigeon
Genus *Petrophassa* rock-pigeons
Genus *Phaps* bronzewings
Genus *Ptilinopus* fruitdoves

ORDER PSITTACIFORMES
Family Cacatuidae
Genus *Cacatua* cockatoos
Genus *Callocephalon* Gang-gang Cockatoo
Genus *Calyptorhynchus* black cockatoos
Genus *Nymphicus* Cockatiel
Genus *Probosciger* Palm Cockatoo
Family Loriidae
Genus *Glossopsitta* lorikeets
Genus *Psitteuteles* Varied Lorikeet
Genus *Trichoglossus* lorikeets
Family Psittacidae
Genus *Alisterus* King Parrot
Genus *Aprosmictus* Red-winged Parrot
Genus *Barnardius* Ringneck
Genus *Eclectus* Electus Parrot
Genus *Geoffroyus* Red-cheeked Parrot
Genus *Geopsittacus* Night Parrot
Genus *Lathamus* Swift Parrot
Genus *Melopsittacus* Budgerigar
Genus *Neophema* parrots
Genus *Northiella* Bluebonnet
Genus *Pezoporus* Ground Parrot
Genus *Platycercus* rosellas
Genus *Polytelis* parrots
Genus *Psephotus* parrots
Genus *Psittaculirostris* figparrots
Genus *Purpureicephalus* Red-capped Parrot

ORDER CUCULIFORMES
Family Cuculidae
Genus *Centropus* coucals
Genus *Chrysococcyx* bronze-cuckoos
Genus *Cuculus* cuckoos
Genus *Eudynamis* koels
Genus *Scythrops* Channel-billed Cuckoo

ORDER STRIGIFORMES
Family Tytonidae
Genus *Tyto* barn owls

Family Strigidae
Genus *Ninox* owls

ORDER CAPRIMULGIFORMES
Family Caprimulgidae
Genus *Caprimulgus* nightjars
Family Podargidae
Genus *Podargus* frogmouths
Family Aegothelidae
Genus *Aegotheles* owlet-nightjars

ORDER APODIFORMES
Family Apodidae
Genus *Apus* swifts
Genus *Collocalia* swiftlets
Genus *Hirundapus* swifts

ORDER CORACIIFORMES
Family Alcedinidae
Genus *Ceyx* kingfishers
Genus *Dacelo* kookaburras
Genus *Halcyon* kingfishers
Genus *Syma* kingfishers
Genus *Tanysiptera* paradise-kingfishers
Family Meropidae
Genus *Merops* bee-eaters
Family Coraciidae
Genus *Eurystomus* rollers, Dollarbird

ORDER PASSERIFORMES
Family Pittidae
Genus *Pitta* pittas
Family Atrichornithidae
Genus *Atrichornis* scrub-birds
Family Menuridae
Genus *Menura* lyrebirds
Family Alaudidae
Genus *Mirafra* larks
Family Motacillidae
Genus *Anthus* pipits
Genus *Motacilla* wagtails
Family Hirundinidae
Genus *Cecropis* martins
Genus *Cheramoeca* White-backed Swallow
Genus *Hirundo* swallows
Family Campephagidae
Genus *Coracina* cuckooshrikes
Genus *Lalage* trillers

Family Turdidae
  Genus *Zoothera* thrushes
Family Orthonychidae
  Genus *Cinclosoma* quailthrushes
  Genus *Orthonyx* Chowchilla, Logrunner
  Genus *Psophodes* wedgebills
Family Timaliidae
  Genus *Pomatostomus* babblers
Family Eopsaltridae
  Genus *Drymodes* scrub-robins
  Genus *Eopsaltria* robins
  Genus *Melanodryas* robins
  Genus *Microeca* flycatchers
  Genus *Petroica* robins
  Genus *Poecilodryas* robins
  Genus *Tregellasia* robins
Family Pachycephalidae
  Genus *Colluricincla* shrike-thrushes
  Genus *Falcunculus* Shriketit
  Genus *Oreoica* Crested Bellbird
  Genus *Pachycephala* whistlers
Family Monarchidae
  Genus *Arses* monarchs
  Genus *Machaerirhynchus* flycatchers
  Genus *Monarcha* monarchs
  Genus *Myiagra* flycatchers
  Genus *Rhipidura* fantails
Family Sylviidae
  Genus *Acrocephalus* reed-warblers
  Genus *Cinclorhamphus* songlarks
  Genus *Cisticola* cisticolas
  Genus *Eremiornis* spinifexbird
  Genus *Megalurus* grassbirds
Family Maluridae
  Genus *Amytornis* grasswrens
  Genus *Malurus* fairywrens
  Genus *Stipiturus* emuwrens
Family Acanthizidae
  Genus *Acanthiza* thornbills
  Genus *Aphelocephala* whitefaces
  Genus *Crateroscelis* fernwrens
  Genus *Dasyornis* bristlebirds
  Genus *Gerygone* warblers
  Genus *Origma* Rock Warbler
  Genus *Pycnoptilus* Pilotbird
  Genus *Sericornis* scrubwrens
  Genus *Smicrornis* Weebill
Family Neosittidae
  Genus *Daphoenositta* sitellas

Family Climacteridae
  Genus *Climacteris* treecreepers
  Genus *Cormobates* White-throated Treecreeper
Family Ephthianuridae
  Genus *Ashbyia* Gibberbird
  Genus *Ephthianura* chats
Family Pardalotidae
  Genus *Pardalotus* pardalotes
Family Meliphagidae
  Genus *Acanthagenys* Spiny-cheeked Honeyeater
  Genus *Acanthorhynchus* spinebills
  Genus *Anthochaera* wattlebirds
  Genus *Certhionyx* honeyeaters
  Genus *Conopophila* honeyeaters
  Genus *Entomyzon* Blue-faced Honeyeater
  Genus *Glycichaera* Green-backed Honeyeater
  Genus *Glyciphila* Tawny-crowned Honeyeater
  Genus *Grantiella* Painted Honeyeater
  Genus *Lichmera* honeyeaters
  Genus *Manorina* miners
  Genus *Meliphaga* honeyeaters
  Genus *Melithreptus* honeyeaters
  Genus *Myzomela* honeyeaters
  Genus *Philemon* friarbirds
  Genus *Phylidonyris* yellow-winged honeyeaters
  Genus *Plectorhyncha* Striped Honeyeater
  Genus *Ramsayornis* Bar-breasted Honeyeater
  Genus *Trichodere* White-streaked Honeyeater
  Genus *Xanthomyza* Regent Honeyeater
  Genus *Xanthotis* honeyeaters
Family Dicaeidae
  Genus *Dicaeum* flower-peckers, Mistletoebird
Family Nectariniidae
  Genus *Nectarinia* Yellow-bellied Sunbird
Family Zosteropidae
  Genus *Zosterops* white-eyes
Family Estrildidae
  Genus *Aidemosyne* Plum-headed Finch
  Genus *Emblema* firetails
  Genus *Erythura* parrotfinches
  Genus *Lonchura* mannikins
  Genus *Neochmia* finches
  Genus *Poephila* finches
Family Oriolidae
  Genus *Oriolus* orioles
  Genus *Sphecotheres* Figbird
Family Sturnidae
  Genus *Aplonis* starlings

Family Dicruridae
    Genus *Dicrurus* drongos
Family Artamidae
    Genus *Artamus* woodswallows
Family Corcoracidae
    Genus *Corcorax* White-winged Chough
    Genus *Struthidea* Apostlebird
Family Grallinidae
    Genus *Grallina* Magpielark
Family Cracticidae
    Genus *Cracticus* butcherbirds
    Genus *Gymnorhina* Australian Magpie
    Genus *Strepera* currawongs
Family Ptilonorhynchidae
    Genus *Ailuroedus* catbirds
    Genus *Chlamydera* bowerbirds
    Genus *Prionodura*
    Genus *Ptilonorhynchus* Satin Bowerbird
    Genus *Sericulus* bowerbirds
Family Paradisaeidae
    Genus *Manucodia* manucodes
    Genus *Ptiloris* riflebirds
Family Corvidae
    Genus *Corvus* crows, ravens

Introduced Species

ORDER ANSERIFORMES
    Family Anatidae
        Genus *Anas* Mallard

ORDER GALLIFORMES
    Family Phasianidae
        Genus *Lophortyx* California Quail
        Genus *Phasianus* Ringneck Pheasant

ORDER COLUMBIFORMES
    Family Columbidae
        Genus *Columba* Rock Pigeon
        Genus *Streptopelia* Laughing Turtledove, Senegal
            Turtledove

ORDER PASSERIFORMES
    Family Muscicapidae
        Genus *Turdus* Blackbird, Song Thrush
    Family Pycnonotidae
        Genus *Pycnonotus* Red-whiskered Bulbul
    Family Estrildidae
        Genus *Lonchura* Nutmeg Mannikin
    Family Passeridae
        Genus *Passer* House, Tree Sparrows
    Family Fringillidae
        Genus *Carduelis* Goldfinch, Greenfinch
    Family Alaudidae
        Genus *Alauda* Skylark
    Family Sturnidae
        Genus *Sturnus* Common Starling
        Genus *Acridotheres* Common Myna

GLOSSARY

Allopatric. Living in different areas. Usually applied to congeneric species. Contrast with **sympatric**.

Altricial. The condition of young birds that are confined to the nest for a long time, being unable to move about independently or feed themselves. Contrast with **precocial**.

Amphibious. Able to live on land and in water.

Anterior. Towards the front end of an animal. Contrast with **posterior**.

Arboreal. Living in trees.

Arthropods. A wide variety of animals with external skeletons and jointed bodies and limbs: eg. crustaceans, millipedes, centipedes, insects, spiders.

Brigalow. An Australian wattle, *Acacia harpophylla*. Also areas of country where this species is dominant in the vegetation.

Caecum. A blind branch of the intestine, often large in herbivorous vertebrates.

Canopy. The upper foliage, usually dense, of a tree or forest.

Carnivorous. Feeding on animals (usually vertebrates).

Caruncle. A flap or appendage of bare skin, often brightly coloured.

Casque. Horny structure on the upper surface of the bill or head of certain birds.

Caudal. Pertaining to the tail.

Cere. Soft skin at the base of the upper bill of certain birds.

Chenopod. Any plant of the family Chenopodiaceae: e.g. saltbush and bluebush.

Cline. A gradual change in the characteristics of a species across its range—usually associated with an environmental gradient: north–south, wet–dry, lowland–mountain, etc. Contrast with **subspecies**.

Cloaca. Terminal part of the gut of vertebrates other than eutherian mammals. In female birds, faeces, urine and eggs pass out of the body via the cloaca. In most male birds, faeces, urine and sperms exit via the cloaca but male ducks, emus and cassowaries have a penis-like organ through which semen passes during copulation.

Congeners. Species that are members of the same genus. Species in the same genus are said to be congeneric.

Conspecific. Belonging to the same species.

Convergent evolution. The evolution of similar body form (or other features) in animals of widely different ancestry: e.g. bats and birds.

Crepuscular. Active around dawn and/or dusk.

Cryptic. Hidden, inconspicuous, skilled at concealment.

Distribution. The maximum extent of the area in which a species is known to occur; also referred to as range. See also **home range, territory**.

Dorsal. Pertaining to the back or upper surface of an animal. Opposite of ventral.

Emergent tree. One that rises well above the surrounding forest canopy.

Endemic. Native to a designated area: e.g. the Budgerigar is endemic to Australia. See also **exotic** and **extralimital**.

Exotic. Foreign, coming from outside a designated area: e.g. the House Sparrow is exotic in respect of Australia.

Extralimital. Living inside and outside of a designated area: e.g. the Koel, which lives in Australia and New Guinea, is extralimital in respect of Australia.

Forb. Any small, non-woody, ground plant which is not a grass. Plants designated as forbs usually have broad-leaves.

Frugivorous. Feeding on fruits.

Gibber desert. Desert with a surface composed largely of smoothly rounded stones (gibbers).

Habitat. The area that provides the physical and biological requirements of a species.

Herbivorous. Feeding on plants.

Home range. The area habitually traversed by an individual animal. It may be exclusive or overlap with the home ranges of other members of the species

Hummock grassland. Areas where grasses of the genus *Triodia* and *Plectrachne* (both often erroneously called "spinifex") are dominant. Soil builds up around each clump of grass, forming a hummock, the ground between the hummocks usually being bare. See also **spinifex**.

Hybrid. The offspring of individuals from two distinct populations. As a general rule, hybrids between related subspecies are viable and fertile; hybrids between different species are rare in nature and seldom fertile.

Insectivorous. Feeding on insects and/or other terrestrial arthropods.

Invertebrate. Any animal that is not a vertebrate: worms, molluscs, arthropods, etc.

Lateral. Referring to the sides.

Lignum. *Muehlenbeckia cunninghami*, a sparse-leafed shrub which grows into a tangled mound of thin canes, often forming dense thickets.

Lingual. Pertaining to the tongue.

Littoral. Pertaining to the edge of a sea or lake.

Mallee. Eucalypt trees with multiple stems arising from a single root stock. Also country in which these trees are dominant in the vegetation.

Melanic, melanotic. Black or darkly pigmented. Usually descriptive of a morph.

Mesic. Refers to an environment intermediate between arid and moist.

Migratory. Moving more or less regularly from one area to another, usually in response to seasonal change. Compare with **sedentary, nomadic**.

Monophyletic. Sharing a common ancestor in the group of organisms under consideration. Contrast with **polyphyletic**.

459

Monsoon rainforest. Low rainforest occurring in small patches across northern Australia from Cape York to the Kimberley. It is lower and less complex than typical tropical rainforest.

Morph. One of two or more distinct forms (usually differing in coloration) which may occur within a freely interbreeding population. See also **polymorphic**.

Nectarivorous. Feeding on nectar (sometimes also pollen).

Nidicolous. Altricial.

Nidifugous. Precocial.

Nomadic. Wandering from place to place, usually from one source of food to another. Contrasts with **sedentary** and **migratory**.

Ocellate. Having a pattern of round 'eye-spots'.

Omnivorous. Feeding on animals and plants.

Opportunistic. In reference to feeding behaviour, eating whatever is available within a wide range of foods. In reference to reproduction, breeding when environmental conditions are favourable, rather than at a particular time of year.

Pelagic. Living at or near the surface of the sea.

Polymorphism. The condition in which a species or population consists of two or more morphs that are distinct in coloration (or other characters). A population with two morphs is said to be **dimorphic**.

Polyphyletic. Not sharing a common ancestor within the group under consideration. Contrast with **monophyletic**.

Posterior. Pertaining to the rear end of an animal. Opposite of anterior.

Precocial. The condition of young birds (such as chickens or emus) that are able to move about and feed themselves shortly after hatching. Contrast with **altricial**.

Proximal. At, or towards the region of attachment of an appendage such as a limb or tail. Opposite of **distal**.

Ratite. Without a keel to the sternum, as in Emu and cassowaries.

Refugial. Referring to a limited area in which a sometimes widespread species survives during unfavourable conditions, such as drought. Such an area is called a refugium.

Relic, relictual. Refers to the isolated remnant population(s) of a species that was once more widely distributed.

Riparian. Pertaining to the land on the sides of a river: riverside.

Sclerophyll. A general term for eucalypt forest or woodland. **Wet sclerophyll** forest is tall and has a dense canopy. **Dry sclerophyll**

forest varies considerably in height and usually has a discontinuous canopy.

Sedentary. Remaining in much the same area throughout the year. Contrast with **migratory** and nomadic.

Senescent. Becoming senile, approaching the end of life.

Spinifex. Strictly, a genus of sea-coast grasses with long, creeping stems. Commonly, and inaccurately, used as a general term for spiky hummock grasses (*Triodia*, *Plectrachne*) of the arid inland. See **hummock grassland**.

Subspecies. One or two or more populations of a species that are recognisably different from each other and usually differ in distribution. Subspecies frequently hybridise where their distributions meet or overlap.

Substrate. The surface—soil, rock, leaf-litter, etc.—on which an animal lives.

Sympatric. Living in the same area. Usually applied to congeneric species. Contrast with **allopatric**.

Syndactyl. (In birds), having the third and fourth toes joined for part of their length.

Taxon. The scientific name of any classificatory group of organisms: e.g., phylum, class, order, genus, species, etc.

Terrestrial. Pertaining to the land. Living on, or mainly on, the ground. Contrast with **arboreal, amphibious, fossorial**.

Territory. An area occupied by one or more individuals and defended against other members of the species.

Torpor. A state of dormancy and diminished temperature regulation over a period of hours or days, in response to cold or food shortage. See also **hibernation**.

Tussock grassland. Areas where Mitchell Grass (*Astrebla*) or bluegrass (*Dicanthium*) are the dominant plants. See also **spinifex** and **hummock grassland**.

Ventral. Pertaining to the belly or under-surface. Opposite to **dorsal**.

Vestigial. Reduced (in the course of evolution) and often without apparent function.

Vine thicket. Rainforest with many hanging vines or lianas.

Wattle. A caruncle, usually on the face or neck, often brightly coloured, and usually of sexual significance.

Zygodactyl. Having two toes directed forward and two back, as in parrots.

INDEX OF COMMON NAMES

INDEX OF SCIENTIFIC NAMES

467